38th UIT Heat Transfer International Conference (UIT 2021)

Journal of Physics: Conference Series Volume 2177

Cassino, Italy
21-23 June 2021

ISBN: 978-1-7138-5476-0
ISSN: 1742-6588

Printed from e-media with permission by:

Curran Associates, Inc.
57 Morehouse Lane
Red Hook, NY 12571

Some format issues inherent in the e-media version may also appear in this print version.

Institute of Physics
Dirac House, Temple Back
Bristol BS1 6BE UK

Phone: 44 1 17 929 7481
Fax: 44 1 17 920 0979

techtracking@iop.org

Additional copies of this publication are available from:

Curran Associates, Inc.
57 Morehouse Lane
Red Hook, NY 12571 USA
Phone: 845-758-0400
Fax: 845-758-2633
Email: curran@proceedings.com
Web: www.proceedings.com

TABLE OF CONTENTS

Preface

The 2021 UIT (Italian Union of Thermo-Fluid Dynamics) International Conference, hereafter referred as 38th UIT 2021, was organized by the Department of Civil and Mechanical Engineering of the University of Cassino (Italy), in collaboration with the UIT, on June 21st-23rd 2021.

The 38th UIT Heat Transfer Conference 2021 was organized by the Organizing Committee of the 38th UIT Heat Transfer Conference, initially scheduled for June 21st-23rd 2021 at the Castello di Gaeta, Italy. Considering the criticalities induced by the spread of COVID-19, difficulties in travelling, social distancing, and strict rules to follow, the Organizing Committee made the difficult decision to organize the 38th UIT Heat Transfer Conference 2021 completely online on June 21st-22nd 2021, using the platform Google Meet, to give the opportunity to the authors to introduce their valuable papers in 11 online parallel technical sessions. A total of 58 abstracts were submitted to the 38th UIT Heat Transfer Conference 2021, 45 of which presented by the Authors in oral sessions, while the remaining13 abstracts were presented in poster sessions.

The 38th UIT Heat Transfer Conference 2021 program scheduled a plenary lecture given by Prof. Giovanni Betta from the University of Cassino and Southern Lazio, while each author confined his talk in 15 minutes, including questions and answers. The majority of participants were from Italy, with a total number of about 70 researchers connected to the 38th UIT 2021 online event.

The 38th UIT Heat Transfer Conference 2021 was a useful occasion to stimulate discussion, further the understanding of heat transfer and related phenomena, present the state-of-the-art of some topics, discuss emerging trends and promote collaborations. The Organizing Committee hope that the event results constituted significant contribution to the knowledge in the following topics: computational fluid dynamic and heat transfer; multi-phase fluid dynamics, heat transfer and interface phenomena; heat transfer in energy systems and nuclear plants; micro and nano scale thermo-fluid dynamics; heat transfer in porous media; measurement techniques for heat and mass transfer.

A special tank to UIT, international advisory committee, local organizing committee and all the participants.

With Kind Regards

Paolo Vigo, Università di Napoli “Parthenope”, Italy
Marco Dell’Isola, Università degli Studi di Cassino e del Lazio Meridionale, Italy
Fausto Arpino, Università degli Studi di Cassino e del Lazio Meridionale, Italy
Alfonso Niro, Politecnico di Milano, Italy

Journal of Physics: Conference Series 2177 (2022) 011001 doi:10.1088/1742-6596/2177/1/011001

Committees

Co-chairs
P. Vigo, Università di Napoli "Parthenope", Italy.
F. Arpino, Università degli Studi di Cassino, Italy

International Advisory Committee
W. Ambrosini, Università degli Studi dell'Aquila, Italy
C. Aprea, Università degli Studi di Salerno II, Italy
F. Asdrubali, Università degli Studi Roma Tre, Italy
P. Asinari, Politecnico di Torino, Italy
P. Baggio, Università degli Studi di Trento, Italy
N. Bianco, Università degli Studi di Napoli "Federico II" , Italy
G.P. Celata, Centro Ricerche ENEA, Italy
M. Costa, Istituto Motori - CNR, Italy
L. De Santoli, Università degli Studi di Roma "La Sapienza", Italy
D. Del Col, Università degli Studi di Padova, Italy
V. Fernicola, Istituto Nazionale di Ricerca Metrologica - INRiM, Italy
O. Manca, Università degli Studi della Campania Luigi Vanvitelli, Italy
N. Massarotti, Università degli Studi di Napoli "Parthenope", Italy
G. Morini, Università degli Studi di Bologna, Italy
V. Naso, Università degli Studi di Napoli "Federico II" , Italy
A. Nicolini, Università degli Studi di Perugia, Italy
A. Niro, Politecnico di Milano, Italy
P. Nithiarasu, Swansea University, Wales
A.J. Nowak, Silesian University of Technology, Poland
B. Panella, Politecnico di Torino, Italy
S. Rainieri, Università degli Studi di Parma, Italy
R. Ricci, Università Politecnica delle Marche, Italy
G. Rizzo, Università degli Studi di Palermo, Italy
B. Sarler,University of Ljubljan, Slovenia
M. Sasso, Università degli Studi del Sannio, Italy
L.A. Tagliafico, Università degli Studi di Genova, Italy
R. Vinuesa, KTH Royal Institute of Technology, Sweden
I. Vušanović, University of Montenegro, Montenegro

Local Organising Committee
L. Canale, Università degli Studi di Cassino, Italy
G. Cortellessa, Università degli Studi di Cassino, Italy
L. Moretti, Università degli Studi di Cassino, Italy
A. Pacitto, Università degli Studi di Cassino, Italy
L. Stabile, Università degli Studi di Cassino, Italy
M. Scungio, Università degli Studi della Tuscia, Italy
F. Zuena, Università degli Studi di Cassino, Italy

Contact details for the declaration

- ✓ Paolo Vigo, Università di Napoli "Parthenope", Italy, paolo.vigo@uniparthenope.it
- ✓ Marco Dell'Isola, Università degli Studi di Cassino e del Lazio Meridionale, Italy, dellisola@unicas.it
- ✓ Fausto Arpino, Università degli Studi di Cassino e del Lazio Meridionale, Italy, f.arpino@unicas.it
- ✓ Alfonso Niro, Politecnico di Milano, Italy, alfonso.niro@polimi.it

Peer review declaration

All conference organisers/editors are required to declare details about their peer review. Therefore, please provide the following information:

- **Type of peer review:** Double-blind review
- **Conference submission management system:** submissions are managed through the website www.uitconference2020.unicas.it
- **Number of submissions received:** 58
- **Number of submissions sent for review:** 49
- **Number of submissions accepted:**48
- **Acceptance Rate (Number of Submissions Accepted / Number of Submissions Received X 100):** 83%
- **Average number of reviews per paper:** 2
- **Total number of reviewers involved:** 60
- **Any additional info on review process:** none
- **Contact person for queries:**

Fausto Arpino
Department of Civil and Mechanical Engineering (DICEM), University of Cassino and Southern Lazio
Email address: f.arpino@unicas.it

A Numerical Approach to Study Shell-Side Fluid Flow in Shell-and-Tube Heat Exchangers

A. Quartararo[1], P.A. Di Maio[1], I. Moscato[2] E. Vallone[1], G. Guagliardo[1]

[1]Department of Engineering, University of Palermo, Viale delle Scienze, Ed. 6, 90128 Palermo, Italy
[2]Fusion Technology Department – Programme Management Unit, EUROfusion Consortium, Boltzmannstraße 2, 85748 Garching, Germany

Corresponding author's e-mail: andrea.quartararo@unipa.it

Abstract. The detailed CFD simulation of the shell-side flow inside shell-and-tube heat exchangers results prohibitive for large-size components, due to the large number of tubes, together with the multi-scale nature of the flow features. To overcome this issue, the porous media approach can be adopted, substituting the tube bundle with a porous domain, where a suitable momentum sink term in the Navier-Stokes equation is provided. In this paper, a novel numerical approach to estimate the crossflow shell-side pressure drop is presented: a complete characterization of the tube bundle was performed by running 2D-CFD steady-state isothermal analyses, collecting pressure gradient magnitudes and directions for different mass flow rates and angles of attack of the incoming flow. The approach was validated against numerical results and compared to the methodologies currently available. Models, assumptions, and boundary conditions are herewith reported and thoroughly discussed, alongside the main results obtained.

1. Introduction

Shell-and-tube Heat eXChangers (HXC) are the most widely used HXCs in process industries, power plants, petroleum refinery and chemical industries. This technology draws its great diffusion from the fairly large ratios of heat transfer area to volume and weight, the easy manufacturability and maintainability and the flexibility to meet a wide range of pressures, temperatures and mass flow rates, as well as the possibility to work with phase-changing and severely fouling and corrosive fluids [1]. The design of shell-and-tube HXCs is based on reliable and well-consolidated methodologies, mainly grounded on experimental data [2]. In fact, a large number of tubes, together with the multi-scale nature of the flow features inside the shell-and-tube HXCs, pose an obstacle to the development of detailed numerical-based methodologies, capable to reproduce the details of the flow field without the need to perform unduly geometrical simplifications.

In fact, despite the great increase of Computational Fluid Dynamics (CFD) codes capabilities, driven by the widespread availability of powerful computing resources, the detailed simulation of large HXCs, generally equipped with several thousand tubes, still results prohibitive. As a consequence, most of the recent research activity on HXCs involving detailed CFD simulations is focussed on small-size equipment [3, 4, 5]. Nevertheless, the level of detail achievable with these simulations allows to gain a deep understanding of the underlying physical processes, and it is of great value to check the occurrence of bypass and recirculation phenomena, responsible for an efficiency reduction of the HXCs.

To overcome this issue, the *porous media concept*, also referred to as *distributed resistance approach*, was firstly introduced by Patankar and Spalding [6], and further developed by other authors [7, 8, 9]. The idea behind this approach is to treat the tube bundle as a porous medium, adopting a modelling technique that correctly reproduces the interaction between fluid and tubes, by means of suitable source terms in the Navier-Stokes equations. The closure problem thus requires additional equations (usually experimental correlations) that relate pressure losses and heat transfer coefficient with the local field variables of the fluid. With regard to pressure loss, it can be obtained by adopting suitable friction factors and geometrical parameters to consider both the bundle porosity and its permeability.

It is moreover important to remark that the friction factor depends on both the shell-side fluid velocity magnitude and its direction. Nevertheless, while the distinction between cross and parallel-flow is always considered within this approach, the friction factor is usually assumed to be independent on the cross-flow angle [7, 10] or calculated separately along the two orthogonal HXC main directions [9].

In this paper, a different implementation of the porous medium approach is proposed: to provide a reliable estimate of the friction factor, an elementary cell of the tube bundle, namely *test bundle*, was simulated by means of 2D-CFD steady-state analyses, following a numerical technique based on the Finite Volume Method (FVM) and adopting the commercial CFD code ANSYS CFX 2020 R2 [11]. The test bundle simulations allowed to obtain a map of pressure gradient magnitudes and directions for different mass flow rates and angles of attack of the incoming flow, and the data acquired were used to define a cross-flow friction factor dependent on the local main flow direction.

To validate this novel approach, a complete 2D-CFD simulation of the inlet section of a large shell-and-tube HXC was performed, with a sufficient level of detail able to capture the flow field around more than 1200 tubes. Furthermore, the same geometry was simulated adopting the porous media approach in ANSYS CFX. The momentum sink term was defined both according to the formulation given in [10] and implementing the data obtained from the test bundle study, in order to assess the advantages of the proposed formulation.

2. Shell-side fluid flow in shell-and-tube HXCs

The shell-side fluid flow inside shell-and-tube HXCs is particularly complex and it is determined by the simultaneous presence of five mixing and interacting flow streams [12]:

- the main flow stream across the bundle;
- the leakage through tubes-to-baffle clearances;
- the flow bypass between the outermost tubes and the inner shell diameter;
- the baffle-to-shell clearance leakages;
- the flow through any channel within the bundle induced by the presence of pass dividers.

The fluid distribution among these five streams is determined by the relative hydraulic resistances of each one of these paths. Therefore, the correct evaluation of the tube bundle friction factor is of the utmost importance both to assess the overall pressure drop of the equipment and to correctly quantify the occurring bypass phenomena, ultimately responsible for a reduction of the HXC heat transfer performance. Among the several experimental correlations available in the literature for shell-side pressure drop in HXCs, Zukauskas' results are considered here [13]. The pressure drop inside a tube bundle can be evaluated by means of the dimensionless Euler number Eu (equivalent to a friction factor per unit row of tubes), defined as

$$Eu = \frac{\Delta p}{\frac{1}{2} N_{rows} \rho v_{max}^2} \tag{1}$$

where Δp is the total pressure drop, N_{rows} is the number of tube rows crossed by the flow, ρ is the fluid density and v_{max} its maximum velocity, calculated at the minimum bundle cross-section. By considering dimensional analysis arguments, Eu can be expressed as a function of Reynolds number Re and few geometrical parameters, i.e. pitch-over-diameter ratios for in-line and transverse directions. Experimental curves of Eu as a function of Re are available, for different bundle geometries both for in-line and staggered tube arrangements, either in graphical [13] and analytical [1] forms.

The adoption of Zukauskas' correlation inside the porous media model needs an additional manipulation: the velocity magnitude required for both Eu and Re numbers is evaluated at the minimum bundle cross-section "seen" by the moving fluid, and it is dependent on the flow main direction. It is thus necessary to re-define both Eu and Re as a function of the superficial velocity, i.e. the average velocity that would occur if the tubes were removed, simply employing geometrical considerations. The superficial velocity formulation enables the definition of a simple $Eu_s = f(Re_s)$ expression that can be adopted inside the porous media approach with no need to consider the bundle porosity and its direction-dependent permeabilities.

3. HXC inlet section 2D-CFD simulation

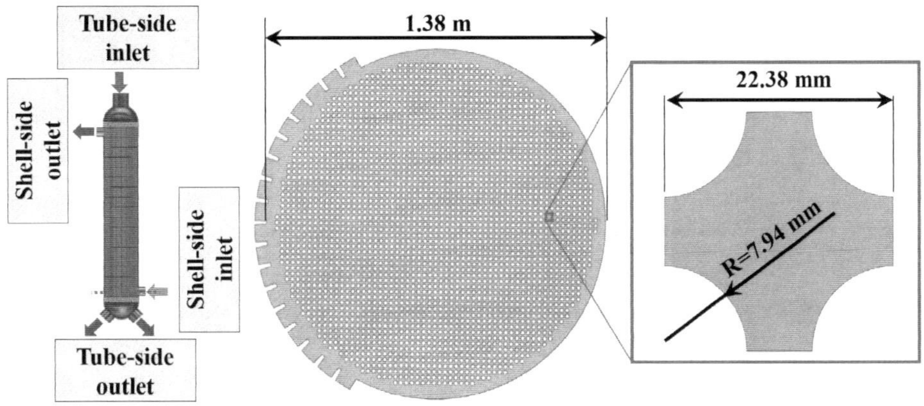

Figure 1. HXC complete geometry (left), 2D inlet section (center) and bundle elementary cell details (right).

The HXC considered for the following simulations is a smaller and simplified version of the DEMO fusion reactor Helium-Molten Salt Intermediate Heat Exchanger (IHX) [14]. With respect to the original DEMO IHX design, the inner shell diameter was roughly halved (from 2.70 to 1.38 m) with the aim to reduce the overall number of tubes (from 10903 to 2643), so to make the simulation manageable with the available computational resources. The details of the resulting geometry are represented in fig. 1. Moreover, the geometry was further simplified, as visible in fig. 2: only half domain was simulated for symmetry reasons, while the tube bundle was cut in correspondence to the baffle plate windows, where the main flow is expected to move along the HXC axial direction. The resulting total number of tubes considered amounts to 1204.

The analyses were performed adopting the Boundary Conditions (BC) shown in fig. 2, while the main assumption and the coolant operating conditions are reported in table 1. Regarding turbulence, ANSYS Generalized $k-\omega$ (GEKO) [11] model with a hybrid wall-function/low-Reynolds wall treatment was adopted.

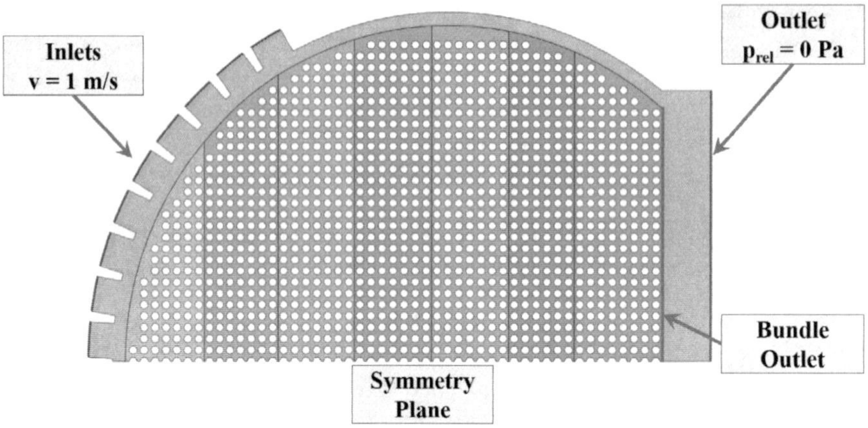

Figure 2. 2D-HXC calculation domain with adopted BCs.

Table 1. Summary of CFD analysis setup.

	Reference Conditions
Analysis Type	Steady-state isothermal
Material Library	Water
Density ρ **[kg/m^3]**	997
Dynamic Viscosity μ **[Pa s]**	$8.9{\cdot}10^{-4}$
Turbulence Model	GEKO
Boundary Layer Modelling	Automatic Wall Functions

Some details of the adopted mesh are depicted in fig. 3, while its main parameters and the quality metrics that are most relevant to ANSYS CFX are summarized in table 2. It is worth mentioning that the average mesh quality metrics are within the acceptable ranges prescribed in [11], and only a small fraction (< 1%) of the overall number of cells is characterized by poor quality. Consequently, the results are not expected to be significantly influenced by mesh quality. Moreover, the grid first layer thickness was chosen so to guarantee y^+ values of $\approx$1.

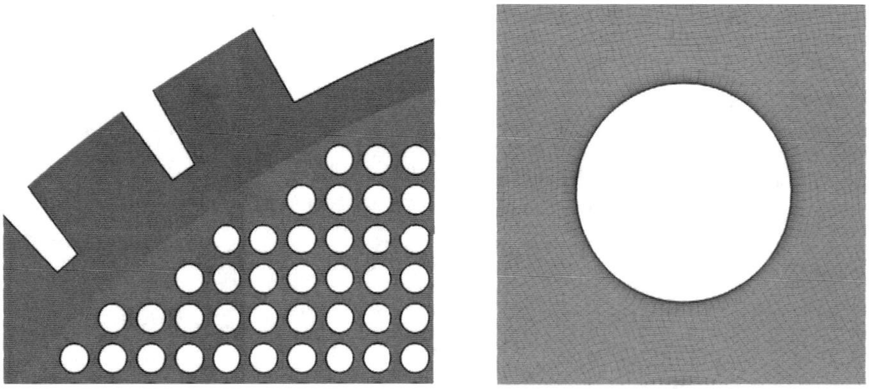

Figure 3. Details of the mesh adopted for the 2D-HXC calculation.

Table 2. Summary of the HXC main mesh parameters and quality metrics.

Mesh Parameter	Value
Nodes/Elements	$5.5{\cdot}10^6/5.4{\cdot}10^6$
Elements Topology	Hybrid (Hexa/Prism)
Inflation Layers Number	30
First Layer Thickness [μm]	2
Layers Growth Rate	1.15
Maximum Element Size [mm]	0.50
Mesh Metric	**Value**
Orthogonality Factor (average/min)	0.99/0.07
Expansion Factor (average/max)	1.1/4.5
Aspect Ratio (average/max)	36/247

Two different figures of merit (FoMs) were selected to assess the HXC performance: the total pressure drop and the bypass flow rate, defined as

$$f_{bypass} = 1 - \frac{\dot{m}_{BundleOutlet}}{\dot{m}_{Outlet}} \tag{2}$$

where the mass flow rates are calculated at the Outlet and Bundle Outlet locations as by fig. 2.

The velocity field inside the HXC is reported in fig. 4, while a total pressure drop of 33.7 kPa and a bypass factor of ≈22% were calculated. It is moreover interesting to observe in fig. 4 that the fluid is partially expelled from the tube bundle in the region circled in red, with a consequent increase of the bypass flow.

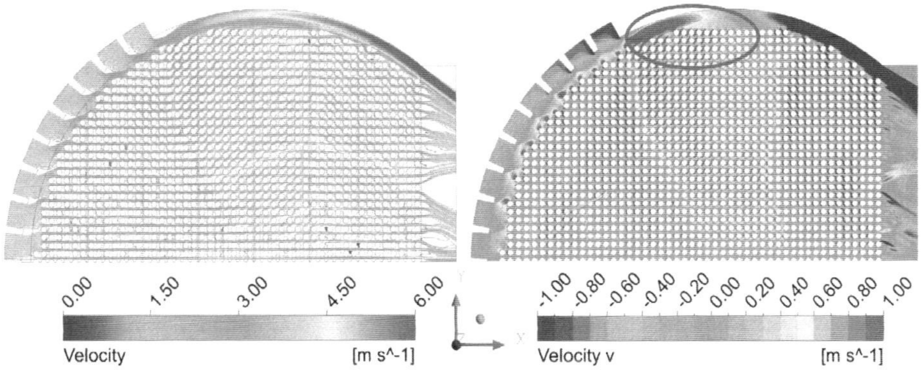

Figure 4. Velocity streamlines (left) and transverse flow velocity (right).

4. 2D-CFD Test Bundle Characterization

The elementary cell of the HXC described in section 3 (visible in fig. 5) was adopted as test bundle, to numerically evaluate the relations between Eu_s and Re_s at different main flow directions by means of 2D-CFD steady-state isothermal analyses, under the same coolant operating conditions of table 1.

With reference to fig. 6, several simulations were performed, imposing the main flow velocity magnitude and direction α. From the analysis of the results obtained, $-\nabla p$ magnitude and direction β were collected, providing a complete map of the test bundle hydraulic resistances. The selection of suitable BCs to be adopted for the simulation resulted challenging, due to the need to precisely control the main flow velocity direction while providing periodic BCs for the Left/Right and Top/Bottom couples of boundaries. The problem was overcome by adopting

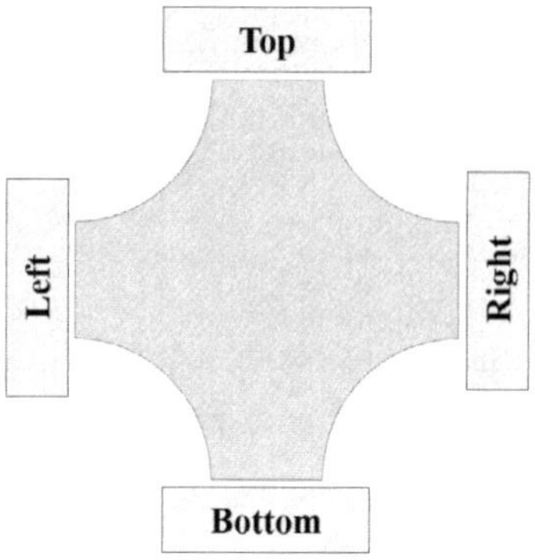

Figure 5. Test bundle geometry and boundary nomenclature.

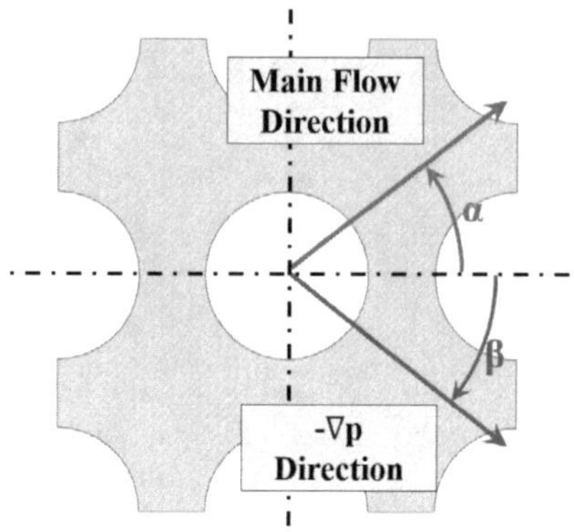

Figure 6. Main velocity and pressure gradient directions with respect to the bundle lattice.

Specified Mass Flow Rate periodic interface conditions (ICs). With this IC model, the solver implements the specified mass flow condition by adjusting the pressure change until the specified mass flow rate is satisfied. Adopting this approach, it is therefore possible to define a specified velocity magnitude taking advantage of the flow incompressibility.

The numerical campaign was conducted considering α values ranging between 0 and 45° at different Re_s values. The results obtained were extended to the range -180 to 180°, exploiting the tube bundle lattice symmetries, and the resulting curves are reported in fig. 7 and fig. 8.

As it may be argued from fig. 7, Eu_s exhibits a quite complex behaviour, characterized by many local minima. At higher values of Re_s, the variations of Eu_s are smaller and the curves tend to flatter. Moreover, it can be observed how the minimum pressure drop is obtained for $\alpha = 45°$ when Re_s is higher than ≈2000, and for $\alpha = 0°$ for lower Re_s numbers. Furthermore, it is interesting to notice in fig. 8 that the pressure gradient is not aligned with the main flow direction.

In particular, β exhibits a pronounced sawtooth shape for high Re_s values, while below $Re_s \approx 4000$, it becomes fairly independent on α.

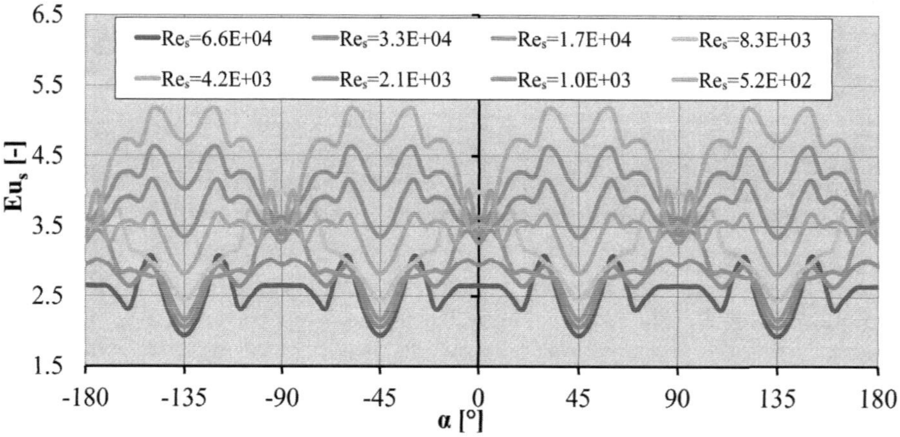

Figure 7. Eu_s as a function of the main flow velocity direction for different Re_s values.

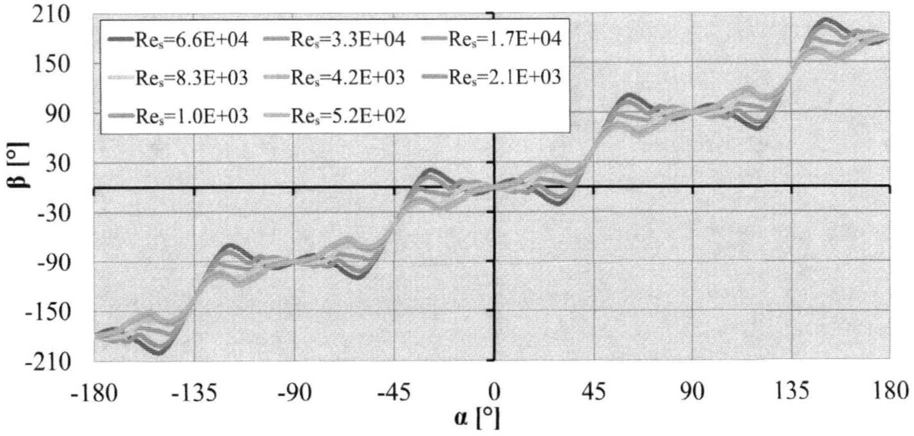

Figure 8. β as a function of the main flow velocity direction for different Re_s values.

Additionally, the results obtained for α=0° and 45° were compared with Zukauskas' experimental curves rescaled according to the superficial velocity formulation, respectively for in-line and staggered tube bundles, as visible in fig. 9. From the analysis of the results obtained, it can be argued that the CFD simulations are able to predict with a good accuracy the experimental results for $Re_s > 10^4$. For Eu_s outside this range, the agreement with experimental data is worse, probably due to turbulence modelling limitations or the assumption of 2D flow, with errors below 30%.

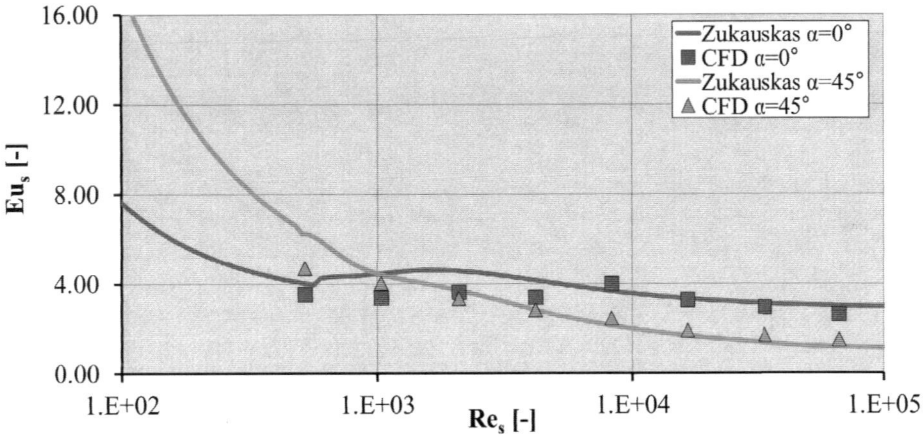

Figure 9. Comparison between Zukauskas' correlation and CFD results.

5. Porous media approach HXC 2D-CFD simulation

The 2D inlet section of the HXC of section 3 was simulated adopting the porous media approach under the same BCs and coolant operating conditions. The geometry was greatly simplified by replacing the tube bundle with a full porous medium, allowing a reduction of the mesh size (whose details are summarized in table 3) by a factor of ≈100, with a consequent remarkable decrease of the computational cost by a factor of ≈20. The momentum sink term was provided starting from the complete maps of Eu_s and β of fig. 7 and fig. 8, supplied in tabular format to the code. It is worth noticing how the reduction of the computational cost related to the lower number of cells is partly lost due to the necessity to interpolate the provided tabular data.

Table 3. Summary of the porous media HXC main mesh parameters and quality metrics.

Mesh Parameter	Value
Nodes/Elements	$5.8{\cdot}10^4/5.6{\cdot}10^4$
Elements Topology	Hybrid (Hexa/Prism)
Inflation Layers Number	30
First Layer Thickness [μm]	10
Layers Growth Rate	1.15
Maximum Element Size [mm]	10 (inside bundle region)
Mesh Metric	**Value**
Orthogonality Factor (average/min)	0.95/0.04
Expansion Factor (average/max)	1.1/84.0
Aspect Ratio (average/max)	36/452

The velocity field inside the HXC is reported in fig. 10, showing a good qualitative agreement with the results of section 3. Regarding the two FoMs, the calculated total pressure drop amounts to 35.7 kPa (+6% with respect to the results of section 3), while the bypass factor is ≈20% (against ≈22%).

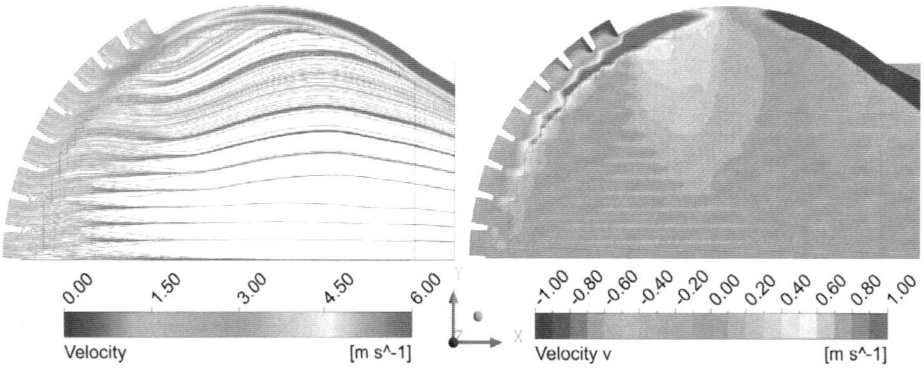

Figure 10. Velocity streamlines (left) and transverse flow velocity (right).

Additionally, to provide a comparison with the existing porous media approaches, the same geometry was tested directly implementing Eu_s from Zukauskas' correlation, following the methodology proposed in [10]. The simulation, whose details are not reported here for the sake of brevity, gave a slightly better prediction of the pressure drop (35.2 kPa, +4%), but considerably underpredicted the bypass flow (≈ 16.7% bypass). The greater bypass factor error may be related to the impossibility of the classical porous medium approach to predict the occurrence of the $\beta = 45°$ low-pressure drop direction at high Re_s, that contributes to the partial flow expulsion from the bundle.

6. Conclusions

A novel numerical methodology to estimate the crossflow shell-side pressure drop has been presented in this paper. To provide a realistic and direction-dependent formulation of the porous media approach, a numerical campaign was performed to study the hydraulic response of a small test bundle at different flow velocity magnitudes and directions. The novel approach was validated with a numerical benchmark, showing a good agreement in terms of pressure drop and bypass prediction. In particular, regarding this latter FoM, the novel methodology indicated better results if compared to the classical porous media formulations.

Disclaimer
This work has been carried out within the framework of the EUROfusion Consortium and has received funding from the Euratom research and training programme 2014-2018 and 2019-2020 under grant agreement No 633053. The views and opinions expressed herein do not necessarily reflect those of the European Commission.

References
[1] E. U. Schlunder, Handbook of Heat Exchager Design, Hemisphere Publishing, 1983.
[2] K. Bell, Delaware method for shell side design, in: S. Kakaç, A. Bergles, F. Mayinger (Eds.), Heat Exchangers: Thermal-Hydraulic Fundamentals and Design, Hemisphere Publishing Corporation, 1981.
[3] E. Ozden, I. Tari, Shell side cfd analysis of a small shell-and-tube heat exchanger, Energy Conversion and Management 51 (5) (2010) 1004–1014.
[4] Q. Wang, et al., Numerical investigation on combined multiple shell-pass shell-and-tube heat exchanger with continuous helical baffles, International Journal of Heat and Mass Transfer 52 (5) (2009) 1214–1222.
[5] J.-F. Zhang, et al., 3d numerical simulation on shell-and-tube heat exchangers with middle-overlapped helical baffles and continuous baffles – part i: Numerical model and results of whole heat exchanger with middle-overlapped helical baffles, International Journal of Heat and Mass Transfer 52 (23) (2009) 5371–5380.
[6] S. Patankar, D. Spalding, Computer analysis of the three-dimensional flow and heat transfer in a steam generator, Numerical Prediction of Flow, Heat Transfer, Turbulence and Combustion (1983) 293–298.
[7] M. Prithiviraj, M. J. Andrews, Three dimensional numerical simulation of shell-and-tube heat exchangers. part i: Foundation and fluid mechanics, Numerical Heat Transfer, Part A: Applications 33 (8) (1998) 799–816.
[8] W. T. Sha, ae al., Multidimensional Numerical Modeling of Heat Exchangers, Journal of Heat Transfer 104 (3) (1982) 417–425.
[9] C. Zhang, et al., Numerical Simulation of Different Types of Steam Surface Condensers, Journal of Energy Resources Technology 113 (2) (1991) 63–70.
[10] Y. He, et al., Numerical simulation and experimental study of flow and heat transfer characteristics of shell side fluid in shell-and-tube heat exchangers, che2005-05, in: Enhanced, Compact and Ultra Compact Heat Exchangers: Science, Engineering and Technology, Hemisphere Publishing Corporation, 2005.
[11] ANSYS Inc., ANSYS CFX-Solver Theory Guide, 2020, Release: 2020 R2.
[12] S. Kakaç, H. Liu, Heat exchangers: Selection, rating, and thermal design, second edition, CRC Press, 2002.
[13] A. Žukauskas, Heat Transfer from Tubes in Crossflow, Advances in Heat Transfer 8 (1972) 93–160.
[14] I. Moscato, Thermal-hydraulic study in support of the design of the DEMO Balance of Plant, Ph.D. thesis, Department of Engineering, University of Palermo, Handle: http://hdl.handle.net/10447/395492 (2020).

On the modelling of tritium transport phenomena at fluid-structure interfaces

F. M. Castrovinci, G. Bongiovì[1], P. Chiovaro[1], P. A. Di Maio[1], F. Franza[2], A. Quartararo[1], G. A. Spagnuolo[3], E. Vallone[1]

[1]Department of Engineering, University of Palermo, Viale delle Scienze, Ed. 6, 90128 Palermo, Italy
[2]Institute for Neutron Physics and Reactor Technology (INR), Karlsruhe Institute of Technology, Hermann-von-Helmholtz-Platz 1, 76344 Eggenstein-Leopoldshafen, Germany
[3]Fusion Technology Department - Programme Management Unit, EUROfusion Consortium, Boltzmannstraße 2, 85748 Garching, Germany

Corresponding author's e-mail: francescamaria.castrovinci@unipa.it

Abstract. One of the main functions of the DEMO Breeding Blanket (BB) system is to ensure the tritium breeding inside the reactor. Tritium is a beta emitter radioactive isotope, subjected to several processes that determine its permeation across materials and its leakage towards the environment, posing potential safety issues in terms of radiological hazard. Thus, the evaluation of tritium inventories inside components and tritium losses towards the environment plays a key role in the fulfilment of the pertinent BB safety requirements. In this regard, a research activity has been carried out, in close cooperation between the University of Palermo and the Karlsruhe Institute of Technology, focussing on the development of a multiphysical model that might realistically simulate 3D tritium transport phenomena across complex fluid-structure interfaces. Models, source terms and boundary conditions assumed for the analyses are herewith reported and critically discussed, together with the main results obtained.

1. Introduction

The European Roadmap, describing the priorities of fusion energy research development towards electricity production, outlines tritium self-sufficiency and management in fusion power plants among the most critical missions [1]. In DEMO, tritium will be produced inside the Breeding Blanket (BB) system and it is subjected to several processes that determine its permeation across materials. Therefore, the prediction of tritium inventories inside the DEMO BB components and tritium losses towards the environment plays a fundamental role to fulfil the relevant safety requirements.

Tritium transport modelling through the BB system is complex and involves a large number of parameters, most of which have not been determined with an adequate level of accuracy yet. Moreover, the applicability of tritium transport models to the DEMO BB requires the physics to be sufficiently comprehensive to take into account all phenomena that can take place under a wide range of operative conditions [2]. Indeed, within the BB, the T generated into the breeder material (e.g., Li-orthosilicate/metatitanate pebble beds or eutectic alloy of PbLi enriched with ^{6}Li) is transported by a T-carrier (He purge gas or the same PbLi) outside the blanket to be extracted. However, the T-carrier is in contact with tens of thousands of square meters of structural material at high temperatures (e.g. 300-500 °C) and with thin thickness. Therefore, the BB conditions are particularly favourable for the T permeation that can pose a safety concern in terms of contamination of the coolant system and releases into the Tokamak building as well as in the environment. Of course, several mitigation measures have been studied to reduce the T permeation (e.g. anti-permeation coatings in alumina) and to control the T inventories.

Due to the extreme complexity of the physics behind tritium permeation and in order to extend the main results of the analyses to more complex geometries, the adoption of a numerical approach is necessary, although it requires strong verification and validation. However, all the currently available tritium transport models (e.g. [3], [4]) refer in most of the cases to system-level or microscopic level, being the reason why a component level model (namely a model referring to an intermediate level) is necessary to improve the accuracy of the predictions in terms of tritium inventories, losses and contaminations of coolant flows [5].

Within this framework, a research campaign has been carried out at the University of Palermo, in close cooperation with the Karlsruhe Institute of Technology, in order to develop a 3D model that takes into account transport phenomena involving tritium transport across fluid-structure interfaces.

The study has been carried out following a theoretical-computational approach based on the Finite Volume Method (FVM) and adopting the commercial Computational Fluid-Dynamic (CFD) code ANSYS CFX 2019 R1 [6].

Tritium dissociative and recombinative phenomena occurring at fluid-solid interfaces (explained in section 2) are governed by a non-linear InterFace Condition (IFC) and the net tritium flux is a function of the local concentration of both diatomic and monoatomic species, which are the variables of the problem. The implementation of non-linear IFCs is not straightforward in the code and, therefore, two novel approaches have been introduced in this work. As a first step, 1D diffusion problems have been considered and the results of their relevant numerical simulations have been compared against the pertaining analytical solutions (section 3) to assess the predictive potential of the methodologies adopted (section 4) and validate the modelling approaches (section 5).

Moreover, parametric analyses have been performed and different cases have been studied by varying the model free parameters, to investigate their impact on permeation rates and concentration profiles.

2. Tritium permeation phenomena

Tritium permeation at fluid-solid interfaces is mainly governed by different surface and bulk processes as schematized in **Figure 1**. In the following, the most significant phenomena will be described.

Surface processes are referred to all the processes occurring on and through the surface of a solid. This category includes adsorption, dissociation, recombination, and desorption, summarized by the following chemical equilibrium expression:

$$T_2 \rightleftarrows 2T \tag{1}$$

In general, surface processes occur in a thick layer, just below the gas-solid interface, which is referred to as a "surface layer". This layer exists due to several causes, such as natural oxide formation, deposition of a coating layer or modification by implantation. It is typically some hundreds Å thick and is not uniform, since the oxygen content of the oxide layer and the density of implanted products changes with depth [7].

Bulk processes comprehend atomic dissolution and interstitial diffusion caused by concentration gradients. It is found experimentally that, from a macroscopic point of view, the flux J [mol m^{-2} s^{-1}] of dissolved tritium atoms inside a solid follows Fick's Law [7]. Two other bulk processes are thermomigration and trapping, the latter occurring when tritium bonds to microstructural features within metals. Both are thermally-driven processes that in this work have been neglected since isothermal conditions have been considered.

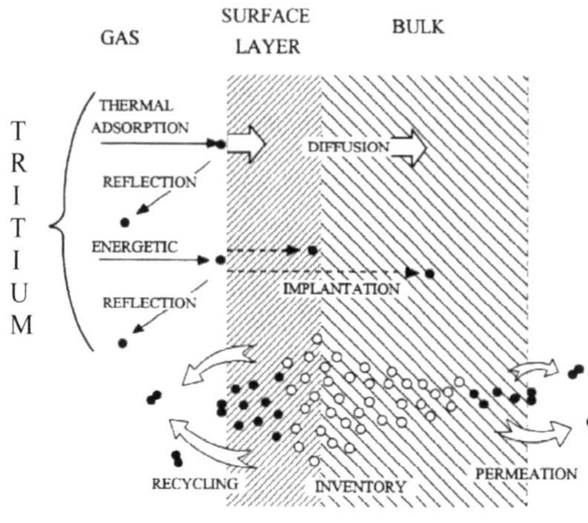

Figure 1. Tritium transport processes across solid boundaries [7].

Permeation can be either limited by surface or diffusion phenomena. The occurring regime is determined by a permeation number W that takes into account tritium partial pressure, surface properties, bulk thickness and permeabilities [8]. Diffusion-limited regime occurs when high partial pressure differences are established across the solid region and, in particular, tritium migration through the metal membrane is limited mainly by diffusion in the metal lattice while the surface processes (such as adsorption, desorption) are significantly faster. On the other hand, when low tritium partial pressure differences occur, a surface-limited regime is established, and the diffusion through the membrane occurs fast enough so that any concentration gradient is cancelled by diffusion [9]. In this case, permeation is limited by the rate at which atoms can recombine back into molecules. Except for extremely high temperatures, this recombination is necessary for tritium to be released from a material.

Tritium dynamics in solid and fluid media is governed by diffusion and/or convective transport phenomena. The set of equations to be solved is strongly simplified by the fact that tritium can be considered a passive scalar since it has no significant influence on the velocity, density and temperature fields. The general transport equation in a generic medium can be derived from the mass continuity equation for the species i in a medium j and it is given by (2):

$$\frac{\partial C_i(\mathbf{x},t)}{\partial t} + \nabla \cdot \mathbf{u}(\mathbf{x},t) C_i(\mathbf{x},t) - \nabla \cdot (D_{ji} \nabla C_i(\mathbf{x},t)) = S_i(\mathbf{x},t) \tag{2}$$

where $\mathbf{x}$ and t denote the space and time coordinates, C_i [mol m^{-3}] is the i-th species concentration field, $\mathbf{u}$ is the vector velocity field, D_{ji} is the diffusivity 2nd order tensor of the i-th species in the j-th medium and S_i is the i-th species source term. Obviously, for the case of a solid medium, the governing equation is further simplified for the absence of the convective term.

The permeation regime mainly depends on the tritium partial pressures in the fluid mean which is strongly related to the examined problem. To avoid any a-priori choice of the permeation regime a more general model derived from classical kinetic theory has been adopted here to calculate the net flux J_i of the atomic species i through a metal surface [10].

Tritium permeation occurs in different ways depending on the nature of the fluid and solid considered. For the aim of this study, a fluid where tritium is present in molecular form has been investigated. Moreover, only the molecular species T_2 has been considered, neglecting for the sake of simplicity the presence of H_2 and HT. Tritium molecules are adsorbed at the interface and dissociated into two T atoms. Then, atoms may desorb from the surface and recombine back into gas as T_2 molecules. The dissociation and recombination fluxes at interfaces (from the fluid side) can be expressed as:

$$J_{T_2,d} = K_d \cdot p_{T_2} \tag{3}$$

$$J_{T_2,r} = K_r \cdot C_T^{\,2} \tag{4}$$

where K_r and K_d are phenomenological constants called recombination and dissociation constant respectively, p_{T_2} is the T_2 partial pressure and C_T is the concentration of the T species.

If the system is not in an equilibrium state, dissociation and recombination fluxes at fluid-structure interfaces are not counterbalanced, and net transport of tritium occurs from one side to the other. The net flux is thus:

$$J_{T_2} = K_d \cdot p_{T_2} - K_r \cdot c_T^{\,2} \tag{5}$$

Considering a gas with a temperature T_g, Dalton's law (6) expresses the relationship between the concentration of T_2 (C_{T_2}) and its partial pressure:

$$C_{T_2} = \frac{1}{RT_g} p_{T_2} = K_g p_{T_2} \tag{6}$$

This relation allows fluxes to be expressed as a direct function of the concentrations as follows:

$$J_{T_2} = K \cdot C_{T_2} - K_r \cdot C_T^{\,2} \tag{7}$$

where K is the ratio of K_d to K_g.

Since the concentration in solid and gas is referred to different species, the continuity of the fluxes at the interface cannot be applied, but looking at the reaction (1), the stoichiometry of the dissociation reaction defines a relationship between the species fluxes, that can be expressed as:

$$J_T = 2 \cdot J_{T_2} \tag{8}$$

Summarizing the conditions at the interface, under the hypothesis of homogeneous and isotropic media, the Cauchy IFCs between gas and solid domains are given below.

$$\begin{cases} -D_{s,T} \nabla C_T \cdot \hat{n}\big|_{g\to s} = 2 \cdot (K \cdot C_{T_2} - K_r \cdot C_T^{\,2}) \\ (-D_{g,T_2} \nabla C_{T_2} + \vec{u}_g \cdot C_{T_2}) \cdot \hat{n}\big|_{s\to g} = K \cdot C_{T_2} - K_r \cdot C_T^{\,2} \end{cases} \tag{9}$$

From the IFCs it is possible to notice how there is neither continuity in the concentration nor in the flux, and furthermore that the fluxes depend non-linearly on the concentrations of the two species. Therefore, these equations significantly differ from those adopted in standard IFCs and their implementation into commercial computational fluid-dynamic codes results quite complex. To this purpose, two novel approaches have been considered and validated against analytical solutions to open the way to the assessment of tritium permeation properly considering 3D transport phenomena across fluid-solid interfaces.

3. Analytical model definition

In mathematical form, the transient mass conservation law (2), resolving the space-time distribution concentration C_i of the i-th species, can be easily solved in one-dimensional geometries, with a reference frame system defined by the space coordinate x (the scalar distribution is supposed to depend only on the x coordinate for the plane geometry and on the radial coordinate r for the cylindrical and the spherical geometries). Note that in the following dimensionless equations the generalized coordinate $\tilde{X}$ has been normalized to the total length of the domain L and, analogously, the problem variables, i.e. concentrations c_i, have been normalized to a reference concentration C_0.

The domain has been split into two parts, a fluid domain from $\tilde{X}=0$ to $\tilde{X}=\Xi$ and a solid domain, from $\tilde{X}=\Xi$ to $\tilde{X}=1$, connected by their common interface.

To obtain an analytical closed-form solution some assumptions have been introduced. The steady-state condition has been considered and a fluid at rest has been studied to focus attention only on the tritium transport phenomena. Under the hypothesis of homogeneous and isotropic media, incompressible fluid and a uniform volumetric source term inside the fluid domain Σ, equation (2) is particularized into two ordinary differential equations (ODEs), i.e. the diffusion equation for fluid domain and the diffusion equation for the solid domain.

The set of Boundary Conditions (BCs) assigned to analytically solve the Cauchy's problem consists of a Neumann BC, imposed at the fluid domain (at $\tilde{X} = 0$) due to symmetry constraint, a Dirichlet BC in the solid domain (at $\tilde{X} = 1$), and of the two non-linear IFCs at Ξ. The resulting set of ODEs together with their BCs is reported in equations (10)

$$\text{a)}\begin{cases} \dfrac{1}{\tilde{X}^m}\dfrac{d}{d\tilde{X}}\left(\tilde{X}^m\dfrac{dc_{T_2}}{d\tilde{X}}\right)+\Sigma=0 \\ \left.\dfrac{dc_{T_2}}{d\tilde{X}}\right|_{\tilde{X}=0}=0 \\ \left.\dfrac{dc_{T_2}}{d\tilde{X}}\right|_{\tilde{X}=\Xi}=-\left(\Lambda c_{T_2}-\Gamma c_T{}^2\right) \end{cases} \qquad \text{b)}\begin{cases} \dfrac{1}{\tilde{X}^m}\dfrac{d}{d\tilde{X}}\left(\tilde{X}^m\dfrac{dc_T}{d\tilde{X}}\right)=0 \\ c_T(\tilde{X}=1)=\dfrac{C_X}{C_0} \\ \left.\dfrac{dc_T}{d\tilde{X}}\right|_{\tilde{X}=\Xi}=-2\delta\cdot\left(\Lambda c_{T_2}-\Gamma c_T{}^2\right) \end{cases} \tag{10}$$

where m = 0, 1, 2 for Cartesian, cylindrical and spherical coordinate systems, respectively.

As regards the sensitivity analyses, an additional set of BCs has been considered for the Cartesian geometry (see equations (11)), maintaining the fluid left boundary at a fixed non-zero concentration (at $\tilde{X} = 0$), without any source term inside the fluid domain.

The set of BCs adopted in equations (10) entails that the tritium flux across the interface is uniquely determined by the imposed volumetric source term, while concentration is unknown and has to be computed. Enforcing Dirichlet-Dirichlet BCs at the ends of the domain, on the contrary, the tritium flux becomes a result of the problem and can be used to assess the quality of the different modelling strategies.

$$\text{a)}\begin{cases} \dfrac{d^2c_{T_2}}{d\tilde{x}^2}=0 \\ c_{T_2}(\tilde{x}=0)=1 \\ \left.\dfrac{dc_{T_2}}{d\tilde{x}}\right|_{\tilde{x}=\xi}=-(\Lambda c_{T_2}-\Gamma c_T{}^2) \end{cases} \qquad \text{b)}\begin{cases} \dfrac{d^2c_T}{d\tilde{x}^2}=0 \\ c_T(\tilde{x}=1)=0 \\ \left.\dfrac{dc_T}{d\tilde{x}}\right|_{\tilde{x}=\xi}=-2\delta\cdot(\Lambda c_{T_2}-\Gamma c_T{}^2) \end{cases} \tag{11}$$

Dimensionless coefficients that figure in the two sets of equations are given in **Table 1**.

Table 1. Coefficients of the dimensionless form equation.

Dimensionless coefficients				
Γ	Λ	δ	Ξ	Σ
$\dfrac{K_r L}{D_{g,T_2}} C_0$	$\dfrac{KL}{D_{g,T_2}}$	$\dfrac{D_{g,T_2}}{D_{s,T}}$	$\dfrac{X_1}{L}$	$\dfrac{S_{T_2} L^2}{D_{g,T_2} C_0}$

Directly integrating the ODEs with constant coefficients - and homogeneous for the solid domain- and applying the related BCs, the concentration distribution profiles c_{T_2} and c_T have been obtained.

4. Numerical model definition

The aforementioned model has been implemented using the commercial code ANSYS CFX which allows solving numerically the tritium transport equation using the FVM approach.

In ANSYS CFX, two adjacent domains can be connected only if the flux continuity condition is imposed. If this condition is not met (8), the two domains are treated as independent, and field variables belonging to one domain cannot be retrieved from the other. Consequently, the imposition of an IFC defined as a function of local concentrations on both sides of the interface is not foreseen by the code. To overcome this issue, two alternative approaches have been conceived and they have been reported in the following paragraph.

The first approach investigated consists of the definition of a thin layer - which, hereafter, is referred to as “dummy layer”- derived from the solid volume, between the fluid and the solid domain. Atomic concentration C_T and molecular concentration C_{T_2} have been defined separately in solid and gas domains as specific passive scalars, while both concentrations have been defined inside the dummy layer.

The dummy layer has to be intended only as a numerical artifact, necessary for the implementation of the correct IFCs. The processes of dissociation and recombination are superficial processes that, due to the solid surface irregularities, occurs in a thin surface layer. Despite this, there is no substantial physical relation between the dummy layer and the physical surface layer.

Moreover, also from a geometric point of view, the dummy layer thickness must be larger than the physical layer thickness for numerical reasons, but it should be significantly thinner than the characteristic size of the domain. The effect of layer thickness assumption has been investigated separately by running a set of dedicated parametric analyses.

Inside this layer, the surface phenomena occurring at fluid-structure interfaces are simulated as bulk phenomena by introducing properly defined effective source terms in the transport equations of the two scalars C_T and C_{T_2}, reported as follows:

$$S_T = \frac{2 \cdot \left(K \cdot C_{T_2} - K_r \cdot C_T^2\right)}{\delta} \cdot \rho_s \tag{12}$$

$$S_{T_2} = -\frac{\left(K \cdot C_{T_2} - K_r \cdot C_T^2\right)}{\delta} \cdot \rho_g \tag{13}$$

where δ is the dummy layer thickness, ρ_s and ρ_g are the solid and the gas density, respectively.

This approach holds because, integrating the above-defined source terms over the dummy layer thickness, the results are equal to the fluxes exchanged at the interface, under the assumption that the concentration of the two species is uniform along the dummy layer width. Consequently, the dummy layer has to be small enough to avoid the development of profile concentrations along its thickness.

Furthermore, different sensitivity analyses have been carried out to verify how the several free parameters of the dummy layer approach may influence the results.

The other approach investigated is based on the implementation of a *Fortran User Routine* in the code ANSYS CFX. The subroutine makes the concentration of the i-th species, initially defined in a domain, available on the interface side belonging to the adjacent domain. Thus, it becomes possible to enforce a flux condition at the interface boundaries as a function of concentrations on both sides (9).

The two novel numerical methods have been tested considering simple model problems, consisting of a fluid and a solid domain, in which tritium diffuses, connected by means of an interface. Steady-state isothermal analyses have been performed considering three elementary geometries - Cartesian, cylindrical and spherical - to validate the numerical results against the relevant analytical models.

Concerning the geometrical features, for each one of the three cases, the solid domain length has been chosen as half of the fluid domain width. In particular, fluid domain width in plane geometry is 1 [m], while the radius of the fluid domain in cylindrical and spherical geometries has been chosen equal to 0.5 [m]. The analysed scenario consists of a fluid at rest with a constant and uniform tritium source (fixed at 1 [$mol \cdot m^{-3} \cdot s^{-1}$]) and a solid external surface maintained at a constant concentration equal to zero.

It has to be highlighted that the geometrical features, the BCs and volume sources are not physically representative of the major engineering applications related to this methodology and, consequently, these model problems have to be intended just for validation purposes.

Journal of Physics: Conference Series 2177 (2022) 012002 doi:10.1088/1742-6596/2177/1/012002

5. Results

Results have been reported in terms of concentration distribution profile along the x-axis for the Cartesian geometry and along the r-axis for the cylindrical and spherical geometries.

The computational results calculated by means of ANSYS CFX code have been compared to their analytical equivalents described in §3. Nodal concentration values obtained with the two approaches are compared to the analytical results and reported for the most significant case, i.e. the cylindrical geometry, in **Figure 2**.

Both concentration and the r coordinate are normalized to the reference variables defined in §3 and for the three geometries studied with a dummy layer thickness of 2 mm, which corresponds to 0.4% of the solid domain thickness.

Maximum and average errors of the two concentration profiles to the local analytical solutions are reported in **Table 2**.

As it is possible to observe in **Figure 2**, numerical results obtained with the two novel methodologies are in good agreement with the analytical results. The maximum error occurs in the proximity of $\tilde{X} = 1$ where the concentration approaches zero due to the Dirichlet BC that has been enforced. In particular, maximum errors of about 1% and average errors lower than 0.5% have been encountered.

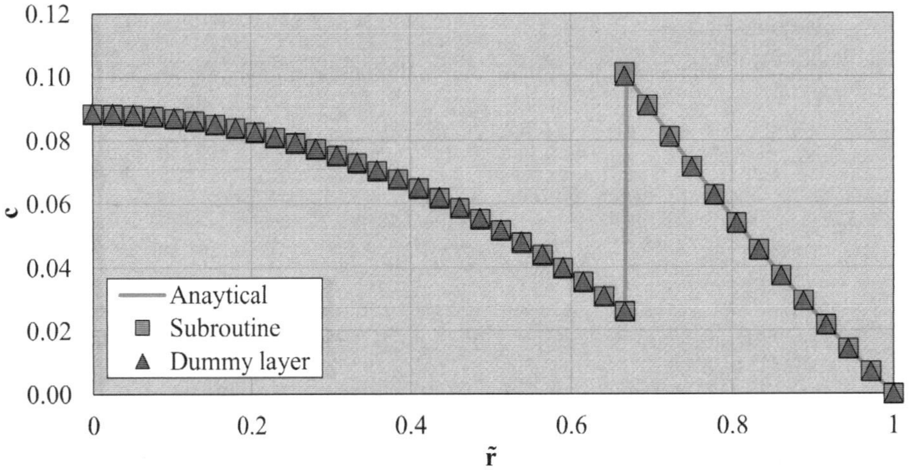

Figure 2. Concentration distribution for cylindrical geometry.

Table 2. Maximum and average errors between numerical and analytical results.

	Dummy layer approach	Subroutine approach
ε_{max}	1.22%	0.76%
ε_{ave}	0.35%	0.24%

The numerical simulations showed that the dummy layer approach requires a lower computational time but its implementation to complex geometries may result more challenging. Indeed, if a detailed geometry has to be assessed, the realization of the dummy layers for each interface can be cumbersome. Conversely, although the subroutine approach is the most promising one because of its easy implementation, it requires a higher number of iterations to converge that may limit its adoption in fusion-relevant cases.

Afterwards, considering the case of the Cartesian geometry, parametric analyses have been performed for different combinations of layer cells number (#Cells), material diffusivity (D) and thickness (S). These are reported in **Table 3** where the layer thickness has been normalized to the solid domain length.

Table 3. Combination of free parameters investigated.

Layer thickness									
3.33E-4		1.67E-3		3.33E-3		1.67E-2		3.33E-2	
D[m²/s]	**#Cells**	**D[m²/s]**	**#Cells**	**D[m²/s]**	**#Cells**	**D[m²/s]**	**#Cells**	**D[m²/s]**	**#Cells**
1E-5	1	1E-5	1	1E-5	1	1E-5	1	1E-5	1
1E-6	2	1E-6	2	1E-6	2	1E-6	2	1E-6	2
1E-7	3	1E-7	3	1E-7	3	1E-7	3	1E-7	3
1E-8	4	1E-8	4	1E-8	4	1E-8	4	1E-8	4

In this case, results have been validated against the Dirichlet-Dirichlet analytical model. In particular, the concentrations and the net flux at the interface on both sides have been compared to the analytical values.

Results have been reported in terms of relative errors on concentration at the fluid and solid sides of the interface and errors on the net flux of T_2. They have been summarized in **Table 4** where cases A, B, C and D stand for dummy layer diffusivities of 1E-5 [m²/s], 1E-6 [m²/s], 1E-7 [m²/s] and 1E-8 [m²/s] respectively.

Table 4. Errors between numerical and analytical models with S = 3.33E-4.

	$\varepsilon(C_{T_2})$ [$\cdot 10^{-3}$ %]				$\varepsilon(C_T)$ [$\cdot 10^{-3}$ %]				$\varepsilon(\varphi_s)$ [$\cdot 10^{-3}$ %]			
# Cells	**A**	**B**	**C**	**D**	**A**	**B**	**C**	**D**	**A**	**B**	**C**	**D**
1	0.2	0.2	0.2	0.1	21.8	22.3	23.4	34.5	5.0	4.8	3.7	7.5
2	0.4	0.4	0.4	0.2	12.3	12.3	12.9	19.2	14.8	14.8	14.1	7.5
3	0.4	0.4	0.4	0.3	10.2	10.2	10.8	16.5	16.7	16.5	16.0	10.4
4	0.4	0.4	0.4	0.3	9.7	9.7	10.2	15.5	17.3	17.3	16.7	11.3

Errors have been reported just for the case with a normalized thickness of 3.33E-4 which represents the best case with the lowest errors. This was expected since, as mentioned before, the dummy layer thickness should be of the same order of magnitude as the surface layer.

For the reasons explained in chapter §3, results have been analysed mainly in terms of errors on the T_2 flux. Looking at the effect of diffusivity, it can be observed that the best configurations are those with the highest diffusivity. The reason is related to the hypothesis of no gradient in concentration across the dummy layer; indeed, if the diffusivity is higher, the profiles inside the layer are flatter, resulting in a better agreement with the analytical models. On the other hand, a decreased diffusivity would play an important role in the case of 3D solutions, since it could be responsible for an increased tritium diffusion perpendicularly to the interface normal direction.

The last parameter is the number of cells across the layer; as it can be observed, just one cell is the optimal configuration, once again for the hypothesis of no concentration gradients, and here, the positive effect is only related to the numerical approximation of the problem in case just one cell is used to discretize the interface, the gradients are inhibited by the averaging properties of the FVM.

Moreover, the presence of a finite-thickness layer, derived from the solid domain, affects the profile that occurs in the solid. In particular, concerning the effects of the diffusivity and number of cells of the dummy layer, the errors exhibit the same trend for the flux and the T_2 concentration at the gas side, while it is the opposite for the T concentration at the solid side. However, the errors remain very low.

To summarize, the best combination of parameters in terms of permeation flux to be adopted is the one with a thin dummy layer, one cell across its length and higher diffusivity. Under these assumptions, it is possible to obtain a configuration where the maximum error is related to the concentration in solid and it is equal to 0.0218%, while the errors on flux and concentration in the gas are practically equal to zero.

As shown in the previous sensitivity analyses, the layer thickness significantly influences the results and, therefore, this parameter has been studied more in detail. In particular, its effect has been evaluated on the cylindrical geometric to see how the curvature of the geometry could affect the results in terms of concentration distribution.

For this reason, six different normalized layer thicknesses, ranging from 8E-3 to 8E-2, have been investigated and results in terms of peaks of concentration distributions compared to analytical solutions of equations (10), are reported in **Figure 3**.

As shown in **Figure 3**, numerical results fit better to the analytical results when the layer width gets thinner and it is possible to see how, increasing the layer thickness, the concentration peak gets lower and far from the analytical results.

Moreover, it can be observed how the errors remain limited up to a normalized dummy layer thickness of 3.2E-2. Above this threshold value, a departure from linearity is observed and the error increases significantly, due to the combined effect of curvature and the development of a concentration profile inside the layer. It can be thus concluded that the effects of curvature remain negligible when the dummy layer thickness is smaller than 1% of the local curvature radius.

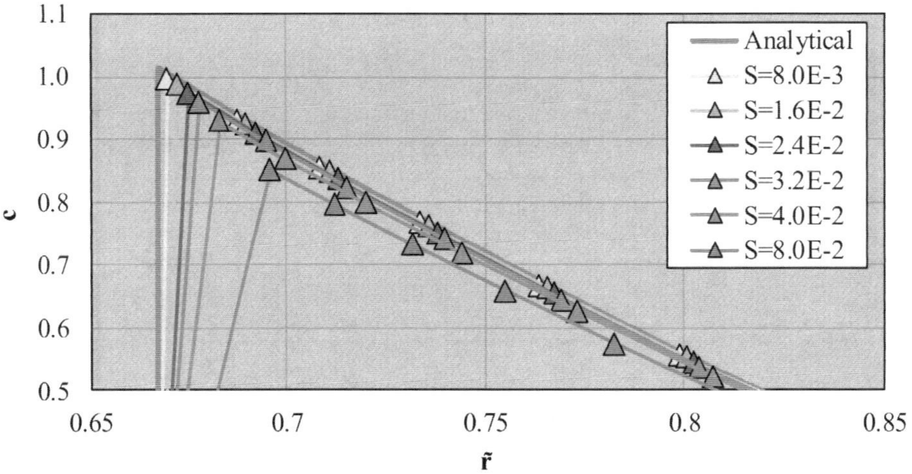

Figure 3. Zoom of concentration peaks near the gas/solid interface for different values of the layer thickness.

Sensitivity analyses reported above have been carried out only for the transition regime between diffusion limited and surface-limited model (with a permeation number $W \approx 1$ [8]) where both processes, diffusion and recombination, are not negligible. Of course, sensitivity analyses on phenomenological constants and studies on the different permeation regimes are outstanding topics for the future.

6. Conclusions

In the frame of the European DEMO BB studies, a research campaign has been carried out at the University of Palermo in cooperation with the Karlsruhe Institute of Technology, aimed at developing a 3D tritium permeation model that could take into account the processes occurring at the fluid-structure interfaces.

A theoretical-computational approach based on the FVM has been followed and the commercial code ANSYS CFX has been adopted. Due to the code limitations for the implementation of non-linear IFCs, two novel approaches have been developed. The first consists of the implementation of surface phenomena in a thin layer (referred to as "dummy layer") as bulk phenomena, while the second approach is based on the implementation of a specific *Fortran User Routine* in the code.

Both methodologies have been fully validated against analytical results, with very small errors, in several diffusion problems. Furthermore, sensitivity analyses have been carried out to assess the impact on the results of the free parameters of the dummy layer approach. In particular, the best configuration (i.e. leading to the smallest numerical errors) comprehends a thin dummy layer, with one cell volume discretization and a higher diffusivity. In the case of cylindrical geometries, it has been found that the effects of curvature remain negligible when the dummy layer thickness is smaller than 1% of the local curvature radius.

Outstanding topics for the future are mainly the integration of the thermofluid-dynamics in the model and its extension to more complex geometries, starting with the implementation of the approaches in 2D and 3D models.

Further developments of this work will aim at the implementation of other phenomena such as trapping, which is significant for fusion-relevant applications. The second step will be the expansion of the model to a multispecies model that includes the hydrogen molecules' transport and their recombination and dissociation at interfaces.

Disclaimer

This work has been carried out within the framework of the EUROfusion Consortium and has received funding from the Euratom research and training programme 2014-2018 and 2019-2020 under grant agreement No 633053. The views and opinions expressed herein do not necessarily reflect those of the European Commission.

References

[1] Donné A J H, The European roadmap towards fusion electricity, *Phil. Trans. R. Soc.*, **377** (2019) 20170432

[2] Ricapito I et al., Tritium transport modelling for breeding blanket: state of the art and strategy for future development in the EU fusion program, *Fus. Eng. and Des.* **87** (2012) 793–797.

[3] Carrella E et al., Tritium modelling in HCPB breeder blanket at a system level, *Fus. Eng. and Des.* **124** (2017) 687–691.

[4] Longhurst G R, *TMAP7: Tritium Migration Analysis Program, User Manual*, Idaho National Laboratory, INEEL/EXT-04-02352, 2004.

[5] Pasler V et al., Development and verification of a component-level hydrogen transport model for a DEMO-like HCPB breeder unit with OpenFOAM, *Fus. Eng. and Des.* **127** (2018) 249-258.

[6] ANSYS Inc., *ANSYS CFX Solver Theory Guide*, Release: 2019 R1, 2019

[7] Serra E, *Hydrogen and Tritium Kinetics in Fusion Reactor Materials*, Joint Research Centre, (1996) EUR 16471 EN

[8] Esteban G A, The surface rate constants of deuterium in the reduced activating martensitic steel OPTIFER-IVb, *Journ. of Nuc. Mat.*, **282** (2000) 89-96

[9] Franza F, *Tritium Transport Analysis in HCPB DEMO Blanket with the FUS-TPC Code*, KIT Scientific Reports, 7642

[10] Zhang H et al., Characterization of Tritium Isotopic Permeation Through ARAA in Diffusion Limited and Surface Limited Regimes, *Fus. Sci. and Tec.* **72**, p. 416–425, 2017

Three-dimensional MHD flow in moderate change ratio orifice

Lorenzo Melchiorri, Alessandro Tassone, Gianfranco Caruso

DIAEE–Sapienza University of Rome, Corso Vittorio Emanuele II, 244, 00186 Roma, Italy

lorenzo.melchiorri@uniroma1.it

Abstract. In fusion reactor blanket design, liquid metals are attractive working fluids since it is possible to combine in a single fluid the functions of coolant, tritium carrier and breeder. These electrically conductive fluids flow in the presence of a strong magnetic field, inducing the appearance of Lorentz forces and magnetohydrodynamic MHD effects. Increased pressure loss, particularly in complex geometry elements, is a critical point for blanket design. The MHD flow through an orifice plate made by electroconductive walls (c = 0.01 ÷ 0.1) has been analysed in this paper using ANSYS CFX in the range Re = 108, and Ha = 0 ÷ 300. A wide recirculation region is detected after the flow exits the orifice, with potentially harmful consequences for efficient tritium removal. Large pressure loss occurs in the orifice due to conductive wall and non-negligible axial length. The 3D pressure drop term is characterized through a local resistance coefficient (k) that is found to be k ≈ 0.205 for well conducting walls (c = 0.1) and k ≈ 0.063 for poorly conducting ones (c = 0.01).

1. Introduction

In fusion reactor blankets, liquid metals are attractive due to the possibility to combine in a single fluid the functions of coolant, tritium carrier and breeder. These electrically conductive fluids interact with the intense magnetic field ($\approx 4 - 8\,T$), used to confine the plasma, causing the appearance of magnetohydrodynamic (MHD) phenomena. Increased pressure drop and flow redistribution are two effects that need to be correctly quantified to support the blanket design [1]. If the fully developed flow behavior is well understood, prediction of MHD pressure losses for developing flows in complex geometrical elements is considerably harder and far from a satisfactory theoretical explanation due to the many governing parameters involved [2]. Experiments and numerical simulations in prototypical configurations are used to estimate these losses and support the blanket design.

Sudden cross-section expansion and contraction are common hydraulic elements that exhibit complex MHD flow features like 3D currents and internal free shear layers originating from the corners [3], [4]. As such, they contribute considerably to the blanket pressure loss and, thus, an accurate estimate is of paramount importance. The most important experimental work dealing with this topic is probably the one performed by Bühler et al. [5]. Numerous simulations have been performed to characterize these components in terms of pressure loss and flow features for both insulating [6]–[9] and electrically conductive [10]–[14] rectangular ducts. A critical point is the orientation of the preferential direction for the duct expansion/contraction regarding the magnetic field, parallel variation being the most pressure loss intensive.

Considerably less studied is the simultaneous effect of a contraction followed by an expansion, as

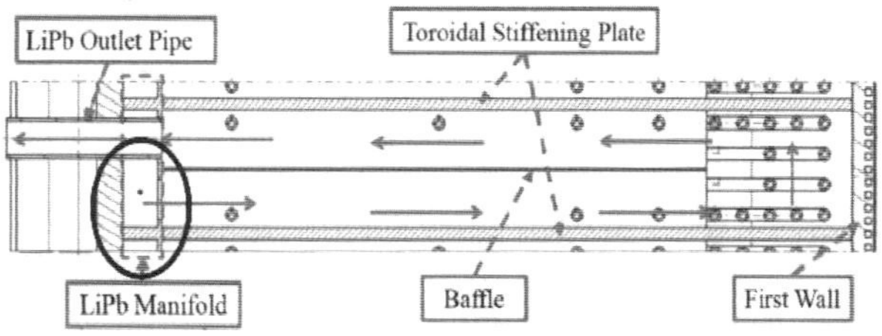

Figure 1: Breeding zone cell: radial-poloidal view, orifice position highlighted [15].

occurs whenever an orifice is met by the fluid. Only two numerical works that have addressed this topic are known to the authors. Singh and Gohil analyzed an insulating duct featuring a rectangular and triangular-shaped orifice by means of a 2D model developed using OpenFOAM; a surprising choice given the inherent 3D flow nature of the chosen geometry [16]. Tassone and Caruso have performed a more detailed study in the framework of the ITER Test Blanket Module development, using a complete 3D model to represent the MHD flow at high magnetic field intensity in a mostly one-sided sudden expansion/contraction from a rectangular orifice [17].

This paper aims to contribute to the effort of characterizing the MHD flow through a rectangular orifice. The geometry considered is a sudden cross-section variation featuring a large restriction in the direction parallel to the magnetic field (z, toroidal) and a small one perpendicular to it (y, poloidal). The main channel is a rectangular duct with electrically conductive walls (c = 0.01 ÷ 0.1). The problem is investigated for Re = 108, and Ha = 0 ÷ 300. The geometry is representative of the orifice connecting the manifold and breeding zone in the Water-Cooled Lithium Lead (WCLL) blanket [18] (Figure 1), whereas scaled-down magnetic field intensity is used to reduce the computational cost.

2. Problem Formulation

The MHD governing equations are obtained by the combination of the hydrodynamic set with Maxwell's one [4]. For an electrically conducting, incompressible, and viscous fluid, these equations can be written for an isothermal flow as

$$\frac{1}{\mathrm{N}}\left(\frac{\partial}{\partial \mathrm{t}}+\mathbf{v}\cdot\nabla\right)\mathbf{v} = -\nabla\mathbf{p}+\frac{1}{\mathrm{Ha}^2}\nabla\mathbf{v}^2+\mathbf{J}\times\mathbf{B} \quad (1)$$

$$\nabla\cdot\mathbf{v}=0, \qquad \nabla\cdot\mathbf{J}=0 \quad (2)$$

$$\mathbf{J}=-\nabla\phi+\mathbf{v}\times\mathbf{B} \quad (3)$$

Where **v**, p, **J**, **B** and ϕ represents velocity, pressure, current density, magnetic field, and electric potential. The variable scales are taken as described in Ref. [10]. The characteristic length scale is the duct half-width in the magnetic field direction (a). The magnetic field is assumed to be constant and aligned to the toroidal direction, $\mathbf{B} = B_0\hat{\mathbf{z}}$, thus neglecting any poloidal field contribution. Using the low magnetic Reynolds number approximation, **B** is independent by **v**, and the set is closed with an additional equation obtained by combining Eqs. (2) and (3):

$$\nabla^2\phi=\nabla\cdot(\mathbf{v}\times\mathbf{B}) \quad (4)$$

The dimensionless parameter governing the flow are the Hartmann number ($Ha = a\ B_0(\sigma/\mu)0.5$), whose square is the ratio between electromagnetic and viscous forces, and the interaction parameter ($N = \sigma a B_0/\rho u_0 = Ha^2/Re$), the ratio of electromagnetic and inertial forces. The problem geometry is shown in Figure 2.

The duct is characterized by an aspect ratio γ and is bounded by uniform thickness walls (t_w). The square orifice is characterized by a toroidal (R_z) and poloidal (R_y) change ratio, as well as an axial shape factor (R_x). A uniform velocity u_0 is assumed at the inlet, which is placed at $X_1 = -9a$, whereas p=0 at the outlet, located at $X_2 = 24a$. These values are chosen to ensure a fully developed state is reached before and after the orifice. In Table 1, geometrical parameters are collected.

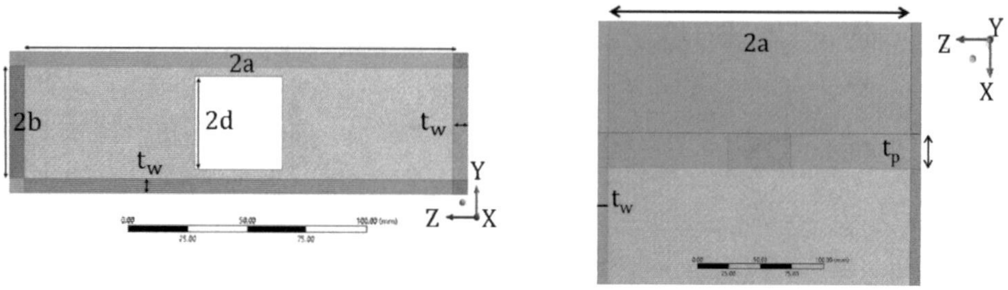

Figure 2: Problem geometry: (left) duct/orifice cross-sectional view; (right) detail of orifice, top view.

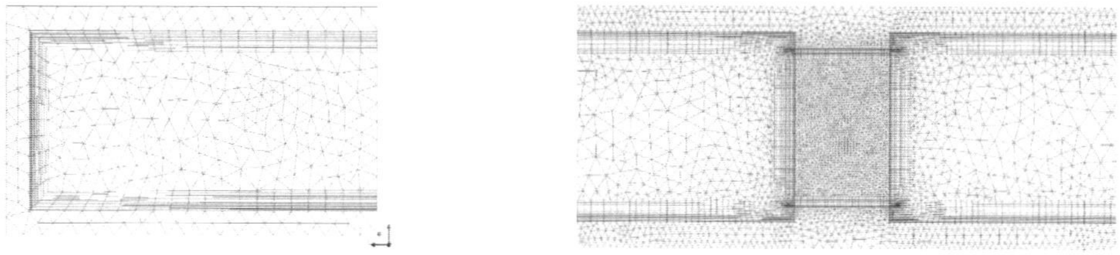

Figure 3: Computational grid: (left) duct mesh, (right) detail of orifice mesh, lateral view.

Table 1: Model geometrical parameters.

Parameter	Symbol	Value (mm)	Parameter	Symbol	Value
Duct toroidal half-length	a	120	**Duct aspect ratio**	γ	a/b = 4
Duct poloidal half-length	b	30	**Toroidal change ratio**	R_z	a/d = 4.8
Orifice half-width	d	25	**Orifice axial shape factor**	R_x	t_p /2d = 0.6
Wall thickness	t_w	8	**Poloidal change ratio**	R_y	b/d = 1.2
Plate radial thickness	t_p	30			

Since the problem features finite conductivity walls ($c = 0.01 \div 0.1$), the relative conductance ratio between wall and fluid is introduced, $c = \sigma_w t_w/(\sigma a)$, and Eq. (4) is a conjugated problem across the fluid and solid domain. At the fluid/solid interface, kinematic no-slip ($\mathbf{v} = 0$), continuity of electric potential ($\phi = \phi_w$), and normal current ($J_n = J_{n,w}$) are the boundary conditions. On the solid external surface, the Neumann boundary condition is used ($\partial\phi/\partial n = 0$) to model the surrounding dielectric medium, as well as for inlet and outlet. The physical properties of both fluid and solid are constant. For the former, it is assumed lithium-lead at 600 K [19], for the latter σ_w is adjusted to obtain c.

3. Numerical Model

The model is solved with ANSYS CFX 18.2. For all the simulations performed, high resolution advection scheme is adopted. The convergence of the solution is controlled with residuals values of mass, velocity and electric potential conservation and monitoring solution variables ($\mathbf{v}$, ϕ, $\mathbf{J}$) at fixed points during the run. Calculations are stopped when the root mean square residuals are lower than 10^{-4} and the monitored variables remain constant. The flow is modelled as laminar and steady.

The model domain is discretized using an unstructured mesh composed by tetrahedrons in the core and solid structures, whereas prismatic elements are used to resolve the boundary layers, as shown in Figure 3. The grid is realized to include at least 2 elements in the Hartmann layer appearing at walls ⊥ **B**, whose thickness is $\delta_H = O(Ha^{-1})$. The mesh axial resolution is increased approaching the orifice

Table 2: Mesh sensitivity on hydrodynamic orifice pressure drop (from Ref. [10]) versus total axial (N_X), orifice axial (N_O), and cross-sectional (N_(Y/Z)) grid resolution.

#	N_X	N_O	N_(Y/Z)	N. of elements	Δp_o (mPa)	$\Delta p_\#$ (mPa)	Error (%)
1	92	6	96 × 24	223 460	6.89	8.289	20.3
2	112	12	144 × 36	646 544		8.003	16.1
3	132	20	192 × 48	1 361 608		7.822	13.5
4	150	30		1 480 380		7.731	12.2
5	171	45		1 575 734		7.675	11.4

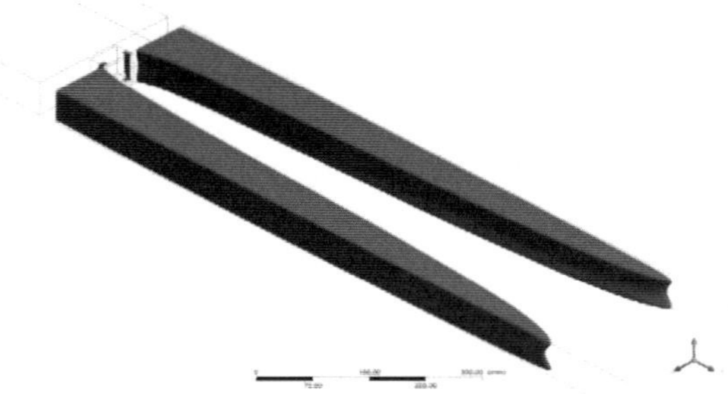

Figure 4: Negative axial velocity iso-surfaces (recirculation regions) for OHD case.

region. A mesh sensitivity study was performed comparing the hydrodynamic orifice Δp calculated by the code with the one reported by Zivkovic et al. [20] for a similar configuration. An overview of the results is presented in Table 2. The grid number 4 was selected as reference for the simulations reported in Section 4. Extensive validation of the CFX electromagnetic model has been performed in the past for pressure-driven flows in finite conductivity ducts. Details can be found in Refs. [11], [12], [15].

4. Results and Discussion

Ordinary hydrodynamic (OHD) simulations are performed to investigate basic flow features and as reference for MHD cases. It is found that contraction ratios R_z and R_y have a significantly different effect on the flow features. The poloidal contraction is relatively small, and it is not accompanied by relevant recirculation regions: the boundary layer reattachment is observed shortly after the orifice exit ($L_y \approx d$). Conversely, the toroidal contraction affects the flow features in a more dramatic fashion. Large recirculation regions are present after the orifice (Figure 4) with maximum toroidal length $L \approx 0.75\,a$. Reattachment points are located much further downstream compared with the poloidal contraction being observed about at $L_z = 40d$. Maximum velocity observed in the contraction is $u_{Max} = 1.49\ mm/s$, which still allows to treat the flow as laminar and steady since $Re_o \approx 900$. Fully developed condition is regained approximately at the axial coordinate $L_x = 72d$.

4.1. MHD Results

In Figure 5, the velocity profiles along poloidal direction within the duct (y = ±b) and inside the orifice (y = ±d) are shown. For a low magnetic field, the velocity profile in the former retains a quasi-hydrodynamic shape but, for increasing Ha, the flow separates into a slug core and jets close to the side walls. Cross-section contraction in the orifice affects the flow features more than by increasing mean velocity: M-shaped profile is observed with quicker jets and enhanced pressure losses compared with duct flow. This phenomenon is explained by the thicker Hartmann walls bounding the orifice flow, c = 6.13 versus c = 0.1. A non-null velocity gradient appears along the duct axis when the flow approaches the orifice contraction that induces an electric potential difference, as shown in Figure 6. This drives 3D currents that are not confined to the cross-section (see Figure 7). These are responsible for non-axial

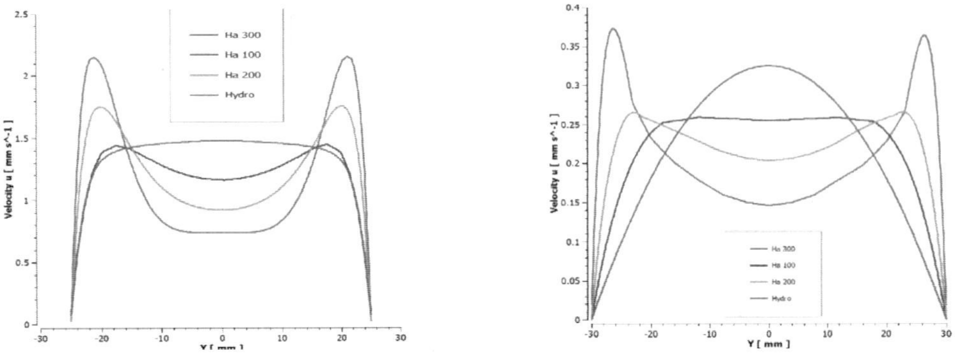

Figure 5: Velocity profile along y-direction (at midplane z = 0) for increasing Ha with c = 0.1 and OHD case: (left) at the orifice center (x=0) and fully developed flow in the duct (right).

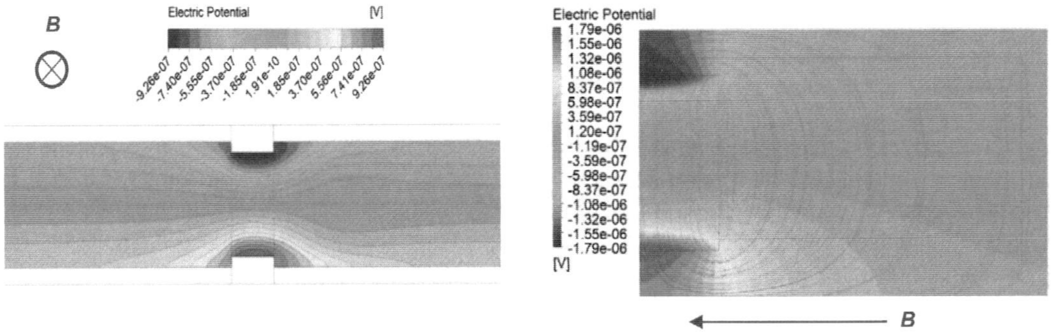

Figure 6: Electric potential contours and current density streamlines: left, z = 0 Ha = 100 c = 0.01; right, x = 0 Ha = 300 c =0.1.

Lorentz forces that causes flow redistribution and additional pressure losses. In Figure 8, it is possible to observe how the interaction between these forces, which push the fluid toward the side walls, and the flow transfer in the internal layer parallel to the magnetic field (not shown) causes core flow reversal at the duct exit. This phenomenon seems to be characteristic of the asymmetric expansion, since it is not reported by Refs. [10], [11], [17], and could lead to tritium accumulation in the manifold. It is interesting to note that a similar phenomenon was described by Rhodes et al. [8] for the sudden expansion from an insulated duct, where a pair of counter-rotating steady vortices were observed downstream of the cross-section variation. To the best of our knowledge, it is the first time that this feature is reported for electrically conductive walls. Smaller flow reversals are observed in the corners after the poloidal expansion. The recirculation region at the egress of the orifice plate appears for all the cases studied but for the lowest magnetic field intensity considered (Ha = 100). The volume of the region is found to increase with c but decreases with Ha. The flow reversal can be explained considering that jets within the orifice are quicker than the ones in the duct and are further enhanced at the orifice exit due to the effect of the axial current: these concurring phenomena cause a flow deficit in the duct central region and the formation of the recirculation. Secondary flow reversals are observed in the corners after the poloidal expansion whereas the Hartmann boundary layer does not undergo any separation. The flow regains fully developed state quicker than in the hydrodynamic simulation thanks to the stabilizing effect of the magnetic field. The orifice effect on the flow can be considered limited to an axial region $L_x = \pm 5\ d$. In Table 3, recirculation zone dimensions are collected.

Figure 7: Current density streamlines Ha = 200 c = 0.1; straight duct (left), upstream and downstream the orifice (right).

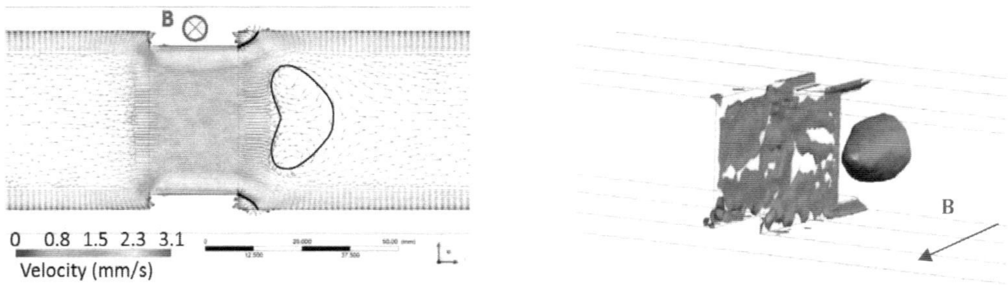

Figure 8: Left, velocity vector on the plane passing through the duct center (z=0) for Ha = 300 and c = 0.1. Flow reversals contoured in black. Right, 3D view of flow reversal for the same case.

Table 3: Downstream primary reverse flow area extension (scaled with orifice half-width d).

		Axial	Toroidal	Poloidal
Ha = 300	**c= 0.01**	0.68	2.75	0.72
	c = 0.1	0.95	4.6	1.24
Ha = 200	**c= 0.01**	0.64	1.91	0.46
	c = 0.1	0.81	3.41	0.8

4.2. Pressure Loss Analysis

If Ha $\gg$ 1, the pressure losses in an MHD flow are dominated by the Lorentz force. In Figure 9, the pressure profile for the flow through the orifice is sketched. The dashed lines represent the pressure profile for a fully developed flow in the main duct, downstream and upstream, and within the orifice region. For the former, it is computed from the area-averaged pressure gradient at the outlet ($\partial_x p_1$), whereas the latter is calculated from the area-averaged pressure gradient at the orifice center ($\partial_x p_2$). As previously noted, 2D MHD flow in the orifice features higher losses than in the duct due to a larger value of wall conductivity. A 3D pressure drop term (Δp_{3D}) is associated to the flow within the orifice due to the appearance of axial currents. It is calculated from the total pressure loss in the model: $\Delta p_{3D} = \Delta p_{tot} - 33a\partial_x p_1 - t_p \partial_x p_2$. Results are presented in Table 4.

In general, Δp_{3D} is influenced by Ha, N, c, and geometrical parameters. For constant c, increasing Ha results in a relative Δp_{3D} decrease over the total one due to a weaker dependence on B compared with 2D MHD flow, for which $\Delta p_{2D} \propto B^2$. A similar trend is observed for increasing c at constant Ha, since an increment of wall conductivity affects more the intensity of the currents closing through it rather than those which are contained within the fluid. The pressure loss can be calculated with $\Delta p_{3D} = 0.5k\sigma u_0 B_0^2 d$, where u_0 and k are the orifice hydrodynamic mean velocity and local MHD resistance coefficient. Regarding this parameter, it converges to $k \approx 0.205$ for c = 0.1, hinting to Δp_{3D} being determined by electromagnetic forces Ha = 200 ÷ 300, whereas inertial forces are still significant for c = 0.01 up to Ha = 300.

5. Conclusions

The 3D MHD flow in an orifice with asymmetrical sudden cross-section variation has been studied

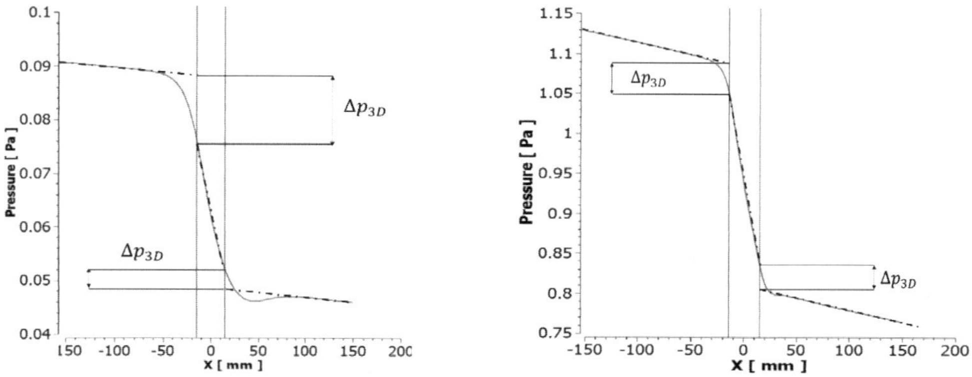

Figure 9: Pressure profiles along the duct centerline for the flow through the orifice: (left) Ha = 100 and c = 0.01, (right) Ha = 300 and c = 0.1. Vertical lines identify the orifice region. Dashed lines mark the pressure distribution for a fully developed flow in the main and orifice channel.

Table 4: 3D orifice pressure loss and MHD local resistance coefficient.

	c=0.01			c=0.1		
Ha	**100**	**200**	**300**	**100**	**200**	**300**
Δp(Pa)	0.04821	0.1279	0.2374	0.1057	0.3727	1.3888
Δp_{3D}(Pa)	0.00784	0.0107	0.0201	0.0134	0.0291	0.0669
k	0.221262	0.075278	0.062891	0.378225	0.204729	0.209324
$\Delta p_{3D}/\Delta p$	16.26%	8.36%	8.48 %	12.72%	7.8 %	4.82 %

using the code ANSYS CFX in the range Ha = 0 ÷ 300 and c = 0.01 ÷ 0.1. The geometry considered is representative of the perforated plate connecting the manifold and breeding zone in the WCLL [15]. The flow main features are described with the most interesting one being the appearance of a wide recirculation region at the exit. The 3D pressure loss for the flow through the orifice can be estimated with $k \approx 0.205$ for c = 0.1 and $k \approx 0.063$ for c = 0.01.

Further work is required to complete the characterization of the MHD flow through a rectangular orifice. It would be desirable to perform simulation closer to blanket conditions, i.e. featuring Ha $\approx 10^4$. An important point that has so far been neglected in the literature is the effect of a skewed magnetic field on the 3D loss. The effect of the orifice shape and axial length should also be investigated and its role in the onset of recirculation regions downstream of it.

Acknowledgments This work has been carried out within the framework of the EUROfusion Consortium and has received funding from the Euratom research and training programme 2014-2018 and 2019-2020 under grant agreement No 633053. The views and opinions expressed herein do not necessarily reflect those of the European Commission.

References

[1] S. Smolentsev, R. Moreau, L. Bühler, and C. Mistrangelo, "MHD thermofluid issues of liquid-metal blankets: Phenomena and advances," *Fusion Eng. Des.*, vol. 85, no. 7–9, pp. 1196–1205, 2010, doi: 10.1016/j.fusengdes.2010.02.038.
[2] S. Smolentsev, "Physical Background, Computations and Practical Issues of the Magnetohydrodynamic Pressure Drop in a Fusion Liquid Metal Blanket," *Fluids*, vol. 6, no. 3, p. 110, 2021, doi: 10.3390/fluids6030110.
[3] J. C. R. Hunt and S. Leibovich, "Magnetohydrodynamic flow in channels of variable cross-section with strong transverse magnetic fields," *J. Fluid Mech.*, vol. 28, no. 2, pp. 241–260, 1967, doi: 10.1017/S0022112067002046.
[4] U. Muller and L. Bhuler, *Magnetofluiddynamics in Channels and Containers*. 1999.
[5] L. Bühler, S. Horanyi, and E. Arbogast, "Experimental investigation of liquid-metal flows through a sudden expansion at fusion-relevant Hartmann numbers," *Fusion Eng. Des.*, vol. 82, no. 15–24, pp. 2239–2245, 2007, doi: 10.1016/j.fusengdes.2007.02.029.
[6] J. Feng, Q. He, H. Chen, and M. Ye, "Numerical Investigation of magnetohydrodynamic flow through Sudden expansion pipes in Liquid Metal Blankets," *Fusion Eng. Des.*, vol. 109–111, pp. 1360–1364, 2016, doi: 10.1016/j.fusengdes.2015.12.023.
[7] H. Kumamaru, "Numerical analyses on liquid-metal magnetohydrodynamic flow in sudden channel contraction," *J. Nucl. Sci. Technol.*, vol. 54, no. 12, pp. 1300–1309, 2017, doi: 10.1080/00223131.2017.1365021.
[8] T. J. Rhodes, S. Smolentsev, and M. Abdou, "Magnetohydrodynamic pressure drop and flow balancing of liquid metal flow in a prototypic fusion blanket manifold," *Phys. Fluids*, vol. 30, no. 5, 2018, doi: 10.1063/1.5026404.
[9] T. Rhodes and S. Smolentsev, "Pressure drop in a prototypical 3D magnetohydrodynamic flow across contraction of a fusion blanket manifold," *J. Nucl. Sci. Technol.*, vol. 00, no. 00, pp. 1–10, 2021, doi: 10.1080/00223131.2021.1892550.
[10] C. Mistrangelo and L. Bühler, "Numerical investigation of liquid metal flows in rectangular sudden expansions," *Fusion Eng. Des.*, vol. 82, no. 15–24, pp. 2176–2182, 2007, doi: 10.1016/j.fusengdes.2007.04.014.
[11] C. N. Kim, "Liquid metal magnetohydrodynamic flows in an electrically conducting rectangular duct with sudden expansion," *Comput. Fluids*, vol. 89, pp. 232–241, 2014, doi: 10.1016/j.compfluid.2013.11.002.
[12] C. N. Kim, "A liquid metal magnetohydrodynamic duct flow with sudden contraction in a direction perpendicular to a magnetic field," *Comput. Fluids*, vol. 108, no. 1000, pp. 156–167, 2015, doi: 10.1016/j.compfluid.2014.12.001.
[13] L. Bühler, "A parametric study of 3D MHD flows in expansions of rectangular ducts," *Fusion Sci. Technol.*, vol. 52, no. 3, pp. 595–602, 2007, doi: 10.13182/FST07-A1553.
[14] X. Xiao and C. N. Kim, "Numerical simulations of MHD flows in a conduit that contracts in the magnetic field direction," *Fusion Eng. Des.*, vol. 160, no. July, p. 111990, 2020, doi: 10.1016/j.fusengdes.2020.111990.
[15] A. Tassone, G. Caruso, A. Del Nevo, and I. Di Piazza, "CFD simulation of the magnetohydrodynamic flow inside the WCLL breeding blanket module," *Fusion Eng. Des.*, vol. 124, pp. 705–709, 2017, doi: 10.1016/j.fusengdes.2017.05.098.
[16] R. J. Singh and T. B. Gohil, "Influence of the presence of the Lorentz force and its direction on the suppression of secondary flow in two different orifices: A numerical study using OpenFOAM," *J. Appl. Fluid Mech.*, vol. 12, no. 3, pp. 751–762, 2019, doi: 10.29252/JAFM.12.03.29242.
[17] A. Tassone and G. Caruso, "Computational MHD analyses in support of the design of the WCLL TBM breeding zone," *Fusion Eng. Des.*, vol. 170, no. February, p. 112535, 2021, doi: 10.1016/j.fusengdes.2021.112535.
[18] A. Del Nevo *et al.*, "Recent progress in developing a feasible and integrated conceptual design

of the WCLL BB in EUROfusion project," *Fusion Eng. Des.*, vol. 146, no. March, pp. 1805–1809, 2019, doi: 10.1016/j.fusengdes.2019.03.040.

[19] D. Martelli, A. Venturini, and M. Utili, "Literature review of lead-lithium thermophysical properties," *Fusion Eng. Des.*, vol. 138, no. November 2018, pp. 183–195, 2019, doi: 10.1016/j.fusengdes.2018.11.028.

[20] V. Zivkovic, P. Zerna, Z. T. Alwahabi, and M. J. Biggs, "A pressure drop correlation for low Reynolds number Newtonian flows through a rectangular orifice in a similarly shaped micro-channel," *Chem. Eng. Res. Des.*, vol. 91, no. 1, pp. 1–6, 2013, doi: 10.1016/j.cherd.2012.05.022.

Coupled Geothermal Energy Simulations

Armando Di Meglio, Nicola Massarotti, Alessandro Mauro, Vincenzo Trombetta*

Department of Engineering, University of Naples "Parthenope" Centro Direzionale, Isola C4, 80133

*vincenzo.trombetta@studenti.uniparthenope.it

Abstract. This paper presents the numerical solution of thermal field of a set of geothermal coaxial probes coupled with the surrounding soil. These probes, used for the excavation of Naples underground's tunnels with the artificial ground freezing technique, are converted for the heat transfer fluid transport for a geothermal plant. One-dimensional energy equation has been used with a single unknown in order to extract the solution of the temperature inside the probe, the temperature of the fluid, which is a boundary condition for calculating the temperature range of the soil surrounding the probes. In turn, the latter is used as a source term for the solution of the energy equation for the fluid. This paper also presents a comparison between two different probe materials i) internal and external stainless-steel crown; ii) internal rigid PVC crown and external stainless-steel crown. Such a numerical scheme allows to take into account the 3D effects of the fluid dynamics in porous media as is the soil, and, at the same time, considering a "reduced order model" inside the geothermal probes, as is the 1D model, for considering the thermal coupling effects.

Nomenclature

A	Morphology porous constant	u_i	Dimensionless velocity components
B_l	Isobaric compressibility of the liquid [K^{-1}]	$\boldsymbol{u}$	Velocity vector [$m\ s^{-1}$]
C	Inertia coefficient	x_i	Dimensionless coordinate axis
c	Specific heat [$J\ kg^{-1}\ K^{-1}$]	Δt	Dimensionless time step
Da	Darcy number	ΔT	Dimensionless temperature difference
g	Gravitational acceleration [$m\ s^{-2}$]	α	Thermal diffusivity [$m^2\ s^{-1}$]
h	Local element size [m]	ε	Porosity
J	Viscosity ratio	κ	Permeability [m^2]
L	Reference length [m]	λ	Thermal conductivity ratio
P	Perimeter [m]	μ	Dynamic viscosity [$N\ s\ m^{-2}$]
p	Dimensionless pressure	υ	Kinematic viscosity [$m^2\ s^{-1}$]
Pr	Prandlt number	ρ	Density [$kg\ m^{-3}$]
Q_t	External heat exchange	Ω	Heat capacity ratio

Q_f	Enternal heat exchange		
Ra	Rayleigh number	**Subscripts**	
T	Dimensionless temperature	*int*	Intern
T	Dimensionless time	*ext*	Extern

1. Introduction

Geothermal energy is the energy contained in the Earth's interior. The term geothermal resources are usually referred to accessible geothermal reservoirs, characterized by the presence of a geothermal fluid. The reservoir temperature is the classification parameter of geothermal resources; Muffler and Cataldi [1] classify the reservoir as low, intermediate, and high enthalpy when the fluid temperature is lower than 90°C, between 90 and 150°, or higher than 150°C, respectively. Other classifications are available in literature [2]-[4].

Regarding the sustainable use of geothermal energy, particular importance is given the coupling between the reservoir and the heat exchange system to transfer energy from and to the ground. Considered the significant risks associated with the geothermal energy source, an accurate modelling of the phenomena is essential [5], [6].

The geothermal energy system studied in this work is obtained from a part of a plant used for soil freezing during the construction of Naples underground [7]. According to this configuration, the geothermal probes are coaxial and horizontal, while the reservoir as low (16 °C).

This paper presents a numerical analysis of a coupled geothermal system that is the interaction of the thermal field between the probes and the soil. The numerical scheme used for the pipe flow is a 1D finite difference method where the unknown is the temperature, while the heat flux from the ground represents a transient source term. Regarding the soil temperature, a 3D quasi-implicit finite element method is adopted, in which the temperature trends of the geothermal probes set up a variable Dirichlet boundary condition. With respect to the available literature, the main novelty lies in the numerical scheme adopted in such a physical system. From one side, the 3D Navier Stokes equations are needed to consider the possible flow regimes in porous media (natural, mixed or forced convection), from the other a 1D finite difference scheme is enough to capture the temperature gradients along the geothermal probes coordinate, because their diameter is much less than their lengths. This allows to significantly the reduce the computational costs which we would considering the 3D geometry for a such coupled thermal problem. The use of a 1D model for the probes requires that the heat transfer coefficients (γ) from the existing literature have to be introduced. A drawback is that in the coaxial probe, the temperature in the connection point between internal/external probe has to prescribe as an additional boundary condition. The goal is to understand if the recovery of these probes for geothermal use has an advantage in energy terms.

2. Theory

2.1 Governing equations

The physical phenomena analysed are convection heat transfer in porous media for surrounding soil, and heat transfer by convection for pipe flow. The dimensionless equations for first phenomenon are the following:

Mass conservation equation:

$$\frac{\partial u_i}{\partial x_i} = 0 \qquad (1)$$

Momentum conservation equation:

$$\frac{1}{\epsilon}\frac{\partial u_i}{\partial t}+\frac{1}{\epsilon^2}u_j\frac{\partial u_i}{\partial x_j}=\frac{\partial p_l}{\partial x_i}+\frac{JPr}{\epsilon}\frac{\partial}{\partial x_j}\left(\frac{\partial u_i}{\partial x_j}\right)-\frac{Pr}{Da}u_i-\frac{C|\mathbf{u}|}{\sqrt{Da}\epsilon^{\frac{3}{2}}}u_i+RaPrT \tag{2}$$

Energy conservation equation:

$$\Omega\frac{\partial T}{\partial t}+u_i\frac{\partial T}{\partial x_i}=\lambda\frac{\partial}{\partial x_i}\left(\frac{\partial T}{\partial x_i}\right) \tag{3}$$

The scales and the parameters used to derive the above dimensionless equations can be derivate from Ref. [8]-[10]. In particular, T is the ground temperature; Ω is the heat capacity; λ is the thermal conductivity and u_i is the fluid velocity.
The equations describing the second phenomenon are considered as follows [11]-[17]:

$$\rho c_p A_{ext}\frac{\partial T_{ext}}{\partial t}+c_p\dot{m}\frac{\partial T_{ext}}{\partial x}=Q_t+Q_f \tag{4}$$

$$\rho c_p A_{int}\frac{\partial T_{int}}{\partial t}+c_p\dot{m}\frac{\partial T_{int}}{\partial x}=-Q_f \tag{5}$$

The equations (4) and (5) describe the fluid in the external and internal crown of the pipe, respectively. Where V is the volume; A_{int} is the cross-section area for the internal pipe, A_{ext} is the cross-section area for the external pipe; T is the temperature of the fluid; $\dot{m}$ is the mass flow; Q_t represent external heat exchange through the pipe wall; Q_f is the internal heat exchange between internal and external fluid. The subscripts *ext* and *int* stand for external and internal fluid.
The Q_t and Q_f terms are detailed below.

$$Q_t=\gamma_{ext}(T_{ext}-T_{fe}) \tag{6}$$

$$Q_f=\gamma_{int}(T_{fi}-T_{fe}) \tag{7}$$

Where:

$$\gamma_{ext}=\frac{1}{\frac{1}{k_t P1_{ext}}+\frac{\ln(r1_{ext}/r0_{ext})}{k_p 2\pi}+\frac{1}{\beta_{ext}P0_{ext}}} \tag{8}$$

$$\gamma_{int}=\frac{1}{\frac{1}{\beta_{int}P0_{int}}+\frac{\ln(r1_{int}/r0_{int})}{k_p 2\pi}+\frac{1}{\beta_{ext}P0_{ext}}} \tag{9}$$

In the above equation, T_{ext} is given as temperature field surrounding the pipe; T_{fe} and T_{fi} are the temperature of external and internal fluid respectively; $r0_{ext}$ and $r0_{int}$ are the internal radius of external and internal crown respectively; $r1_{ext}$ and $r1_{int}$ are the external radius of external and internal crown respectively; the same assessments made for the radius must be made for the perimeter $P0_{ext}$, $P0_{int}$, $P1_{ext}$, $P1_{int}$; k_p is the thermal conductivity of the pipe; β_{int} and β_{ext} are:

$$\beta=Nu\frac{k_f}{d_h} \tag{10}$$

Where k_f is the thermal conductivity of the fluid; d_h is the hydraulic diameter and Nu is the Nusselt number. Figure 1 resumes the equations 4 and 5.

Journal of Physics: Conference Series 2177 (2022) 012004 doi:10.1088/1742-6596/2177/1/012004

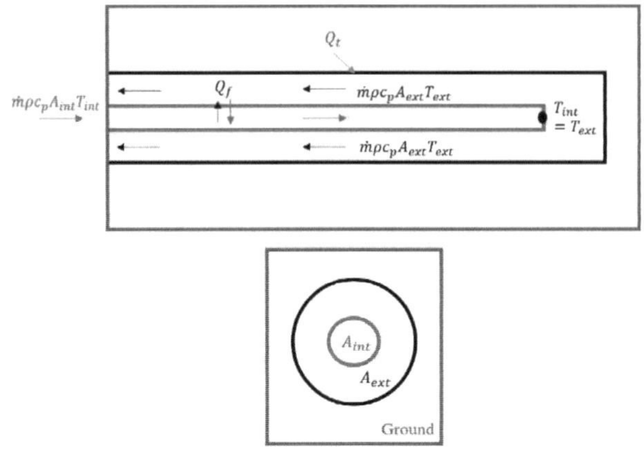

Figure 1. Coaxial Pipe energy scheme.

2.2 Numerical scheme

The numerical procedure used to solve the equations (1-3) is Characteristic Based Split (CBS), a finite element procedure based on a fractional time-stepping algorithm where the convective terms are treated using the Characteristics-Galerkin method [8]. The CBS can be implemented with different solution schemes [18]-[21], each one is subject proper time step restrictions [8]. In particular, in this paper has been used a quasi-implicit scheme. The numerical step of CBS can be written as follow:

Step 1: intermediate velocity calculation

$$\left(CO\boldsymbol{M} + \frac{JPr}{\epsilon}\boldsymbol{K_m}\right)\tilde{\boldsymbol{u}}_i = \left(\frac{1}{\Delta t}\boldsymbol{M} - \frac{1}{\epsilon^2}\boldsymbol{C}\right)\boldsymbol{u}_i^n + RaPr\boldsymbol{MT^n} + \boldsymbol{f} \tag{11}$$

Step 2: pressure calculation

$$\boldsymbol{K_p p^{n+1}} = -\left(\frac{1}{\epsilon\Delta t} + \frac{Pr}{Da} + \frac{C|\boldsymbol{u}|}{\epsilon^{3/2}\sqrt{Da}}\right)\boldsymbol{G_i}\tilde{\boldsymbol{u}}_i \tag{12}$$

Step 3: velocity correction

$$M\boldsymbol{u}_i^{n+1} = \boldsymbol{M}\tilde{\boldsymbol{u}}_i - \frac{1}{CO}\boldsymbol{G_i p^{n+1}} \tag{13}$$

Step 4: temperature calculation

$$\left(\frac{\Omega}{\Delta t}\boldsymbol{M} + \lambda\boldsymbol{K_d}\right)\boldsymbol{T^{n+1}} = \left(\frac{\Omega}{\Delta t}\boldsymbol{M} - \boldsymbol{C}\right)\boldsymbol{T^n} + \boldsymbol{f_d} \tag{14}$$

Where:

$$CO = \left(\frac{1}{\epsilon\Delta t} + \frac{Pr}{Da} + \frac{C|\boldsymbol{u}|}{\epsilon^{3/2}\sqrt{Da}}\right) \tag{15}$$

Where **M** is the mass matrix (lumped using a standard row-summing approach), **C** is the convection matrix, $\boldsymbol{K_d}$ is the diffusion matrix, **G** is the gradient matrix, $\boldsymbol{K_p}$ is the stiffness matrix, $\boldsymbol{K_m}$ is the momentum matrix, and $\boldsymbol{f}$ and $\boldsymbol{f_d}$ are the boundary vectors from momentum and temperature equations. Further details of all terms presented in the above equations are given in [8], [18], [21].

For the equations (4) and (5) has been used a finite difference method for space and time [22]-[26], for a generic spatial point *i*, and temporal point *j* it is possible to write as follow:

$$T_{ext}(i,j) = T_{ext}(i,j-1) + \frac{\dot{m}\Delta t}{\rho A_{ext}\Delta x}\left(T_{ext}(i-1,j) - T_{ext}(i,j)\right) + \frac{\Delta t}{\rho c_p A_{ext}}\left(Q_t + Q_f\right) \tag{16}$$

$$T_{int}(i,j) = T_{int}(i,j-1) + \frac{\dot{m}\Delta t}{\rho A_{int}\Delta x}\left(T_{int}(i-1,j) - T_{int}(i,j)\right) - \frac{\Delta t}{\rho c_p A_{int}}\left(Q_f\right) \tag{17}$$

The equation (16) and (17) don't need to be numerically stabilized, and this numerical scheme results very simple to implement and it has a highly computational efficiency. The out internal temperature is the inlet external temperature.

The equations in this section are coupled, because the temperature computed in equation (14) is used to calculate Q_t and the temperatures computed in the equation (16) is used as Dirichlet boundary condition in equation (14).
Numerical model is validated by Massarotti et al. [27].

3. Results

In this section, the numerical results calculated with the above numerical scheme are described.
The results are shown for the following three cases:

1. Steel-Steel Coaxial pipe flow (1 Day);
2. Steel-PVC Coaxial pipe flow (1Day);
3. Steel-PVC Coaxial pipe flow (5 Days).

The logical and geometric scheme of the coaxial pipe is shown in Figure 2, the length in z-direction is 40 meters and the external and internal radius of internal tube are 0.026 and 0.028 meters, respectively. The external and internal radius of external tube are 0.078 and 0.080 meters.
The ground around the pipes is Neapolitan yellow tuff.

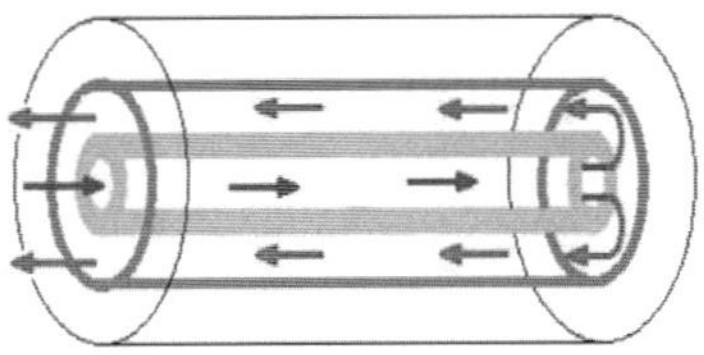

Figure 2. Logical and Geometric scheme of coaxial pipe.

The cases have a constant mass flow inlet. For all cases the inlet temperature is 35 °C and the initial ground temperature is equal to 16 °C. The ground geometry and mesh used is pictured in Figure 3.
The Figure 4 and Figure 5 show the temperature profile of internal and external tube for two different inlet mass flow for both cases 1 and 2. The figures show a big difference of the temperature profiles for the two cases analyzed, justified by the wide difference of the thermal properties of steel and PVC. Despite this difference, the outlet temperatures for case 1 and 2 are almost identical for the flow rate of 0.5 kg/s, but the case 2 has more energy advantage when the flow rate is 0.05 kg/s.

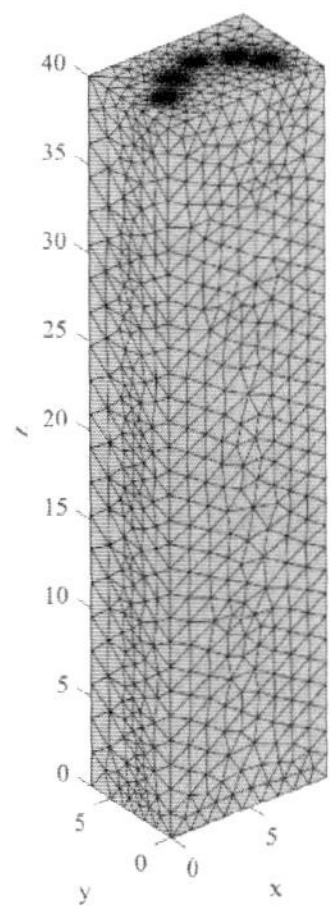

Figure 3. Geometry and Mesh.

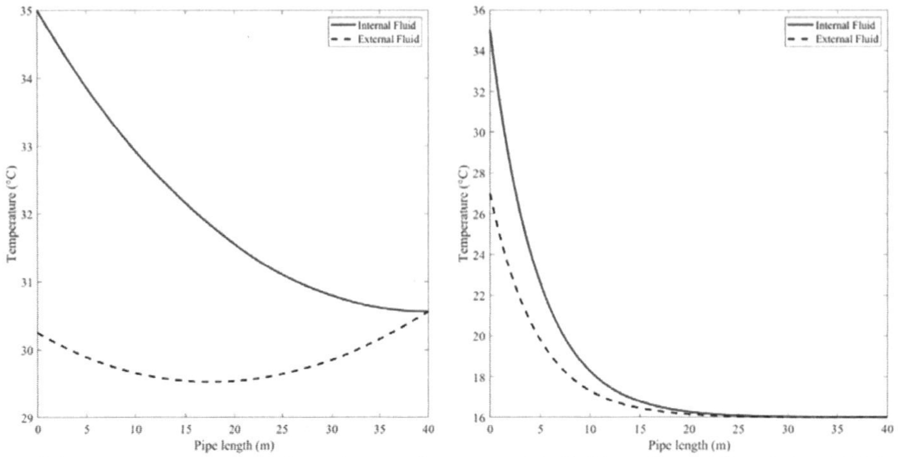

Figure 4. Temperature profile Case 1, mass flow inlet 0.5 (left) and mass flow inlet 0.05 (right).

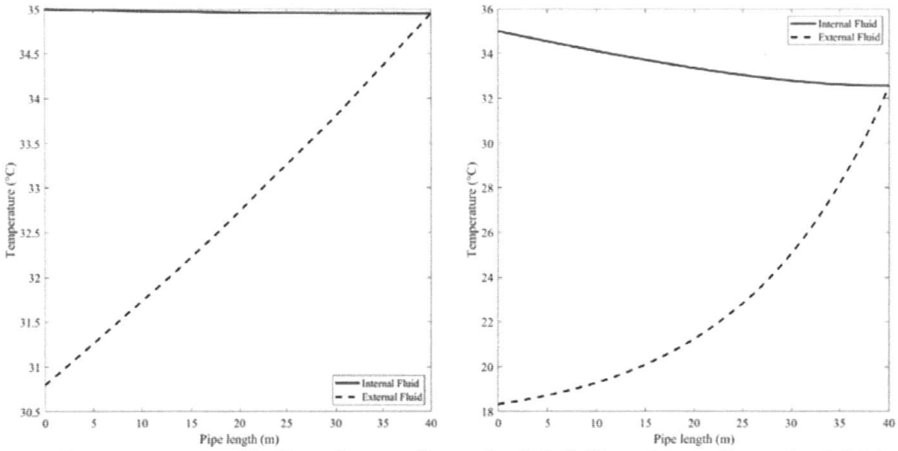

Figure 5. Temperature profile Case 2, mass flow inlet 0.5 (left) and mass flow inlet 0.05 (right).

Regarding the ground temperature, the Figure 6 shows the results for the case 1 at 24 hours.

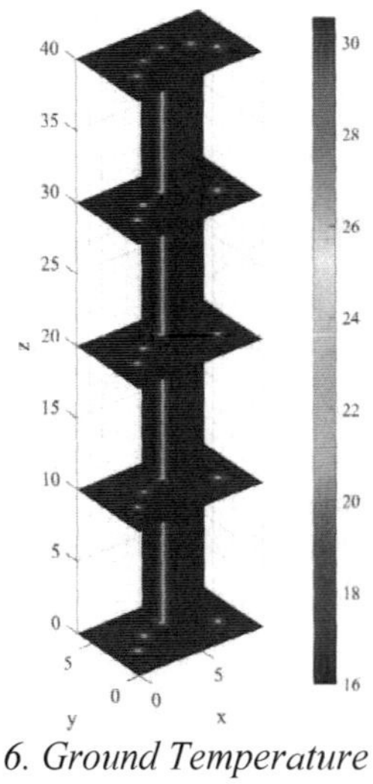

Figure 6. Ground Temperature Case 1.

Journal of Physics: Conference Series 2177 (2022) 012004 doi:10.1088/1742-6596/2177/1/012004

The third case reproduces the case for five days with a mass flow rate equal to 0.5 kg/s. Figure 7 and Figure 8 show the ground temperature and temperature profile of pipe flow, respectively.

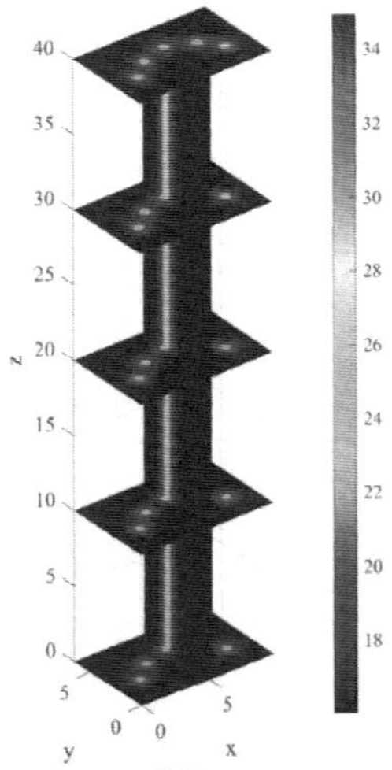

Figure 7. Ground Temperature Case 3.

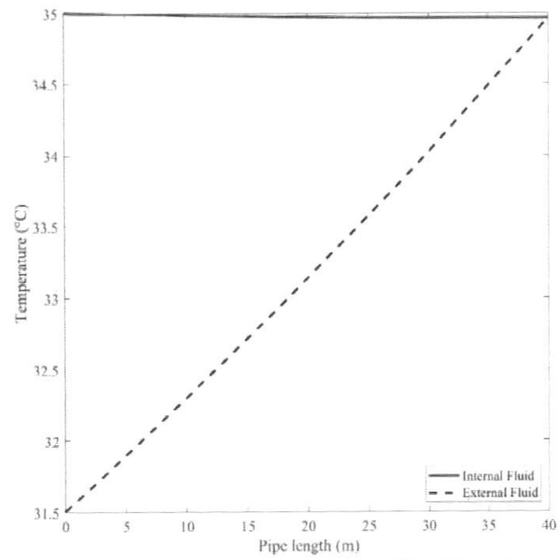

Figure 8. Temperature profile Case3.

4. Conclusion

In this paper, a predictive analysis has been made suitable for the construction of a geothermal plant obtained from a conversion of a plant for the artificial ground freezing for the construction of the Naples underground. The simulations have been made by considering the thermal interaction between the geothermal probes and the soil. In particular, the finite element method has been adopted to simulate the natural convection phenomena in the soil, in which the solution from the advection-diffusion in the coaxial pipe, solved by finite difference method, sets up a variable Dirichlet boundary conditions. In turn, the temperature along the probes becomes a source term for the pipe-flow process.

The results show greater thermal energy recovery for case number 2 compared to case 1 when the mass flow is 0.05 kg/s. In particular, the temperature difference between inlet and outlet is more than doubled. Instead, the cases 1 and 2 are similar when mass flow is 0.5 kg/s.

The results are based on a real life example in which the natural convection solved by 3D FEM-based simulation affects the heat transfer process between the soil and the probes. This numerical scheme can be applied in the future for other geothermal systems for studying how much is convenient, from an energy point of view, the exploiting of geothermal energy.

References

[1] P. Muffler, R. Cataldi, Methods for regional assessment of geothermal resources. Geothermal 1978, 53-89.
[2] MP. Hochstein, Classification and assessment of geothermal resources, In: Dickson MH, Fanelli M, editors. Small geothermal resources. Rome, Italy: Centre for small energy resources; 1990, 31-59.
[3] Y. Benderitter, G. Cormy, Possible approach to geothermal research and relative cost estimate, In: Dickson MH, Fanelli M, editors. Small geothermal resources. Rome, Italy: Centre for small energy resources; 1990, 61-67.
[4] R. Haenel, L. Rybach, L. Stegena, Handbook of terrestrial heat-flow density determination. Kluwer Academic; 1988, 9-57.
[5] A. Carotenuto, M. Ciccolella, N. Massarotti, A. Mauro, Models for thermo-fluid dynamics phenomena in low enthalpy geothermal energy systems: A review. Renewable and Sustainable Energy Reviews; 2016, 331-355.
[6] N. Massarotti, A. Mauro, V. Trombetta, Proper generalized decomposition for geothermal applications, Thermal Science and Engineering Progress, 2021.
[7] A. Mauro, G. Normino, F. Cavuoto, P. Marotta, N. Massarotti, Modelling artificial ground freezing for construction of two tunnels of a metro station in Napoli (Italy), Energies 2020, 13, 1272.
[8] O. C. Zienkiewicz, R. L. Taylor, and P. Nithiarasu, The Finite Element Method for Fluid Dynamics, 6th ed., Elsevier Butterworth-Heinemann, Oxford, UK, 2005.
[9] K. Vafai, Handbook of Porous Media, 2nd ed., Taylor & Francis, New York, 2005.
[10] F. Arpino, A. Carotenuto, M. Ciccolella, G. Cortellessa, N. Massarotti, A. Mauro, Transient natural convection in partially porous vertical annuli, International Journal of Heat and Technology (2016), 512-518.
[11] F. P. Incropera, D. P. DeWitt, Fundamentals of heat and mass transfer, 5th ed., John Wiley & Sons, 2002, 480-490.
[12] Shames IH, Mechanics of fluids. McGraw-Hill, 1982.
[13] M. S. Bhatti and R. K. Shah. Turbulent and Transition Flow Convective Heat Transfer in Ducts, In Handbook of Single-Phase Convective Heat Transfer, ed. S. Kakaç, R. K. Shah, and W. Aung. Wiley Interscience, 1987.
[14] S. E. Haaland. Simple and Explicit Formulas for the Friction Factor in Turbulent Pipe Flow, Journal of Fluids Engineering, 1983, 89-90.
[15] R. K. Shah and M. S. Bhatti, Laminar Convective Heat Transfer in Ducts. In Handbook of Single-Phase Convective Heat Transfer, ed. S. Kakaç, R. K. Shah, and W. Aung. New York: Wiley Interscience, 1987.
[16] P. K. Swamee and A. K. Jain. Explicit Equations for Pipe-Flow Problems, Journal of the Hydraulics Division ASCE 102, no. HY5 1976, 657-664.
[17] R. W. Lewis, P. Nithiarasu, and K. N. Seetharamu, Fundamentals of the Finite Element Method for Heat and Fluid Flow, Wiley, Chichester, UK, (2004).
[18] O.C. Zienkiewicz and R. Codina, A general algorithm for compressible and incompressible flow, Part I. The split characteristic based scheme, Int. J. Num. Meth Fluids (1995), 869-885.
[19] P. Nithiarasu, An efficient artificial compressibility (AC) scheme based on the characteristic based split (CBS) method for incompressible flows, Int. J. Numerical Methods Eng. (2003), 1815-1845.
[20] P. Nithiarasu, R. Codina, O.C. Zienkiewicz, Characteristic based split (CBS) scheme - a unified approach to fluid dynamics, Int. J. Numer. Methods Eng. (2006), 1514-1546.

[21] N. Massarotti, F. Arpino, R. W. Lewis, and P. Nithiarasu, Explicit and Semi-Implicit CBS Procedures for Incompressible Viscous Flows, Int. J. Numer. Meth. Eng. (2006), 1618-1640.
[22] A. Quarteroni, Numerical models for differential problems, 3th ed., Springer, 2017.
[23] J. Smoller, Shock waves and reaction-diffusion equations. Springer-Verlag, 1983.
[24] MN. Ozisik, Mn. Czisik, Finite difference methods in heat transfer, CRC press, 1994.
[25] R. J. LeVeque, Finite Difference Methods for Ordinary and Partial Differential Equations: Steady-State and Time-dependent Problems, Society for Industrial and Applied Mathematics, 2007.
[26] G. H. Golub and J. M. Ortega. Scientific computing and differential equations: An introduction to numerical methods. Academic Press, 1992.
[27] N. Massarotti, A. Mauro, V. Trombetta, A general numerical procedure for solidification and melting in porous media and free fluids, International Journal of Thermal Sciences, 2021.

Journal of Physics: Conference Series 2177 (2022) 012005 doi:10.1088/1742-6596/2177/1/012005

Numerical simulation of a low Prandtl number flow with a four-parameters turbulence model through an explicit algebraic definition of Reynolds stress and turbulent heat flux

G Barbi[1] A Chierici[1], L Chirco[1], V Giovacchini[1], S Manservisi[1] and L Sirotti[1]

[1] University of Bologna - Department of Industrial Engineering DIN
Laboratory of Montecuccolino - Via dei Colli 16, Bologna 40136 (BO), Italy

E-mail: giacomo.barbi3@unibo.it

Abstract. Computational Fluid Dynamics codes usually adopt the Reynolds analogy in order to simulate dynamic and thermal flow fields for ordinary fluids like water and air. On the other hand, in low Prandtl fluids, such as heavy liquid metals like Lead-Bismuth Eutectic (LBE), the time scales of temperature and velocity fields are rather different and therefore similarity hypothesis cannot be used. Furthermore, to properly predict a complex flow field characterized by anisotropic behavior, it is necessary to overcome eddy-viscosity models and move to more advanced turbulence models. In the present work, we propose a nonlinear method for the computation of the Reynolds stress tensor and of the turbulent heat flux. Explicit algebraic models (EAM) and new time scales have been implemented using a logarithmic four parameters turbulence model (i.e. K-Ω-K_θ-Ω_θ). This new model is validated through the simulation of plane channel and cylinder flows and results are compared with DNS data.

1. Introduction

Turbulence modeling has been so far an important issue through engineering design for industrial applications. For this reason, Computational Fluid Dynamics has become a high-performance tool to predict fluid flow behavior and to understand essential physical quantities as heat transfer and pressure drop. Several turbulence models are available in CFD codes and produce accurate results for flow simulations of fluids like water or air. However, traditional turbulence models are not adequate to simulate liquid metals [1] which are more and more considered for different engineering applications characterized by large heat load, e.g. in IV generation nuclear reactors and solar power plant [2].

The similarity hypothesis between dynamic and thermal fields, which is generally accepted for fluids with $Pr \simeq 1$ and is widely implemented in commercial codes, does not hold for low Prandtl number fluids, i.e. liquid metals. The similarity hypothesis and the assumption of a constant turbulent Prandtl number (0.85 - 0.9) are not able to predict liquid metal behavior leading to numerical results in contradiction with heat transfer experimental correlations [3, 4]. Moreover, conventional eddy viscosity models based on Boussinesq approximation are not able to appropriately predict the wall-limiting and flow anisotropic behavior [5]. To overcome the Reynolds analogy, the use of higher-order approximations for the Reynolds stress tensor and turbulent heat flux is recommended. As a consequence, Reynolds Stress Models are recognized as a more accurate alternative to the linear viscosity model. From the pioneering work of Pope [6], many non-linear formulations have been developed in what is now known as Explicit Algebraic Stress Model (EASM), Explicit Algebraic Heat-Flux Model (EAHFM), or similarly Non-Linear Eddy Viscosity Model (NLEDM) [7, 8].

In this work, we propose a four-parameter logarithmic model K-Ω-K_θ-Ω_θ with an Explicit Algebraic Model for Reynolds stress tensor and turbulent heat flux. The logarithmic model is chosen for its numerical stability since the four turbulent variables are enforced to be positive. In addition, attention is put into the modeling of time scales present in $\overline{\mathbf{u}'\mathbf{u}'}$ and $\overline{\mathbf{u}'T'}$ explicit algebraic expressions [9, 10], with the aim of predicting dynamic and thermal variables in both near-wall and bulk regions. To validate these models, by referring to benchmark data obtained by DNS simulations, two geometries have been investigated i.e. plane channel and cylinder geometry.

In the next section, the conservation equations and the turbulence model are presented, focusing mainly on the stress tensor and turbulent heat flux formulation. The boundary conditions set for each turbulent transport equation are presented. In the third section, the simulation parameters are introduced and the numerical results for each geometry are reported and compared with DNS data. Finally, conclusions about the accuracy of the implemented model are drawn and discussed.

2. Transport equations and turbulence model

Fluid flows, such as heavy liquid metals, can be described by the incompressible form of the Averaged Navier-Stokes Equations, here presented

$$\nabla \cdot \mathbf{u} = 0\,, \tag{1}$$

$$\frac{\partial \mathbf{u}}{\partial t} + \mathbf{u}\cdot\nabla\mathbf{u} = -\frac{1}{\rho}\nabla P + \nabla\cdot\left[\nu\left(\nabla\mathbf{u}+\nabla\mathbf{u}^T\right) - \overline{\mathbf{u}'\mathbf{u}'}\right] + \mathbf{g}\,, \tag{2}$$

$$\frac{\partial T}{\partial t} + \mathbf{u}\cdot\nabla T = \nabla\cdot\left[\alpha\nabla T - \overline{\mathbf{u}'T'}\right] + \frac{Q}{\rho C_p}\,, \tag{3}$$

where $\mathbf{u}$ is the average fluid velocity, T is the average temperature, P is the pressure and Q is the volumetric heat source. After the Reynolds averaging process, in equations (2) and (3) two new terms appear called Reynolds stress tensor $\overline{\mathbf{u}'\mathbf{u}'}$ and turbulent heat flux $\overline{\mathbf{u}'T'}$. In these terms, $\mathbf{u}'$ and T' represent the fluctuating components of velocity and temperature, respectively. According to Nagano [5, 11, 9], in this work a second-order closure model is implemented in order to properly predict the Reynolds stress tensor. The generic component is expressed as

$$\overline{u_i'u_j'} = \frac{2}{3}k\delta_{ij} - \frac{2\nu_t}{f_{R1}}S_{ij} - \frac{4C_D k f_\tau}{f_{R1}}(S_{ik}\Omega_{kj} - \Omega_{ik}S_{kj}) + \frac{4C_D k f_\tau}{f_{R1}}\left(S_{ik}S_{kj} - \frac{1}{3}S_{mn}S_{mn}\delta_{ij}\right), \tag{4}$$

where the strain tensor S_{ij} is equal to $\frac{1}{2}\left(\frac{\partial u_i}{\partial x_j} + \frac{\partial u_j}{\partial x_i}\right)$ and the vorticity tensor Ω_{ij} is defined as $\frac{1}{2}\left(\frac{\partial u_i}{\partial x_j} - \frac{\partial u_j}{\partial x_i}\right)$. In equation (4), δ_{ij} represents the Kronecker delta, and the model functions f_τ and f_{R1} are the following

$$f_\tau = \tau_{R_0}^2 + \tau_{R_w}^2, \qquad f_{R1} = 1 + \frac{22}{3}\left(C_D\tau_{R_0}\right)^2\Omega^2 + \frac{2}{3}\left(C_D\tau_{R_0}\right)^2\left(\Omega^2 - S^2\right)f_B\,,$$

where $S^2 = S_{ij}S_{ij}$ and $\Omega^2 = \Omega_{ij}\Omega_{ij}$. It is important to underline that the function f_τ is the sum of the squares of the characteristic time scale $\tau_{R_0} = \nu_t/k$ and the wall-reflection time scale given by $\tau_{R_w} = \sqrt{\frac{f_{R1}/C_D}{f_{S\Omega}}}\left(1 - \frac{3C_{v1}f_{v2}}{8}\right)f_{v1}^2$.

Concerning the ν_t modeling, new variables are introduced to account for the near-wall behavior of turbulent quantities. In particular, considering the wall distance δ and the

Kolmogorov length scale $(\nu^3/\varepsilon)^{1/4}$, it is possible to define the turbulent Reynolds number R_t and the non-dimensional distance from wall surface R_d as

$$R_t := \frac{k^2}{\nu\varepsilon}, \qquad R_d := \frac{\delta}{(\nu^3/\varepsilon)^{1/4}}. \tag{5}$$

With these new quantities, the eddy viscosity can be modeled as

$$\nu_t = C_\mu f_\mu \frac{k^2}{\varepsilon}, \tag{6}$$

where f_μ is the model damping function and C_μ is a constant equal to 0.12. In this paper, it has been used the one proposed by Abe [5] defined as

$$f_\mu = \left\{1 - e^{\left[-(R_d/26)^2\right]}\right\}\left\{1 + \frac{35}{R_t^{3/4}} e^{\left[-(R_t/30)^{3/4}\right]}\right\}. \tag{7}$$

The other constant and model functions are indicated in Table 1.

C_D	C_η	C_{v1}	C_{v2}	C_{tm}	$f_w(\xi)$	R_{tm}	f_B
0.8	5.0	0.4	2000	130	$\exp\left[-(R_{tm}/\xi)^2\right]$	$\frac{C_{tm} R_d R_t^{1/4}}{C_{tm} R_t^{1/4} + R_d}$	$1 + C_\eta (C_D \tau_{R_0})^2 (\Omega^2 - S^2)$

S_f	f_{v1}	f_{v2}	$f_{S\Omega}$
$1 + \sqrt{S^2}\tau_u$	$\exp\left[-\frac{(R_{tm}/52)^2 S_f}{(R_{tm}/52)^2 + S_f}\right]$	$1 - \exp\left(-\frac{\sqrt{R_t}}{C_{v2}}\right)$	$\frac{\Omega^2}{2} + \frac{S^2}{3} - \left[\left(\sqrt{\frac{S^2}{2}} - \sqrt{\frac{\Omega^2}{2}}\right) f_w(1)\right]^2$

Table 1: Constant and model functions for equation (4).

For the Explicit Algebraic Heat-Flux Model the formulation of Abe [12] has been used, with a different characteristic time scale for turbulent heat transfer. The turbulent heat flux components are explicitly calculated as

$$\overline{u'_j T'} = -\frac{C_{T1}}{f_{RT}} \tau_m \overline{u'_j u'_k} \frac{\partial T}{\partial x_k} + \frac{C_{T1}}{f_{RT}} \tau_m^2 \left[(C_{T2} - C_{T3}) S_{jl} \overline{u'_l u'_k} + (C_{T2} + C_{T3}) \Omega_{jl} \overline{u'_l u'_k}\right] \frac{\partial T}{\partial x_k}, \tag{8}$$

where f_{RT} is given by

$$f_{RT} = 1 + \frac{1}{2}\tau_m^2 \left[(C_{T2} + C_{T3})^2 \Omega^2 - (C_{T2} - C_{T3})^2 S^2\right],$$

where we remark that $\Omega^2 = \Omega_{ij}\Omega_{ij}$. According to the isotropic four-parameter turbulence model [10], τ_m can be expressed as

$$\tau_m = \tau_u \left(f_{1\theta} B_{1\theta} + f_{2\theta} B_{2\theta} + f_{3\theta} B_{3\theta}\right), \tag{9}$$

where τ_u is the characteristic dynamical time scale defined as $\tau_u = k/\varepsilon = 1/(C_\mu e^\Omega)$. The thermal characteristic time scale is defined as the sum of three contributions: $B_{1\theta} = 1/Pr_{t,\infty}$ is the correction term for the bulk region with $Pr_{t,\infty} = 1.3$; $B_{2\theta} = 2R/(R + C_\gamma)$ is the contribution for the intermediate region with $R = \tau_u/\tau_t = \omega_\theta/\omega$ the thermal to mechanical turbulent time scale ratio and $C_\gamma = 0.3$; finally, $B_{3\theta} = 1.3\sqrt{2R}/(Pr R_t^{3/4})$ is introduced as a correction term for the near-wall region. The blending functions appearing in (9) are $f_{1\theta} = (1 - e^{\sqrt{Pr}R_d/14})(1 - e^{R_d/14})$, $f_{2\theta} = f_{1\theta}e^{-(R_t/500)^2}$ and $f_{3\theta} = f_{1\theta}e^{-(R_t/200)^2}$. The other model constants in (8) are $C_{T1} = 0.18$, $C_{T2} = 0.18$, and $C_{T3} = 0.02$.

In this paper we overcome the Boussinesq hypothesis, in fact the Reynolds stress tensor is not considered proportional to strain tensor through the eddy viscosity ν_t and in the same way the turbulent heat flux is not parallel to the temperature gradient by the turbulent thermal diffusivity α_t.

The closure of $\overline{\mathbf{u}'\mathbf{u}'}$ and $\overline{\mathbf{u}'T'}$ requires the evaluation of the turbulent kinetic energy k, its dissipation ε, the mean-squared temperature fluctuations value k_θ and its dissipation ε_θ define as

$$k := \frac{1}{2}\overline{\mathbf{u}'^2}, \qquad \varepsilon := \nu\overline{||\nabla\mathbf{u}'||^2}, \qquad k_\theta := \frac{1}{2}\overline{T'^2}, \qquad \varepsilon_\theta := \alpha\overline{||\nabla T'||^2}. \tag{10}$$

We now introduce the specific dissipation rate $\omega = \varepsilon/C_\mu k$ and $\omega_\theta = \varepsilon_\theta/C_\mu k_\theta$, where C_μ is a model constant. To achieve more numerical stability, in this work a logarithmic version of the previous variables has been chosen [13, 14]. In fact, with this approach the variables are defined always positive avoiding negative values with no physical meaning and consequently problems of numerical instability. The new variables are defined as

$$K := \ln(k), \qquad \Omega := \ln(\omega), \qquad K_\theta := \ln(k_\theta), \qquad \Omega_\theta := \ln(\omega_\theta). \tag{11}$$

The transport equations are the following

$$\frac{\partial K}{\partial t} + \mathbf{u}\cdot\nabla K = \nabla\cdot\left[\nu^K_{eff}\nabla K\right] + \nu^K_{eff}\nabla K\cdot\nabla K + \frac{P_k}{e^K} - C_\mu e^\Omega, \tag{12}$$

$$\frac{\partial \Omega}{\partial t} + \mathbf{u}\cdot\nabla \Omega = \nabla\cdot\left[\nu^\Omega_{eff}\nabla \Omega\right] + 2\nu^\Omega_{eff}\nabla K\cdot\nabla\Omega + \nu^\Omega_{eff}\nabla\Omega\cdot\nabla\Omega + \frac{c_{\varepsilon1}-1}{e^K}P_k + - C_\mu\left(c_{\varepsilon2}f_\varepsilon - 1\right)e^\Omega, \tag{13}$$

$$\frac{\partial K_\theta}{\partial t} + \mathbf{u}\cdot\nabla K_\theta = \nabla\cdot\left[\alpha^{K_\theta}_{eff}\nabla K_\theta\right] + \alpha^{K_\theta}_{eff}\nabla K_\theta\cdot\nabla K_\theta + \frac{P_\theta}{e^{K_\theta}} - C_\mu e^{\Omega_\theta}, \tag{14}$$

$$\frac{\partial \Omega_\theta}{\partial t} + \mathbf{u}\cdot\nabla \Omega_\theta = \nabla\cdot\left[\alpha^{\Omega_\theta}_{eff}\nabla \Omega_\theta\right] + 2\alpha^{\Omega_\theta}_{eff}\nabla K_\theta\cdot\nabla\Omega_\theta + \alpha^{\Omega_\theta}_{eff}\nabla\Omega_\theta\cdot\nabla\Omega_\theta + + \frac{c_{p1}-1}{e^{K_\theta}}P_\theta + \frac{c_{p2}}{e^K}P_k - \left(c_{d1}-1\right)C_\mu e^{\Omega_\theta} - c_{d2}C_\mu e^\Omega, \tag{15}$$

where the production terms of the turbulent kinetic energy and the temperature fluctuations are defined as

$$P_k := -\overline{u_i'u_j'}\frac{\partial u_i}{\partial x_j}, \qquad P_\theta := -\overline{u_j'T'}\frac{\partial T}{\partial x_j}. \tag{16}$$

In the previous equations a simple gradient diffusion hypothesis (SGDH) model is considered to model the pressure and turbulent diffusion, and therefore ν^i_{eff} and α^i_{eff} are equal to

$$\nu^i_{eff} = \nu + \frac{\nu_t}{\sigma_i}, \qquad \alpha^i_{eff} = \alpha + \frac{\alpha_t}{\sigma_i}. \tag{17}$$

To model the near-wall behavior two blending functions are introduced, respectively for the Ω and Ω_θ equations [15, 16], as

$$f_\varepsilon = \left[1 - e^{(R_d/3.1)}\right]^2\left\{1 - 0.3e^{-(R_t/6.5)^2}\right\}, \quad c_{d2} = \left[c_{\varepsilon2}f_2 - 1\right]\left[1 - e^{(-R_d/5.7)}\right]^2, \tag{18}$$

where $f_2 = 1 - 0.3e^{-(R_t/6.5)^2}$. In Table 2 the model constants are summarized.

For the eddy thermal diffusivity appearing in diffusive terms of equations (14)-(15) we use the model presented in [10]. It is possible to define α_t as

$$\alpha_t = C_\theta k\tau_m\,, \tag{19}$$

where $C_\theta = 0.1$ and τ_m is given by equation (9).

$c_{\varepsilon 1}$	$c_{\varepsilon 2}$	C_μ	σ_k	σ_ε	c_{p1}	c_{p2}	c_{d1}	$\sigma_{k\theta}$	$\sigma_{\varepsilon\theta}$
1.5	1.9	0.12	1.4	1.4	1.025	0.9	1.1	1.4	1.4

Table 2: Constant values for the equations (12)-(15).

3. Numerical results

In this section we present and discuss the numerical results obtained with the anisotropic four-parameters turbulence model for fully developed turbulent flows in two different geometries, i.e. plane channel and cylindrical pipe. A fully developed flow along a plane channel or a cylindrical pipe is driven by a pressure drop force $F \propto Re_\tau^2$, where $Re_\tau = u_\tau \delta/\nu$ is the turbulent Reynolds number and u_τ is the friction velocity defined as $u_\tau = \sqrt{\tau_w/\rho}$ with τ_w the wall shear stress. DNS data for turbulent fully developed flow plane over a plane channel and pipe are available for different values of Re_τ and Pr number [4, 17]. Numerical simulations are performed for a flow of Lead-Bismuth-Eutectic with physical properties reported in Table 3 and $Pr = 0.025$ for different cases of Re_τ, i.e. $Re_\tau = 180, 395$ for plane channel and $Re_\tau = 186, 360$ for pipe. The

Viscosity	ν	0.001844	$Pa\, s$
Density	ρ	10340	kg/m^3
Thermal conductivity	λ	10.72	$W/(mK)$
Specific heat	C_p	145.75	$J/(kgK)$

Table 3: Physical properties employed for the numerical simulations.

numerical solution of the Navier-Stokes equations and the system (12)-(15) is obtained with the finite element multigrid code FEMuS [18]. In particular, we use quadratic Taylor-Hood finite elements for all the variables but the pressure, for which we use linear finite elements in order to satisfy the BBL inf-sup condition needed for the stability of the discrete Navier-Stokes approximation. The SUPG stabilization term is used in all the equations [19]. The code provides solutions of the thermal fields once the dynamical one is obtained.

3.1. Plane channel

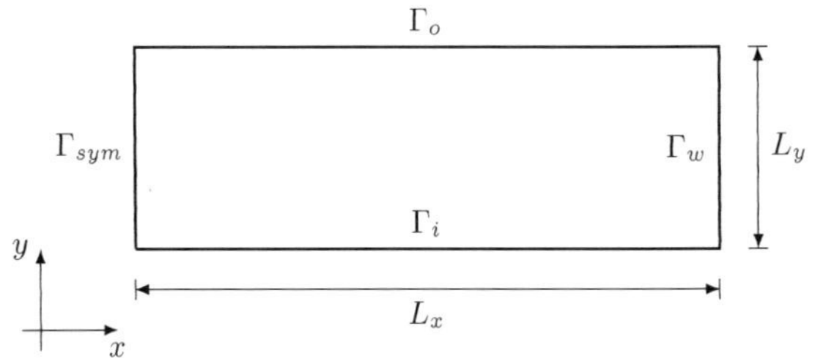

Figure 1: Sketch of the computational domain.

Consistent with available DNS results [4] we adopt a plane channel geometry to validate the model implemented in the code. A representative sketch of the computational domain is reported in Figure 1. This geometry is outlined with two plates located at a distance $L = 2L_x = 0.0605$ m. A general mesh refinement is performed near wall boundaries to obtain for the first mesh a non dimensional distance from the wall $x^+ = xu_\tau/\nu < 1$. Let Γ_i, Γ_o ,Γ_w and Γ_{sym} be

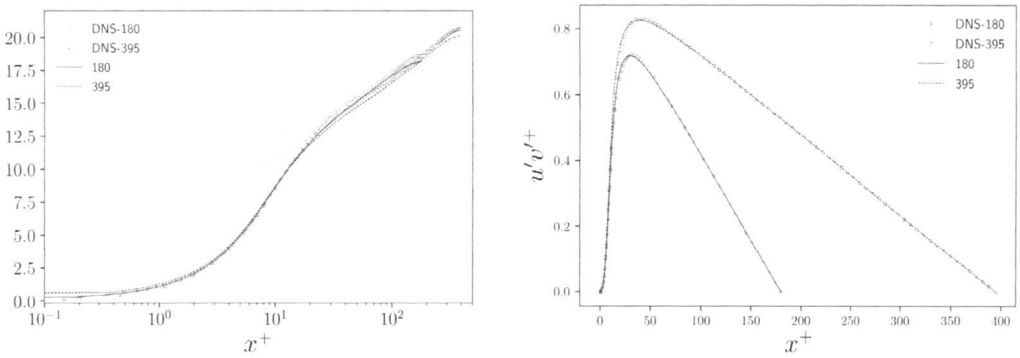

Figure 2: Dynamical fields for the plane channel geometry: non-dimensional velocity field on the left and non-dimensional turbulent shear stress on the right.

the inlet, the outlet, the wall and the symmetry surface, respectively. On the inlet and outlet surface, we apply periodic boundary conditions for the dynamic field due to the fully developed flow. We impose a uniform heat flux of $360000\,W/m^2$ on the wall surface. Therefore, we can redefine the temperature field by introducing the variable $\theta = T - T_{w0} - L_y \Delta T_b$, where ΔT_b is the normal temperature difference, T_{w0} is a constant value on Γ_w and L_y is the axial length of the computational domain. We can apply periodic conditions also on θ. Considering a fully developed turbulent flow, the resulting system is therefore formed by the simplified averaged Navier-Stokes equations with the boundary conditions reported in Table 4, where v is the streamwise component of velocity, u the wall-normal component (u_r for the cylindrical case), $\frac{\partial}{\partial n}$ indicates the normal derivative and $\frac{\partial}{\partial y}$ the streamwise one.

	u/u_r	v	θ	K	Ω	K_θ	Ω_θ
Γ_i	$\frac{\partial u}{\partial y} = 0$	$\frac{\partial v}{\partial y} = 0$	$\frac{\partial \theta}{\partial y} = 0$	$\frac{\partial K}{\partial y} = 0$	$\frac{\partial \Omega}{\partial y} = 0$	$\frac{\partial K_\theta}{\partial y} = 0$	$\frac{\partial \Omega_\theta}{\partial y} = 0$
Γ_o	$\frac{\partial u}{\partial y} = 0$	$\frac{\partial v}{\partial y} = 0$	$\frac{\partial \theta}{\partial y} = 0$	$\frac{\partial K}{\partial y} = 0$	$\frac{\partial \Omega}{\partial y} = 0$	$\frac{\partial K_\theta}{\partial y} = 0$	$\frac{\partial \Omega_\theta}{\partial y} = 0$
Γ_w	$u = 0$	$\frac{\partial v}{\partial n} = \frac{\mu}{\delta}$	$\theta = 0$	$\frac{\partial K}{\partial n} = \frac{2}{\delta}$	$\Omega = \ln(\frac{2\nu}{C_\mu \delta^2})$	$\frac{\partial K_\theta}{\partial n} = \frac{2}{\delta}$	$\Omega_\theta = \ln(\frac{2\nu}{C_\mu \delta^2})$
Γ_{sym}	$u = 0$	$\frac{\partial v}{\partial n} = 0$	$\frac{\partial \theta}{\partial n} = 0$	$\frac{\partial K}{\partial n} = 0$	$\frac{\partial \Omega}{\partial n} = 0$	$\frac{\partial K_\theta}{\partial n} = 0$	$\frac{\partial \Omega_\theta}{\partial n} = 0$

Table 4: Boundary conditions for the state variables.

It has been noticed that the velocity field depends on the pressure gradient which is a function of the friction Reynolds number. In particular, for a fully developed turbulent flow, the existing DNS data are referred to different friction Reynolds numbers, therefore the simulations results are provided for the cases of $Re_\tau = 180$ and $Re_\tau = 395$. To compare the simulation results with DNS data, the quantities are plotted in non-dimensional form. Both the dynamical and the thermal fields are reported as functions of the dimensionless wall distance x^+. The non-dimensional mean velocity field $u^+ = u/u_\tau$ is shown on the left of the Figure 2. A good prediction of both linear and logarithmic behavior can be observed from a comparison with the DNS data. The Reynolds shear stress is correctly predicted for both values of Re_τ as shown on the right of Figure 2.

The non-dimensional mean temperature profile $\theta^+ = \theta/T_\tau = \theta u_\tau \rho C_p/q$ is reported on the left of Figure 3. An overall good agreement with DNS results can be observed for $Re_\tau = 180$, while some discrepancies can be noticed for the case with $Re_\tau = 395$. In particular, the simulated

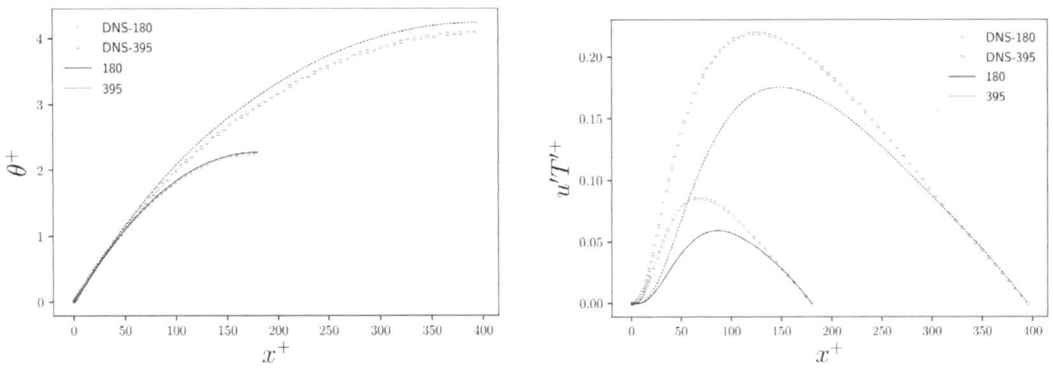

Figure 3: Thermal fields for the plane channel geometry: non-dimensional temperature field on the left and non-dimensional wall-normal turbulent heat flux on the right.

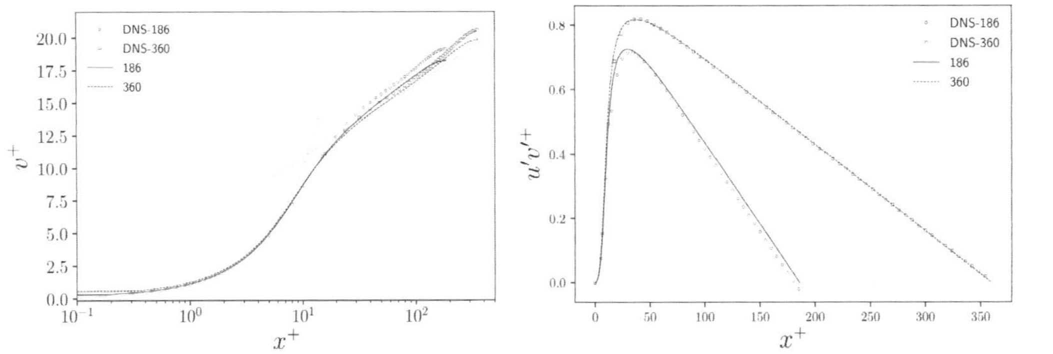

Figure 4: Dynamical fields for the cylindrical pipe.

temperature profile is slightly overestimated compared with DNS results. On the right of Figure 3 the normal component of the turbulence heat flux is reported. This component is underestimated for both Re_τ values. A possible reason for this behaviour, could be a not perfect estimation of the thermal time scale τ_m and the choice of constant models, i.e. C_μ is set equal to 0.12 instead of the standard value 0.09. However, this underestimation does not affect strongly the prediction of the temperature field.

3.2. Cylindrical geometry

The cylindrical geometry results are referred to a pipe with diameter $D = 0.0605$ m and with $Re_\tau = 186, 360$, according to the available DNS data [17]. The boundary conditions are reported in Table 4 with the normal direction represented by the radial component u_r.

In Figure 4 on the left the non-dimensional mean velocity profile is provided as a function of the non-dimensional wall distance x^+. The simulation results show a slight underestimation of the velocity field in the logarithmic region. The non-dimensional shear stress $u'v'^+$ prediction is in good agreement with DNS data, as shown on the right of Figure 4. DNS data of thermal fields are not available for $Re_\tau = 360$ and $Pr = 0.025$, therefore the temperature field and wall-normal turbulent heat flux are compared with reference data only for $Re_\tau = 186$. The temperature field prediction is in good agreement with DNS data while the normal component of the turbulent heat flux is underestimated.

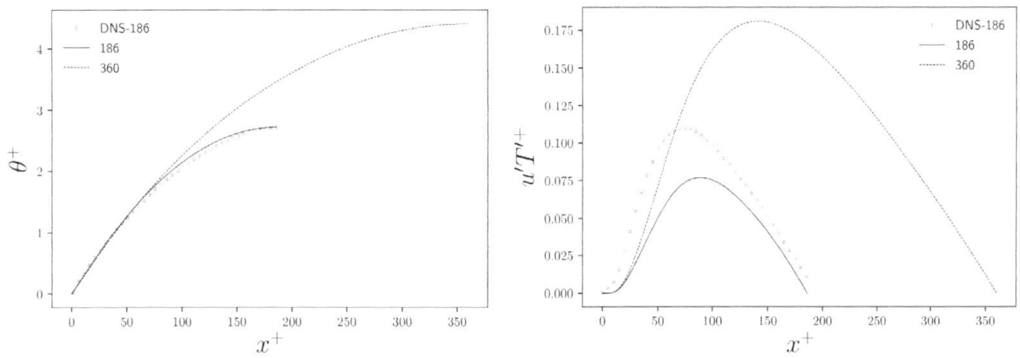

Figure 5: Thermal fields for the cylindrical pipe.

4. Conclusions

In the present work, the simulation of low Prandtl number fluids has been performed with the use of an Explicit Algebraic Stress Model (EASM) for the Reynolds stress tensor, and an Explicit Algebraic Heat Flux Model (EAHFM) for the turbulent heat flux components, overcoming the Boussinesq hypothesis not suitable for anisotropic flows [9, 12]. Moreover, a four-parameter logarithmic model K-Ω-K_θ-Ω_θ has been implemented to close the equations system and compute all the present variables. Two different time scales have been adopted for the evaluation of turbulent viscosity and turbulent diffusivity [5, 10]. Also, a plane channel and a cylindrical geometry have been simulated with a fully developed turbulent flow characterized by a constant heat flux on the wall. The dynamic fields profiles show a good agreement with DNS simulations for both geometries. On the other hand, the thermal variables differ slightly from DNS results. Further improvements to this model will be investigated in future works to better understand the physical behavior of liquid metals in several engineering applications.

References

[1] Grötzbach G 2013 *Nuclear Engineering and Design* **264** 41–55
[2] Heinzel A, Hering W, Konys J, Marocco L, Litfin K, Müller G, Pacio J, Schroer C, Stieglitz R, Stoppel L *et al.* 2017 *Energy Technology* **5** 1026–1036
[3] Cheng X and Tak N 2006 *Nuclear Engineering and Design* **236** 1874–1885
[4] Kawamura H, Abe H and Matsuo Y 1999 *International Journal of Heat and Fluid Flow* **20** 196–207
[5] Abe K, Kondoh T and Nagano Y 1997 *International Journal of Heat and Fluid Flow* **18** 266–282
[6] Pope S 1975 *Journal of Fluid Mechanics* **72** 331–340
[7] Suga K and Abe K 2000 *International Journal of Heat and Fluid Flow* **21** 37–48
[8] Gatski T and Jongen T 2000 *Progress in Aerospace Sciences* **36** 655–682
[9] Hattori H, Morita A and Nagano Y 2006 *International Journal of Heat and Fluid Flow* **27** 671–683
[10] Manservisi S and Menghini F 2014 *International Journal of Heat and Mass Transfer* **69** 312–326
[11] Hattori H and Nagano Y 2004 *Theoretical and Computational Fluid Dynamics* **17** 313–330
[12] Abe K i, Kondoh T and Nagano Y 1996 *International Journal of Heat and Fluid Flow* **17** 228–237
[13] Ilinca F, Pelletier D *et al.* 1998 *Computers & Fluids* **27** 291–310
[14] Da Via R, Manservisi S and Menghini F 2016 *International Journal of Heat and Mass Transfer* **101** 1030–1041
[15] Abe K, Kondoh T and Nagano Y 1994 *International Journal of Heat and Mass Transfer* **37** 139–151
[16] Abe K, Kondoh T and Nagano Y 1995 *International Journal of Heat and Mass Transfer* **38** 1467–1481
[17] Redjem-Saad L, Ould-Rouiss M and Lauriat G 2007 *International Journal of Heat and Fluid Flow* **28** 847–861
[18] Chierici A, Barbi G, Bornia G, Cerroni D, Chirco L, Da Vià R, Giovacchini V, Manservisi S, Scardovelli R and Cervone A 2021 *9th edition of the International Conference on Computational Methods for Coupled Problems in Science and Engineering (COUPLED PROBLEMS 2021)*
[19] Brooks A N and Hughes T J 1982 *Computer Methods in Applied Mechanics and Engineering* **32** 199–259

Journal of Physics: Conference Series 2177 (2022) 012006 doi:10.1088/1742-6596/2177/1/012006

Numerical simulation of a low Prandtl number flow over a backward facing step with an anisotropic four-equation turbulence model

G. Barbi[1], **A. Chierici**[1], **V. Giovacchini**[1], **F. Quarta**[1] **and S. Manservisi**[1]

[1] University of Bologna - Department of Industrial Engineering DIN
Laboratory of Montecuccolino - Via dei Colli 16, Bologna 40136 (BO), Italy

E-mail: valentin.giovacchin2@unibo.it

Abstract. In recent years the use of liquid metals has become more and more popular for heat transfer applications in many fields ranging from IV generation fast nuclear reactors to solar power plants. Due to their low Prandtl number values, the similarity between dynamical and thermal fields cannot be assumed and sophisticated heat turbulence models are required to take into account the anisotropy of the turbulent heat transfer involving liquid metals. In the present work, we solve an anisotropic four-equation turbulence model coupled with the Reynolds Averaged Navier Stokes system of equations to simulate a turbulent flow of liquid sodium over a vertical backward-facing step. We implement an explicit algebraic model for Reynolds stress tensor and turbulent heat flux that takes into account flow anisotropic behavior. We study forced and mixed convection regimes when a uniform heat flux is applied on the wall behind the step. Linear isotropic approximations for eddy viscosity and eddy thermal diffusivity underestimate the turbulent heat flux components while this anisotropic model shows a better agreement with DNS results.

Key Words: four-equation turbulence model, explicit algebraic model, Reynolds-stress tensor, turbulent heat flux, anisotropy, liquid metals.

1. Introduction

Liquid metals with their low Prandtl number have gained increasing attention in recent years. Compared with other coolant fluids, such as air or water, liquid metals provide larger heat fluxes and can withstand higher temperatures. Moreover, some liquid metals, like sodium, can flow in the liquid phase on a wide range of temperature without the need for high pressurized systems [1]. Due to these properties, liquid metals are currently considered in a broad range of industrial applications, including the production of steel and semiconductors, thermal solar plants [2, 3] and Generation IV nuclear power plants [4, 5], i.e. the Lead Fast Reactor (LFR) and the Sodium Fast Reactor (SFR).

In the nuclear context, thermal-hydraulics is recognized as one of the key issues in the design and construction of liquid metal cooled reactors and Computational Fluid Dynamics (CFD) is regarded as a valuable tool to analyze the thermal-hydraulics behavior of nuclear systems. In the Reynolds-Averaged Navier-Stokes (RANS) framework, a large number of models have been developed in past decades for the closure of the Reynolds stress tensor $\overline{u_i'u_j'}$. The first-order models are based on the isotropic eddy diffusivity ν_t, while the second-order models use transport equations for each component of Reynolds stress tensor. From an academic point of view, the second-order models would be the best modeling of flows with anisotropic momentum transfer, but the usage of these techniques requires a considerably increased numerical effort [6]. In this work, for the closure of the momentum equation, we propose an Explicit Algebraic Stress Model (EASM) which belongs to a class of models between first and second-order. This class of models is derived from the second-order transport equation with the hypothesis of local equilibrium between production and dissipation, therefore convection and diffusion terms are neglected. The remaining closure terms are formulated in terms of the turbulent kinetic energy k and its dissipation rate ε.

For the turbulent heat flux $\overline{u_i'T'}$, only a restricted number of models have been developed and validated. Most of them are first-order models based on the Simple Gradient Diffusion Hypothesis (SGDH) which assumes the similarity between the turbulent heat flux and the molecular heat conduction introducing the turbulent thermal diffusivity α_t. Furthermore, the Reynolds analogy is almost universally applied. This approach assumes a similarity between α_t and ν_t introducing the turbulent Prandtl number Pr_t which is often set equal to a constant value in the range 0.8-1. This concept can reproduce reasonable results in the forced convection regime and for fluids with $Pr \approx 1$, whereas it is inadequate for applications involving non-unity Prandtl number fluids like liquid metals and/or non negligible buoyancy effects [7]. For these applications, the most promising models require the introduction of additional transport equations. In [8, 9] an implicit Algebraic Heat Flux Model (AHFM) model has been proposed and implemented in STAR-CCM+. Its closure requires one additional transport equation for the evaluation of the temperature variance $\overline{T'^2}$ and the assumption of a constant thermal to mechanical time scale ratio R. In [10, 11, 12, 13] an isotropic four-equation model has been proposed. The model introduces two additional thermal transport equations for the evaluation of the squared temperature fluctuations k_θ and its dissipation ε_θ. In the original formulation of this model, the turbulent heat flux is evaluated with an SGDH approach. In this work, we propose an anisotropic version of the above-mentioned four-equation model and we suggest for the turbulent heat flux an Explicit Algebraic Heat Flux Model (EAHFM). The thermal model is coupled to an Explicit Algebraic Stress Model (EASM). For the closure of $\overline{u_i'u_j'}$ and $\overline{u_i'T'}$ we solve four transport equations for the variables k-ε-k_θ-ε_θ.

In this work, we study liquid sodium turbulent flows over a vertical backward-facing step using the anisotropic four-equation turbulence model. This flow has been extensively studied in the literature for different flow regimes. Several DNS simulations with different Richardson and Reynolds numbers are available for the comparison and validation of turbulence models [14]. In Section 2 the anisotropic four-equation model is presented, while in Section 3 results for forced and mixed convection are discussed in comparison with literature data.

2. Mathematical Model

The turbulent flow is simulated using a RANS set of equations with the assumption of incompressible flow and the Oberbeck-Boussinesq approximation, namely

$$\frac{\partial u_i}{\partial x_i} = 0\,, \tag{1}$$

$$\frac{Du_i}{Dt} = -\frac{1}{\rho}\frac{\partial p}{\partial x_i} + \frac{\partial}{\partial x_j}\left[\nu\left(\frac{\partial u_i}{\partial x_j} + \frac{\partial u_j}{\partial x_i}\right) - \overline{u_i'u_j'}\right] - g_i\beta(T - T_0)\,, \tag{2}$$

$$\frac{DT}{Dt} = \frac{\partial}{\partial x_i}\left(\alpha\frac{\partial T}{\partial x_i} - \overline{u_i'T'}\right), \tag{3}$$

where u_i, p and T are the mean velocity components, pressure and temperature, while u_i' and T' are the fluctuating velocity components and temperature. The constant ν is the kinematic viscosity, α is the thermal diffusivity, ρ is the fluid density and β is the thermal expansion coefficient. The operator D/Dt indicates the substantial derivative $D/Dt = \partial/\partial t + u_j\partial/\partial x_j$. The Reynolds stress tensor $\overline{u_i'u_j'}$ is the averaged product of velocity fluctuations and the turbulent heat flux $\overline{u_i'T'}$ is the averaged product of velocity and temperature fluctuations. The Explicit Algebraic Shear stress Model (EASM) derives from the full transport equation for the Reynolds stress tensor under the hypothesis of local equilibrium between production and dissipation [15]. The resulting algebraic expression contains products of strain-rate tensor $S_{ij} = \frac{1}{2}(\frac{\partial u_i}{\partial x_j} + \frac{\partial u_j}{\partial x_i})$, vorticity tensor $\Omega_{ij} = \frac{1}{2}(\frac{\partial u_i}{\partial x_j} - \frac{\partial u_j}{\partial x_i})$ and eddy viscosity ν_t as follows

$$\overline{u_i'u_j'} = \frac{2}{3}k\delta_{ij} - \frac{2\nu_t}{f_R}S_{ij} - \frac{4C_D k f_\tau}{f_R}\left(S_{ik}\Omega_{kj} - \Omega_{ik}S_{kj} - S_{ik}S_{kj} + \frac{1}{3}S^2\delta_{ij}\right), \tag{4}$$

where $S^2 = S_{mn}S_{mn}$. The eddy viscosity ν_t is given by

$$\nu_t = C_\mu f_\mu k \tau_u \,, \tag{5}$$

where C_μ and f_μ are the model constant and function, τ_u denotes the characteristic time scale defined as $\tau_u = k/\varepsilon$, with k the turbulent kinetic energy and ε its dissipation rate

$$k = \frac{1}{2}\overline{u_i'u_i'} \,, \tag{6}$$

$$\varepsilon = \nu\overline{\frac{\partial u_i'}{\partial x_k}\frac{\partial u_i'}{\partial x_k}} \,. \tag{7}$$

The function f_μ plays a key role in the eddy viscosity description and in turbulence modeling. The modeling of f_μ is performed using the non-dimensional wall-distance $R_d = \upsilon y_d/\nu$ and the turbulent Reynolds number $R_t = k^2/(\nu\varepsilon)$, with $\upsilon = (\nu\varepsilon)^{\frac{1}{4}}$ the Kolmogorov velocity scale and y_d the wall distance at a point, i.e. the distance between that point and nearest point on the wall surfaces. According to [15], we define the function f_μ as follows

$$f_\mu = \left\{1 - \exp\left[-\left(\frac{R_d}{26}\right)^2\right]\right\}\left\{1 + \frac{35}{R_t^{\frac{3}{4}}}\exp\left[-\left(\frac{R_t}{30}\right)^{\frac{3}{4}}\right]\right\}. \tag{8}$$

The function f_τ in (4) reproduces the wall-limiting behavior and anisotropy of the Reynolds normal stress components near the wall and it is defined as

$$f_\tau = \tau_{R_0}^2 + \tau_{R_W}^2 \,, \tag{9}$$

where $\tau_{R_0} = \nu_t/k$ is the characteristic time scale and τ_{R_W} is the wall reflection time scale, defined by the expression [16]

$$\tau_{R_W} = \sqrt{\frac{f_R}{6C_D f_{S\Omega}}}\left(1 - \frac{3C_{v1}f_{v2}}{8}\right)f_{v1}^2 \,. \tag{10}$$

The model constants and functions appearing in (4) and (10) are reported in Table 1. For the closure of the EASM model and the evaluation of k and ε, we propose a logarithmic turbulence model K-Ω, which improves the stability of a standard k-ω model [11, 13]. This logarithmic model has been proven to be more robust [11] due to the positiveness of the solution variables and the logarithmic variables have smoother profiles than the ones of the natural variables [17]. The

C_D	C_μ	C_η	C_{v1}	C_{v2}	C_{tm}	R_{tm}
0.8	0.09	5	0.4	2000	130	$\frac{C_{tm}R_dR_t^{0.25}}{C_{tm}R_t^{0.25}+R_d}$

f_{v2}	f_{v1}	$f_{S\Omega}$
$1-\exp(-\sqrt{R_t/C_{v2}})$	$\exp(-R_{tm}^2/2025)$	$\frac{\Omega^2}{2}+\frac{S^2}{3}-\left\lvert\sqrt{\frac{S^2}{2}}-\sqrt{\frac{\Omega^2}{2}}\right\rvert\exp(-R_{tm}^2)$

f_R	f_B
$1+\frac{22}{3}(C_D\tau_{R_0})^2\Omega^2+\frac{2}{3}(C_D\tau_{R_0})^2(\Omega^2-S^2)f_B$	$1+C_\eta(C_D\tau_{R_0})^2(\Omega^2-S^2)$

Table 1: Model constants and functions for the Explicit Algebraic Stress Model (4),(10).

transport equations for the logarithmic variables $K=\ln(k)$ and $\Omega=\ln(\omega)$, where $\omega=\varepsilon/(C_\mu k)$ is the specific dissipation rate of turbulent kinetic energy, are the following

$$\frac{DK}{Dt}=\frac{\partial}{\partial x_i}\left[\left(\nu+\frac{\nu_t}{\sigma_k}\right)\frac{\partial K}{\partial x_i}\right]+\left(\nu+\frac{\nu_t}{\sigma_k}\right)\frac{\partial K}{\partial x_i}\frac{\partial K}{\partial x_i}+\frac{P_k}{e^K}+C_b\frac{P_b}{e^K}-C_\mu e^\Omega\,, \tag{11}$$

$$\begin{aligned}\frac{D\Omega}{Dt}=\frac{\partial}{\partial x_i}\left[\left(\nu+\frac{\nu_t}{\sigma_\omega}\right)\frac{\partial \Omega}{\partial x_i}\right]+\left(\nu+\frac{\nu_t}{\sigma_\omega}\right)\frac{\partial \Omega}{\partial x_i}\frac{\partial \Omega}{\partial x_i}+2\left(\nu+\frac{\nu_t}{\sigma_\omega}\right)\frac{\partial K}{\partial x_i}\frac{\partial \Omega}{\partial x_i}+\\+\frac{P_k}{e^K}(C_{\varepsilon1}-1)+(C_b-1)\frac{P_b}{e^K}-C_\mu(C_{\varepsilon2}f_\varepsilon-1)e^\Omega\,.\end{aligned} \tag{12}$$

where $P_k=-\overline{u_i'u_j'}\partial u_i/\partial x_j$ is the production rate of turbulent kinetic energy and $P_b=-\beta g_i\overline{u_i'T'}$ is the source term for buoyant flows. The model constants are $\sigma_k=\sigma_\omega=1.4$, $C_{\varepsilon1}=1.9$, $C_{\varepsilon2}=1.5$, $C_b=1.44$, while the model function f_ε is given by

$$f_\varepsilon=\left\{1-0.3\exp\left[-\left(\frac{R_t}{6.5}\right)^2\right]\right\}\left\{1-\exp\left[-\left(\frac{R_d}{3.7}\right)^2\right]\right\}. \tag{13}$$

The explicit Algebraic Heat Flux Model (EAHFM) derives from the transport equation for the turbulent heat flux in local equilibrium state neglecting the diffusive term [18]. Under this hypothesis, the explicit algebraic expression for the turbulent heat flux can be written as

$$\overline{u_i'T'}=-\frac{C_{t1}}{f_{RT}}\tau_m\overline{u_i'u_j'}\frac{\partial T}{\partial x_j}+\frac{C_{t1}}{f_{RT}}\tau_m^2[(C_{t2}-C_{t3})S_{ij}+(C_{t2}-C_{t3})\Omega_{ij}]\overline{u_j'u_k'}\frac{\partial T}{\partial x_k}\,. \tag{14}$$

In contrast with SGDH models, the components of turbulent heat flux $\overline{u_i'T'}$ and the mean temperature gradient are not necessarily in alignment due to the effects of the mean-velocity gradient and the Reynolds stress tensor $\overline{u_i'u_j'}$. To predict the heat transfer in wall flows, the characteristic time scale τ_m plays a key role since the expression (14) derives from the local equilibrium hypothesis, which does not hold in the near-wall region. According to [10, 11], the characteristic thermal time scale τ_m is modeled as

$$\tau_m=\tau_u f_{1t}\left(\frac{1}{Pr_t}+\frac{2R}{R+C_\gamma}f_{2t}+1.3\frac{\sqrt{2R}}{PrR_t^{\frac{3}{4}}}f_{3t}\right), \tag{15}$$

where $R=(\varepsilon/k)/(\varepsilon_\theta/k_\theta)$ is the thermal to mechanical turbulent time scale ratio. Model constant and model functions are reported in Table 2. The closure of the turbulent heat flux requires two

C_γ	C_{t1}	C_{t2}	C_{t3}	Pr_t
$0.25/Pr^{\frac{1}{4}}$	0.18	0.18	0.02	1.3
f_{1t}			f_{2t}	f_{3t}
$[1-\exp(-\frac{R_d}{14})][1-\exp(-\frac{\sqrt{Pr}R_d}{14})]$			$\exp[-(\frac{R_t}{500})^2]$	$\exp[-(\frac{R_t}{200})^2]$
f_{RT}				
$1+0.5\tau_m^2[(C_{t2}+C_{t3})^2\Omega^2-(C_{t2}-C_{t3})^2S^2]$				

Table 2: Model constants and functions for Explicit Algebraic Heat Flux Model (14)-(15).

additional transport equations for the evaluation of k_θ and ε_θ. We propose a logarithmic K_θ-Ω_θ turbulence model, where K_θ and Ω_θ represent the logarithmic values of mean temperature fluctuactions k_θ and its dissipation rate $\omega_\theta = \varepsilon_\theta/(C_\mu k_\theta)$. The transport equations for the logarithmic quantities are expressed as

$$\frac{DK_\theta}{Dt} = \frac{\partial}{\partial x_i}\left[\left(\alpha+\frac{\alpha_t}{\sigma_{k\theta}}\right)\frac{\partial K_\theta}{\partial x_i}\right] + \left(\alpha+\frac{\alpha_t}{\sigma_{k\theta}}\right)\frac{\partial K_\theta}{\partial x_i}\frac{\partial K_\theta}{\partial x_i} + \frac{P_{k_\theta}}{e^{K_\theta}} - C_\mu e^{\Omega_\theta}\,, \tag{16}$$

$$\begin{aligned}\frac{D\Omega_\theta}{Dt} &= \frac{\partial}{\partial x_i}\left[\left(\alpha+\frac{\alpha_t}{\sigma_{\omega_\theta}}\right)\frac{\partial \Omega_\theta}{\partial x_i}\right] + \left(\alpha+\frac{\alpha_t}{\sigma_{\omega_\theta}}\right)\frac{\partial \Omega_\theta}{\partial x_i}\frac{\partial \Omega_\theta}{\partial x_i} + 2\left(\alpha+\frac{\alpha_t}{\sigma_{\omega_\theta}}\right)\frac{\partial K_\theta}{\partial x_i}\frac{\partial \Omega_\theta}{\partial x_i} + \\ &\quad + \frac{P_{k_\theta}}{e^{K_\theta}}(C_{p1}-1) + C_{p2}\frac{P_k}{e^K} - (C_{d1}-1)C_\mu e^{\Omega_\theta} - C_{d2}C_\mu e^{\Omega}\,,\end{aligned} \tag{17}$$

where $P_{k_\theta} = -\overline{u_j' T'}\partial T/\partial x_j$ and C_{d2} is the following model function

$$C_{d2} = \left\{1.9\left[1-0.3\exp\left(-\frac{R_t^2}{42.25}\right)\right]-1\right\}\left[1-\exp\left(-\frac{R_d^2}{25}\right)\right], \tag{18}$$

while $C_{p1} = 1.025$, $C_{p2} = 0.9$, $C_{d1} = 1.1$, $\sigma_{k_\theta} = \sigma_{\omega_\theta} = 1.4$. The eddy thermal diffusivity appearing in the diffusive terms of (16)-(17) is simplified as the scalar quantity $\alpha_t = C_\theta k \tau_m$, with $C_\theta = 0.1$.

3. Numerical Results

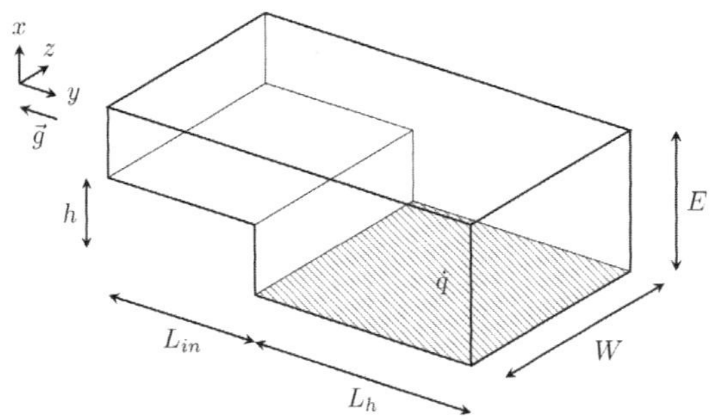

Figure 1: Representative sketch of backward-facing step geometry.

L_{in}/h	L_h/h	W	E_r	Re	Ri	Pr
2	20	0	1.5	4805	$0 - 0.338$	0.0088

Table 3: Geometrical parameters of the simulated domain.

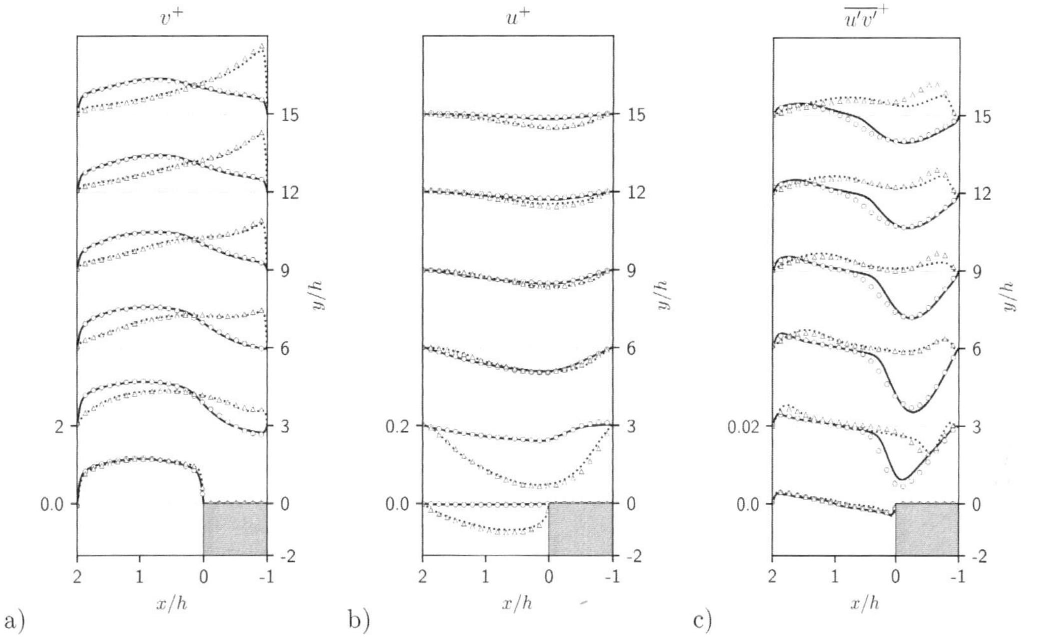

Figure 2: Profile of dynamical fields: a) mean streamwise velocity v^+ b) mean wall-normal velocity u^+ c) shear stress $\overline{u'v'}^+$. —: Simulation results with $Ri = 0$; --: Simulation results with $Ri = 0.338$. △: DNS data with $Ri = 0.338$; ∘ : DNS data with $Ri = 0$.

In this section, we report the results obtained for the simulation of a turbulent flow of liquid sodium over a vertical backward-facing step. The computational domain reproduces the reference DNS domain [14] and a representative sketch is reported in Figure 1. The inlet section length is L_{in}, the step height is h, the domain width is W and the downstream channel height is E. The expansion ratio is $E_r = E/(E-h)$. The geometrical parameters of the simulated domain are reported in Table 3. The system of equations is solved with the finite element code FEMuS [19]. The mesh consists of 27820 cells with 107411 biquadratic nodes. Mesh cells are clustered near the corner step and reattachment zone. Mesh refinement is performed near walls.

We have considered two different values of Richardson number $Ri = g\beta\Delta T h/U_b^2$, $Ri = 0$ corresponding to forced convection and $Ri = 0.338$ corresponding to mixed convection regime. In terms of boundary conditions, a fully developed inflow condition has been set for the velocity field and turbulent variables. For the temperature, a uniform value is set, i.e. $T_{in} = 423.15K$. The same temperature is used as the reference value for the evaluation of liquid sodium properties obtaining $Pr = 0.0088$. At the outlet section, an outflow boundary condition is imposed on the velocity field and for all the other variables homogeneous Neumann conditions are set. All the remaining boundaries have been treated as adiabatic no-slip walls, except for the wall behind

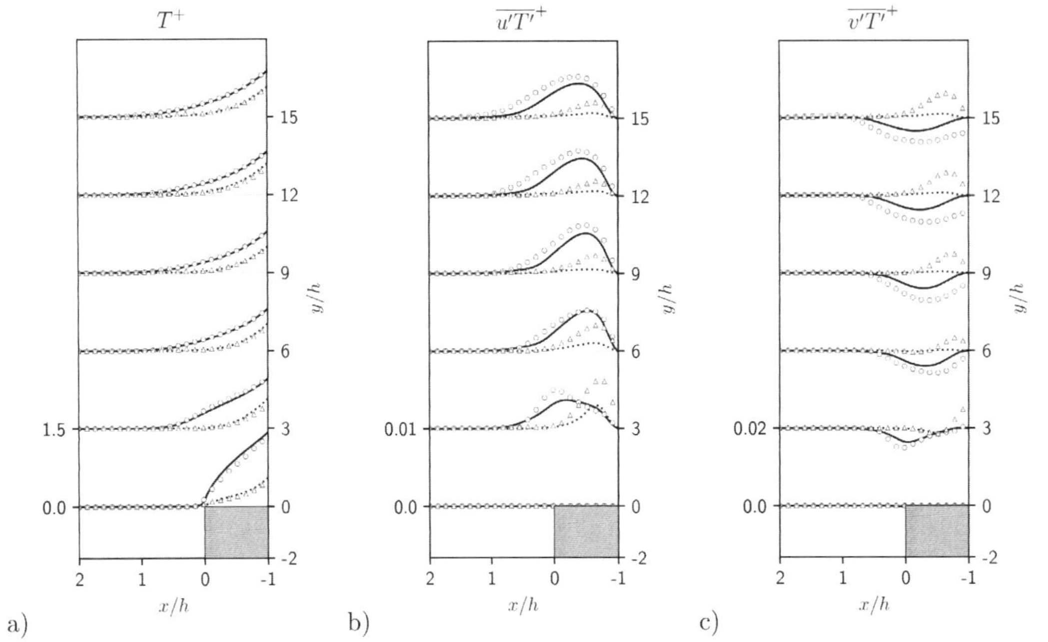

Figure 3: Profile of thermal fields: a) mean temperature T^+ b) mean wall-normal turbulent heat flux $\overline{u'T'}^+$ c) mean streamwise turbulent heat flux $\overline{v'T'}^+$. —: Simulation results with $Ri = 0$; ---: Simulation results with $Ri = 0.338$. △: DNS data with $Ri = 0.338$; ◦ : DNS data with $Ri = 0$.

the step where a uniform heat flux $\dot{q}$ is imposed.

3.1. *Dynamical fields*

In this section the results obtained for flow fields are compared with DNS data. Non-dimensional profiles of velocity, taken on channel cross-section planes, are reported for several streamwise coordinate y/h values, for forced and mixed convection regimes. The streamwise positions included in these plots correspond to the locations where DNS data are available [14]. The streamwise $v^+ = v/U_b$ and wall-normal velocity component $u^+ = u/U_b$ are reported respectively in Figure 2a) and Fig 2b). The velocity field prediction is in good agreement with DNS results for both forced and mixed convection. In the forced convection case ($Ri = 0$), the typical features of the flow over a backward-facing step, i.e. the separation and reattachment behind the step, can be observed from the plot taken at $y/h = 3$ where v^+ assumes negative values close to the heated wall. In the $Ri = 0.338$ case, a jet flow develops near the heated wall due to the acceleration of the fluid caused by buoyancy. As a consequence, away from the heated wall the streamwise velocity is significantly reduced. In Figure 2c) the non-dimensional shear stress component $\overline{u'v'}^+ = \overline{u'v'}/U_b^2$ is reported in comparison with DNS data for forced and mixed convection regimes. In both cases, the prediction of the shear stress is in good agreement with reference data but some discrepancies in the mixed convection case are observed near the heated wall in the locations $y/h > 9$.

3.2. *Thermal Fields*

Non-dimensional profiles of temperature, taken on channel cross-section planes, are reported in Figure 3 for different values of streamwise coordinate y/h, for forced and mixed cases. An overall good agreement with DNS data is obtained for both $Ri = 0$ and $Ri = 0.338$. The discrepancies with DNS results in the forced convection regime are limited to the plot taken at $y/h = 0$ where the temperature is slightly overestimated, while in mixed regime a deviation from DNS results is observed on the plot at $y/h = 3$. The turbulent heat flux components along wall-normal and streamwise directions are reported respectively in Figure 3b) and Figure 3c). With the proposed anisotropic model, the wall-normal component shows a good agreement with DNS data and the streamwise component results only slightly underestimated with $Ri = 0$. Despite the discrepancies in the prediction of both components of the turbulent heat flux, the temperature profiles are in perfect agreement with DNS data. In the mixed convection regime both components are strongly underestimated without consequences on the prediction of the temperature field over the domain since the heat transfer in this case is dominated by the diffusive heat transfer.

4. Conclusion

In this paper, an anisotropic four-equation turbulence model has been applied to the modeling of turbulent liquid sodium flow over a vertical backward-facing step in both forced and mixed convection regimes. An Explicit Algebraic Stress Model (EASM) and an Explicit Algebraic Heat Flux Model (EAHFM) have been used to model the Reynolds stress and turbulent heat flux components. The closure has been performed with four transport equations for the logarithmic variable K-Ω-K_θ-Ω_θ. A reasonable prediction of the velocity and temperature fields has been obtained in forced and mixed convection regimes. The shear stress component shows a good agreement with DNS results in the forced convection case while in the mixed convection it is slightly underestimated near to the outlet section. The prediction of both turbulent heat flux components is significantly improved with respect to the isotropic model [13] in the forced convection case. This work represents the first application of the anisotropic four-equation turbulence model to a complex flow configuration at low Prandtl number and the results are very promising. Further simulations including complex configurations and buoyancy effects will be performed and presented in future works.

References

[1] Heinzel A, Hering W, Konys J, Marocco L, Litfin K, Müller G, Pacio J, Schroer C, Stieglitz R, Stoppel L *et al.* 2017 *Energy Technology* **5** 1026–1036
[2] Marocco L, Cammi G, Flesch J and Wetzel T 2016 *International Journal of Thermal Sciences* **105** 22–35
[3] Frazer D, Stergar E, Cionea C and Hosemann P 2014 *Energy Procedia* **49** 627–636
[4] Manservisi S and Menghini F 2014 *Nuclear Engineering and Design* **273** 251–270
[5] Cheng X and Tak N i 2006 *Nuclear Engineering and Design* **236** 385–393
[6] Grötzbach G 2007 *Anisotropy and Buoyancy in Nuclear Turbulent Heat Transfer: Critical Assessment and Needs for Modelling* (Citeseer)
[7] Shams A, De Santis A, Koloszar L, Ortiz A V and Narayanan C 2019 *Nuclear Engineering and Design* **353** 110220
[8] De Santis A and Shams A 2018 *Annals of Nuclear Energy* **117** 32–44
[9] Shams A and De Santis A 2019 *International Journal of Heat and Mass Transfer* **130** 290–303
[10] Manservisi S and Menghini F 2014 *International Journal of Heat and Mass Transfer* **69** 312–326
[11] Da Via R, Manservisi S and Menghini F 2016 *International Journal of Heat and Mass Transfer* **101** 1030–1041
[12] Da Vià R, Giovacchini V and Manservisi S 2020 *Applied Sciences* **10** 4337
[13] Da Vià R and Manservisi S 2019 *International Journal of Heat and Mass Transfer* **135** 591–603
[14] Niemann M and Fröhlich J 2016 *International Journal of Heat and Mass Transfer* **101** 1237–1250
[15] Abe K, Kondoh T and Nagano Y 1997 *International Journal of Heat and Fluid Flow* **18** 266–282
[16] Hattori H, Morita A and Nagano Y 2006 *International Journal of Heat and Fluid Flow* **27** 671–683
[17] Ilinca F, Pelletier D *et al.* 1998 *Computers & fluids* **27** 291–310
[18] Abe K i, Kondoh T and Nagano Y 1996 *International journal of heat and fluid flow* **17** 228–237
[19] Chierici A, Barbi G, Bornia G, Cerroni D, Chirco L, Da Vià R, Giovacchini V, Manservisi S, Scardovelli R and Cervone A 2021 Femus-platform: a numerical platform for multiscale and multiphysics code coupling *9th edition of the International Conference on Computational Methods for Coupled Problems in Science and Engineering (COUPLED PROBLEMS 2021)*

Journal of Physics: Conference Series 2177 (2022) 012007 doi:10.1088/1742-6596/2177/1/012007

Numerical analysis of thermo-fluid problems in 3D domains by means of the RBF-FD meshless method

Riccardo Zamolo, Davide Miotti and Enrico Nobile

Dipartimento di Ingegneria e Architettura, Università degli Studi di Trieste, via A. Valerio 10, 34127 Trieste, Italy

E-mail: rzamolo@units.it

Abstract. The use of CAE (Computer Aided Engineering) software, commonly applied to the design and verification of a great variety of manufactured products, is totally reliant on accurate numerical simulations. Classic mesh-based methods, e.g., Finite Element (FEM) and Finite Volume (FVM), are usually employed for such simulations, where the role of the mesh is crucial for both accuracy and time consumption issues. This is especially true for complex 3D domains which are typically encountered in most practical problems. Meshless, or meshfree, methods have been recently introduced in order to replace the usual mesh with much simpler node distributions, thus purifying the data structures of any additional geometric information. Radial Basis Function-Finite Difference (RBF-FD) meshless methods have been shown to be able to easily solve problems of engineering relevance over complex-shaped domains with great accuracy, with particular reference to fluid flow and heat transfer problems. In this paper the RBF-FD method is employed to solve heat transfer problems with incompressible, steady-state laminar flow over 3D complex-shaped domains. The required node distributions are automatically generated by using a meshless node generation algorithm, which has been specifically developed to produce high quality node arrangements over arbitrary 3D geometries. The presented strategy represents therefore a fully-meshless approach for the accurate and automatic simulation of thermo-fluid problems over 3D domains of practical interest.

1. Introduction

Traditional CAE (Computer Aided Engineering) software usually rely on classic mesh-based methods, e.g., Finite Element (FEM) [1] and Finite Volume (FVM)[2] for solving boundary value problems over generic 3D domains. These methods are based on well understood mathematical models, which have proved to be reliable and accurate in the hands of adequately experienced users. With time, their widespread adoption has brought to the surface a number of inherent limitations, for example the need for processing and storage of the so called connectivity information. The main consequences are [3]:

- the cost of the mesh creation and the size of the consequent data structure,
- the inability to allow large geometric deformation of the domain.

The first one refers to both computational and economical costs; indeed the generation of a mesh often require the intervention of an experienced human worker, and operator costs now outweigh those of CPU time for the computer [3]. Such an operation becomes necessary for instance when high level of accuracy is demanded and therefore some local refinement of the mesh is needed at certain spots of the domain. The second weakness becomes of interest in many practical cases, for example when the solver is paired with some optimization algorithm and the geometry is deformed multiple times.

For those reasons the so-called meshless or meshfree methods have recently been proposed as an alternative to the mesh-based ones for the solution of partial differential equations over complex-shaped domains [3–6]. The common characteristic is the use of the so-called field nodes: at first a set of nodes is scattered over the domain, then shape functions are constructed for each internal node.

In this paper an implementation of the Radial Basis Function-generated Finite Difference (RBF-FD) meshless method [4, 7–11] is used to solve thermo-fluid problems over 3D geometries. The main advantages of this method is that no connectivity information is stored or processed, this implies the overcoming of all the limitations of the mesh-based methods enumerated so far. Furthermore, the employed node generation process lends itself to being automatically executed in parallel and therefore is particularly suitable for High Performance Computing applications. The employed code is developed using Julia programming language [12], which allows extensive code reuse and excellent computational performance already at the development stage.

In order to illustrate the numerical properties of the proposed method and its usability in the field of CFD simulations, several computations for a natural convection problem over a complex 3D geometry at a Rayleigh number $\mathrm{Ra}_D = 10^3$ will be presented.

2. Governing equations

Let us consider an incompressible fluid with density ρ, kinematic viscosity ν, thermal diffusivity α, thermal conductivity k and volumetric temperature expansion coefficient β. The resulting nondimentional conservation equations of mass, momentum and energy are

$$\nabla \cdot \mathbf{u} = 0, \tag{1}$$

$$\frac{\partial \mathbf{u}}{\partial t} + (\mathbf{u} \cdot \nabla)\mathbf{u} = -\nabla p + \sqrt{\frac{\mathrm{Pr}}{\mathrm{Ra}_D}}\nabla^2 \mathbf{u} + \mathbf{k}T, \tag{2}$$

$$\frac{\partial T}{\partial t} + \mathbf{u} \cdot \nabla T = \frac{1}{\sqrt{\mathrm{Pr}\cdot\mathrm{Ra}_D}}\nabla^2 T, \tag{3}$$

where $\mathbf{k}$ is the unit vector along the vertical direction z. In the above equations, length, velocity $\mathbf{u} = (u, v, w)$, time t, pressure p and temperature T are made dimensionless by taking respectively D, $U_0 = \sqrt{g\beta D\Delta T}$, D/U_0, ρU_0^2 and ΔT as reference quantities, where g is the gravitational acceleration. $\mathrm{Pr} = \nu/\alpha$ is the Prandtl number and $\mathrm{Ra}_D = \mathrm{Pr}\cdot g\beta D^3\Delta T/\nu^2$ is the Rayleigh number. The chosen values for the presented computations are $\mathrm{Pr} = 0.71$ and $\mathrm{Ra}_D = 10^3$, leading to a steady-state solution with the presented geometry.

3. Numerical method

In order provide a better explanation of the RBF-FD meshless method, we can describe it as the succesive application of four steps:

(i) generation of node distributions,
(ii) RBF interpolation or function approximation,
(iii) collocation technique,
(iv) solution procedure.

Each of the aforementioned steps will be briefly presented as follows. Interested readers might find a more detailed discussion on the very same method in [13].

3.1. Node distributions

Node distributions are represented by a certain number N of 3D points fulfilling a prescribed spacing function s and which are required to be scattered within the domain Ω and on its boundary $\partial\Omega$, where the latter is assumed to be concretely available in the form of a `.stl` (stereolithography) file. This process can be further divided in two stages:

- generation of a volumetric node distribution within Ω (nodes are not placed on $\partial\Omega$ yet) that satisfies a certain spacing s on average,
- iterative refinement of the initial node distribution through a node-repel approach that provides a suitable node distribution also on $\partial\Omega$.

In the second step the nodes are moved according to mutual repulsion forces that arise between neighbors. Whenever one of them is pushed outside Ω, it is projected onto the nearest point on the boundary $\partial\Omega$. This projection operation is efficiently performed by exploiting an octree data structure [14, 15] for the boundary triangles of the stereolithography surface.

3.2. RBF Interpolation

The value of a generic function ζ at any point is approximated applying the Radial Point Interpolation Method (RPIM) using Radial Basis Functions (RBFs) augmented with polynomial terms [3]. This means that the value $\zeta(x)$ taken by such function at the point $x \in \Omega$ can be locally approximated by the quantity $\zeta^h(x)$, defined as follows:

$$\zeta(x) \approx \zeta^h(x) := \sum_{i=1}^{\bar{n}} \alpha_i \varphi(\|x - x_i\|_2) + \sum_{j=1}^{m} \beta_j p_j(x), \tag{4}$$

where:

- $\{\varphi(\|x - x_i\|_2) : i = 1, \dots, \bar{n}\}$ is a set of $\bar{n}$ RBFs, each of which is associated to the corresponding node x_i. The set of the considered nodes is local, i.e. $x_1, \dots, x_{\bar{n}}$ are the $\bar{n}$ nearest nodes to x. For brevity of notation, $\varphi(x, x_i)$ will be used instead of $\varphi(\|x - x_i\|_2)$.
- the multiquadratic Radial Basis Functions [16] are used, defined as $\varphi(r) := \sqrt{1 + (\varepsilon r)^2}$
- $\{p_j : j = 1, \dots, m\}$ is a complete polynomial basis of degree P.

The calculation of the weights $\{\alpha_1, \dots, \alpha_{\bar{n}}\}$ and $\{\beta_1, \dots, \beta_m\}$ depends on the position of the stencil, that is, if some nodes of the stencil belongs to the boundary $\partial\Omega$, then boundary conditions must be taken into account. When none of the nodes belonging to the stencil lies on the boundary $\partial\Omega$, then the weights are estimated assuming that the values of the function u at the nodes of the stencil are known, this translates into a linear system of $\bar{n}$ equations of the form:

$$\zeta^h(x_i) = \zeta(x_i), \quad i = 1, \dots, \bar{n} \tag{5}$$

where the definition of $\zeta^h(x)$ can be substituted at each row. In order to make the system solvable, other m equations are attained by enforcing the following additional orthogonality conditions:

$$\sum_{i=1}^{\bar{n}} \alpha_i p_j(x_i) = 0, \quad j = 1, \dots, m \tag{6}$$

On the other hand, when some boundary nodes are included in the stencil, instead of assuming the true value of the function ζ in the corresponding equation, the appropriate boundary condition are imposed. In the case of Robin b.c. (i.e. $a\zeta + b\,{}^{\partial\zeta}/_{\partial n} = g$) applied to the ith node of the stencil, the corresponding equation becomes:

$$a(x_i)\,\zeta^h(x_i) + b(x_i)\,\frac{\partial\zeta^h}{\partial n}(x_i) = g(x_i), \quad x_i \in \partial\Omega \tag{7}$$

where a, b, g are all known functions defined on the boundary. By substituting the definition of ζ^h given in Equation (4) and reordering the terms in order to isolate the unknown coefficients α's and β's, we get:

$$\sum_{k=1}^{\bar{n}} \alpha_k \left(a(x_i)\varphi(x_i, x_k) + b(x_i)\frac{\partial \varphi}{\partial n}(x_i, x_k) \right) + \sum_{j=1}^{m} \beta_j \left(a(x_i)p_j(x_i) + b(x_i)\frac{\partial p_j}{\partial n}(x_i) \right) = g(x_i) \quad (8)$$

We can now enforce the orthogonality conditions of Equation (6) and thus obtain a linear system of the form:

$$\underbrace{\begin{bmatrix} \boldsymbol{\Phi_{BC}} & \boldsymbol{P_{BC}} \\ \boldsymbol{P^T} & \mathbf{0} \end{bmatrix}}_{\boldsymbol{M}} \begin{Bmatrix} \boldsymbol{\alpha} \\ \boldsymbol{\beta} \end{Bmatrix} = \begin{Bmatrix} \zeta \\ \boldsymbol{g} \\ 0 \end{Bmatrix} \quad (9)$$

where the block matrix $\boldsymbol{P}^T$ comes from Equation (6), $\{\alpha_1, \ldots, \alpha_{\bar{n}}, \beta_1, \ldots, \beta_m\}$ is the vector of unknowns and both $\boldsymbol{\Phi_{BC}}$ and $\boldsymbol{P_{BC}}$ come from Equation (4) but have boundary conditions (8) enforced at the appropriate rows.

3.3. *Collocation Techinque*

Given a linear partial differential equation $\mathcal{L}(\zeta) = f$ in the unknown field ζ, the RBF expansion (4) is made valid at each node x_i which does not lie on the boundary. The following sparse linear system is obtained:

$$[\boldsymbol{C}]\,\{\boldsymbol{\zeta}^h\} = \boldsymbol{q} - \boldsymbol{f} \quad (10)$$

where $\boldsymbol{\zeta}^h = \{\zeta^h(x_1), \ldots, \zeta^h(x_{N_I})\}$ is the vector of unknown scalar values $\zeta^h(x_i)$ evaluated at all N_I inner nodes, i.e. those contained in Ω, while the vector $\boldsymbol{q}$ comes from the enforcement of boundary conditions at the interpolant level.

3.4. *Solution Procedure*

At each time step, the computation of velocity, pressure and temperature through Equations (1)-(3) is decoupled using a projection scheme with a three-level Gear scheme for the time discretization while the buoyancy term is modeled with the Boussinesq approximation. A tentative velocity $\mathbf{u}^*$ is computed from the linearized nondimensional momentum equation:

$$\frac{3\mathbf{u}^* - 4\mathbf{u}^l + \mathbf{u}^{l-1}}{2\Delta t} + \mathbf{u}^l \nabla \mathbf{u}^* = -\nabla p^l + \sqrt{\frac{\mathrm{Pr}}{\mathrm{Ra}_D}} \nabla^2 \mathbf{u}^* + \mathbf{k}T^l, \quad (11)$$

where l is the time level and Δt is the nondimensional time step size.

The tentative velocity $\mathbf{u}^*$ is then forced to satisfy the continuity equation (1) by means of an irrotational correction $\mathbf{u}^{l+1} = \mathbf{u}^* - \nabla\Phi$, leading to the Poisson equation $\nabla^2\Phi = \nabla \cdot \mathbf{u}^*$ in the auxiliary variable Φ. The pressure is then updated as $p^{l+1} = p^l + \Phi/\Delta t$ and the temperature is computed from the discretized nondimensional energy equation:

$$\frac{3T^{l+1} - 4T^l + T^{l-1}}{2\Delta t} + \mathbf{u}^{l+1}\nabla T^{l+1} = \frac{1}{\sqrt{\mathrm{Pr}\cdot\mathrm{Ra}_D}} \nabla^2 T^{l+1}. \quad (12)$$

The previous equations are discretized in space with the same RBF-FD scheme presented in Section 3.2. Sparse linear systems like the one in Equation (10) are attained for each component of velocity and for temperature.

In the current implementation such linear systems are preconditioned with an Incomplete LU factorization (ILU) [17] (package `IncompleteLU` in `Julia`) of the sparse matrix $\boldsymbol{C}$ whenever

Journal of Physics: Conference Series 2177 (2022) 012007 doi:10.1088/1742-6596/2177/1/012007

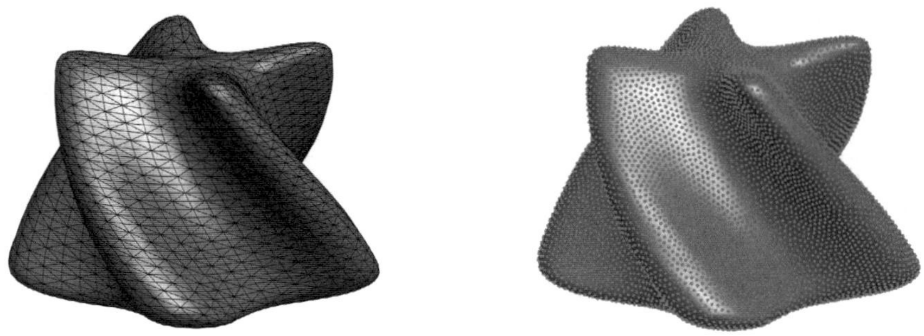

Figure 1. `.stl` surface of the 3D object (left) and distribution of the boundary nodes on the same object (right).

is needed. The Poisson equation for the velocity correction is also preconditioned through an ILU factorization which can be performed only once at the beginning of the simulation. ILU preconditioning is then followed by the application of the Biconjugate Gradient Stabilized Method [18] (package `IterativeSolvers` in `Julia`) using a relative error of 10^{-10}. The computational time required for each time step ranges from 2 seconds for $N \approx 100,000$ nodes and $P = 2$ to 18 seconds for $N \approx 500,000$ nodes and $P = 4$ on a laptop equipped with a quad-core Intel i7-6700HQ 2.6GHz processor with 16 GB of RAM memory.

3.5. Convergence Criterion

The convergence to steady-state is declared when the RMS time derivative for the computed velocity fields $\mathbf{u} = \{u, v, w\}$ and temperature field T becomes smaller than $5 \cdot 10^{-5}$, where the RMS time derivative $\dot{F}_{RMS}$ for a generic field F over the domain Ω is defined by:

$$\dot{F}_{RMS} = \sqrt{\frac{1}{\mu(\Omega)} \int_{\Omega} \left(\frac{\partial F}{\partial t} \right)^2 d\Omega} \approx \sqrt{\frac{1}{\sum_i s_i^3} \sum_i \left(\frac{F_i^{l+1} - F_i^l}{\Delta t} \right)^2 s_i^3} \tag{13}$$

where $\mu(\Omega)$ is the measure of the domain Ω and the index i runs over each internal node.

4. Geometry, boundary conditions and auxiliary computations

A complex-shaped 3D object with characteristic length D, represented in Figure 1, is enclosed in a cubic cavity with side length $W = 5D$. The 3D object is centered with respect to the cavity in the $x - y$ plane, while the distance between the bottom cavity wall and the lower surface of the object is D, as visible in Figure 4. The following boundary conditions are enforced:

- no-slip condition $\mathbf{u} = \mathbf{0}$ on all boundary surfaces;
- $T = 1$ on the surface of the 3D object;
- $T = 0$ on the inner walls of the cubic cavity;
- $\dfrac{\partial \Phi}{\partial \mathbf{n}} = 0$ on all boundary surfaces;

In order to ensure the uniqueness of the calculated auxiliary variable Φ, an additional condition is imposed in the form of a Lagrange multiplier. The complete problem for Φ assumes therefore the following form:

$$\begin{cases} \nabla^2\Phi + \lambda = \nabla \cdot \mathbf{u}^* & in\ \Omega \\ \dfrac{\partial \Phi}{\partial \mathbf{n}} = 0 & on\ \partial\Omega \\ \displaystyle\int_\Omega \Phi = 0 \end{cases} \tag{14}$$

Since boundary layers are expected at the surface of the object, the following spacing function is employed to obtain refined node distributions in the proximity of such geometry:

$$s(\mathbf{x}) = s_{max} - \frac{s_{max} - s_{min,j}}{\frac{d(\mathbf{x})}{\lambda} + 1} \tag{15}$$

where s_{max} is the asymptotic maximum spacing and $s_{min,j}$ is a minimum spacing attained on each triangle of the original `.stl` surface. $d(\mathbf{x})$ is the distance of point $\mathbf{x}$ from the closest `.stl` triangle, denoted in Equation (15) with the index j, and λ is a parameter used for controlling the transition from $s_{min,j}$ to s_{max}.
In the presented case the value of s_{max} is chosen in order to attain a sought number of nodes, then s_{min} is chosen to be $s_{min} = 0.8s_{max}$ on vertical and bottom walls of the domain, $s_{min} = 0.65s_{max}$ on the top wall and $s_{min} = 0.1s_{max}$ on the surface of the object, while $\lambda = 0.1$. An example of a boundary node distribution over the object is depicted in Figure 1 for a total number of nodes $N \approx 185,000$.

The mean Nusselt number over the surface Γ of the considered object is given by:

$$\overline{\mathrm{Nu}}_D = \frac{1}{\mu(\Gamma)} \int_\Gamma \frac{\partial T}{\partial \mathbf{n}} d\Gamma \approx \frac{1}{\sum_b s_b^2} \sum_b \left.\frac{\partial T}{\partial \mathbf{n}}\right|_b s_b^2 \tag{16}$$

where the index b runs over each boundary node on Γ.

5. Results

In order to highlight the convergence properties of the employed RBF-FD approach, several simulations are carried out for different node distributions using a total number of nodes ranging from $N \approx 100,000$ nodes to $N \approx 740,000$ nodes. Three polynomial degrees $P = 2, 3, 4$ are employed. Convergence curves for $P = 2$ and $P = 3$ are shown in Figure 2 in terms of normalized RMS errors for the flow variables $\mathbf{u} = \{u, v, w\}$, T and p. The normalized RMS error $err(F)$ of a generic field F is defined as:

$$err(F) = \sqrt{\frac{1}{\mu(\Omega)} \int_\Omega \left(\frac{F - \bar{F}}{\max(F)} \right)^2 d\Omega} \tag{17}$$

where the reference solution $\bar{F}$ is a highly accurate solution obtained with polynomial degree $P = 4$ on a fine node distribution with $N \approx 545,000$ nodes.

The convergence curves depicted in Figure 2 show that the resulting order of accuracy for $P = 2$ ranges from 2.1 to 2.5 for each flow variable, while for $P = 3$ the resulting order of accuracy ranges from 3.2 to 3.7, as expected. The orders of accuracy for larger polynomial degrees $P > 3$ are therefore expected to be greater than or equal to P for the present problem, although this assumption could not be verified due to the large memory requirements for $P \geq 4$ on 3D problems with the present RBF-FD implementation, which makes it unsuitable for a personal (laptop) computer.

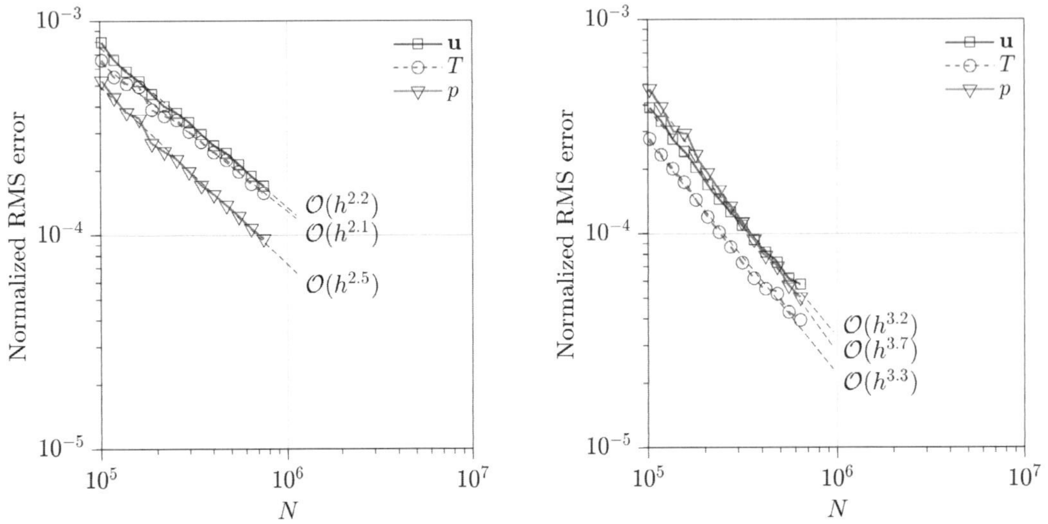

Figure 2. Convergence curves for the flow variables $\mathbf{u} = \{u, v, w\}$, T and p for $P = 2$ (left) and $P = 3$ (right).

Table 1. Mean Nusselt number.

	$\overline{\mathrm{Nu}}_D$			
N	$P = 2$	$P = 3$	$P = 4$	Fluent
100,000	3.85	3.91	3.87	-
116,666	3.88	3.92	3.89	4.36*
136,110	3.89	3.95	3.92	-
158,794	3.95	4.01	3.98	-
185,258	3.97	4.02	4.00	-
216,133	4.00	4.06	4.03	4.36*
252,154	4.03	4.08	4.06	-
294,178	4.05	4.10	4.08	-
343,206	4.08	4.13	4.12	-
400,404	4.10	4.15	4.14	-
467,136	4.12	4.17	4.16	4.37*
544,989	4.13	4.18	4.17	-
635,816	4.15	4.20	-	-
741,782	4.18	-	-	-

* using inflation layers with patch-conforming mesh from .stl model

The computed values of the mean Nusselt number $\overline{\mathrm{Nu}}_D$ for different number of nodes N and polynomial degrees $P = 2, 3, 4$ are reported in Table 1, where mean Nusselt numbers computed with Ansys Fluent 2020R1 are also reported for comparison. In the latter case inflation layers (Wed6 elements) via patch-conforming meshing are employed at the object surface. Despite the refined node distribution, i.e., reduced node spacing near the object surface, the $\overline{\mathrm{Nu}}_D$ values

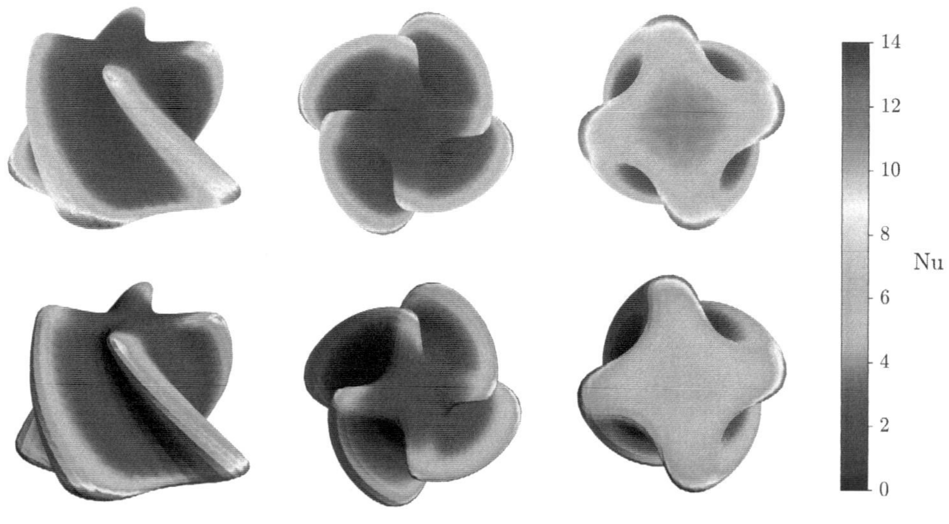

Figure 3. Visual comparison of contour plots of local Nusselt number at the object surface: RBF-FD (top row) vs. Ansys Fluent (bottom row).

show a slow convergence to the reference value $\overline{\mathrm{Nu}}_D = 4.37$, regardless of the polynomial degree P. This is mainly due to the fact that the generated node distributions are isotropic, i.e., node spacing is independent upon the spatial directions, therefore boundary layers are not efficiently solved as with inflation layers.

A graphical, qualitative comparison between RBF-FD ($P = 3$, $N \approx 640,000$ nodes) and Ansys Fluent ($470,000$ elements) is depicted in Figure 3 in terms of local Nusselt number at the object surface. The comparison highlights a very good agreement, to graphical accuracy, in the surface distribution of the local Nu number between the computed results and the reference ones. Figure 4 depicts the computed temperature field on the $x - z$ plane at $y = W/2 = 2.5$ together with the velocity vector plot projected on the same plane, highlighting the classic natural convection flow originated by the temperature differential between the object surface and the cavity walls.

At last, due to its relevance for practical applications, we report a comparison between the computational times required by the geometrical discretizations of the employed computational domain. The node generation for the RBF-FD meshless approach, as described in subsection 3.1, required 4s for the octree phase and an average of $5\mathrm{s}/10^5$ nodes for the iterative refinement phase, while the average time for the mesh generation in ANSYS Meshing was $6\mathrm{s}/10^5$ cells.

6. Conclusions

In this work the RBF-FD meshless method is applied to the simulation of heat transfer problems with incompressible laminar flow over a 3D complex-shaped domain.

The major strength of the proposed method lies in its potential convenience for the end user. The node generation algorithm is capable of coping with very complex geometries without any need for human intervention, furthermore, by choosing the order of the polynomial basis P, it is possible to decide the order of accuracy of the method (see Figure 2). While the code still is at the development stage, the good level of parallelization allowed by the Julia programming language already allows an efficient solution of large systems of equations (e.g. $N \approx 750,000$),

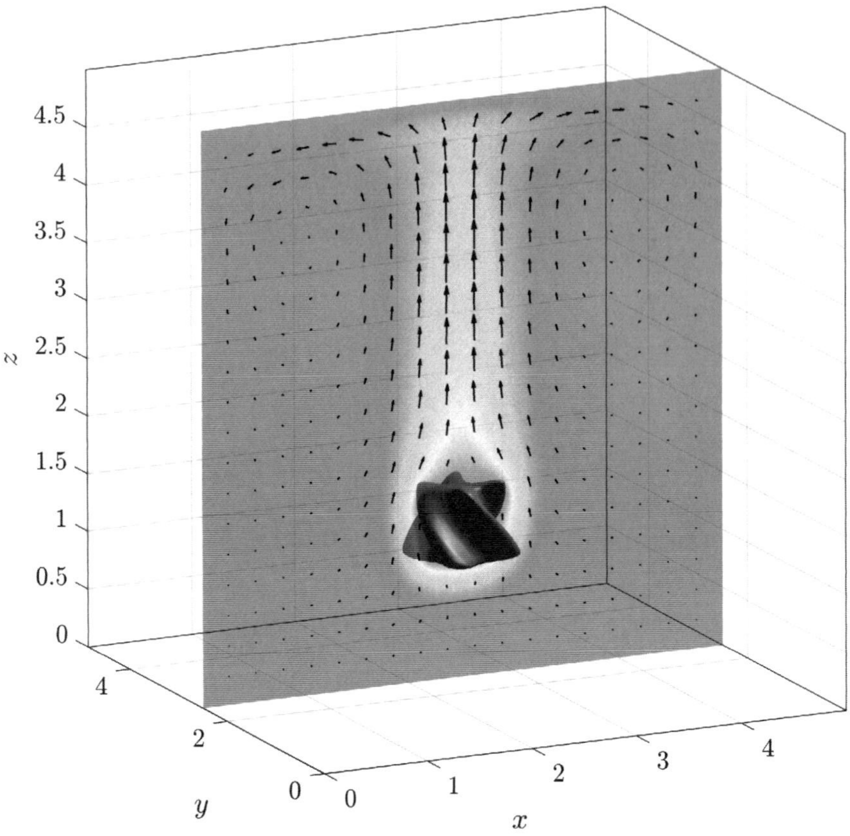

Figure 4. Temperature contour plot and velocity vector plot projected on the $x - z$ plane (T ranges between $T = 0$ on the cube faces and $T = 1$ on the object surface).

thus leading to reliable results in terms of accuracy also on a personal computer.

Future developments of the code will be aimed at addressing its limitations with the presented implementation, such as the need for directional node refinement close to boundary walls, and adding important features such as turbulence models.

We believe that the RBF-FD method has the potential to reach and outperform the accuracy allowed by the best traditional CAE solvers and to make the design process far more convenient and democratic by overcoming the need to generate a mesh.

References

[1] Zienkiewicz O C 1977 *The finite element method* vol 3 (McGraw-hill London)
[2] Versteeg H K and Malalasekera W 2007 *An introduction to computational fluid dynamics: the finite volume method* (Pearson education)
[3] Liu G R and Gu Y T 2005 *An introduction to meshfree methods and their programming* (Springer Science & Business Media)
[4] Fornberg B and Flyer N 2015 *A primer on radial basis functions with applications to the geosciences* (SIAM)
[5] Liu G R 2009 *Meshfree methods: moving beyond the finite element method* (Taylor & Francis)
[6] Li H and Mulay S S 2013 *Meshless methods and their numerical properties* (CRC press)
[7] Divo E and Kassab A J 2007 *J. Heat Transf.* **129** 124–136
[8] Zamolo R and Nobile E 2019 *Numer. Heat Tr. B-Fund.* **75** 19–42

[9] Šarler B and Vertnik R 2006 *Comput. Math. Appl.* **51** 1269–1282
[10] Waters J and Pepper D W 2015 *Numer. Heat Tr. B-Fund.* **68** 185–203
[11] Kosec G and Slak J 2020 *AIP Conference Proceedings* **2293** 420094
[12] Bezanson J, Edelman A, Karpinski S and Shah V B 2017 *SIAM Rev.* **59** 65–98
[13] Miotti D, Zamolo R and Nobile E 2021 *Energies* **14** 1351
[14] Frey P and George P 2013 *Mesh Generation: Application to Finite Elements* ISTE (Wiley) ISBN 9781118623824 URL https://books.google.it/books?id=TZQokVbCwGIC
[15] de Berg M, Cheong O, van Kreveld M and Overmars M 2008 *Computational Geometry: Algorithms and Applications* (Springer Berlin Heidelberg) ISBN 9783540779742 URL https://books.google.it/books?id=9nhHUZpKzeEC
[16] Hardy R L 1971 *J. Geophys. Res.* **76** 1905–1915
[17] Saad Y 2003 *Iterative Methods for Sparse Linear Systems* (Philadelphia, Pennsylvania: SIAM) chap 10: Preconditioning Techniques, pp 297–368 2nd ed
[18] van der Vorst H 1992 *SIAM J. Sci. Comput.* **13**(2) 631–644

Assessment of operating temperature within the new pavilion for slag management in Terni

Mattia Manni[1,*], Claudia Fabiani[2], Andrea Nicolini[2], Anna Laura Pisello[2], Federico Rossi[2], Franco Cotana[1,2]

[1] CIRIAF - Interuniversity Research Centre on Pollution and Environment "M. Felli", Via G. Duranti 67, 06125 Perugia (Italy)

[2] Department of Engineering, University of Perugia, Via G. Duranti 93, 06125 Perugia (Italy)

Corresponding author's e-mail address: manni@crbnet.it.

Abstract. The whole metallurgic sector produces up to 200 million tons of slag, which are tapped from the blast furnace (at a temperature of 1,500°C), and then need to be cooled down before disposal. These cooling processes are generally conducted open-air, significantly affecting local environmental quality of the surroundings. The present study aims at investigating the potential of an innovative slag cooling system housed within a pavilion, designed in order to minimize the emission of dust and pollutants out from the metallurgic plant. Such a system consists of a depressurized environment whose top surface is treated with black pigments and cooled down by water streams above it. Air is continuously extracted and then adequately filtered before being released outdoor. A numerical model was elaborated for evaluating the main heat flows developed within and through the pavilion's envelope for the case study in Terni, central Italy. Once the physical and geometrical properties of the slag and the pavilion were defined, the heat exchanged with the air and water due to convection, as well as the latent heat dissipated through water evaporation was quantified. Results demonstrated the effectiveness of the water-based cooling system in keeping the roof temperature lower than 328 K without compromising the mechanical properties of the material. The evaporated water mass ranged between 4.2 kg h^{-1} and 79.6 kg h^{-1} and was strongly influenced by seasonal weather conditions.

1. Introduction

The metallurgic sector can produce up to 200 million tons of slag worldwide throughout the year, by about 300 kg per tons of metal produced [1]. When tapped from the blast furnace, slag show a temperature value near 1,773 K, therefore they must be cooled down before disposal. Slag are generally named depending on the furnace used for their production [2]. The main typologies of slag (from iron and steelmaking industries) are blast-furnace slag (ironmaking slag) and steel-furnace slag. The latter is further divided into basic-oxygen-furnace slag, electric-arc-furnace slag, and ladle slag. Different sub-products can be obtained from the slag depending on the applied cooling process [3]. Rapidly cooling the slag with air and steam produces slag-wool, while a spreading water for cooling the material generates particles of solidified slag with a maximum diameter of about 5 mm. Finally, slag-pellets and slag-foams are produced by rapidly cooling the slag with air and slowly cooling the slag with water, respectively.

These cooling processes are generally performed open-air, and therefore they are responsible for compromising local environmental quality of the surroundings. In fact, large amounts of dust and pollutants are released in the atmosphere and moved by the intense convective flows caused by the large temperature gradients with the adjacent air domain. In recent years, several innovative slag cooling techniques making use of dedicated pavilions with high-efficiency filtering systems have been proposed in the attempt to stem local environmental degradation [4]. However, despite their promising potential, these solutions require the optimization of several factors right from the preliminary stage of the design process. In particular, temperature values reached by the structural elements as well as the air volume beneath them have to be quantified for avoiding the load-bearing structure to collapse and the filtering system to be damaged.

Within this framework, the present study aims at investigating the applicability of an innovative slag cooling system housed within a pavilion (Figure 1), specifically design to face this environmental challenge, i.e. to minimize the worsening of local air quality and overheating phenomenon compromising microclimate wellbeing in the surroundings of the industrial plants. These surroundings are indeed characterized by urban areas, with also residential areas in close proximity, facing dangerous emissions of dust and pollutants. The proposed system consists of a depressurized environment whose top surface is treated with black pigments (to enhance its absorptivity in the side facing the slag) and cooled down by water streams above it. Air is also continuously extracted from the centre of the room and then adequately filtered before being released in the outdoors.

The paper is structured as follows. The Materials and methods section (section 2) describes the case study, the numerical model, and the investigated scenarios (sub-sections 2.1, 2.2, and 2.3). The Result and Discussion section (section 3), presents and discusses the outcomes from the numerical simulations (sub-section 3.1), the correlation analyses (sub-section 3.2), as well as the main limitations of the study (sub-section 3.3). Finally, the Conclusions and future developments section summarizes the results and the implications of this work (section 4).

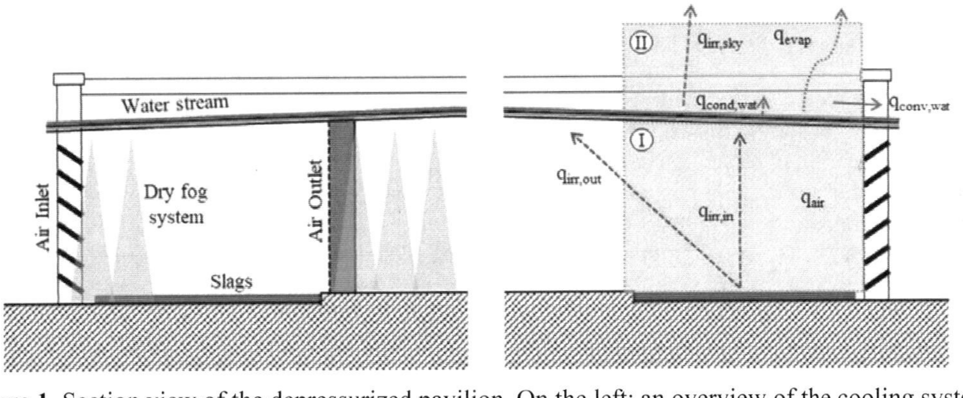

Figure 1. Section view of the depressurized pavilion. On the left: an overview of the cooling systems' layout (i.e. dry-fog system, water stream, mechanical ventilation). On the right: heat fluxes considered in the two computational domains.

2. Materials and methods

2.1. Case study

The case study is the new pavilion realized by the Acciai Speciali Terni (AST) company in Terni, Italy, to reduce the emissions due to the cooling of basic-oxygen-furnace slag. The pavilion covers an area of 7,400 m^2 and it is around 10 m high. A 5,700 m^2 surface is arranged into 28 cooling ramps where the fluid slag are poured. The outer walls of the pavilion are made of blinds that enable air flowing between them and in addition to orient it towards the slag for improving convection efficiency. Two fans allow depressurization and extract an air volume by 565,000 m^3 per hour (7.5 air changes per hour) from the centre of the room. The corresponding air velocity which can be measured at the outlet is 1.3 m/s. The highest operating temperature achievable by the air filtering system is 373 K. A dry-fog system is installed in close proximity to the air inlets (blinds integrated to the façade) as well as in the central zone where slag-handling vehicles are driven. Up to 611 nozzles nebulized the water over a circular area having a diameter of 3 m. The flow rate corresponds to 4 l/min, and the size of water particles is lower than 200 μm. Furthermore, the dry-fog system contributes to cooling the vehicles as well as to clean them from the aforementioned dust. The roof cover of the pavilion is made of steel sheets which are cooled down by a water mass (3 mm thick) flowing on them. The velocity of the water stream is 0.3 m/s, while the flow rate equals 600 m^3/h. The water flows along the tilted surface of the roof and then is collected into a tank where the absorbed thermal energy is then stored. The initial water volume is reinstated by adding demineralized water to compensate the amount of water previously evaporated. The demineralization process is fundamental to minimize the deposit of minerals on the rooftop layer.

2.2. *Numerical model*
Based on the above behaviour, a 3D numerical model was implemented in MatLab environment and used for evaluating the main heat flows developed within and through the previously described pavilion's envelope. Once the physical and geometrical properties of the slag and the pavilion are defined, the numerical model is capable of quantifying the amount of heat exchanged with the air and water due to convection, as well as the latent heat dissipated through water evaporation. In more detail, the most important contributions in terms of local heat transfer phenomena are analytically described and used for investigating the thermal response of the system in stationary conditions, considering the scenarios described in the specific section. The heat transfer due to irradiation are evaluated by exploiting a Monte Carlo-based routine for calculating the view factors among the various surfaces, similar to the one exploited in [5]. The amount of radiative exchanges between slag and covering, and between slag and façades are therefore analysed. Finally, the heat loss due to irradiation between the water flowing on the pavilion's top and the sky-dome is estimated. Reasonable simplification hypotheses were made for analysing fluid turbulences, but consistently with the aim of the numerical model which is supposed to be utilized during the preliminary stages of the design process. By doing so, the effectiveness of the newly proposed cooling process based on air and water forced convection, as well as water evaporation is assessed, and the operating temperature of indoor air and load-bearing elements of the pavilion structure (i.e. covering) is estimated. In order to do so, the temperature of the poured slag is defined as an input variable together with the weather conditions. At a later stage, the overall heat dissipation capability of the cooling system is also investigated considering different operational scenarios, i.e. varying the effective flow rate of the water stream, as well as water and air temperature conditions.

The modelled space consists of a parallelepiped volume ($20.5 \cdot 5.5 \cdot 10.5$ m^3) above one of the ramps (20.5 m long, 5.5 wide), as representative of the whole pavilion. Such a volume represents the air mass between the bottom surface, which simulates the ramp where slag were poured, and the top surface, the covering. This is consequently organized into two layers: the aluminium roof structure and the water streams. No air turbulence is modelled, thus the air mass is considered as static to be representative of the worse operating condition in which the mechanical ventilation is not working, and the heat can be dissipated only by the water-based cooling system installed on the covering.

Energy balance equations
The energy balance equations implemented in the numerical model permit to evaluate heat transfer due to conduction, convection, and irradiation phenomena. Two domains were considered which corresponded to the indoor and the outdoor environments, respectively. The surface of the ramp where tapped slag are poured was included in the former. The heat flow due to the slag (Q_{slag}) was estimated as:

$$Q_{slag} = A_{slag} \cdot \sigma \cdot T_{slag}{}^{4} \tag{1}$$

where A_{slag} is the area covered by the slag, σ is the Stefan-Boltzmann constant, and T_{slag} is the slag temperature.

When it came to the heat flows exiting the indoor environment, the radiative exchange between slag and covering ($Q_{irr,in}$) was estimated as:

$$Q_{irr,in} = VF_{slag-to-roof} \cdot A_{slag} \cdot \sigma \cdot \varepsilon_{slag} \cdot \left(T_{slag}{}^{4} - T_{roof}{}^{4}\right) \tag{2}$$

where $VF_{slag\text{-}to\text{-}roof}$ is the view factor between the slag and the roof, ε_{slag} is the slag emissivity, and T_{roof} is the roof temperature.

Similarly, the radiative heat flows ($Q_{irr,out}$) exiting the indoor domain through the lateral surface boundaries and reaching the pavilion's covering were calculated by using the same equation as before (2), but with a different view factor. It is worth highlighting that such lateral surface boundaries are the three virtual surfaces separating the cooling ramp from the rest of the indoor space. The lateral surface boundary corresponding to the outer wall was not considered.

Moreover, the heat absorbed by the indoor air volume (Q_{air}) was quantified as:

$$Q_{air} = m_{air} \cdot c_{p} \cdot \left(T_{air,out} - T_{air}\right) \tag{3}$$

where m_{air} is the air mass, c_p is the specific heat capacity of the air (varies with the temperature), and $T_{air,out}$ and T_{air} are the air temperature outside and inside the pavilion, respectively.

According to the energy conservation law in stationary regime, the energy balance of the first domain (indoor environment) can be expressed as it follows:

$$Q_{slag} = Q_{irr,in} + Q_{irr,out} + Q_{air} \tag{4}$$

As far as the second domain is concerned (outdoor environment), the thermal energy entering the system was considered as high as $Q_{irr,in}$, while the exiting heat flows were due to the convection of the water ($Q_{conv,wat}$), the conduction through the water layer ($Q_{cond,wat}$), the irradiation from the roof towards the sky ($Q_{irr,sky}$), and the latent heat of water evaporation (Q_{evap}).

The $Q_{conv,wat}$ was calculated as:

$$Q_{conv,wat} = h_{wat} \cdot A_{roof} \cdot \left(T_{roof} - T_{wat}\right) \tag{5}$$

where h_{wat} is the convective heat transfer coefficient of the water flowing on the roof, A_{roof} is the roof surface and T_{wat} is the water temperature.

In particular, the h_{wat} coefficient was evaluated for the case of a fluid that is forced to flow over a surface. Firstly, the Reynolds (Re) number was quantified for the water as:

$$Re_{wat} = \rho_{wat}(T_{wat}) \cdot v_{wat} \cdot L \cdot \nu_{wat}{}^{-1} \tag{6}$$

where ρ_{wat} is the water density which depends on the water temperature, v_{wat} is the speed of the water, L is the geometric parameter, and ν_{wat} is the water viscosity.

Following this, the Nusselt number (Nu) was estimated as described in [6].

$$Nu_{wat} = 0.036 \cdot [Pr_{wat}(T_{wat})]^{1/3} \cdot Re_{wat}{}^{0.8} \tag{7}$$

where Pr_{wat} is the Prandtl number and it varies with the water temperature, while Re_{wat} is the Reynolds number for water.

Finally, the convective heat transfer coefficient for water was calculated as:

$$h_{wat} = Nu_{wat} \cdot [\lambda_{wat}(T_{wat})]^{-1} \tag{8}$$

where Nu_{wat} is the Nusselt number of water, and λ_{wat} is the thermal conductivity of water (it depends on the temperature).

To estimate the $Q_{cond,wat}$ amount, the (9) was used:

$$Q_{cond,wat} = \lambda_{wat} \cdot dz_{wat}{}^{-1} \cdot A_{roof} \cdot \left(T_{roof} - T_{air,out}\right) \tag{9}$$

where λ_{wat} is the thermal conductivity of water, dz_{wat} is the thickness of the water layer, and A_{roof} is the area of the roof surface.

The irradiative exchange between the roof surface and the sky-dome takes place through the water layer which is almost transparent to these wavelengths. The $Q_{irr,sky}$ was calculated as:

$$Q_{irr,sky} = A_{roof} \cdot \sigma \cdot \varepsilon_{roof} \cdot \left(T_{roof}{}^{4} - T_{air,out}{}^{4}\right) \tag{10}$$

where A_{roof} is the area of the roof surface, σ is the Stefan-Boltzmann constant, ε_{roof} is the roof emissivity, and T_{roof} and $T_{air,out}$ are the temperature of the roof and outdoor air, respectively. In this case, the view of the sky-dome from the roof surface was unobstructed, thus the view factor was set equal to one. Although using the sky temperature instead of the air temperature would have been more accurate, the latter was considered since the calculation of the sky temperature requires additional information such as cloudiness level which were not available for the case study. Nonetheless, this simplified approach cab be considered as adequate at this stage of the design process as explained in the section about limitations of the study (sub-section 3.3).

The latent heat of evaporation (Q_{evap}) was assessed as reported in the following equation.

$$Q_{evap} = h_{evap}(T_{wat}) \cdot g_h \tag{11}$$

where h_{evap} is the evaporation heat of water (varying with water temperature) and g_h is the amount of evaporated water per hour. However, a detailed description of the routine describing the evaporative phenomenon is provided in the specific subsection.

Water evaporation calculation for latent heat estimation

The quantification of the evaporated water amount from the roof cooling system was conducted by following the approach outlined in [7]. According to this, the water mass evaporated per hour was calculated as:

$$g_h = \left(25 + 19 \cdot v_{air,out}\right) \cdot A_{roof} \cdot (x_s - x) \tag{12}$$

where $v_{air,out}$ is the velocity of the air above the water surface, x_s is the maximum humidity ratio of saturated air at the same temperature as the water surface, and x is the humidity ratio of air.

Based on the ideal gas law, the humidity ratio air can be expressed as:

$$x = 0.62198 \cdot P_w \cdot (P_a - P_w)^{-1} \tag{13}$$

where P_w is the partial pressure of water vapour in moist air, and P_a is the atmospheric pressure of moist air.

Following this, the maximum amount of water vapour in the air is observed when P_w equals the saturation pressure of water vapour (P_{ws}).

$$x_s = 0.62198 \cdot P_{ws} \cdot (P_a - P_{ws})^{-1} \tag{14}$$

The P_{ws} was estimated according to the equation reported in [8]:

$$\log\left(P_{ws} \cdot P_c^{-1}\right) = T_c \cdot T^{-1} \cdot (a_1 \cdot \mu + a_2 \cdot \mu^{1.5} + a_3 \cdot \mu^3 + a_4 \cdot \mu^{3.5} + a_5 \cdot \mu^4 + a_6 \cdot \mu^{7.5}) \tag{15}$$

where P_c is the critical pressure (22.064 MPa), T_c is the critical temperature (647.096 K), T is the outdoor air temperature, and μ is equal to:

$$\mu = \left(1 - T \cdot T_c^{-1}\right) \tag{16}$$

Finally, the six coefficients (a_i) were set equal to the values in Table 1.

Table 1. Numerical values set for the coefficient in the (15).

a_1	a_2	a_3	a_4	a_5	a_6
−7.85951783	1.84408259	−11.7866497	22.6807411	−15.9618719	1.80122502

Then, P_w was calculated considering that:

$$P_w = 0.01 \cdot RH_{\%} \cdot P_{ws} \tag{17}$$

where $RH_{\%}$ is the relative humidity of the outdoor air expressed as percentage.

Once the evaporated water mass is calculated, the h_{wat} corresponding to the water temperature set as input parameter is multiplied to g_h to estimate the latent heat of evaporation.

2.3. *Investigated scenarios*

The effectiveness of the investigated cooling system was assessed by modelling scenarios with different boundary conditions concerning the temperature of slag (T_{slag}), the characteristics of the outdoor environment (T_{air}, $RH_{\%}$,), and the operating parameters of the water-based cooling system above the pavilion (layer thickness, water stream velocity).

When it came to the slag temperature, the values to be simulated were chosen according to the values coming from the field transient monitoring (i.e. 450 minute- transient after the pouring of fluid slag on the cooling ramp reported in Figure 2a). In particular, the highest (1,005 K) and the lowest temperature (420 K) were considered together with the average value (585 K).

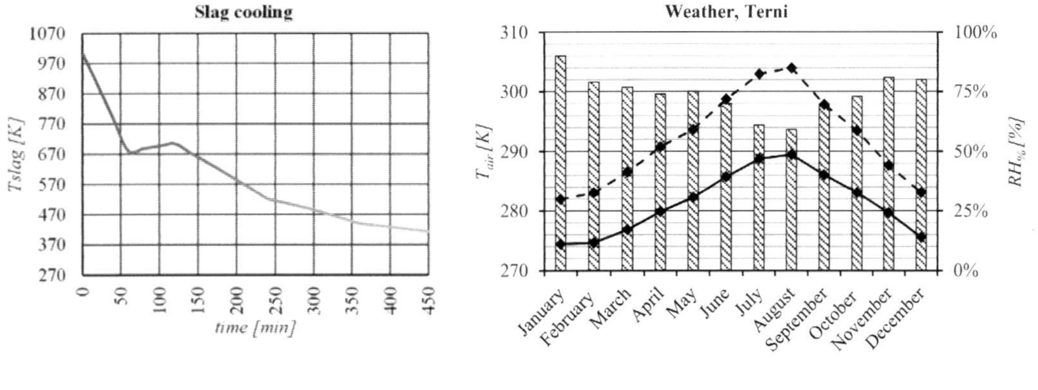

Figure 2. From the left: (a) monitored temperature values of the slag when poured on the ramp; (b) maximum and minimum monthly values of T_{air} (lines graph) and average monthly values of $RH_{\%}$ (bars graph) in Terni, throughout the year.

Environmental conditions representative of summer and winter seasons, respectively, were considered. On the one hand, air temperature as high as 278 K was associated to an air relative humidity of 80% for simulating the cooling effect of the water mass during the winter. On the other hand, an air temperature of 313 K was considered along to an air relative humidity of 60% for the summer. The wind speed was equal to 0.5 m s^{-1}, and it was kept as constant throughout the study.

Finally, different operating parameters of the cooling system were assessed by varying the layer thickness of the water mass flowing above the pavilion from 0.003 m to 0.005 m, and to 0.010 m; while the water stream velocity ranged between 0.3 m s^{-1} and 1.0 m s^{-1}.

3. Results and discussion

3.1. Numerical simulations' results

The outcomes from the numerical simulations in the investigated scenarios are reported in Table 2 along to the input parameters. The Q_{slag} entering the domain corresponding to the indoor environment was considered as high as $6.5 \cdot 10^6$ W, $7.5 \cdot 10^5$ W, and $2.0 \cdot 10^5$ W for a T_{slag} of 1,005 K, 585 K, and 420 K, respectively.

The indoor air temperature was up to 2 K higher than the outdoor one, and generally the highest increments were estimated for the lowest slag temperature (420 K). On the contrary, the temperature of the roof was higher (up to 328 K) when the slag was just poured on the ramp and its temperature was 1,005 K, during summer (T_{air} equal to 313 K).

When it came to the evaporated water mass, two different amounts were estimated throughout the 36 investigated scenarios. This mainly depends on the season environmental condition (T_{air} and $RH_{\%}$) and it corresponded to 4.2 kg h^{-1} and 79.6 kg h^{-1} in winter and in summer, respectively.

Table 2. Overview of the input and output parameters for each simulated scenario.

ID	Input							Output										
	T_{air} [K]	$RH_{\%}$ [%]	T_{slag} [K]	$v_{air,out}$ [m·s^{-1}]	dz_{wat} [m]	v_{wat} [m·s^{-1}]	T_{wat} [K]	$Q_{irr,in}$ [W]	$Q_{irr,out}$ [W]	Q_{air} [W]	$Q_{conv,wat}$ [W]	$Q_{cond,wat}$ [W]	$Q_{irr,sky}$ [W]	Q_{evap} [W]	h_{wat} [W m^{-2} K^{-1}]	g_h [kg h^{-1}]	$T_{air,in}$ [K]	T_{roof} [K]
1	278	80	1,005	0.5	0.003	0.3	278	$2.6 \cdot 10^6$	$3.2 \cdot 10^6$	$7.2 \cdot 10^5$	$1.7 \cdot 10^6$	$7.6 \cdot 10^5$	$2.2 \cdot 10^4$	$1.1 \cdot 10^4$	432	4.2	278.6	313.5
2	278	80	585	0.5	0.003	0.3	278	$2.8 \cdot 10^5$	$3.6 \cdot 10^5$	$1.5 \cdot 10^6$	$1.2 \cdot 10^5$	$5.2 \cdot 10^4$	$1.3 \cdot 10^3$	$1.1 \cdot 10^4$	432	4.2	279.1	280.5

3	278	80	420	0.5	0.003	0.3	278	$6.4\cdot10^{4}$	$8.0\cdot10^{4}$	$2.5\cdot10^{6}$	$9.7\cdot10^{4}$	$4.3\cdot10^{4}$	$1.1\cdot10^{3}$	$1.1\cdot10^{4}$	432	4.2	279.8	280.0
4	313	60	1,005	0.5	0.003	0.3	288	$2.6\cdot10^{6}$	$3.2\cdot10^{6}$	$7.2\cdot10^{5}$	$2.0\cdot10^{6}$	$2.5\cdot10^{5}$	$9.0\cdot10^{3}$	$2.0\cdot10^{5}$	492	79.6	313.6	324.5
5	313	60	585	0.5	0.003	0.3	288	$2.8\cdot10^{5}$	$3.5\cdot10^{5}$	$3.8\cdot10^{5}$	$5.3\cdot10^{5}$	$-3.4\cdot10^{5}$	$-1.1\cdot10^{4}$	$2.0\cdot10^{5}$	492	79.6	313.3	297.3
6	313	60	420	0.5	0.003	0.3	288	$6.0\cdot10^{4}$	$7.6\cdot10^{4}$	$2.2\cdot10^{5}$	$3.7\cdot10^{5}$	$-4.1\cdot10^{5}$	$-1.3\cdot10^{4}$	$2.0\cdot10^{5}$	492	79.6	313.2	294.5
7	278	80	1,005	0.5	0.003	0.6	278	$2.6\cdot10^{6}$	$3.2\cdot10^{6}$	$7.5\cdot10^{5}$	$2.0\cdot10^{6}$	$5.0\cdot10^{5}$	$1.4\cdot10^{4}$	$1.1\cdot10^{4}$	752	4.2	278.6	301.5
8	278	80	585	0.5	0.003	0.6	278	$2.8\cdot10^{5}$	$3.6\cdot10^{5}$	$2.5\cdot10^{6}$	$1.7\cdot10^{5}$	$4.3\cdot10^{4}$	$1.1\cdot10^{3}$	$1.1\cdot10^{4}$	752	4.2	279.8	280.0
9	278	80	420	0.5	0.003	0.6	278	$6.4\cdot10^{4}$	$8.0\cdot10^{4}$	$6.0\cdot10^{5}$	$1.2\cdot10^{5}$	$3.0\cdot10^{4}$	$7.4\cdot10^{2}$	$1.1\cdot10^{4}$	752	4.2	278.4	279.4
10	313	60	1,005	0.5	0.003	0.6	288	$2.6\cdot10^{6}$	$3.3\cdot10^{6}$	$7.1\cdot10^{5}$	$2.3\cdot10^{6}$	$-2.6\cdot10^{4}$	$-8.8\cdot10^{2}$	$2.0\cdot10^{5}$	856	79.6	313.6	311.8
11	313	60	585	0.5	0.003	0.6	288	$2.8\cdot10^{5}$	$3.5\cdot10^{5}$	$2.5\cdot10^{6}$	$4.8\cdot10^{5}$	$-4.4\cdot10^{5}$	$-1.4\cdot10^{4}$	$2.0\cdot10^{5}$	856	79.6	314.8	292.8
12	313	60	420	0.5	0.003	0.6	288	$6.1\cdot10^{4}$	$7.7\cdot10^{4}$	$2.0\cdot10^{6}$	$4.3\cdot10^{5}$	$-4.6\cdot10^{5}$	$-1.4\cdot10^{4}$	$2.0\cdot10^{5}$	856	79.6	314.5	292.2
13	278	80	1,005	0.5	0.005	0.3	278	$2.6\cdot10^{6}$	$3.2\cdot10^{6}$	$7.4\cdot10^{5}$	$2.0\cdot10^{6}$	$5.2\cdot10^{5}$	$2.6\cdot10^{4}$	$1.1\cdot10^{4}$	432	4.2	278.6	318.6
14	278	80	585	0.5	0.005	0.3	278	$2.8\cdot10^{5}$	$3.6\cdot10^{5}$	$1.2\cdot10^{6}$	$1.4\cdot10^{5}$	$3.7\cdot10^{4}$	$1.5\cdot10^{3}$	$1.1\cdot10^{4}$	432	4.2	278.9	280.9
15	278	80	420	0.5	0.005	0.3	278	$6.3\cdot10^{4}$	$8.0\cdot10^{4}$	$2.5\cdot10^{6}$	$9.7\cdot10^{4}$	$2.6\cdot10^{4}$	$1.1\cdot10^{3}$	$1.1\cdot10^{4}$	432	4.2	279.8	280.0
16	313	60	1,005	0.5	0.005	0.3	288	$2.6\cdot10^{6}$	$3.2\cdot10^{6}$	$7.4\cdot10^{5}$	$2.1\cdot10^{6}$	$1.7\cdot10^{5}$	$1.0\cdot10^{4}$	$2.0\cdot10^{5}$	492	79.6	313.6	326.1
17	313	60	585	0.5	0.005	0.3	288	$2.8\cdot10^{5}$	$3.5\cdot10^{5}$	$4.4\cdot10^{5}$	$4.2\cdot10^{5}$	$-2.3\cdot10^{5}$	$-1.2\cdot10^{4}$	$2.0\cdot10^{5}$	492	79.6	313.3	295.4
18	313	60	420	0.5	0.005	0.3	288	$6.1\cdot10^{4}$	$7.6\cdot10^{4}$	$3.0\cdot10^{5}$	$2.5\cdot10^{5}$	$-2.7\cdot10^{5}$	$-1.4\cdot10^{4}$	$2.0\cdot10^{5}$	492	79.6	313.2	292.2
19	278	80	1,005	0.5	0.005	0.6	278	$2.6\cdot10^{6}$	$3.2\cdot10^{6}$	$8.2\cdot10^{5}$	$2.1\cdot10^{6}$	$3.3\cdot10^{5}$	$1.5\cdot10^{4}$	$1.1\cdot10^{4}$	752	4.2	278.7	303.6
20	278	80	585	0.5	0.005	0.6	278	$2.8\cdot10^{5}$	$3.6\cdot10^{5}$	$2.5\cdot10^{6}$	$1.7\cdot10^{5}$	$2.6\cdot10^{4}$	$1.1\cdot10^{3}$	$1.1\cdot10^{4}$	752	4.2	279.8	280.1
21	278	80	420	0.5	0.005	0.6	278	$6.4\cdot10^{4}$	$8.1\cdot10^{4}$	$1.3\cdot10^{6}$	$1.3\cdot10^{5}$	$2.0\cdot10^{4}$	$8.1\cdot10^{2}$	$1.1\cdot10^{4}$	752	4.2	279.0	279.5
22	313	60	1,005	0.5	0.005	0.6	288	$2.6\cdot10^{6}$	$3.2\cdot10^{6}$	$7.3\cdot10^{5}$	$2.3\cdot10^{6}$	$-1.6\cdot10^{4}$	$-9.1\cdot10^{2}$	$2.0\cdot10^{5}$	856	79.6	313.6	311.8
23	313	60	585	0.5	0.005	0.6	288	$2.8\cdot10^{5}$	$3.5\cdot10^{5}$	$2.2\cdot10^{6}$	$4.5\cdot10^{5}$	$-2.7\cdot10^{5}$	$-1.4\cdot10^{4}$	$2.0\cdot10^{5}$	856	79.6	314.6	292.4
24	313	60	420	0.5	0.005	0.6	288	$6.1\cdot10^{4}$	$7.7\cdot10^{4}$	$1.3\cdot10^{6}$	$2.6\cdot10^{5}$	$-3.0\cdot10^{5}$	$-1.5\cdot10^{4}$	$2.0\cdot10^{5}$	856	79.6	314.0	290.4
25	278	80	1,005	0.5	0.01	0.3	278	$2.6\cdot10^{6}$	$3.3\cdot10^{6}$	$8.6\cdot10^{5}$	$2.1\cdot10^{6}$	$2.8\cdot10^{5}$	$2.9\cdot10^{4}$	$1.1\cdot10^{4}$	432	4.2	278.7	322.5
26	278	80	585	0.5	0.01	0.3	278	$2.8\cdot10^{5}$	$3.5\cdot10^{5}$	$1.5\cdot10^{6}$	$1.5\cdot10^{5}$	$2.0\cdot10^{4}$	$1.7\cdot10^{3}$	$1.1\cdot10^{4}$	432	4.2	279.1	281.3
27	278	80	420	0.5	0.01	0.3	278	$6.4\cdot10^{4}$	$8.0\cdot10^{4}$	$2.5\cdot10^{6}$	$9.7\cdot10^{4}$	$1.3\cdot10^{4}$	$1.1\cdot10^{3}$	$1.1\cdot10^{4}$	432	4.2	279.8	279.9
28	313	60	1,005	0.5	0.01	0.3	288	$2.6\cdot10^{6}$	$3.2\cdot10^{6}$	$8.5\cdot10^{5}$	$2.2\cdot10^{6}$	$9.5\cdot10^{4}$	$1.1\cdot10^{4}$	$2.0\cdot10^{5}$	492	79.6	313.7	327.5
29	313	60	585	0.5	0.01	0.3	288	$2.8\cdot10^{5}$	$3.5\cdot10^{5}$	$6.5\cdot10^{5}$	$3.1\cdot10^{5}$	$-1.3\cdot10^{5}$	$-1.3\cdot10^{4}$	$2.0\cdot10^{5}$	492	79.6	313.5	293.3
30	313	60	420	0.5	0.01	0.3	288	$6.1\cdot10^{4}$	$7.7\cdot10^{4}$	$5.1\cdot10^{5}$	$1.2\cdot10^{5}$	$-1.5\cdot10^{5}$	$-1.5\cdot10^{4}$	$2.0\cdot10^{5}$	492	79.6	313.4	289.7
31	278	80	1,005	0.5	0.01	0.6	278	$2.6\cdot10^{6}$	$3.3\cdot10^{6}$	$1.1\cdot10^{6}$	$2.3\cdot10^{6}$	$1.7\cdot10^{5}$	$1.6\cdot10^{4}$	$1.1\cdot10^{4}$	752	4.2	278.8	305.1
32	278	80	585	0.5	0.01	0.6	278	$2.8\cdot10^{5}$	$3.6\cdot10^{5}$	$2.5\cdot10^{6}$	$1.7\cdot10^{5}$	$1.3\cdot10^{4}$	$1.1\cdot10^{3}$	$1.1\cdot10^{4}$	752	4.2	279.8	280.2
33	278	80	420	0.5	0.01	0.6	278	$6.4\cdot10^{4}$	$8.0\cdot10^{4}$	$2.0\cdot10^{6}$	$1.4\cdot10^{5}$	$1.1\cdot10^{4}$	$8.8\cdot10^{2}$	$1.1\cdot10^{4}$	752	4.2	279.5	279.5
34	313	60	1,005	0.5	0.01	0.6	288	$2.6\cdot10^{6}$	$3.3\cdot10^{6}$	$9.4\cdot10^{5}$	$2.3\cdot10^{6}$	$-8.8\cdot10^{3}$	$-9.8\cdot10^{2}$	$2.0\cdot10^{5}$	856	79.6	313.7	311.8
35	313	60	585	0.5	0.01	0.6	288	$2.8\cdot10^{5}$	$3.5\cdot10^{5}$	$1.8\cdot10^{6}$	$3.4\cdot10^{5}$	$-1.4\cdot10^{5}$	$-1.4\cdot10^{4}$	$2.0\cdot10^{5}$	856	79.6	314.3	291.1
36	313	60	420	0.5	0.01	0.6	288	$6.2\cdot10^{4}$	$7.7\cdot10^{4}$	$1.3\cdot10^{6}$	$1.2\cdot10^{5}$	$-1.6\cdot10^{5}$	$-1.6\cdot10^{4}$	$2.0\cdot10^{5}$	856	79.6	314.0	288.8

3.2. Correlation analyses

Correlations among the estimated heat flows were assessed through the Pearson coefficient. Pearson coefficient measures the relationship between two variables and it ranges from -1 to +1, with +1 representing a positive correlation, -1 representing a negative correlation, and 0 representing no relationship. The outcomes from this analysis are reported in Table 3.

The $Q_{irr,in}$ and the $Q_{irr,out}$ were perfectly correlated between them and with the $Q_{conv,wat}$, while they showed high degrees of correlation with the $Q_{cond,wat}$ and the $Q_{irr,sky}$. Conversely, their correlations with the Q_{air} were moderate, and no correlation was observed with the Q_{evap}. Similar coefficients were calculated for the $Q_{conv,wat}$, while both the $Q_{cond,wat}$ and the $Q_{irr,sky}$ were characterized by lower correlation degrees when it came to the $Q_{irr,in}$, $Q_{irr,out}$, Q_{air}, and $Q_{conv,wat}$. On the other hand, they presented stronger correlations between them and with the Q_{evap}. Finally, the Q_{evap} was only correlated to the Q_{air} (medium degree), $Q_{cond,wat}$ (high degree), and $Q_{irr,sky}$ (high degree).

Table 3. Correlation coefficients calculated for each couple of estimated heat flows.

	$Q_{irr,in}$	$Q_{irr,out}$	Q_{air}	$Q_{conv,wat}$	$Q_{cond,wat}$	$Q_{irr,sky}$	Q_{evap}
$Q_{irr,in}$	1.0000	1.0000	-0.4454	0.9864	0.6753	0.7311	-0.0004
$Q_{irr,out}$	1.0000	1.0000	-0.4447	0.9869	0.6733	0.7316	-0.0011
Q_{air}	-0.4454	-0.4447	1.0000	-0.4710	-0.1628	-0.1368	-0.3525
$Q_{conv,wat}$	0.9864	0.9869	-0.4710	1.0000	0.5553	0.6470	0.1074
$Q_{cond,wat}$	0.6753	0.6733	-0.1628	0.5553	1.0000	0.8992	-0.6056
$Q_{irr,sky}$	0.7311	0.7316	-0.1368	0.6470	0.8992	1.0000	-0.6229
Q_{evap}	-0.0004	-0.0011	-0.3525	0.1074	-0.6056	-0.6229	1.0000

Similarly, inter-correlations among the g_h, $T_{air,in}$, and T_{roof} output parameters as well as correlations between these and the main numerical model variables were investigated. The estimated correlation coefficients are presented in Table 4.

Table 4. Inter-correlations among the g_h, $T_{air,in}$, and T_{roof} output parameters and correlations between them and the main numerical model variables.

	T_{air}	$RH_{\%}$	T_{slag}	dz_{wat}	v_{wat}	T_{wat}	h_{wat}	g_h	$T_{air,in}$	T_{roof}
g_h	1.0000	-1.0000	0.0000	0.0000	0.0000	1.0000	0.2322	1.0000	0.9996	0.3600
$T_{air,in}$	0.9996	-0.9996	-0.0115	0.0024	0.0103	0.9996	0.2428	0.9996	1.0000	0.3472
T_{roof}	0.3600	-0.3600	0.8595	0.0052	-0.1916	0.3600	-0.1034	0.3600	0.3472	1.0000

The g_h was perfectly correlated to the outdoor air properties (i.e. T_{air}, $RH_{\%}$), and no correlation was found with the characteristic of the water flow (i.e. water layer thickness, water velocity), except with the water temperature (perfect correlation). As far as the $T_{air,in}$ is concerned, it was found to be strongly correlated to T_{air}, $RH_{\%}$, T_{wat}, and $T_{air,in}$, while its correlations to T_{roof} (medium degree) and h_{wat} (low degree) were weaker. Finally, a high correlation coefficient was calculated between the T_{roof} and the T_{slag}. Conversely, lower correlations were observed with the other data set concerning outdoor air temperature and relative humidity, indoor air temperature, water temperature, and the evaporated water mass. Finally, no correlation was observed between T_{roof} and the dz_{wat} parameter, while a low degree-correlation was found when it came to v_{wat}.

3.3. Limitations

The main limitations of the study are reported and discussed in this section. The first important limitation lies in the fact that the contribution of solar radiation was not considered in the energy balance equations. This simplification allowed to minimize the computational time and can be justified by considering that variations in output parameters (i.e., g_h, T_{roof}, $T_{air,in}$) due to solar radiation are much smaller compared to those produced by the radiative flux emitted by the slag, that exhibits a temperature of up to 1,005 K.

A second limitation, concerns the air volume which was modelled as a solid without any turbulence as well as neglecting heat transfer through the outer walls and may result in a systematic overestimation of the operational temperatures. Considering the outdoor air temperature instead of the sky temperature also leads to higher temperatures. However, such an approach permits to design a more resistant load-bearing structure and a more efficient filtering system.

Thirdly, a simplification was made regarding the roof's temperature. In fact, the outer and inner temperature of the roof are considered as equal. This was considered as acceptable since the covering of the pavilion is made of steel single-layer panels which are 1 mm-thick and show a relatively high thermal conductivity value (54 W m^{-1} K^{-1}).

4. Conclusions and future developments

The present work investigated the effectiveness of a water-based cooling system for the new pavilion for slag management in Italy, by assessing the operating temperature of the indoor air and the roof as well as the amount of evaporated water.

The main findings were summarized as follows:

- The roof temperature did not exceed the 328 K, hence the mechanical properties of the material were not compromise;
- The average air temperature was up to 2 K higher than the outdoor one;
- The evaporated water mass was strongly influenced by seasonality, and it ranged between 4.2 kg h^{-1} and 79.6 kg h^{-1};
- High and medium correlations were estimated between T_{roof} and T_{slag}, T_{air}, $RH_{\%}$, T_{wat}, g_h, and $T_{air,in}$, while low and null correlations were found with dz_{wat} and v_{wat}.

In conclusion, the water-based cooling system was demonstrated to be effective in maintaining the operating temperature values below the warning levels, and avoiding the collapsing of pavilion load-bearing elements. However, a more detailed investigation which also considers transient regime simulations is needed to better evaluate the influences of the cooling system on both the indoor environment and the pavilion envelope. Alongside this, the future developments may concern the analysis of the cooling potential of different systems integrated, such as the water-based one from this work and some other standard strategies (forced ventilation, water spreading on the slag).

References

[1] T.S. Naidu, C.M. Sheridan, L.D. van Dyk, Basic oxygen furnace slag: Review of current and potential uses, Miner. Eng. 149 (2020) 106234. https://doi.org/https://doi.org/10.1016/j.mineng.2020.106234.

[2] W. Zhang, F. He, Y. Xiao, M. Xie, F. Li, J. Xie, H. Yang, Z. Li, Structure, viscosity, and crystallization of glass melt from molten blast furnace slag, Int. J. Appl. Glas. Sci. 11 (2020) 676–684. https://doi.org/https://doi.org/10.1111/ijag.15054.

[3] J. Guo, Y. Bao, M. Wang, Steel slag in China: Treatment, recycling, and management, Waste Manag. 78 (2018) 318–330. https://doi.org/https://doi.org/10.1016/j.wasman.2018.04.045.

[4] European Confederation of Iron and Steel Industries, EUROFER Annual Report, Brussels, 2007.

[5] M. Manni, E. Bonamente, G. Lobaccaro, F. Goia, A. Nicolini, E. Bozonnet, F. Rossi, Development and validation of a Monte Carlo-based numerical model for solar analyses in urban canyon configurations, Build. Environ. 170 (2020). https://doi.org/10.1016/j.buildenv.2019.106638.

[6] Y.A. Cengel, A.J. Ghajar, Heat and mass transfer: Fundamentals and applications, 5th ed., McGraw-Hill Professional, New York, 2014.

[7] Engineering Toolbox, Evaporation from Water Surface, (2004). https://www.engineeringtoolbox.com/evaporation-water-surface-d_690.html (accessed April 28, 2021).

[8] W. Wagner, A. Pruß, The IAPWS Formulation 1995 for the Thermodynamic Properties of Ordinary Water Substance for General and Scientific Use, J. Phys. Chem. Ref. Data. 31 (2002) 387–535. https://doi.org/10.1063/1.1461829.

Numerical Model of an Evaporative Condenser for the Food Refrigeration Industry

Michael Giovannini, Marco Lorenzini

Università di Bologna, DIN – Dipartimento di Ingegneria Industriale, Via Fontanelle 40, I-47121, Forlì (FC) Italy,

marco.lorenzini@unibo.it

Abstract. Evaporative condensers play a key role in refrigeration plants for industrial applications, as they are needed to dispose of the thermal and electric power required by the process. Modelling them as lumped capacitance systems is very expedient when trying to simulate e. g. the steady-state and dynamic behaviour of the whole refrigeration plant, but at the expenses of loss of information and often oversimplifying assumptions. This paper analyses two models for heat and mass transfer in cooling equipment, Poppe's and its earlier, simpler version, Merkel's. The two approaches are applied to describe the transport phenomena in an evaporative condenser and the predicted cooling power compared to the data declared by the manufacturer under certain operating conditions. It is demonstrated that predictions are accurate within less than 10%.

1. Introduction

Evaporative condensers have come into use at food processing and storage plants comparatively recently. This piece of heat transfer equipment basically combines an air-cooled and a water-cooled condenser together, with benefits in terms of overall costs.

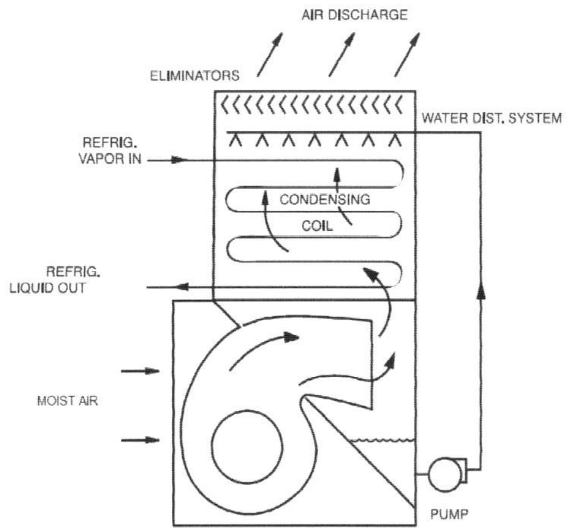

Figure 1: Schematic drawing of an evaporative condenser.

In this type of condenser the heat is transferred from the refrigerant through tube walls to a thin film of water, which flows over the outer surface of the tubes and then to air, mainly due to partial evaporation of water. These condensers can be profitably employed especially in dry and hot climates. Water consumption in the condenser is moderate and can be estimated to amount to 0.4 kg per 1000 kJ of energy transferred as heat, in the case evaporation is the only means of heat removal. During summertime up to 80% of the thermal energy removed may be due to evaporation. As sketched in Fig.1, water is sprayed onto the tubes by means of nozzles and partly evaporates, but mostly flows by gravity into a sump, where it is collected and pumped back to the nozzles above the tube bank. The evaporative process is enhanced by radial or axial fans, which either blow or draft air over the tubes.

On one hand, there is no longer the need to treat and circulate large amounts of water, on the other, fans needed to maintain the airflow are smaller and require significant less power than in air-cooled systems. The most prominent feature, however, is that evaporative condensers can be designed for a lower condensation temperature than single-fluid condensers, which results in lower power consumption and smaller operation pressure designs for the compressors. In the winter months, evaporative condensers may be operated as air-cooled machines, thus eliminating water pumping and reintegration costs. Also, the use of a fluid collector and separator downstream of the condenser ensures a vapor quality of the refrigerant at the outlet close to zero (saturated liquid). The disadvantages of evaporative condensers are related to fouling issues, as scales tends to form rather quickly on the outside of the tubes, and some cleaning difficulties of the inside of the ducts may also be encountered [1,2].

Given the wide range of transport phenomena involved and the practical significance of the problem, the mechanisms of heat and mass transfer occurring have been the subject of several investigations in the past: the earliest studies, which all successive research refers to, are those of Merkel [3] and Poppe [4], and they will be further detailed in the following. Merkel's model was applied with additional simplifying assumptions by Baker and Shyrock [5] to study a condenser. Extensive work has been carried out by Dreyer, starting from his Doctoral Thesis [6], in which some calculation methods for heat transfer are outlined; Poppe's model is the starting point, from which a simplified formulation is obtained, close to Merkel's, and a lumped-parameter model is also hinted at. In this work [6] a list of correlation to compute the heat and mass transfer coefficients is given, according to the type of evaporative condenser considered: the models differ for the fluid flow direction, layout and presence or lack of additional heat transfer blocks to cool the water dripping from the tubes, known as fill packs. Further explanations on the calculation processes in [6] are given in a later work [7], which examines the modes of computation best suited for a given layout and geometry of the tubes.

A rather involved, if general, methodology to determine evaporative heat transfer for different arrangements and two or three participating fluids is given by Halàsz [8].

Merkel's model was used by Hasan and Siren [9] to describe evaporative heat transfer in closed wet cooling towers used for climatization of office buildings; in [10] the same subject was further investigated employing simplified analytical formulations in CFD simulations.

Air-cooled heat exchangers and cooling towers are treated extensively in the book by Kröger [11], which describes both Merkel's and Poppe's approaches and considers them for several practical applications. Heat transfer and pressure drop in heat exchangers are discussed, and a wide selection of applicable correlations offered, also regarding the heat and mass transfer coefficients in the liquid film around tubes.

In a further work, [12], Kloppers and Kröger discuss the effect of the Lewis factor (which had already been related to the Lewis number in [4]) on the heat exchanger performance for different layouts of cooling towers.

Facao and Oliveira, [13], discuss the possibility of employing computer simulations to obtain mass transfer correlations based on their comparison of experimental data for an indirect contact cooling tower with the results of CFD simulations.

Expanding on the work of Dreyer, [4], Yilmaz [14] has suggested more recently an integral analytical approach to obtain the performance of wet cooling towers.

The literature survey demonstrates that the modelling of evaporative coolers, wet cooling towers and air-cooled condenser continues to be a lively area of interest, given the wide range of modelling issues which are also dependent on the details required by the application for which they are devised.

This work intends to investigate the behaviour of an evaporative condenser without fill pack to use it as a simulation block in the framework of a larger, mostly lumped-parameter, model of a refrigeration plant for food (fruit and vegetables) storage located in northern Italy. The full model is aimed towards analysis of plant performance when condensation temperature is lowered following more favourable ambient conditions, that is cooler weather. Evaporative condensers allow this, which has benefits in terms of energy consumption and service life, but there is a strong resistance from the conductors, lest storage conditions in the cells should fail to ensure perfect conservation of the goods. Simulation thus becomes a way to demonstrate the feasibility and profitability of this choice.

2. Modelling approaches: Poppe's and Merkel's

The choice of the model for the evaporative condenser was dictated, as mentioned above, by the final use of the model itself, i.e. the simulation of the behaviour of a refrigeration plant for fruit and vegetable storage. A full CFD model would be computationally too intensive and difficult to integrate into a larger framework, potential candidates remain therefore lumped-parameter and distributed-parameter models. The former would be a computationally better choice, as quantities are at most time-dependent only, and the model becomes algebraic for steady state conditions. Unfortunately, the underlying assumptions required to use a lumped parameter formulation (as discussed by Kröger in [11]), especially those concerning water temperature distribution throughout the tube banks are not verified in practice. For this reason, the choice fell on a distributed-parameter model, in which the quantities not only depend on time, but also on some spatial coordinate, which is determined by the physical process to be described.

Several models found in literature, [3-14] have been considered, all dealing with three fluids, of which two (air and water) interact directly, whereas the third (a refrigerant of some kind) undergoes condensation behind a physical barrier (the pipe wall). The governing equations can be applied to co-flow counterflow and crossflow arrangements, which determine the way in which numerical integration of the equations is carried out.

2.1. Merkel's simplified model

The more extensive version of the model employed, the one given by Poppe [4] will be detailed in the following subsection. Here the simplifying assumptions underlying it are recalled, as they are shared by Merkel's model as well. These are:

- negligible radiation heat transfer;
- uniform distribution of recirculated water, both over the coil and the fill-pack, if present;
- uniform temperature in the water film over the coil, owing to its small thickness;
- area of the heat transfer surface between liquid film and water equals that of the outer pipe wall;
- contributions to heat transfer by end-elements of coil is negligible;
- the enthalpy increase in the water due to circulation pump is neglected.

In order to reduce the number of equations to be solved, further simplifying assumptions may be made, as suggested by Merkel, namely:

- variations in the mass flowrate of water through the coil due to evaporation of the liquid film are neglected;
- the heat transfer rate during the evaporative process is comparable to the mass transfer rate, which is the same as taking the Lewis factor Le_f, i.e. the ratio of the Stanton number for mass transfer to the Stanton number for heat transfer close to unity.

Within this framework, the governing equations that describe the problem are all derived from heat and mass transfer balances over a control volume and are listed below for dry air specific enthalpy, water temperature and refrigerant specific enthalpy.

$$dh_a=\frac{\beta}{\dot{m}_a}\left(h_{a,sw}-h_a\right)dA_o \tag{1}$$

$$dT_w=\frac{1}{\dot{m}_w\cdot c_{p,w}}\left(\dot{m}_a dh_a+\dot{m}_r dh_r\right) \tag{2}$$

$$dh_r=\frac{U}{\dot{m}_r}\left(T_r-T_w\right)dA_o \tag{3}$$

To which a further equation is added, in case the refrigerant is superheated:

$$dT_r=\frac{1}{c_{p,r}}dh_r \tag{4}$$

The subscripts 'w', 'r' and 'a' refer to the three fluids, namely water, refrigerant and air, whilst 's' stands for 'saturation and 'o' for 'outer'. The mass transfer coefficient is β, c_p is the specific thermal capacity at constant pressure, T is the temperature and A the outer area of the pipe. $\dot{m}$ is the mass flowrate and U is the overall heat transfer coefficient, as defined by Eq. (5):

$$U=\left(\frac{1}{\alpha_r}\left(\frac{d_o}{d_i}\right)+\frac{d_o}{2\lambda}\ln\left(\frac{d_o}{d_i}\right)+\frac{1}{\alpha_w}\right)^{-1} \tag{5}$$

In Eq. (5) d is the pipe diameter, the subscript 'i' is for 'inner', λ is the thermal conductivity of the pipe and α the heat transfer coefficient. The pressure drop of the refrigerant flowing within the cooling coil is neglected and all thermodynamic and thermophysical properties needed in the computations are obtained from data fits.

Several correlations for the heat and mass transfer coefficients are available in the literature [11], and an extensive discussion on them for a similar case, albeit with a fill pack has been given in [15]. In the calculations, for the refrigerant, Shah's correlation is used during condensation, and Dittus-Boelter for the superheated fluid [11,15]. The mass transfer coefficient b is computed according to the correlations by either Parker and Treybald [11,15] or Nitsu [11,15], Eqs. (6) and (7) respectively:

$$\beta=0.04935\cdot\left(\frac{\dot{m}_a}{A_c}\right)^{0.905} \tag{6}$$

$$\beta=0.076\cdot\left(\frac{\dot{m}_a}{A_c}\right)^{0.8} \tag{7}$$

where the subscript 'c' is for 'cross-section'. For the heat transfer in the water film, the operating conditions, which are detailed below, fell out of the bonds of the available correlations. It was therefore chosen to use the one closest to the operating conditions, i.e. Erens and Dreyer's [11,15]:

$$\alpha_w=2843\cdot\left(\frac{\Gamma_m}{d_o}\right)^{0.384} \tag{8}$$

Where Γ_m is the average mass flowrate of water over the pipes per unit length.

2.2. Poppe's extended model

The more complex formulation of the governing equations proposed by Poppe [4] was also implemented. This is obtained by dropping the last two assumptions mentioned in the previous subsection. Equations (3) and (4) remain unchanged, whilst Eqs. (1) and (2) become:

$$dh_a=\frac{\beta}{\dot{m}_a}\left(Le_f\left(h_{a,sw}-h_a\right)-(Le_f-1)\left(w_{a,sw}-w_a\right)h_v\right)dA_o \tag{9}$$

$$dT_w=\frac{1}{\dot{m}_w\cdot c_{p,w}}\left(\dot{m}_a dh_a+c_{p,w}T_w d\dot{m}_w+\dot{m}_r dh_r\right) \tag{10}$$

To which two more equations are added:

$$dw_a=\frac{\beta}{\dot{m}_a}\left(w_{a,sw}-w_a\right)dA_o \tag{11}$$

$$d\dot{m}_w=-\dot{m}_a dw_a \tag{12}$$

In Eqs. (11) and (12) w is the humidity ratio and the superscript 'v' is for 'vapour'. The set of equations constituted by Eq. (3) with Eqs. (9)-(12) is valid for unsaturated air; if this is not the case, Eqs. (9) and (11) are replaced by Eqs. (13) and (14) respectively:

$$dh_a=\frac{\beta}{\dot{m}_a}\left(Le_f\left(h_{a,sw}-h_a\right)-(Le_f-1)\left(w_{a,sw}-w_{as}\right)h_v+Le_f(w_a-w_{as})c_{p,w}T_w\right)dA_o \tag{13}$$

$$dw_a=\frac{\beta}{\dot{m}_a}\left(w_{a,sw}-w_{as}\right)dA_o \tag{11}$$

The Lewis factor is computed for both unsaturated and supersaturated conditions using Bosnjacovic's equations [11].

3. Model implementation

The model was implemented for the case in which no fill pack is installed, and the results of the simulation were checked against the data declared by one manufacturer (BAC-Baltimore, USA) for their VXC model; in [15] the results for an evaporative condenser with fill pack (BAC – CXVE) are reported and the influence of the choice of the correlations for heat and mass transfer coefficients are thoroughly discussed.

Figure 2 shows how each rank, that constitutes a separate circuit, through which the condensing refrigerant flows, is subdivided in control volumes, together with the flow paths for the refrigerant, air and water. Balance equations are applied for each control volume and are discretised with fixed steps through a forward scheme. Integration is carried out using an in-house code developed in Python, starting from the topmost element where the superheated ammonia enters the circuit and following the path of the refrigerant. The outputs for Poppe's model are air enthalpy, humidity ratio and temperature, water temperature over the coil and in the sump, and refrigerant temperature, which are mapped unto the elements of the domain (hence the distributed-parameter nature of the method). If Merkel's approach is considered, no information on air temperature and humidity ratio is obtained.

The heat transfer from the refrigerant to the cooling fluid is also computed. A predictor-corrector approach is used, whilst an iterative (control) is employed to obtain the inlet values of enthalpy for the air in the element corresponding to the refrigerant inlet. When Poppe's model is used, the temperature of the air must at the inlet element must be determined likewise.

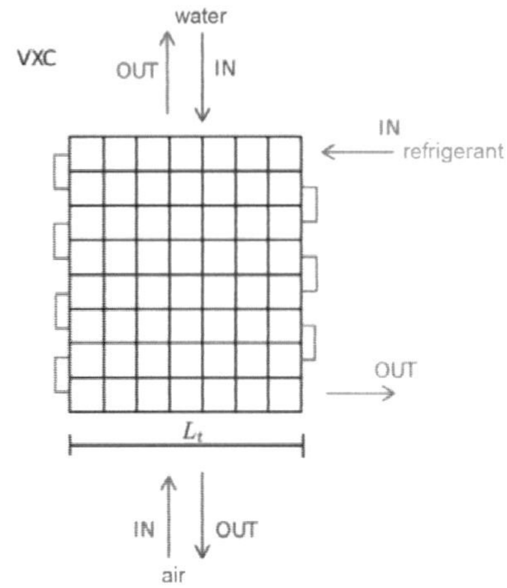

Figure 2. Flow directions for the fluids and element discretisation.

4. Results and discussion

The operating conditions against which the model is tested are air dry bulb and wet bulb temperatures, T_{db}=37 °C, T_{wb}=25.3 °C, corresponding to a relative humidity φ=38 %, and condensation pressure for ammonia $p_{r,s}$=13.50 bar (which corresponds to a saturation temperature $T_{r,s}$=35.0 °C). The cooling power stated by the manufacturer under these operating conditions (steady state) is P_c=1410 kW.

Figure 3 shows the plots for Merkel's model, whilst those for Poppe's are shown in Fig. 4.

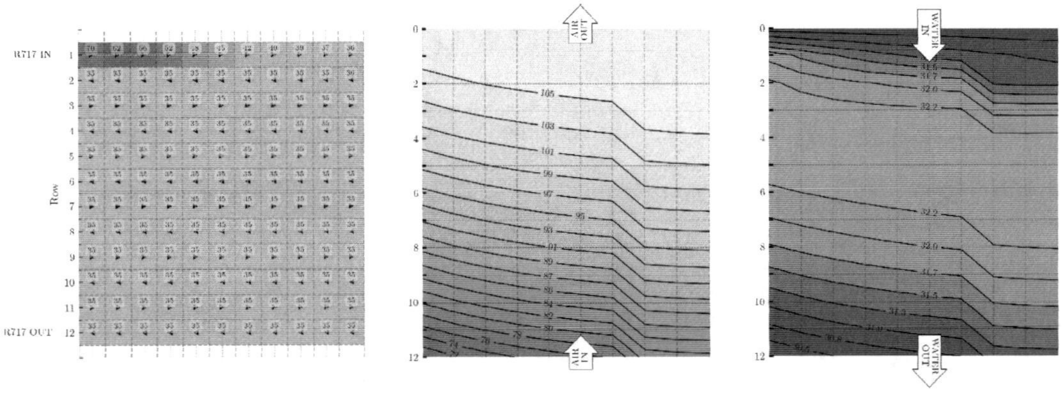

(a) (b) (c)

Figure 3 Refrigerant temperature (°C) (a), enthalpy of air (kJ·kg^{-1}) (b) and water temperature (°C) (c), Merkel's model.

The refrigerant temperature is plotted in Fig. 3a: as is consistent with the operation mode of evaporative condensers, the fluid enters at superheated state (70 °C, topmost left element) and quickly reaches saturation temperature, exiting at a quality close to zero (0.01 in this case). The thermal power removed from R717 is P_c=1374.1 kW, less than 2.5% below the value declared by the manufacturer. Figure 3b is the plot of air enthalpy: as expected, isolines run almost horizontally, as is the case for the temperature of the water dripping from the tubes, Fig. 3c. It can be noticed that the higher refrigerant temperature at the inlet creates higher temperature gradients in the water, and the interaction of the liquid dripping downwards with the air being blown from below is responsible for the ensuing pattern. A distributed parameter formulation is thus demonstrated as being necessary, since the water temperature over the tubes is not uniform, and in [15] it was proven that this approach allows to discard correlations for heat and mass transfer which apparently yield even better results, but which would cause the temperature of the water in some volumes to fall several degrees below the wet bulb temperature, which is unphysical. The model also yields the amount of heat transfer to the air due to evaporation, which amounts to around 98% of the total, again as expected for an evaporative condenser. It is also to be remarked that the convective heat transfer coefficient in the water film, given by Eq. (8) is obtained outside the bounds of the correlation, and that the value obtained α_w=2769 W m^{-2} K^{-1} is significantly lower than the values obtained experimentally in [6].

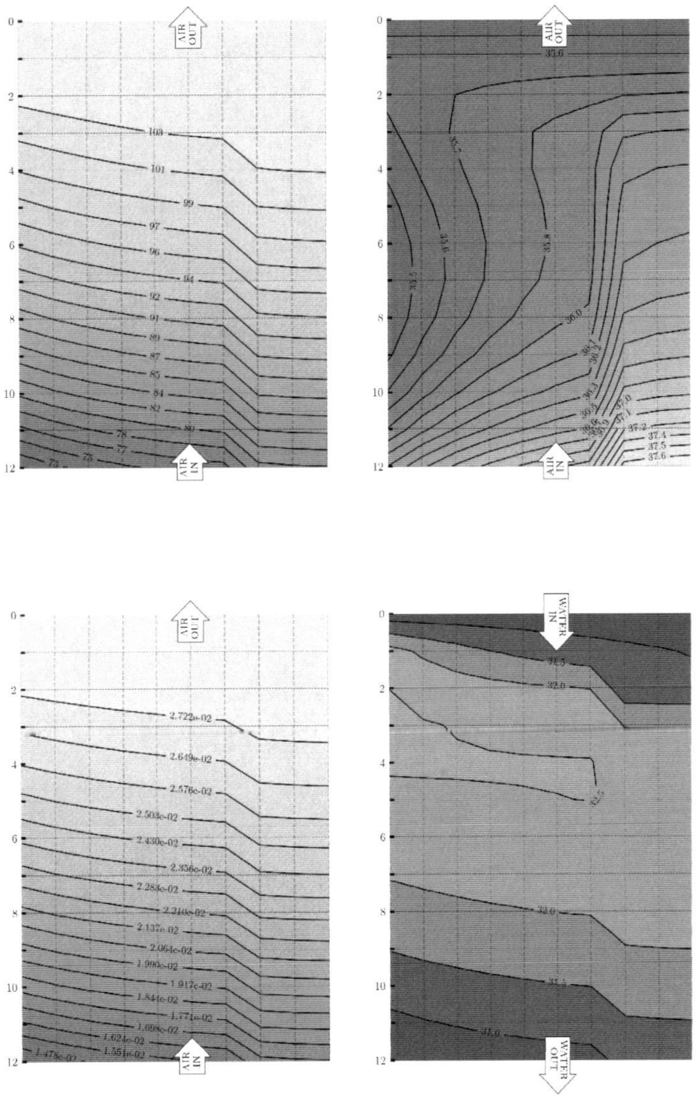

Figure 4. Air enthalpy (kJ·kg^{-1}) (top left), temperature (°C) (top right), humidity ratio (kg_w·kg_a^{-1}) (bottom left) and water temperature (°C) (bottom right) over the cooling coil, Poppe's model.

The results of the simulation with Poppe's model are shown in Fig. 4 (the temperature distribution for the refrigerant is omitted, since it is the same as Fig.3): in this case the temperature of moist air and the humidity ratio over the cooling coil are also obtained. The air enthalpy and water temperature follow a similar pattern to those found with Merkel's model. The water temperature in the sump is T_s=30.8 °C whilst Merkel's model would yield T_s=30.6 °C: even if the difference is small, it has nonetheless an influence on the cooling capability, which, in this case is P_c=1282.6 kW, that is, about 9% below the reference value. Poppe's model, although more complex and richer in output information, appears to perform slightly worse than Merkel's in predicting the cooling performance of the evaporative condenser. One element which deserves further analysis, though, is the actual value of the heat transfer coefficient in the film, which was obtained outside of the bounds for the correlation, and which might be somewhat higher: in this case, Merkel's model might overestimate the results and Poppe's prove to be more accurate.

5. Conclusions

Two different models to capture the transport phenomena occurring in evaporative condensers were described and implemented to check their capability to correctly predict the cooling performance of a piece of equipment as declared by the manufacturer. Both models were accurate within 10% and but the simpler (Merkel's) came closer, underestimating by little more than 2% the cooling power, whereas the more complex formulation (Poppe's) underestimated the quantity by about 9%. An element of uncertainty turned out to be the convective heat transfer coefficient in the film, for which no suitable correlation for the operating conditions of the problem was found. As the value extrapolated from Eq. (8) seems to be lower than those reported from experiments in [6], the matter deserves further investigation. To this aim, some field measurements on units installed in a food storage facility are planned.

The model of the evaporative condenser is thought as a block in a larger system which simulates the behaviour of a refrigeration loop for industrial applications, especially the savings in terms of energy which can be obtained by letting the condenser operate at temperatures closer to the wet bulb, especially on colder days. Although developed for steady state situations, the model can easily be adapted to transient conditions by introducing the storage term in the energy and mass balance equations.

References

[1] Lebedev V, *Refrigerating engineering*, 1986, (MIR Moscow, Russia)
[2] Bonauguri E, Miari D, *Tecnica del freddo* 1988, (Hoepli, Milan, IT)
[3] Merkel F, *1925 VDI-Zeitchrift* **70** pp 123–128
[4] Poppe M., and Rögener H., 1991 *VDI-Wärmeatlas*, pp Mi 1–Mi 15.
[5] Baker D R and Shryock H A, *1961 Journal of Heat Transfer*, **83**(3) pp 339–349
[6] Dreyer A A, *PhD thesis,* 1988 (Stellenbosch University, Stellenbosch, RSA).
[7] Erens P J, and Dreyer A A, *1989 N and O Journal*, **5**(1) pp 1–10
[8] Halasz B, *1998 Revue generale de thermique*, **37**(4) pp 245–255
[9] Hasan A A, and Siren K, *2002 Energy and buildings*, **34**(5) pp 477–486
[10] Hasan A A, *PhD thesis*, 2005 (Helsinki University of Technology, Helsinki, Finland)
[11] Kröger D G, *Air-cooled heat exchangers and cooling towers, v.1.* 2004 (Pennwell Books, Tulsa OK, USA).
[12] Kloppers J C, and Kröger D G, *2005 Int. journal of thermal sciences*, **44**(9) pp 879–884
[13] Facao J, and Oliveira A C, *2008 Num. Heat Transfer, Part A: Applications*, **54**(10) pp 933–944
[14] Yilmaz A., *2010 Journal of Thermal Science and Technology*, **30**(2) pp 45–56
[15] Giovannini M, Lorenzini M *2021 Proc. of ATE-Hefat 2021* (Amsterdam, The Netheralands).

Journal of Physics: Conference Series 2177 (2022) 012010 doi:10.1088/1742-6596/2177/1/012010

The influence of Variable Geometry Control on a R290 Ejector Refrigeration System

Giorgio Besagni[1], Fabio Inzoli[1]

[1]Politecnico di Milano, Department of Energy, via Lambruschini 4a, 20156, Milano (Italy)

giorgio.besagni@polimi.it

Abstract. The large-scale deployment of ejector refrigeration systems (i.e., solar-based ejector refrigeration systems), although representing a promising alternative compared with mechanical compressor ones, is hindered due to limitations regarding ejector control modes. Indeed, ejectors are fluid-dynamics controlled devices and, because of their fixed geometry, they operate at their highest efficiency in a narrow range of operating conditions, which is in contrast with the dynamic pressure and temperature levels characterizing real applications. In this context, variable geometry ejectors (*VGE*) represent a promising solution to increase the flexibility and operation range of this component. The present study aims to extend the present body of knowledge regarding *VGE* systems, evaluating the impact of a spindle-provided ejector operated with R290 on the performance of the refrigeration system. The analysis has been carried out using an integrated lumped parameter/*CFD* approach, thus linking the local flow properties and global performances. Different spindle positions have been tested to assess how the different nozzle area ratios affect both the entrainment ratio and the critical pressure. Results showed that increasing primary nozzle area ratio the system can effectively reduce the thermal input, increasing the average *COP* at the expanse of a lower critical pressure. In conclusion, using a moving spindle control system might ensure an improvement of the ejector performance.

1. Introduction

Ejector is a device constituted by a primary nozzle, a suction chamber, a mixing chamber, and a diffuser, whose main geometrical parameters are presented in Figure 1. The "*high energy*" primary flow accelerates and expands through the primary nozzle creating a low-pressure region nearby the nozzle exit; subsequently, the secondary flow is entrained into the mixing chamber because of the vacuum-effect and the shear action between the primary and secondary flows. The primary and the secondary flows mix within the mixing chamber and the resulting stream moves into a diffuser where the high velocity fluid is gradually decelerated and increased in pressure due to subsonic conditions.

The entrainment, the pressure recovery, and the mixing effects provided by the ejector, makes it suitable to be employed in ejector refrigeration systems (*ERSs*; Figure 2a presents the layout of a standard ejector refrigeration system, *SERS*) [1]. In general, *ERSs* are promising alternative compared with compressor-based technologies owing to their reliability, limited maintenance, low initial and operational costs, and no working fluid limitations [2]; also, the generator of an *ERS* might exploit low-grade heat energy, making *ERSs* valuable in contributing towards reducing electricity consumption in the residential sector [3].

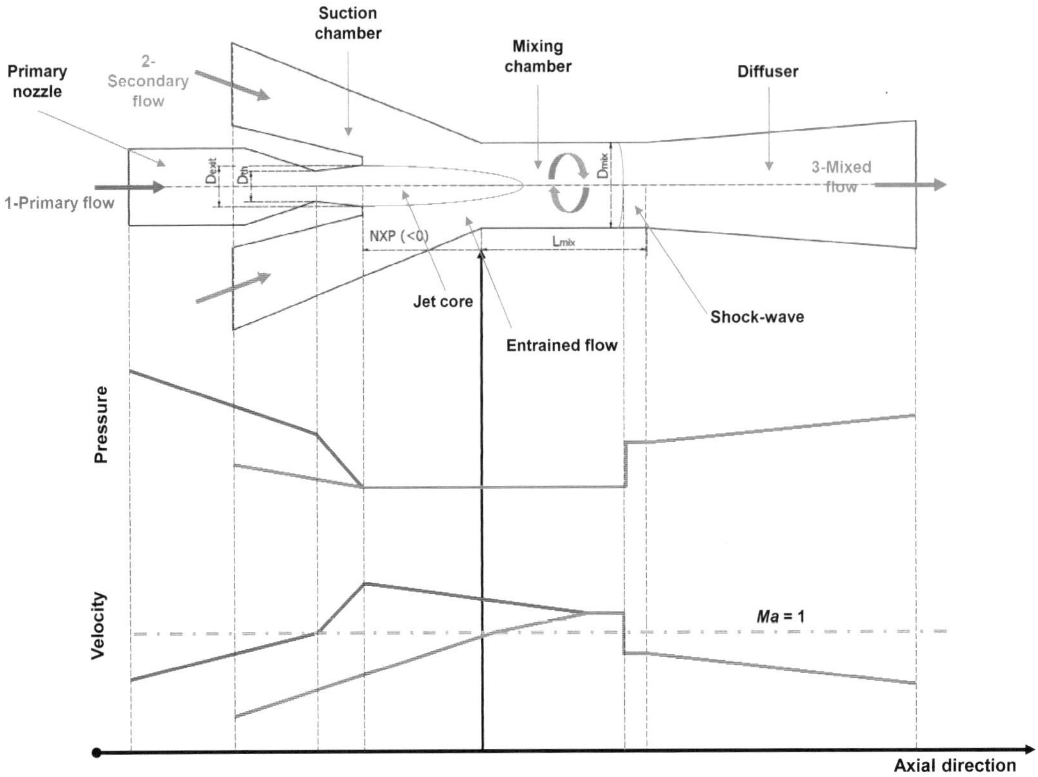

Figure 1. Ejector design and qualitative axial pressure and velocity trends

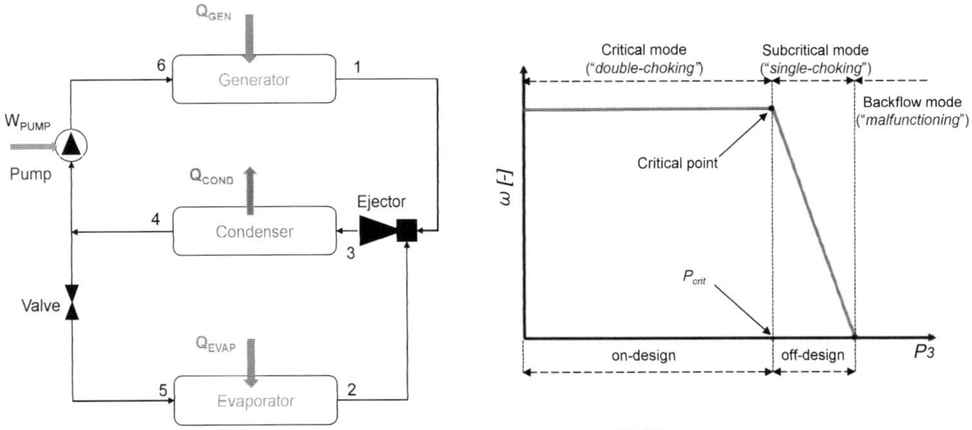

(a) Layout of a standard ejector refrigeration system *(b) Ejector operating curve*

Figure 2. Ejector component (a) and ejector refrigeration system (b)

Unfortunately, the large-scale deployment of *ERSs* is hindered owing to two main drawbacks: the low coefficient of performance (in the range of 0.1–0.7) and the relevant influence of ejector operation on the performance of the whole system. The latter can be easily explained by the fact that the ejector is a fluid-dynamics controlled device, where the fluid-dynamic interactions at the "*local-scale*" impact on the performances at the "*component-scale*", namely the entrainment ratio ω (viz- the ratio between secondary and primary mass flow rates, Figure 2b). These multi-scale relationships can be easily explained considering that the ejector operation relies on two concurrent physical phenomena: (i) the low-pressure fluid entrainment process caused by the primary flow expansion and (ii) the compression effect, provided by the diffuser, which raises the secondary flow pressure from the evaporator to the condenser pressure. These two effects are contrasting, and an improvement of the former would deteriorate the latter: hence, for a given nozzle area ratio, the ejector operating curve is imposed by the primary and secondary flow boundary conditions. On the other hand, Variable Geometry Ejectors (*VGE*) represent a promising solution to increase the flexibility and operation range of this component. In a *VGE*, the spindle acts on the nozzle area, and primary flow rate, and changes the ejector's response by adjusting its entrainment ratio accordingly with the requirements of the system. (i.e., temperature set point, thermal load, …).

The present-day discussion regarding ejector technology is even more challenging owing to the current transition in refrigerants; given the European regulation aimed to limit the emission of fluorinated greenhouse gases, the existing market is expected to change sharply in the next years, and 3rd generation refrigerants (i.e., R134a) will be most likely replaced by natural (i.e., hydrocarbons) and 4th generation fluids (i.e., hydrofluoroolefins) [4]. Among the new refrigerants, *R290* (propane) is promising owing to its favourable thermodynamic properties, which make it suitable for refrigeration applications in the medium-long views.

This paper aims at extending knowledge on *VGE* systems, evaluating the impact of a spindle-provided ejector operated with R290 on the performance of an *ERS*. The present paper employs a multi-scale concept, by using an integrated Computational Fluid Dynamic (*CFD*) - Lumped Parameter Model (*LPM*) of the *ERS*. The *CFD* approach solves the fluid-dynamics within the ejector ("*local-scale*") and provides the ejector entrainment ratio ("*component-scale*"). Conversely, the refrigeration cycle ("*system-scale*") has been modelled by a *LPM* approach using as input data *CFD* outcomes.

2. Methods

2.1. Test case

The tested ejector has been derived from Del Valle et al. [5] and has been already employed in the previous screening of new refrigerants and ejector design sensitivity analysis in Besagni et al. [6]. It should be noted how, compared with the latter reference, the mixing chamber diameter has been reduced to 4 mm to extend the on-design operating mode. The present ejector has been provided of a spindle (Figure 3) which can move axially within the primary nozzle. The neutral spindle position (*SP*) in which the tip of the spindle is placed right in the nozzle throat has been named as *SP#0*. Moving the spindle towards the nozzle exit, the nozzle throat area is reduced, and the area ratio is increased. The effects of the *SP* have been analyzed in the range 0-7 mm, with a discretization of 1 mm each position, whose code name will be referred as *SP#0..7*, depending on the spindle position. A 2D axysimmetric geometrical model has been designed and discretized according to the mesh criteria defined in ref. [7], with an average cell number of 228,000 after the double cycle mesh adaption.

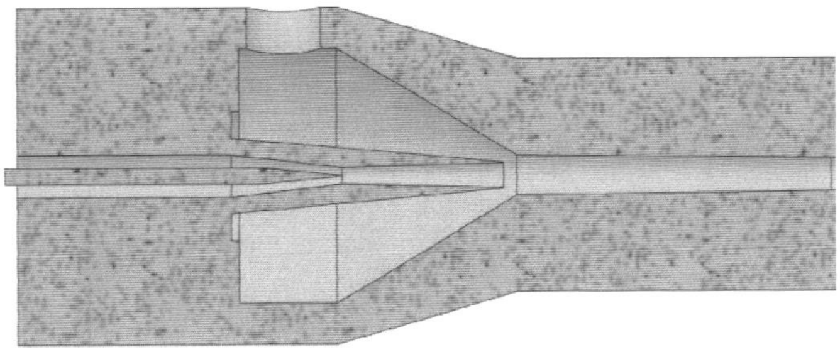

Figure 3. Ejector primary nozzle equipped with the spindle

Propane (*R290*) has been selected as refrigerant and the ejector has been tested with fixed thermodynamic conditions (temperature and pressure) at primary and secondary inlet. Primary inlet superheating has been set equal to 10°C, while secondary superheating has been set equal to 4°C to avoid condensation inside the ejector.

Table 1 summarizes the inlet boundary conditions (T and P), the max value of outlet pressure (P_{3max}) and the critical pressure (P_{3crit}). Modifying the spindle position and varying the pressure outlet condition the VGE's operating curves has been obtained. The critical conditions has been obtained with an iterative approach.

Table 1. Boundary conditions of the simulated cases; the primary flow is labelled as 1, the secondary flow 2 and the outlet stream 3 (Figure 2a)

Code name (spindle position)	T_1 [°C]	P_1 [kPa]	T_2 [°C]	P_2 [kPa]	$P_{3,crit}$ [kPa]	$P_{3,max}$ [kPa]
SP#0	84.2	2,813.5	14	636.6	1,014.1	1,052.7
SP#1	84.2	2,813.5	14	636.6	1,009.0	1,039.7
SP#2	84.2	2,813.5	14	636.6	988.9	1,026.8
SP#3	84.2	2,813.5	14	636.6	961.8	1,014.0
SP#4	84.2	2,813.5	14	636.6	918.6	964.2
SP#5	84.2	2,813.5	14	636.6	869.9	916.2
SP#6	84.2	2,813.5	14	636.6	806.0	836.5
SP#7	84.2	2,813.5	14	636.6	731.5	772.3

2.2. Numerical modelling

The finite volume code ANSYS Fluent (*Release 2020 – R1*) has been used to solve the 2-D axysimmetric steady-state Reynolds Averaged Navier-Stokes (*RANS*) equations for the turbulent compressible Newtonian flow, employing *k-ω SST* as turbulence model [7]. To limit the numerical diffusion, second-order upwind numerical schemes have been used for the spatial discretization, except for the pressure equation. In this case, PRESTO! scheme has been chosen since it is designed for flows involving steep pressure gradients. Second-order upwind schemes also for the turbulence model variables have been used. Gradients are evaluated by a least-squares approach. The initialization has been performed by a two-step approach: (i) a hybrid initialization followed by a (ii) full multi-grid (*FMG*) scheme. Pressure-based solver with Coupled algorithm has been adopted since it is described by Croquer et al. [8] as far more stable than density based-solver nevertheless sufficiently accurate and suitable for high-velocity compressible flows. Pseudo-transient option was enabled, which was found to speed up the steady-state solution. R290 properties have been evaluated with the real-gas NIST database. Ejector inlets boundary conditions are prescribed in terms of total pressure and temperature, while the turbulence boundary conditions have been implemented as hydraulic diameter and the turbulent intensity (5% for the primary flows and 2% for the secondary one) [7]. Outlet condition has been modelled as a pressure boundary condition.

2.3. Multi-scale modelling

The above-described *CFD* model is used to solve the "*component-scale*" (viz., the entrainment ratio ω) and the fluid-dynamics ("*local-scale*"). Once such information is derived, the ejector component is then included within a refrigeration cycle, which is modelled by a *LPM* approach, to estimate the "*system-scale*" performances. The considered cycle has been the *SERS* architecture (Figure 2a) since it is the elementary scheme and, thus, its results could be extended to any other derived system. The advantage of such approach is to consider the fluid dynamics phenomena within the cycle performances. *SERS* input data concern (i) P_1, T_1, P_2, T_2 and P_3, which act as boundary conditions for the *CFD* simulations, and (ii) the mass flow rates $\dot{m}_1$ and $\dot{m}_2$ and T_3, which are the *CFD* model output. To model the *SERS*, the prevailing assumption concerns the absence of pressure losses within the heat exchangers, so to relate ejector boundary conditions to the cycle pressure and pressure levels (Figure 2a). Given the ejector boundary conditions, the thermodynamic points of the *SERS* cycle have been obtained; in particular, in this process, the pump has been modelled considering an isentropic efficiency equal to 0.9.[6]. After all thermodynamic points of the *SERS* cycle have bene obtained, the cooling power ($\dot{Q}_{evap}$), input thermal power ($\dot{Q}_{gen}$) and electrical power required by the pump ($\dot{W}_{pump}$) are computed using the mass flow rates $\dot{m}_1$ and $\dot{m}_2$, which are *CFD* model output:

$$\dot{Q}_{evap}=\dot{m}_2(h_2\text{-}h_5) \quad (1)$$

$$\dot{Q}_{gen}=\dot{m}_1(h_1\text{-}h_6) \quad (2)$$

$$\dot{W}_{pump}=\dot{m}_1(h_6\text{-}h_4) \quad (3)$$

Then the cycle *COP* is computed as:

$$COP=\frac{\dot{Q}_{evap}}{\dot{Q}_{gen}+\dot{W}_{pump}} \quad (4)$$

Coupling Eqs. (1-3) with Eq. (4), it follows:

$$COP=\omega\frac{h_2\text{-}h_5}{h_1\text{-}h_4} \quad (5)$$

Based on Eq. (5) it is worth noting that COP is related to ω, thus linking ejector performances ("*component-scale*") and the system *COP* ("*system-scale*").

3. Results

SERS performances have been computed based on the modelling approach described in Section 2.3, whereas the outcomes are presented in Figures 4-9. Figure 4 displays the ejector operating curves for the different spindle positions and the critical point operation is highlighted in red. Figure 5 displays the relationship between the spindle position and the primary and secondary mass flow rates as well as the generator and evaporator thermal powers (Eqs. (1-2)), for the ejector operated in the critical conditions. Figure 6 displays the relationship between the spindle position and ejector ω and P_{crit}, and system *COP* are presented. for the ejector operated in the critical conditions. Figure 7 displays the *SERS P-h* representations and Figures 8 and 9 display and the Mach contours of the flow fields within the ejector at the critical conditions.

Looking at Figure 4, it is noted that when moving the spindle towards the mixing chamber, ω is increased, but P_{crit} is decreased, limiting the maximum allowable condenser temperature. This effect can be explaining by considering that the spindle position affects the primary and secondary mass flow rates: the former decreases as the spindle moves towards the nozzle exit, whereas the latter is slightly increased (Figure 5). Indeed, shifting the spindle inside the nozzle, the nozzle throat area is reduced and so does the primary mass flow rate (related to the generator thermal power); the trend is not linear with the spindle position, as the nozzle throat area is ring-shaped, hence the first positions of the spindle have a smaller impact on ejector performances than the following ones. Although less pronounced than the primary mass flow rate change, the increase in secondary mass flow rates is caused by the different expansion of the primary jet (Figure 8 and Figure 9). An increase in the nozzle area ratio leads the primary flow to a more pronounced contraction of the jet core, which frees more cross-section area to secondary flow entrainment. The concurrent increase of secondary flow rate and decrease of the motive one results in a noticeable improvement of the ejector entrainment ratio (Figure 6). It should be noted that the higher entrainment performance is countered by a lower P_{crit}, (Figure 4 and Figure 6) which is found to be reduced when the spindle is moved towards the nozzle exit position: moving the spindle from 0 to 7 mm ω is increased by 85.8% while the P_{crit} is reduced by 283 kPa. The above-reported improvement of the ejector entrainment ratio, moving the spindle toward the nozzle exit position, can also be observed by considering the changes in the thermodynamic representation of the cycle (Figure 7, comparing *SERS P-h* at critical conditions for *SP#0* and *SP#7*).

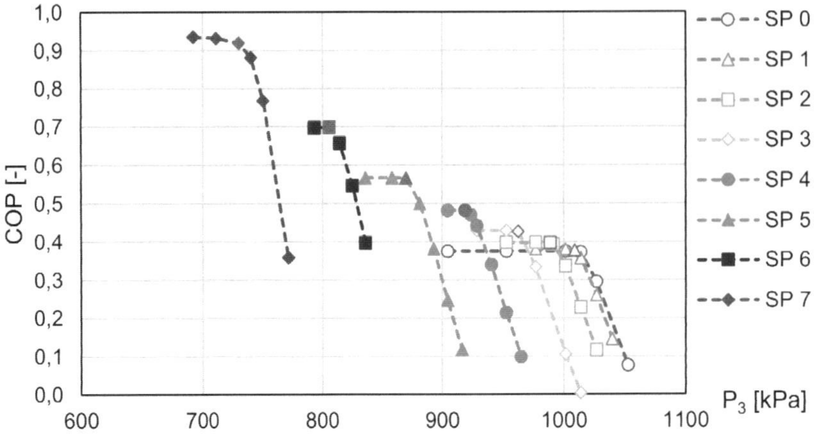

Figure 4. Ejector operating curves for the different spindle positions

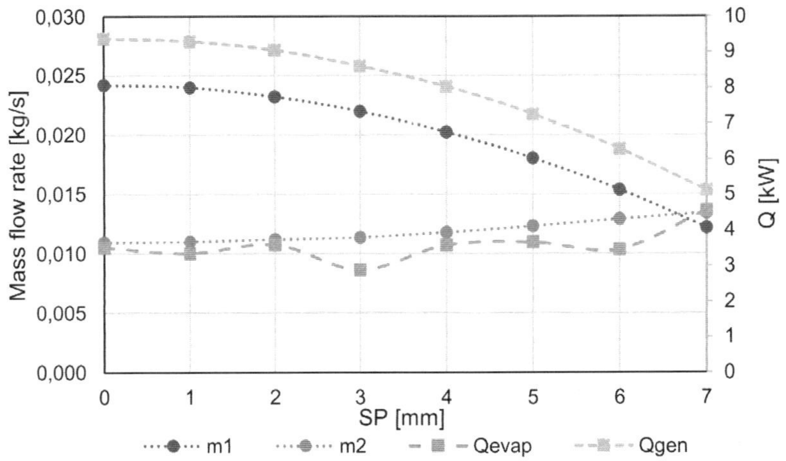

Figure 5. The effect of the Spindle position on mass flow rates and thermal power (critical condition)

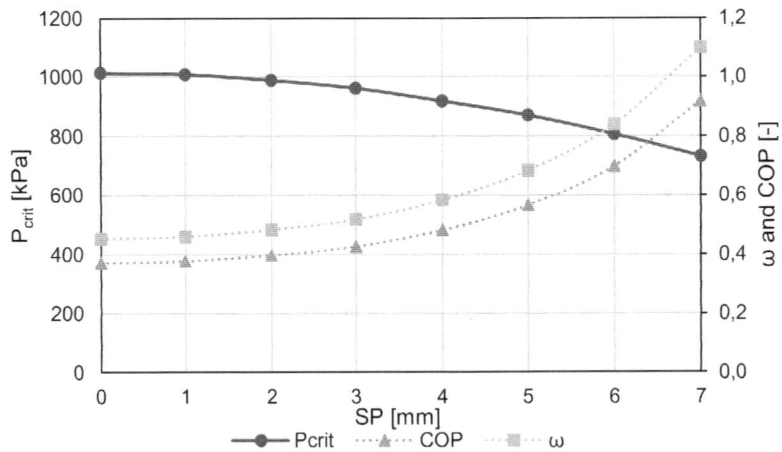

Figure 6. The effect of the Spindle position on P_{crit}, ω and COP (critical condition)

The different spindle positions affect the two main performance parameters of ejectors, namely ω and $P_{crit,}$ which influence the system performance, in turn. *ω* directly influences the *COP* of the cycle, as it can be concluded by observing the *COP* definition as a function of the cycle properties. P_{crit}, instead, not only affects the maximum allowable condenser temperature of the system for which the ejector works at its best efficiency in the double-chocking operating mode but has a lower impact on the cooling capacity, too (Figure 7), the lower condenser temperature for the ejector used with *SP#7* mm (owing to the lower P_{crit}), results in a different positioning of point# 4,5,6 which are shifted toward the left part of the *P-h* diagram. This causes the specific generator input and the specific cooling capacity to increase with the result that their ratio; It should also be noted that the cooling power, is also increased because of the higher secondary mass flow rate $\dot{m}_2$.

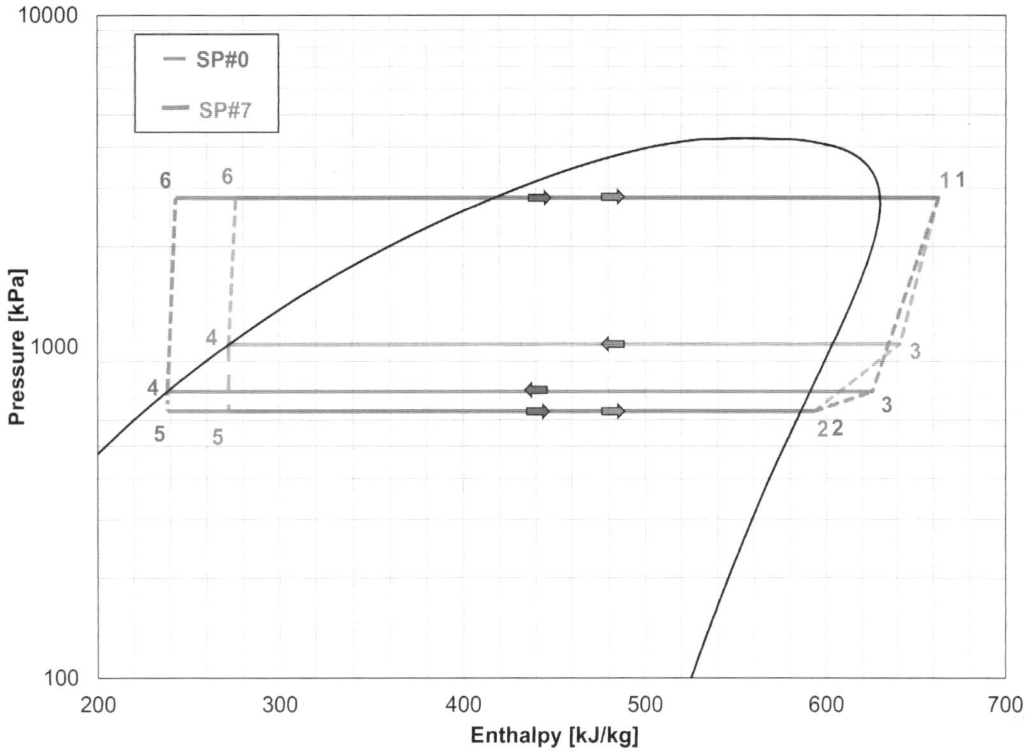

Figure 7. *SERS P-h* representations of *SP#0* and *SP#7* at the ejector critical conditions (Figure *4*)

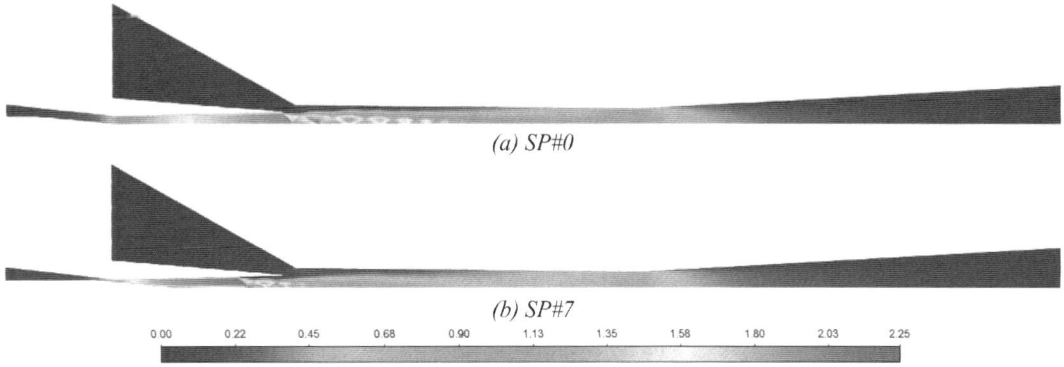

Figure 8. Ejector flow fields (Mach number contours) *of SP#0 and* SP#7 at the ejector critical conditions (Figure 4)

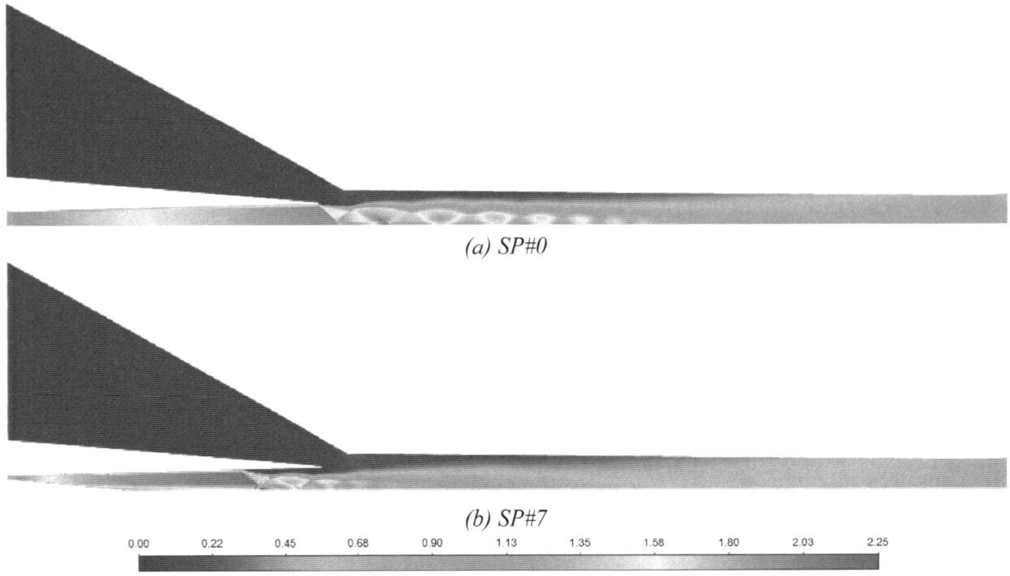

(a) SP#0

(b) SP#7

Figure 9. Ejector flow fields (Mach number contours) *of SP#0 and* SP#7 at the ejector critical conditions (Figure 4) – closer look at the nozzle exit position

4. Conclusions

In this paper, a multi-scale numerical study of a *VGE* operated with R290 has been presented. A *CFD* model has been used to solve the fluid-dynamics within the ejector ("*local-scale*") and to provide the ejector performances ("*component-scale*"). Conversely, the *SERS* performance ("*system-scale*") has been modelled by a *LPM* approach using as input data *CFD* outcomes. It is noted that, when moving the spindle towards the mixing chamber, ω is increased because of the primary flow rate reduction, but P_{crit} is decreased, limiting the maximum allowable condenser temperature. In conclusion, using a moving spindle control system might ensure an improvement of the ejector performance; this is of practical use, for example, when the condenser saturation pressure decreases (i.e., owing to a lower solar heat source). Future works will be devoted to developing dynamic look-up tables to be implemented in a variable ejector control system.

5. Acknowledgment

Authors are grateful to Nicolò Cristiani, for the meaningful discussions and for his valuable support.

6. References

[1] Aidoun Z, Ameur K, Falsafioon M and Badache M 2019 *Inventions* **4** 15
[2] Besagni G, Mereu R and Inzoli F 2016 *Renew. Sust. Energ. Rev.* **53** 373-407
[3] Knobloch F, Pollitt H, Chewpreecha U, Daioglou V and Mercure J F 2019 *Energ. Effic.* **12**(2) 521-550
[4] Cavallini A 2020 *J. Phys. Conf. Ser.* **1599** 012001
[5] Del Valle J G, Jabardo J S, Ruiz F C and Alonso J S J 2014. *Int. J. Refrig.* **46** 105-113
[6] Besagni G, Cristiani N, Croci L, Guédon G R and Inzoli F 2021 *Appl. Therm. Eng.* **186** 116502
[7] Besagni G, Cristiani N, Croci L, Guédon G R and Inzoli F 2021 *Appl. Therm. Eng.* **186** 116431
[8] Croquer S, Poncet S and Aidoun Z 2016 *Int. J. Refrig.* **61** 140-52

Convective heat transfer CFD analysis of forced flow through a half-stadium serpentine mini-channel at low Reynolds number

M CORTI[1], C LA TERRA[1,2], C FANCIULLI[2] and A NIRO[3]

(1,3) Politecnico di Milano, Department of Energy,
Via Lambruschini 4, 20156 Milano, Italy
(2) CNR – ICMATE, Via Previati 1/E, 23900 Lecco, Italy

(3) Corresponding author e-mail: alfonso.niro@polimi.it

Abstract. Spiral and serpentine channels are a very interesting solution to enhance heat transfer in small or miniaturized heat exchangers. In order to properly design these devices, of course, it is essential having a good knowledge of heat transfer characteristics of forced flow through this kind of ducts. However, the data base in open literature is rather lacking being investigations concerning flow mainly restricted to Dean number less than 500 and specific geometric parameters. In order to overcome these limits, a CFD analysis of a laminar forced flow through serpentine mini-channels has been started using a commercial code at finite volumes, namely Ansys Fluent. In this paper, after discussing the preliminary CFD analysis carried out on a straight pipe to tune the tool, we present the results on a flow through a half-stadium serpentine mini-channel with Reynolds number ranging between 200 and 2000. As thermal boundary conditions, a uniform wall temperature is assumed on the flat wall of the half-stadium duct, whereas the others are adiabatic. Finally, simultaneously developing flow condition is adopted. Heat transfer performances and pressure drops have been compared with ones in the straight tube, with the same length and hydraulic diameter as the serpentine duct. The results show heat transfer capability increases due to the presence of curvature as well as an earlier transition from laminar to turbulent flow.

Key Words: *forced convection, curved ducts, serpentine, mini-channels*

1. Introduction

Heat transfer enhancement is mainly driven by the need of saving energy as well as of making heat exchangers more efficient or with more compact size. This paper presents a CFD analysis of a laminar forced flow through serpentine mini-channels (Figure 1) applied in a miniaturized thermoelectric generator (TEG) [1]. Among various developments of this technology there is the will to make these TEGs portable. The size reduction involves the reduction of the surfaces designed for heat transfer, therefore arises the need to find geometries that allow to optimize the overall dimensions and, at the same time, allow an adequate heat transfer. A solution can be found using curved mini-channels, in fact the presence of the curvature plays an important role in increasing heat transfer [2].

Figure 1: Serpentine mini-channels applied in TEG

As is known in literature [2], when a fluid enters a curved tube, it is pushed from the centre towards the walls of the duct due to centrifugal forces as a results of the presence of the curvature itself. The fluid consequently creates a pair of secondary flows, commonly known as Dean vortices [3]. It is thanks to the presence of these secondary flows, which promote the mixing of the fluid between the bulk region and the layers close to the walls, that an increase in the heat transfer is obtained even in laminar flows. However, the fluid motion in the curved ducts causes, in addition to an increase in heat transfer, an increase in pressure drops. It is necessary to evaluate an acceptable compromise between these two factors.

Curved channels have been studied for a long time in literature regarding classical geometries such as helical coiled tubes, sinusoidal serpentine tubes, Archimedean spirals, twisted tubes etc. but they can also be miniaturized for in-plane applications [4] and assume various shapes for which there are however no generic correlation in literature. Correlations are linked to strict constraints both geometric and related to the working fluid, and range of applicability [5], [6].

Given our field of application with Dean numbers $De = Re\sqrt{D/2R_c}$ between about 200 and 2000, there are no suitable correlations literature. Therefore, a numerical model has been developed to conduct this characterization.

2. Problem Configuration and Setup

Numerical simulations have been carried out using the software Ansys Fluent based on finite volume method. This allowed an accurate reproduction of the studied geometry as can be seen in Figure 2.

The solution domain is a 3D serpentine with a half-stadium cross section, it has been analysed with Re ranging between 200 and 2000. The working fluid is water, incompressible, with constant specific heat being far from the boiling point.

As the value of the critical Reynolds number in the present work is not known a priori, emerged the need for a mesh suitable for the study of both laminar and turbulent flow, that could capture the characteristics in both flow regimes.

The mesh created is a butterfly grid, it is a structured grid in which are simultaneously applied a Cartesian grid in the centre of the duct and a cylindrical O-grid, i.e. axial symmetric, close to the wall, around the Cartesian. Its effectiveness for modelling fluid motion within circular cross section ducts has been demonstrated in the literature [7]–[9] and this mesh has been adapted to suite an half-stadium section (Figure 3). This mesh ensures a y^+ equal to 1 at the wall to solve the laminar sublayer. To obtain a grid independent solution has been carried out a sensitivity analysis, choosing a grid with $1.6 * 10^6$ elements.

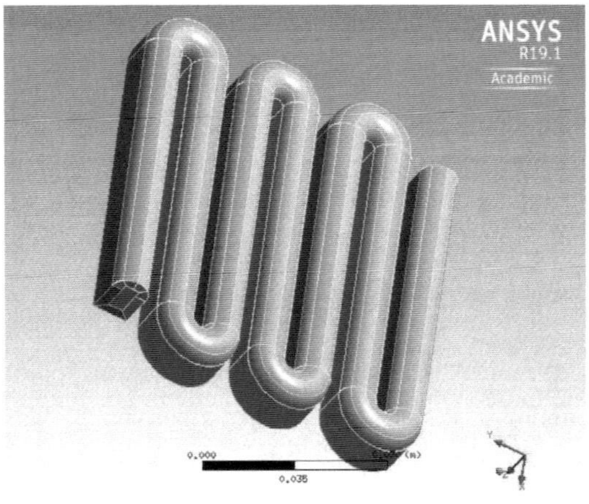

Figure 2: Serpentine mini-channel domain

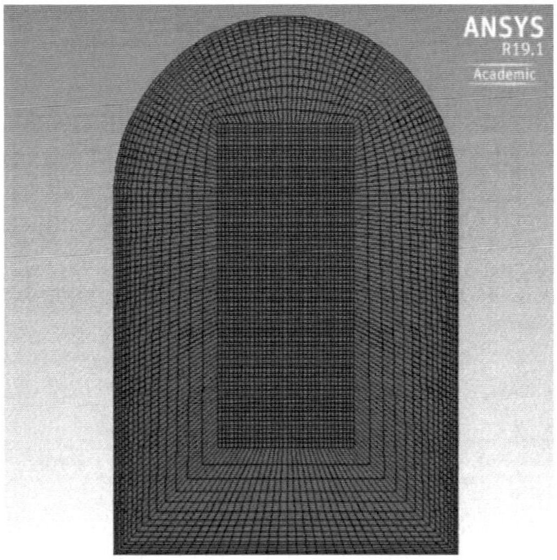

Figure 3: Adapted Butterfly grid for half-stadium cross section serpentine

The adopted mesh has the following quality parameters (Table 1).

	Element Quality	*Aspect Ratio*	*Orthogonal Quality*	*Skewness*
Mean value	0.810	0.980	0.949	0.166

Table 1: Mesh quality parameters for half-stadium cross section serpentine

2.1. Boundary Conditions

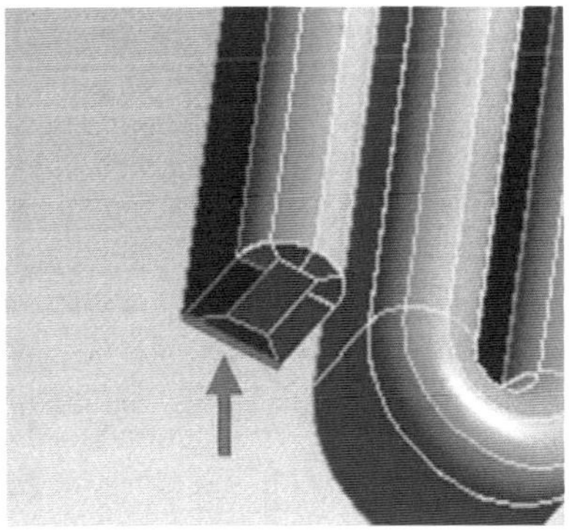

Figure 4: Heated wall at constant temperature

Inlet: mass flow inlet basing on the selected Reynolds number (for half-stadium cross section), $\dot{m} = f(Re)$. The working fluid, water, enters the channel at uniform temperature, $T_{in} = 293.15\ K$.

Outlet: pressure outlet $P_{out} = 0$ Pa .

Wall: no-slip velocity condition is applied for all the walls, moreover one wall (Figure 4) is heated at constant temperature, imposing $T_{wall} = 323.15\ K$, the other walls are diabatic, imposing $\dot{q} = 0\ W/m^2$.

2.2. *Solver setting and models*

The Coupled algorithm is used for pressure-velocity coupling in the case of simulations with laminar flow, while the Semi-Implicit Method for Pressure Linked Equation (SIMPLE) algorithm is used for simulations with turbulent flow.

Laminar flow has been solved with the laminar viscous model proposed by Fluent, while for turbulent flow the k-ω-SST model has been chosen between RANS turbulence models. It uses as default the Enhanced Wall Treatment (a wall function for the near-wall region i.e., viscous sublayer, buffer region and fully-turbulent outer region) [11].

Using from 15 to 30 cores Intel(R) Xeon(R) Gold 6148 CPU @ 2.40GHz, computational time has been of few hours per calculation.

Simulations stop at the level-out point in case of laminar model and at the round-off point (residuals lower than 10^{-6}) for RANS.

2.3. *Preliminary CFD analysis*

A preliminary CFD analysis on a straight pipe has been carried out to tune the tool.

For a straight pipe with circular cross section, the local Nusselt numbers for not fully developed laminar flow for all Reynolds analysed in the study have been numerically calculated and these values have been compared with the Nusselt numbers obtained through the Graetz problem at $Pr = \infty$ [12].

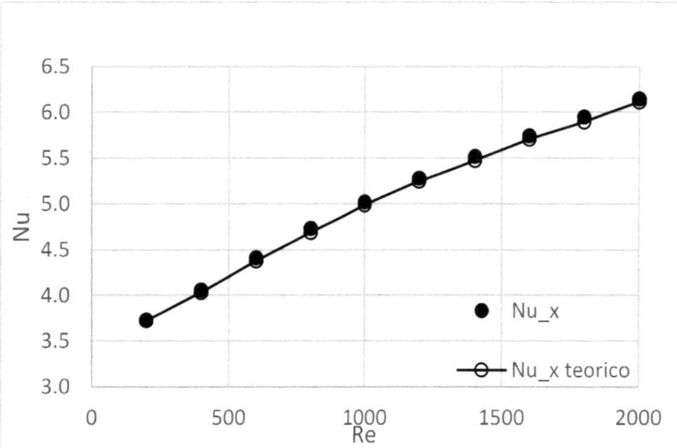

Figure 5: Local Nusselt number comparison for straight tube

The errors remained within 1% as can be seen in Figure 5. A similar procedure has been also followed in the case of a laminar fully developed flow, obtaining good agreement with the expected theoretical results.

Regarding the comparison of the average Nusselt numbers in laminar flow regime, the average Nusselt calculated through the numerical simulation has been compared with the Nusselt number calculated with the Shah-Hausen correlation [12] showing errors between 5 and 10%. Finally, a simulation for turbulent flow has been carried out and the errors on the Nusselt number resulted between 1 and 10%.

The friction factors $f_0 = (\Delta P_0/\rho)(D/L)(2/u^2)$ have been also compared with the theoretical values, obtained from the pressure drops deriving from the simulations. For laminar flow the errors remained within 1% basing on Darcy correlation [13] while for turbulent flow the errors resulted within 10% basing both on Blasius correlation and Kakac-Shah [14]. The only suitable correlation founded in literature for pressure drop for geometries similar to ours is for a circular section serpentine [15], the errors with respect to our results (for a circular section serpentine too) have been lower than 10%.

3. Results

Using Fluent the bulk temperature T_b and the heat flux q have been calculated as mean section variables, from which can be calculated the local convective heat transfer coefficient $h_x = q_x/(T_w - T_b)$ and the related Nusselt number at any section $Nu = h_x D/k$.

3.1. Earlier transition from laminar to turbulent

Regarding Nusselt number from the outlet section, the Figure 6 show $Log(Nu)$ vs $Log(Re)$ for every Reynolds number analysed.

At a certain Reynolds number, i.e. 1000, the values are more dispersed with respect to the previous ones. This lead to assume a transition in the flow regime. Another evidence was given by velocity and temperature profiles at the outlet sections too, at Re 1000 they change trend compared to the previous ones.

For Reynolds numbers affected by this phenomenon, the simulations have then been carried out according to the turbulence model k-ω-SST.

By plotting the new values of the Nusselt number in the previous graph, these points align themselves along a line with an angular coefficient different from the one of the line that interpolates only the Nusselt number calculated from laminar simulations (Figure 7).

A transition in the flow regime is confirmed, passing from laminar to turbulent around Re 1000.

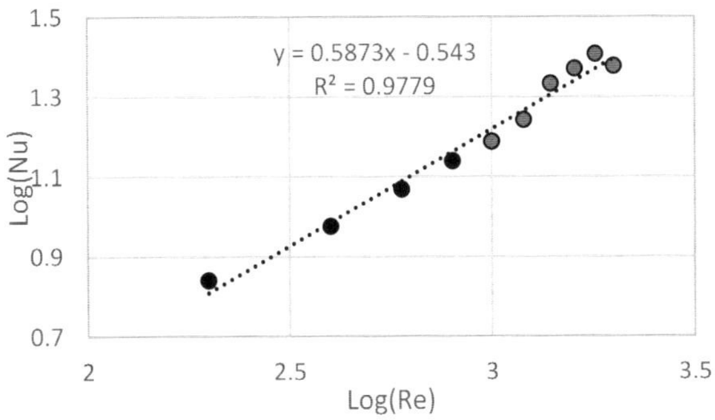

Figure 7: Nusselt numbers calculated from laminar simulation

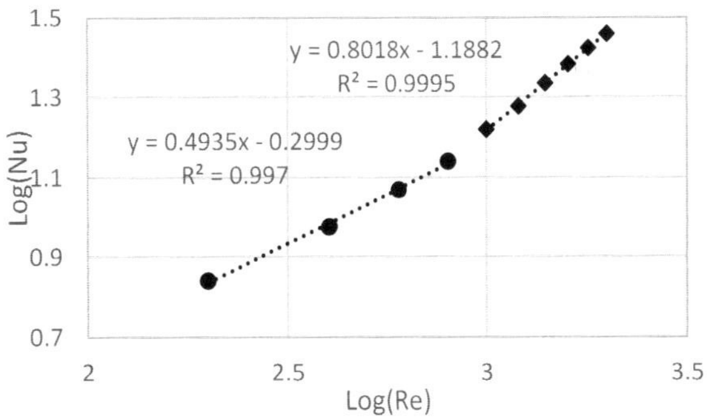

Figure 7: Nusselt numbers calculated with laminar (Re 200 - 800) and with k-ω-SST (Re 1000 - 2000)

3.2. Convective Heat Transfer Coefficient

For Reynolds number equal to 2000 the Figure 8 shows the temperature contour (taken on a plane close to the heated wall) of the fluid. It is clearly higher in the straight ducts and lower in the curved ducts of the serpentine: this implies a lower heat transfer coefficient for the straight ducts and a higher heat transfer coefficient for the curved ones. Some sections have been built along the entire coil to extract the local heat transfer coefficient (Figure 9).

Plotting the values of the local heat transfer coefficient calculated at any sections of the serpentine as a function of the length of the duct, it is visible the presence of a trend of peaks and troughs (Figure 10). In presence of curvatures the heat transfer is greater while in the straight duct parts it is lower.

Furthermore, as regards the first straight duct, away from the curvature, the value of its local convective heat transfer coefficient is lower compared to the value assumed in the subsequent straight sections as it has not yet been affected by the effects of the curvature.

This highlights that the curvature improves heat transfer.

Journal of Physics: Conference Series 2177 (2022) 012011 doi:10.1088/1742-6596/2177/1/012011

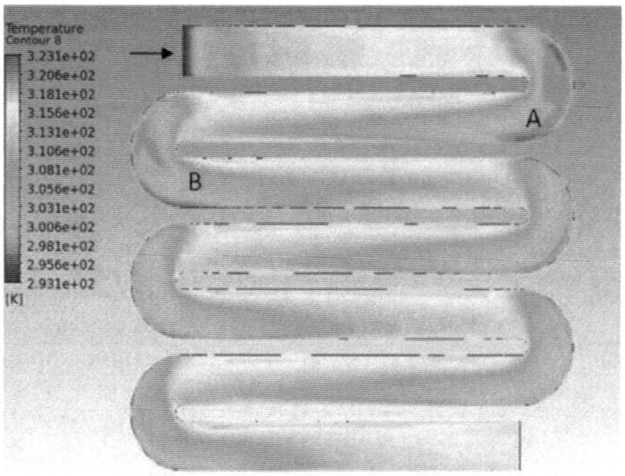

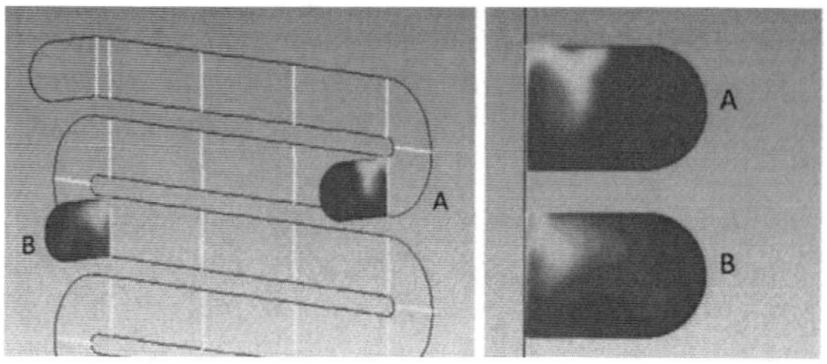

Figure 8: Sections and temperature contour

Figure 9: Temperature Contour

Given the fluctuating trend of the local convective heat transfer coefficient, an attempt has been made to provide a trend starting from the quantities of temperature and heat flux.

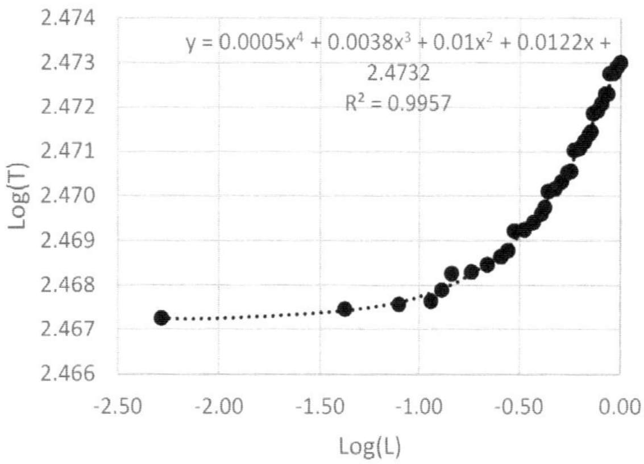

Figure 11: Local Temperature distribution

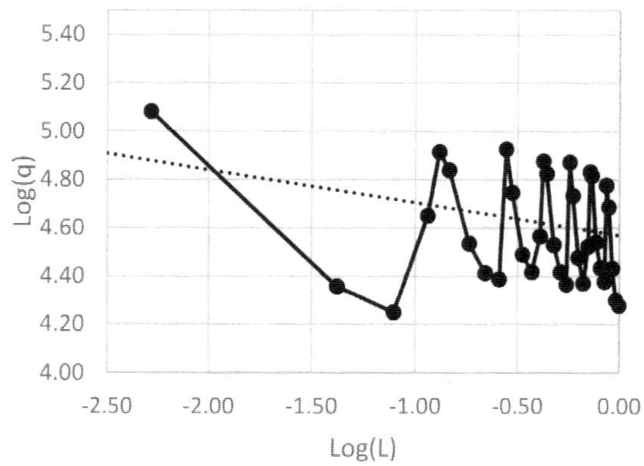

Figure 11: Local Wall Heat Flux distribution

The temperature distribution (Figure 11) in a Log-Log graph has been interpolated with a degree 4 polynomial law so as to maximize the coefficient of determination R^2.

The heat flux (Figure 12) assumes a non-monotonous trend due to the presence of curvatures; it cannot be interpolated with a polynomial function, therefore a linear interpolation has been used, albeit with a low coefficient of determination in order to provide an average trend.

By calculating the bulk temperatures and the heat flux using the trend provided, the interpolated convective heat transfer coefficient is calculated. Plotting this quantity in function of the length of the duct, a regular average trend has been obtained (Figure 10), gradually asymptotic as the x coordinate of the graph increases.

The condition of fully developed flow is reached in the duct.

Obviously it could not be possible to insert a straight duct in the TEG instead of the serpentine, but the values of the local convective heat transfer coefficient of a straight duct of same length as serpentine, same diameter and boundary conditions have been plotted in the graph (Figure 10). The local convective heat transfer coefficient of the straight tube is always smaller than the one of the serpentine.

Using the serpentine is achieved an advantage both in geometric and thermal terms.

Furthermore, by calculating the average convective heat transfer coefficient section by section and plotting the data series obtained (Figure 10), it can be seen how the values of the average convective heat transfer coefficient tend to overlap those of the local convective heat transfer coefficient deriving from the trend.

The average convective heat transfer coefficient has been computed as follows:

$$\bar{h} = \frac{1}{\sum \Delta A_x}\left(\sum_{x=1}^{n_{sez}} h_x * \Delta A_x\right) \quad .$$

As approximations, the local convective heat transfer coefficient interpolated at the outlet can be used as an estimation of the average convective heat transfer coefficient of the duct.

4. Conclusion

In conclusion, the curvature increases the heat transfer and anticipates the transition from the laminar to turbulent flow regime. Using curved mini-channels is possible to optimize heat transfer following the need of reducing size.

The serpentine mini-channel is an adequate solution for the cooling of the TEG, without affecting the electronic device.

The local convective heat transfer coefficient interpolated at the outlet of the serpentine mini-channel can be used as an estimation of the average convective heat transfer coefficient of the duct.

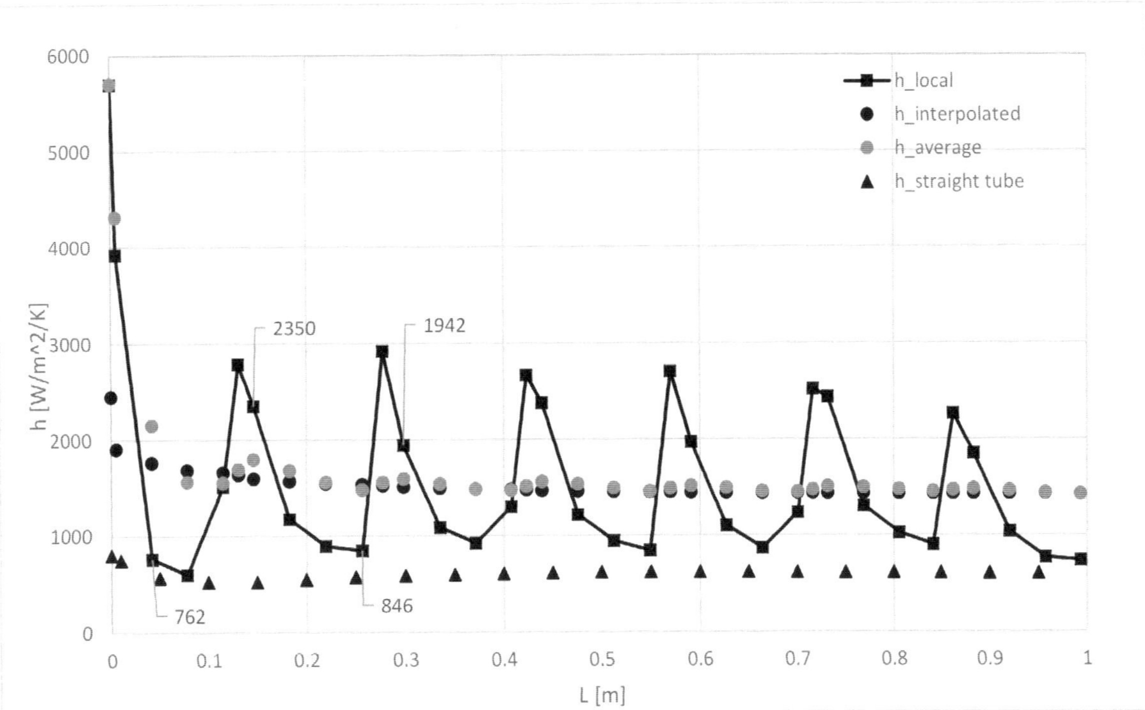

Figure 12: Local and Average Convective Heat Transfer Coefficient

Nomenclature

Symbol	Quantity	SI Unit	Symbol	Quantity	SI Unit
D	Equivalent Diameter	m	Rc	Curvature Radius	m
De	Dean Number	-	Re	Reynolds number	-
f	Friction factor	-	T_b	Bulk temperature	K
h	Convective heat transfer coefficient	W/m^2K	T_w	Wall temperature	K
k	Thermal conductivity	W/m/K	u	Bulk velocity (average)	m/s
$\dot{m}$	Mass flow rate	kg/s	y+	Dimensionless wall distance $y^+ = \Delta_{y_P}/v\sqrt{\tau_w/\rho}$	-
Nu	Nusselt number	-	τ_w	Wall shear stress	Pa
P	Pressure	Pa	ρ	Density	kg/m^3
$\dot{q}$	Wall heat flux	W/m^2	ν	Kinematic viscosity	m^2/s

References

[1] C. Fanciulli, H. Abedi, L. Merotto, R. Dondè, S. De Iuliis, and F. Passaretti, "Portable thermoelectric power generation based on catalytic combustor for low power electronic equipment," *Appl. Energy*, vol. 215, no. February, pp. 300–308, 2018.

[2] M. Ghobadi and Y. S. Muzychka, "A Review of Heat Transfer and Pressure Drop Correlations for Laminar Flow in Curved Circular Ducts," *Heat Transf. Eng.*, vol. 37, no. 10, pp. 815–839, 2016.

[3] W. R. Dean, "The stream-line motion of fluid in a curved pipe," *Philos. Mag.*, vol. 5, no. 7, pp. 208–223, 1928.

[4] P. Poredoš, U. Tomc, N. Petelin, B. Vidrih, U. Flisar, and A. Kitanovski, "Numerical and experimental investigation of the energy and exergy performance of solar thermal, photovoltaic and photovoltaic-thermal modules based on roll-bond heat exchangers," *Energy Convers. Manag.*, vol. 210, no. October 2019, p. 112674, 2020.

[5] P. L. Spedding, E. Benard, and G. M. McNally, "Fluid flow through 90 degree bends," *Dev. Chem. Eng. Miner. Process.*, vol. 12, no. 1–2, pp. 107–128, 2004.

[6] P. Naphon and S. Wongwises, "A review of flow and heat transfer characteristics in curved tubes," *Renew. Sustain. Energy Rev.*, vol. 10, no. 5, pp. 463–490, 2006.

[7] V. Hernandez-Perez, M. Abdulkadir, and B. J. Azzopardi, "Grid generation issues in the CFD modelling of two-phase flow in a pipe," *J. Comput. Multiph. Flows*, vol. 3, no. 1, pp. 13–26, 2011.

[8] V. Hernandez-Perez, "Gas-liquid two-phase flow in inclined pipes.," University of Nottingham, 2008.

[9] M. Colombo, A. Cammi, G. R. Guédon, F. Inzoli, and M. E. Ricotti, "CFD study of an air-water flow inside helically coiled pipes," *Prog. Nucl. Energy*, vol. 85, pp. 462–472, 2015.

[10] S. V Patankar and D. B. Spalding, "A calculation procedure for heat, mass and momentum transfer in three-dimensional parabolic flows," *Int. J. Heat Mass Transf.*, vol. I, pp. 1787–1806, 1972.

[11] I. ANSYS, *ANSYS Meshing User's Guide*. ANSYS, Inc., 2013.

[12] T. L. Bergman, A. S. Lavine, F. P. Incropera, and D. . Dewitt, *Fundamentals of Heat and Mass Transfer*. John Wiley & Sons, 2011.

[13] F. M. White, *Viscous Fluid Flow*. McGraw-Hill, 2006.

[14] S. Kakac, R. K. Shah, and W. Aung, *Handbook of Single-Phase Convective Heat Transfer*. John Wiley & Sons, 1987.

[15] M. Ciofalo and M. Di Liberto, "Fully developed laminar flow and heat transfer in serpentine pipes," *Int. J. Therm. Sci.*, vol. 96, pp. 248–266, 2015.

Macroscopic model for porous media in oscillating flow for CFD applications

Armando Di Meglio*, Nicola Massarotti, Vincenzo Trombetta

Department of Engineering, University of Naples "Parthenope" Centro Direzionale, Isola C4, 80133

*armando.dimeglio001@studenti.uniparthenope.it

Abstract. In this paper a macroscopic model for porous media in oscillating flow is derived for CFD applications. In non -linear modelling of thermoacoustic devices, their cores, useful for the energy conversion, cannot be, in general, modelled like a traditional porous media through friction factor and Nusselt number correlations, recovered from steady state. In this work the analytical solutions in the frequency domain of the linearized Navier-Stokes equations are spatially averaged for an oscillating flow between parallel plates case to build a macroscopic model of an equivalent porous media through a complex permeability and Nusselt number. The results, obtained by assuming a zero mean temperature gradient, show that there is a good agreement between macroscopic solution of a porous box (representing a parallel plates stack) and the average of the solutions of oscillating flow between parallel plates, both in terms of temperature and velocity fields.

Introduction

The porous media have large range of application in real life and industrial applications. Among them, there is the field of thermo-acoustics, where a porous core plays the role of energy conversion from acoustic energy into heat and vice-versa [1]. Generally, because of random and complex physical geometries of porous media, a local solution of velocity and temperature field prohibitively expensive. A macroscopic approach is generally used to model the porous media, even, theoretically, with unsteady flow [2]. This method allows to calculate the average velocity field and temperature field on a representative elementary volume (REV)[3]. The average of the exact governing equations produces new unknowns (spatial fluctuations) which must be expressed as functions of the average field itself, through coefficients fitted from experimental data or numerical simulations of the microscopic scale. Most of macroscopic model for porous media applies to steady state flow. Instead in thermo-acoustics, theoretically, this model cannot be ever applied because flow is oscillating. Macroscopic model resumed from steady flow data have been applied to porous cores (regenerator) of the so called "travelling wave" thermoacoustic devices where the hypothesis of quasi-steady state is assumed [4],[5],[6],[7]. In other word, friction factor (for pressure drop) and Nusselt number (for heat transfer) correlations are considered valid at each instant of time [8]. In the core of "standing wave" thermoacoustic devices, the above assumptions cannot be considered as valid. In fact, in CFD applications, the velocity and temperature fields are solved from a microscopic point a view [9],[10], [11], [12], [13].

In this paper, the authors want to build a first macroscopic model that can be ever applied (for example to thermoacoustic stacks) in oscillating flow regime, resuming it from analytical solutions of linearized Navier-Stokes equations in the frequency domain. The proposed model, compared to those available in the literature, that are recovered from the steady-state flows, is based on frequency-dependence correlations for pressure drop and heat transfer terms. They intrinsically take into account a possible time phase-shift between the variables. A verification of the macroscopic approach is proposed by comparing its solution, in terms of velocity and fluid temperature field, with the corresponding average of the same fields, gained from microscopic solutions of oscillating flow between parallel plates in 2D, under the assumption of mean axial temperature gradient and isothermal walls. The main advantage is related to the possible application of such a model in more complex setups (for example a full standing wave device, where the interactions of different components like heat exchangers, resonators, full stack want to be studied) where the use of a microscopic approach is not feasible due to the much higher computational costs.

1. Theory

1.1. Governing equations of the microscopic scale

The microscopic scale of a fluid flow problem coupled with heat transfer is governed, in general, by the unsteady and compressible Navier Stokes system of equations, which are continuity (1), momentum (2) and energy equation (3), written as follows [14]:

$$\frac{\partial \rho}{\partial t} + \nabla \cdot (\rho \boldsymbol{v}) = 0 \tag{1}$$

$$\frac{\partial (\rho \boldsymbol{v})}{\partial t} + \nabla \cdot (\rho \boldsymbol{v}\boldsymbol{v}) = -\nabla p + \mu \nabla^2 \boldsymbol{v} + \frac{\mu \nabla (\nabla \cdot \boldsymbol{v})}{3} \tag{2}$$

$$\frac{\partial (\rho e + \frac{1}{2}\rho |\boldsymbol{v}|^2)}{\partial t} = -\nabla \cdot \left[\rho \boldsymbol{v} \left(\frac{1}{2} |\boldsymbol{v}|^2 + h \right) - k_f \nabla T - \boldsymbol{v} \cdot \boldsymbol{\tau} \right] \tag{3}$$

where $\rho, \boldsymbol{v}, \mu, e, h, \boldsymbol{\tau}, t$ are respectively the density, velocity, dynamic viscosity, internal energy, enthalpy, viscous stress tensor and time. The FEM (Finite Element Method) microscopic solution of the oscillating flow problem between parallel plates is clearly based on the equations system written above, where any sub-model for the porous medium is not required.

1.1.1. Linearized equations and solutions. To linearize the previous system of equations, each variable has to be written as the sum of a mean time average, indicated with 0 subscript, and fluctuation part (whose mean time average is zero by definition) indicated with 1 subscript. The time-derivative operator is replaced with $i\omega$. In linear thermoacoustics, the mean time of the velocity is zero. The air particles, with reference to average on an acoustic cycle, are at rest. Therefore, the velocity has only the fluctuation term, while the mean pressure term is spatially uniform (if the pressure was not uniform there would be an average component of velocity). By substituting each variable thus written in (2), (3), neglecting the product between two fluctuations (valid until the amplitude of each first order variable is much less than the mean time value) the previous system and the derivative along the propagation axis x (for theory of the order magnitude the length scale λ, acoustic wavelength, is much smaller than oscillating viscous and thermal penetration depths δ) is written as follows [15]:

$$i\omega \rho_0 u_1 = -\frac{dp_1}{dx} + \mu \frac{\partial^2 u_1}{\partial y^2} \tag{4}$$

$$\rho_0 c_p \left(i\omega T_1 + u_1 \frac{dT_0}{dx} \right) = i\omega p_1 + k \frac{\partial^2 T_1}{\partial y^2} \tag{5}$$

where p and T are pressure and temperature field. The solutions of (4) and (5) are obtained using zero wall boundary conditions at the parallel walls [15]:

$$u_1(y) = \frac{i}{\rho_0 \omega}\frac{dp_1}{dx}(1-h_v) \tag{6}$$

$$T_1(y) = \frac{p_1}{\rho_0 c_p}(1-h_k) - \frac{1}{\rho_0\,\omega^2}\frac{dp_1}{dx}\frac{dT_0}{dx}\frac{(1-h_k)-\mathrm{Pr}(1-h_v)}{1-Pr} \tag{7}$$

where h_v and h_k are "shape" functions, also known as viscous and thermal functions, depending on the geometry and fluid properties; Pr is Prandtl number. In the case of parallel plate [15]:

$$h_v = \frac{\cosh\left[\frac{(1+\mathrm{i})\mathrm{y}}{\delta_\mathrm{v}}\right]}{\cosh\left[\frac{(1+\mathrm{i})\mathrm{y}_0}{\delta_\mathrm{v}}\right]}, \qquad h_k = \frac{\cosh\left[\frac{(1+\mathrm{i})\mathrm{y}}{\delta_k}\right]}{\cosh\left[\frac{(1+\mathrm{i})\mathrm{y}_0}{\delta_k}\right]} \tag{8}$$

where δ_v and δ_k are the viscous and thermal penetration depths:

$$\delta_v = \sqrt{\frac{2v}{\omega}}, \qquad \delta_k = \sqrt{\frac{2k}{\rho_0 c_p \omega}} \tag{9}$$

1.2. Macroscopic approach

1.2.1. Momentum equation. The most generalized macroscopic version of the momentum equation (2) corresponds to the (2.51) equation of the Vafai's Handbook of Porous Media [3]. The interfacial force in terms of Darcy velocity, involves contributions from both shear forces and hydrodynamic pressure. Most of research efforts focuses on the determination of the closure coefficients for interfacial forces to obtain the permeability at low Reynolds number (Darcy's law) and the inertial coefficient (Forchheimer effects) in steady flow condition. In oscillating flow, only few experiments were conducted on packed column at low frequency, where all the non-dimensional curves collapse on the steady state correlation. At high frequency, velocity and pressure drop are not in phase because there is pressure gradient component shifted by 90°respect to velocity. The focus of the authors is on the Darcy friction term, where the relation between pressure drop and velocity field is linear. That means the focus is on the determination of the permeability.

In a steady state flow, according to the Darcy law the relation between the pressure gradient and the Darcy velocity ($\boldsymbol{v_D} = \phi\langle \boldsymbol{v}\rangle$, with ϕ *porosity*) is:

$$\nabla p = -\frac{\mu}{\kappa}\boldsymbol{v_D} \tag{10}$$

In frequency domain, two observations are worth making. The first, pressure gradient and velocity may not be phased (as eq. (6) suggests), so the previous equation cannot be directly applied for oscillating quantities at each instant of time. The second, in frequency domain the pressure gradient includes a purely inertial contribution due to oscillating flow, which is not caused by viscosity. Therefore, it is useful to write the pressure gradient, by inverting (6) and isolating the term $i\omega\rho_0\langle u_1\rangle$ (purely inertial pressure gradient) from the Darcy term.

$$\frac{dp_1}{dx} = -i\omega\rho_m\langle u_1\rangle - \frac{\mu}{\kappa_c}\langle u_1\rangle\phi \tag{11}$$

where the subscript c is used to distinguish the complex permeability used in frequency domain from the classical permeability. The comparison between (11) and the spatial average of (6) along the y coordinate allows to obtain the complex permeability.

$$\langle u_1 \rangle = \frac{i}{\rho_0 \omega} \frac{dp_1}{dx} (1 - f_v) \tag{12}$$

where f_v is the spatial average of h_v, $f_v = \tanh[(1+i)y_0/\delta_v]/[(1+i)y_0/\delta_v]$ for parallel plate stack. Therefore:

$$\kappa_c = \phi \delta_v^2 \frac{1 - f_v}{2i\, f_v} \tag{13}$$

So, the real part of the reciprocal of (13) is the corresponding permeability of a steady state flow, where pressure gradient and velocity are 180° degrees out of phase (opposite sign), while the imaginary part represents a new additional source term which takes into account the phase shift between pressure gradient and velocity. The "dynamic" permeability has been treated, among others, by Lafarge *et al.* [16].

$$DARCY\ FRICTION\ TERM = -\mu \Re\left[\frac{1}{\kappa_c}\right] u_D,\ NEW\ TERM = \frac{\mu}{\omega} \Im\left[\frac{1}{\kappa_c}\right] \frac{\partial u_D}{\partial t} \tag{14}$$

The time-domain macroscopic momentum equation, suitable for CFD-based simulations in porous media, where a Darcy velocity formulation is used, has to consider the classical Darcy friction term like in (10) and an additive "inertial term" (velocity and pressure gradient 90° degrees out of phase) which origins from the imaginary part of complex permeability. The latter is known as a "virtual mass term" in the Handbook of porous media. So, the x component of the macroscopic momentum equation inspired by (2.51) equation of the Vafai's Handbook of Porous Media [3] and "corrected" with the new term on the left of (15), market in red, with is:

$$\begin{gathered} \frac{1}{\phi} \rho \frac{\partial u_D}{\partial t} \left\{ \phi \Im \left[\frac{\delta_v^2}{2\kappa_c}\right] + 1 \right\} + \frac{1}{\phi^2} (\boldsymbol{v_D} \cdot \nabla) u_D + \nabla \cdot \left[-\langle p \rangle \boldsymbol{I} + \frac{\langle K \rangle}{\varphi} \right] + \Re\left[-\frac{\mu}{\kappa}\right] u_D, \\ K = \mu\left(\nabla u_D + \nabla u_D{}^{\mathrm{T}}\right) - \frac{2}{3} \mu \,(\nabla \cdot \boldsymbol{v_D}), \end{gathered} \tag{15}$$

1.2.2. Energy equation. A macroscopic non-equilibrium thermal model must deal with thermal tortuosity, thermal dispersion and interfacial heat transfer [3]. The thermal dispersion, is a second order term, neglected a priori in the linear thermoacoustic theory for stack/regenerators. The thermal tortuosity is zero, by definition, for a parallel plate configuration and therefore is outside our scope. The interfacial heat transfer is instead present in a non-equilibrium thermal model for porous media, as in our case because the walls of the solid matrix are considered isothermal. In this context, authors want to calculate the dimensionless interfacial heat transfer coefficient-Nusselt number- suitable to setup the macroscopic energy equation, resuming it from analytical solution in frequency domain.

By invoking the assumption of zero mean temperature gradient, equation (5) is:

$$\rho_0 c_p i\omega T_1 - i\omega p_1 = k \frac{\partial^2 T_1}{\partial y^2} \tag{16}$$

The previous equation (16) is integrated over the fluid domain and applying the diverge theorem to the right side, as made by Liu and Garret [17]:

$$V(\rho_0 c_p i\omega \langle T_1\rangle - i\omega p_1) = \int_{A_{fs}} k\nabla T \, dA_{fs} \tag{17}$$

Defining the hydraulic ratio as $V/A_{fs} = r_h$ and observing that the right-hand side is the heat flux that can be written as $Nu_c \frac{k}{r_h}(\langle T_{1s}\rangle - \langle T_1\rangle)A_{fs}$

$$(\rho_0 c_p i\omega \langle T_1\rangle - i\omega p_1) = Nu_c \frac{k}{r_h^2}(\langle T_{1s}\rangle - \langle T_1\rangle) \tag{18}$$

By spatially averaging (7):

$$\langle T_1\rangle = \frac{p_1}{\rho_0 c_p}(1 - f_k) \tag{19}$$

where f_k is thermal average function of h_k, as f_v for h_v. By replacing (19) in (18), and invoking the isothermal condition for solid walls $\langle T_{1s}\rangle = 0$, the complex Nusselt number Nu can be obtained:

$$Nu_c = 2i\left(\frac{y_0}{\delta_k}\right)^2 \frac{f_k}{(1 - f_k)} \tag{20}$$

So, the real part of complex Nusselt is the actual interfacial heat transfer coefficient used for modeling irreversible heat transfer between fluid and solid matrix in steady flow, while the complex part takes into account that heat flux $-k\partial T/\partial y|_{wall}$ and temperature difference $\langle T_{1s}\rangle - \langle T_1\rangle$ may not be in phase. The macroscopic version of energy equation has to include the classical term of interfacial heat transfer and another one coming from the phase shift between heat flux and temperature difference.

$$q_{sf}(INTER.HEAT.FLUX) = \frac{k}{y0^2}\Re[Nu](T - T_s), NEW\ TERM = \frac{k}{\omega y0^2}\Im[Nu]\frac{\partial(T)}{\partial t} \tag{21}$$

In (21) note T_s is a constant, by assumption. In the time-domain, the (2.56) of Handbook of Porous Media [3], the energy equation is "corrected" and made suitable for CFD applications in oscillating flow as follows, where the new term is marked in red:

$$\phi\rho_f c_p\left(1 + \frac{\delta_k^2\Im[Nu_c]}{2\phi y_0^2}\right)\frac{\partial\langle T\rangle}{\partial t} + \rho_f c_p \boldsymbol{v_D}\cdot\nabla\langle T\rangle - \nabla\cdot(\phi k\nabla\langle T\rangle) = \frac{\partial\langle p\rangle}{\partial t} + \boldsymbol{v_D}\cdot\nabla p + q_{sf} \tag{22}$$

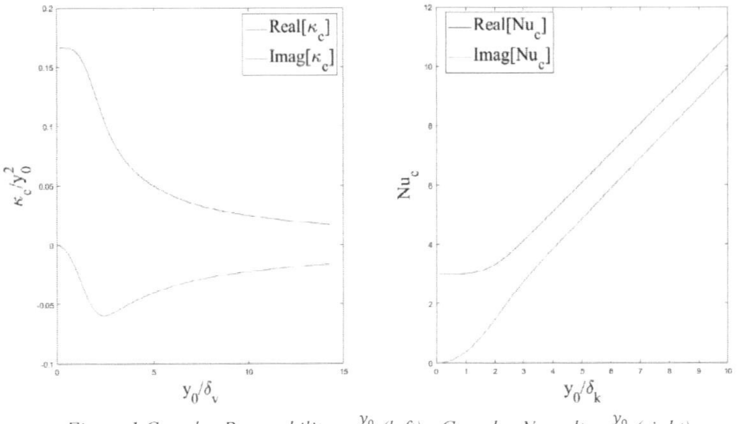

Figure 1 Complex Permeability vs $\frac{y_0}{\delta_v}$ (left) - Complex Nusselt vs $\frac{y_0}{\delta_k}$ (right)

The trends of (13) and (20) versus the frequency are respectively pictured on the left and on the right of the Figure 1. It is noted that for frequency tending to zero (steady-flows like condition), both the imaginary parts of the numbers, therefore also the red marked terms in the macroscopic equations, disappear, as it is expected.

2. Results

In the first part, the solution of microscopic scale in elementary cell of a parallel plate stack is compared with analytical solutions provided in the theoretical section 1.1.1. In the second part the macroscopic field of the equivalent porous media model described in section 2.2 is compared with the spatial average of the above microscopic solution. The computational domains of micro and macro scale are shown in the following Figure 2, while the boundary conditions in Table 1 Initial conditions are zero velocity, uniform temperature $T_0 = 300K$ and an initial pressure distribution $p = p_1 \cos(k_w x)$, with $k_w = 3.62m^{-1}, p_1 = 100Pa.$ 2) A commercial code based on FEM is adopted. In particular, a seconder scheme and a BDF scheme (backward differentiation formulas, second order) have been respectively employed for the spatial and time discretization. In Figure 3 the computational grids used for the calculations are pictured. For the microscopic configurations, the mesh size at the fluid-solid interface must be smaller than the viscous and thermal penetrations depths in eq. (9). A detailed mesh sensitivity analysis can be found in a previous work, even if for another scope, the same setup is analyzed [18].

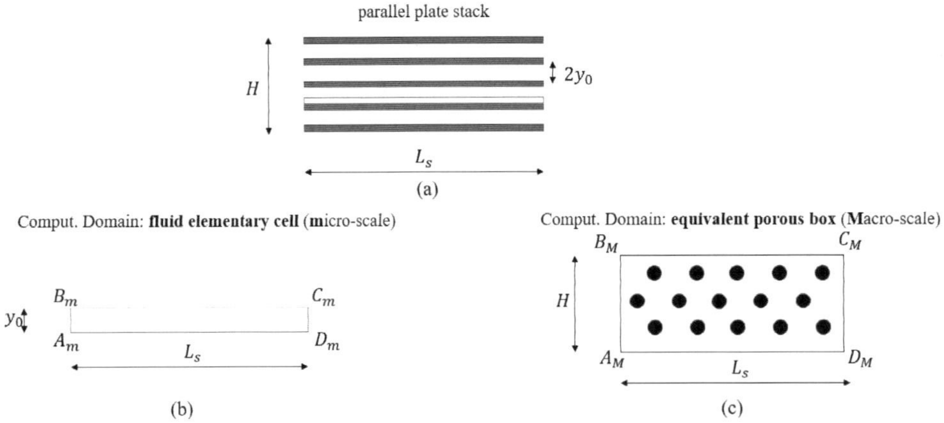

Figure 2 Physical domain (a), microscopic computational domain (b), macroscopic computational domain (c)

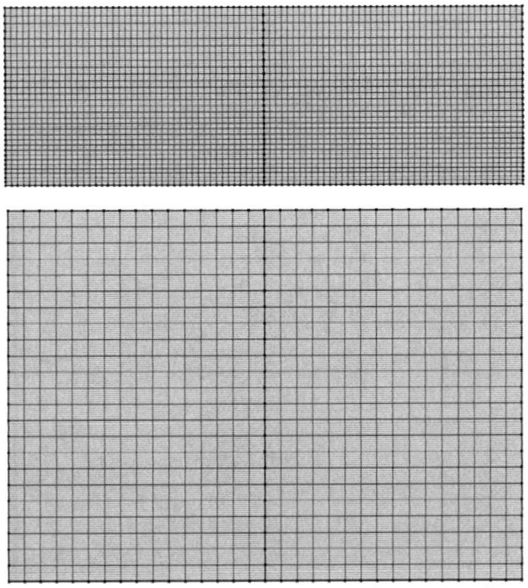

Figure 3 Computational grids for macroscopic (750 elements) and microscopic (2500 elements) setups

Table 1 Boundary conditions for microscopic e and macroscopic setups

	MICRO-SCALE	MACRO-SCALE
AB	Adiabatic, $p = p_1 \cos(\omega t)$	Adiabatic, $p = p_1 \cos(\omega t)$
BC	symmetry	symmetry
CD	Adiabatic, $p = p_1 \cos(\omega t) \cos(k_w L_s)$	Adiabatic, $p = p_1 \cos(\omega t) \cos(k_w L_s)$
DA	Isothermal wall	symmetry

2.1. Microscopic scale vs Analytical

In the following Figure 3 there is the comparison between velocity and analytical solution calculated along the height of half a channel ($y_0 = 5 \cdot 10^{-4} m$), for five different instants of time in the period $T = 1/f = 10^{-2} s$ ($f = 100Hz$). The fluid is air, considered as an ideal gas, with properties calculated at $300K$.

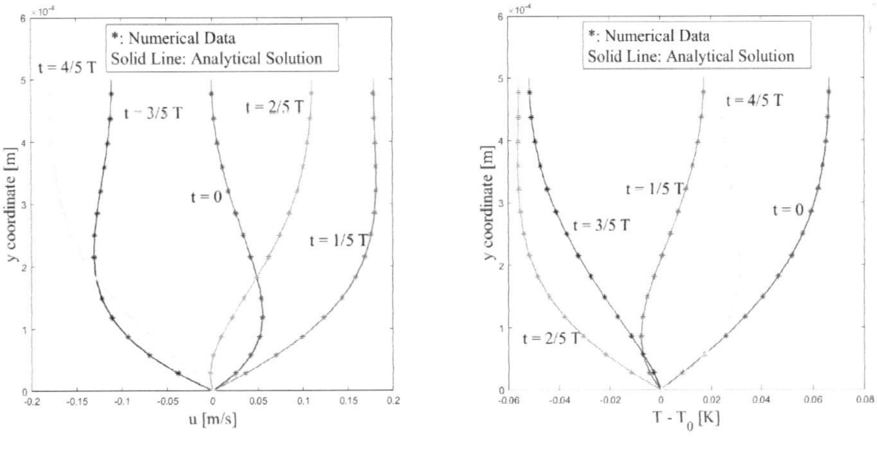

Figure 4 Comparison between analytical solutions and numerical solutions (micro-scale): velocity (left)-temperature (right)

2.2. *Macroscopic scale vs Microscopic scale*

After verifying the validity of microscopic model with analytical solution, the microscopic solution is spatially averaged (along the y coordinate) and therefore compared with the numerical solution of the macroscopic computational domain, modeled as "porous box" ($L_s = 0.03m$, $H = 0.019m$) with porosity $\phi = 0.5$, permeability ad Nusselt number calculated as in (13) and (20), being equal frequency, air properties, and length scale y_0 to the previous microscopic setup.

The agreement between the micro and macro scale solutions is very good. Therefore, in a similar problem, where the non-linear effects are negligible, the solution of the averaged allow the significantly reduce the computational costs, especially in presence of more REVs. On the other side, it is important to highlight that the proposed model is valid until the linearity assumption is valid (each first order variable subscripted with 1 much lower than the time average value subscripted with 0). Moreover, the mathematical closure of the non-linear macroscopic model is not a trivial issue especially when the system has not reached the so called the periodic oscillating flow.

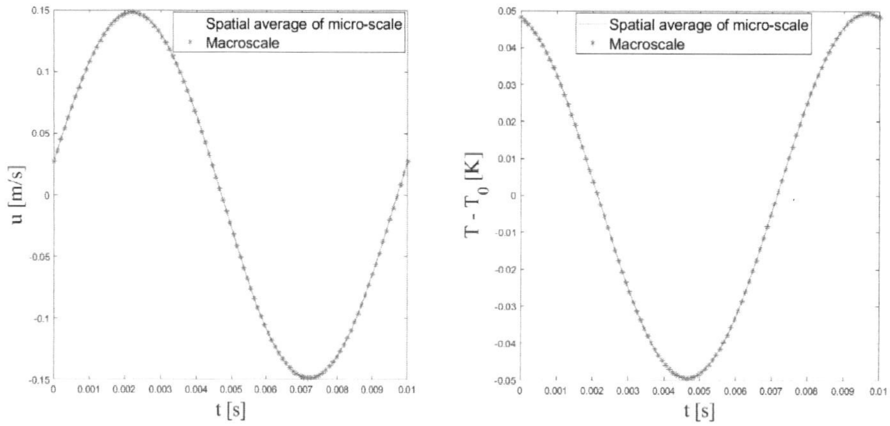

Figure 5 Comparison between macro and micro scale: velocity (left)-temperature (right)

Conclusion

In this paper a macroscopic model for porous media in oscillating flow is derived. By assuming the assumption of linearity, the Navier Stokes equation can be simplified in the frequency domain at the first order. Due to phase shift between pressure gradient and velocity, a complex permeability is needed to represent the effect of the forces exchanged between fluid and solid matrix. The real part of the inverse of permeability represents the classical Darcy term used in steady state modeling, while the imaginary part is an added contribution to the inertial forces (unsteady term of momentum equation). In the case of mean temperature gradient, the complex Nusselt number is derived and used to build the interfacial heat transfer term generally used in the energy equation for porous media with the assumption of non-thermal equilibrium. By analogy, the real part of complex Nusselt is used to the describe the classical interfacial heat transfer in the steady state modelling, while the imaginary part contributes the thermal inertia. The complex permeability and Nusselt number are specifically derived for an oscillating flow between parallel plates, but their formulation can be extended by changing the (f_v, f_k) functions for other geometries. The numerical solutions of the microscale, spatially averaged along the height of the channel and verified with analytical solutions, is compared to the solution of an equivalent porous box. The percentage error between microscale-numerical solutions and between micro and macro-scale is less than 1%. Further efforts will focus in the generalization of the macroscopic model, by including a mean temperature gradient, relaxing the isothermal hypothesis for the solid medium, and taking into account the Forchheimer effects end effects of the porous media, in order to apply it to a real thermoacoustic device.

References

[1] M.A.G. Timmer, K. De Blok, T.H. Van Der Meer, Review on the conversion of thermoacoustic power into electricity, Cit. J. Acoust. Soc. Am. 143 (2018) 841. https://doi.org/10.1121/1.5023395.
[2] D. Lasseux, F.J. Valdés-Parada, F. Bellet, Macroscopic model for unsteady flow in porous media, J. Fluid Mech. 862 (2019) 283–311. https://doi.org/10.1017/jfm.2018.878.
[3] K. Vafai, Handbook of porous media, second edition, CRC Press, 2005. https://doi.org/10.1201/9780824741501.
[4] G.W. Swift, W.C. Ward, Simple harmonic analysis of regenerators, J. Thermophys. Heat Transf. 10 (1996) 652–662. https://doi.org/10.2514/3.842.
[5] C. Scalo, S.K. Lele, L. Hesselink, Linear and nonlinear modelling of a theoretical travelling-wave thermoacoustic heat engine, J. Fluid Mech. 766 (2015) 368–404. https://doi.org/10.1017/jfm.2014.745.
[6] G.Y. Yu, E.C. Luo, W. Dai, J.Y. Hu, Study of nonlinear processes of a large experimental thermoacoustic- Stirling heat engine by using computational fluid dynamics, J. Appl. Phys. 102 (2007). https://doi.org/10.1063/1.2784997.
[7] J.A. Lycklama à Nijeholt, M.E.H. Tijani, S. Spoelstra, Simulation of a traveling-wave thermoacoustic engine using computational fluid dynamics, J. Acoust. Soc. Am. 118 (2005) 2265–2270. https://doi.org/10.1121/1.2035567.
[8] N.A.A. Qasem, S.M. Zubair, Compact and microchannel heat exchangers: A comprehensive review of air-side friction factor and heat transfer correlations, Energy Convers. Manag. 173 (2018) 555–601. https://doi.org/10.1016/j.enconman.2018.06.104.
[9] J. Lin, C. Scalo, L. Hesselink, High-fidelity simulation of a standing-wave thermoacoustic-piezoelectric engine, J. Fluid Mech. 808 (2016) 19–60. https://doi.org/10.1017/jfm.2016.609.
[10] G. Chen, L. Tang, B.R. Mace, Bistability and triggering in a thermoacoustic engine: A numerical study, Int. J. Heat Mass Transf. 157 (2020) 119951. https://doi.org/10.1016/j.ijheatmasstransfer.2020.119951.
[11] M.E.H. Tijani, S. Spoelstra, P.W. Bach, Thermal-relaxation dissipation in thermoacoustic systems, Appl. Acoust. 65 (2004) 1–13. https://doi.org/10.1016/j.apacoust.2003.07.002.
[12] F. Zink, J. Vipperman, L. Schaefer, CFD simulation of thermoacoustic cooling, Int. J. Heat Mass Transf. 53 (2010) 3940–3946. https://doi.org/10.1016/j.ijheatmasstransfer.2010.05.012.
[13] Z. Bouramdane, H. Hafs, A. Bah, M. Alaoui, N. Martaj, S. Savarese, Standing wave thermoacoustic refrigerator: The principle of thermally driven cooling, Proc. 2018 6th Int. Renew. Sustain. Energy Conf. IRSEC 2018. (2018) 1–6. https://doi.org/10.1109/IRSEC.2018.8702929.
[14] G.W. Swift, Thermoacoustics: A Unifying Perspective for Some Engines and Refrigerators, n.d.
[15] G.W. Swift, Thermoacoustic engines, J. Acoust. Soc. Am. 84 (1988) 1145–1180. https://doi.org/10.1121/1.396617.
[16] D. Lafarge, P. Lemarinier, J.F. Allard, V. Tarnow, Dynamic compressibility of air in porous structures at audible frequencies, J. Acoust. Soc. Am. 102 (1997) 1995–2006. https://doi.org/10.1121/1.419690.
[17] J. Liu, S.L. Garrett, Relationship between Nusselt number and the thermoviscous (Rott) functions, J. Acoust. Soc. Am. 119 (2006) 1457–1462. https://doi.org/10.1121/1.2165000.
[18] A. Di Meglio, E. Di Giulio, R. Dragonetti, N. Massarotti, Analysis of heat capacity ratio on porous media in oscillating flow, Int. J. Heat Mass Transf. (2021). https://doi.org/10.1016/j.ijheatmasstransfer.2021.121724.

A numerical approach to the fractional Laplacian operator with applications to quasi-geostrophic flows

G Barbi[1], D Capacci[1], A Chierici[1]*, L Chirco[1], V Giovacchini[1], S Manservisi[1]

[1]University of Bologna, DIN, Lab. of Montecuccolino, Via dei Colli 16, Bologna 40136, Italy

E-mail: andrea.chierici4@unibo.it

Abstract. The quasi-geostrophic models have been very successful for the study of oceanic and atmospheric dynamics in the mid-to-high latitude region of the earth where the Coriolis effect is significant. The governing equation of a quasi-geostrophic model is a transport equation with a fractional dissipation term. Although fractional operators have become a topic of great interest in the research community, the numerical discretization of such operators is very challenging due to their non-local behavior. In this work, we propose the numerical approximations of the bounded fractional Laplacian on a finite element discretization. In particular, we rely on the Riesz method and use a semi-analytical technique to approximate all the integrals involving the interaction between the inner local and the outer region. We test the implemented algorithm with numerical benchmarks, and we apply it to quasi-geostrophic flows. All the presented simulations are computationally expensive, due to the non-local behavior of the fractional Laplacian. For this reason, a parallel implementation of the numerical code has been developed.

1. Introduction

The quasi-geostrophic model is normally used to study oceanic and atmospheric dynamics in the mid-to-high latitude region of the earth when the Coriolis effect may be significant. Charney introduced the three-dimensional model which has been developed by several other authors, for example see [1, 2, 3] and references therein.

The main assumption of the quasi-geostrophic model is that the potential vorticity is uniform along with the geostrophic flow [4, 5]. This assumption, given x_1, x_2, y_1, y_2, z_1 and a generic surface $\hat{\mathcal{S}} = \{(x, y, z) \in \mathbb{R}^3 : x \in [x_1, x_2], y \in [y_1, y_2], z = z_1\}$, implies that the vorticity $\hat{\psi}(x, y, z, t)$ evaluated above $\hat{\mathcal{S}}$ satisfies $\Delta\hat{\psi} = 0$ for $z > 0$ and $\lim_{z\to\infty} \hat{\psi}(x, y, z, t) = 0$. The potential temperature on $\hat{\mathcal{S}}$ is defined as $\theta = \partial_z \hat{\psi}|_{z=0}$.

By considering the restriction of the domain to a finite two-dimensional surface, the potential vorticity on the surface $\psi(x, y, t) = \hat{\psi}(x, y, 0, t)$ and potential temperature $\theta(x, y, t)$ conservation, the model corresponds to the surface quasi-geostrophic (SQG) equation [6]. If we restrict this to the finite surface $\mathcal{S}$, the model reduces to the elliptic partial differential equation $(-\Delta)^{\frac{1}{2}}\psi = \theta$.

The fractional Laplace operator is currently used in many scientific and engineering applications, including fracture mechanics, turbulence, image processing and stochastic processes. Recently, many works have been published on the nontrivial task of the numerical implementation of fractional operators, see for example [7, 8, 9, 10] and references therein. For example in [10], the reader can find the numerical simulation of quasi-geostrophic flows using a spectral approach.

In this paper, we present a numerical approach to the quasi-geostrophic flows for the above model equation that involves a non-local operator in the form of a Riesz fractional Laplace operator where we introduce a semi-analytical technique and calculate directly the double integral over a bounded domain [11].

This article is organized as follows. We introduce the numerical technique and simulate the fractional Laplace operator in section 2. In section 3, we describe the quasi-geostrophic physical model and apply the numerical techniques introduced in the section before. In section 4, some simple numerical results of quasi-geostrophic flows with homogeneous Dirichlet boundary conditions are presented for different imposed initial conditions.

2. The fractional Laplace operator

In this section, we introduce the fractional operators and the fractional Sobolev spaces. Given a generic n-dimensional domain $\Omega \subset \mathbb{R}^n$, the Sobolev space of order $s \in (0,1)$ restricted to the domain Ω is denoted by

$$H^s(\Omega) = W^{s,2}(\Omega) = \left\{ u \in L^2(\Omega) : \frac{|u(\boldsymbol{x}) - u(\boldsymbol{y})|}{|\boldsymbol{x}-\boldsymbol{y}|^{\frac{n}{2}+s}} \in L^2(\Omega \times \Omega) \right\} . \tag{1}$$

The fractional Laplacian operator $(-\Delta)^s$ on the introduced Sobolev space can be written as

$$(-\Delta)^s u(\boldsymbol{x}) = 2^{2s} \frac{s\,\Gamma(\frac{n+2s}{2})}{\pi^{n/2}\,\Gamma(1-s)} \; P.V. \int_{\mathbb{R}^n} \frac{u(\boldsymbol{x}) - u(\boldsymbol{y})}{|\boldsymbol{x}-\boldsymbol{y}|^{n+2s}} d\boldsymbol{y}, \qquad 0 < s < 1 \,, \tag{2}$$

where $P.V.$ means "in the principle value sense" [7] and $\Gamma(\cdot)$ is the Gamma function, defined as $\Gamma(k) = \int_0^\infty t^{k-1} e^{-t} dt$ [8]. We consider the steady-state non-local diffusion equation

$$(-\Delta)^s u(\boldsymbol{x}) = f(\boldsymbol{x}) \qquad \text{on } \Omega \,, \tag{3}$$

with homogeneous Dirichlet boundary conditions $u(\boldsymbol{x}) = 0$ for $\boldsymbol{x} \in \partial\Omega$. The weak formulation of the introduced fractional diffusive problem can be written as

$$2^{2s} \frac{s\,\Gamma(\frac{n+2s}{2})}{2\pi^{n/2}\,\Gamma(1-s)} \int_{\mathbb{R}^n} \int_{\mathbb{R}^n} \frac{u(\boldsymbol{y}) - u(\boldsymbol{x})}{|\boldsymbol{x}-\boldsymbol{y}|^{n+2s}} \left(v(\boldsymbol{y}) - v(\boldsymbol{x})\right) d\boldsymbol{y}\, d\boldsymbol{x} = \int_\Omega f v \, d\boldsymbol{x} \,, \quad \forall v \in H^s_\Omega(\mathbb{R}^n) \,. \tag{4}$$

To restrict this system on a bounded domain Ω we set $u(\boldsymbol{x}) = 0$ and $v(\boldsymbol{x}) = 0$ for all $\boldsymbol{x} \in \mathbb{R}^n \setminus \Omega$.

2.1. Numerical modeling of Riesz fractional Laplacian

We introduce now a finite element based numerical approach to the fractional Laplacian operator, and in particular to the diffusive problem in (4). We first define the kernel

$$\mathcal{K}(\boldsymbol{x},\boldsymbol{y}) = \frac{u(\boldsymbol{y}) - u(\boldsymbol{x})}{|\boldsymbol{x}-\boldsymbol{y}|^{n+2s}} \left(v(\boldsymbol{y}) - v(\boldsymbol{x})\right) ,$$

and then split the integrals over $\mathbb{R}^n$ as

$$\begin{aligned} \int_{\mathbb{R}^n} \int_{\mathbb{R}^n} \mathcal{K}(\boldsymbol{x},\boldsymbol{y}) \; d\boldsymbol{y} d\boldsymbol{x} &= \int_\Omega \int_\Omega \mathcal{K}(\boldsymbol{x},\boldsymbol{y}) \; d\boldsymbol{y} d\boldsymbol{x} + \\ &\int_{\mathbb{R}^n \setminus \Omega} \int_\Omega \mathcal{K}(\boldsymbol{x},\boldsymbol{y}) \; d\boldsymbol{y} d\boldsymbol{x} \; + \int_\Omega \int_{\mathbb{R}^n \setminus \Omega} \mathcal{K}(\boldsymbol{x},\boldsymbol{y}) \; d\boldsymbol{y} d\boldsymbol{x} \,, \end{aligned} \tag{5}$$

since over $\mathbb{R}^n \setminus \Omega$ we have $\int_{\mathbb{R}^n \setminus \Omega} \int_{\mathbb{R}^n \setminus \Omega} \mathcal{K}(\boldsymbol{x}, \boldsymbol{y})\, d\boldsymbol{y} d\boldsymbol{x} = 0$. Moreover, for the integrand symmetry, $\int_{\Omega} \int_{\mathbb{R}^n \setminus \Omega} \mathcal{K}(\boldsymbol{x}, \boldsymbol{y})\, d\boldsymbol{y} d\boldsymbol{x} = \int_{\mathbb{R}^n \setminus \Omega} \int_{\Omega} \mathcal{K}(\boldsymbol{x}, \boldsymbol{y})\, d\boldsymbol{y} d\boldsymbol{x}$, so that (4) becomes

$$\begin{aligned} & 2^{2s} \frac{s\,\Gamma(\frac{n+2s}{2})}{2\pi^{n/2}\,\Gamma(1-s)} \int_{\Omega}\int_{\Omega} \frac{u(\boldsymbol{y}) - u(\boldsymbol{x})}{|\boldsymbol{x}-\boldsymbol{y}|^{n+2s}} \big(v(\boldsymbol{y}) - v(\boldsymbol{x})\big)\, d\boldsymbol{y} d\boldsymbol{x} + \\ & + 2^{2s} \frac{s\,\Gamma(\frac{n+2s}{2})}{\pi^{n/2}\,\Gamma(1-s)} \int_{\Omega} u(\boldsymbol{x}) v(\boldsymbol{x}) \int_{\mathbb{R}^n \setminus \Omega} \frac{1}{|\boldsymbol{x}-\boldsymbol{y}|^{n+2s}}\, d\boldsymbol{y} d\boldsymbol{x} = \int_{\Omega} f v \, d\boldsymbol{x}\,. \end{aligned} \tag{6}$$

The first term of (6) leads to a numerical system characterized by a dense matrix. We split this term into its four different contributions. The stiffness matrix A_{ij} can be written as $A_{ij} = A_{ij}^{11} + A_{ij}^{12} + A_{ij}^{21} + A_{ij}^{22}$ to ease the numerical assemblying of the stiffness matrix. We remark that, since the kernel is a symmetric function in both $\boldsymbol{x}$ and $\boldsymbol{y}$, $A_{ij}^{11} = A_{ij}^{22}$ and $A_{ij}^{12} = A_{ij}^{21}$ [12]. Thus, for the stiffness matrix we have $A_{ij} = 2A_{ij}^{11} + 2A_{ij}^{12}$.

The second term of (6) can be implemented in many different ways. An approach based on the restriction of the integral over $\mathbb{R}^n \setminus \Omega$ to the boundary $\partial\Omega$ can be found in [11] and a similar approach is used in this work, based on both a numerical and an analytical approach to the mixed integral. In particular, over a fixed Gauss node $\boldsymbol{x}_g$ for the integration of the outer integral on Ω we compute a loop over the entire boundary of the numerical problem ($\partial\Omega$). For each $\boldsymbol{x}_g$, the integral over $\mathbb{R}^n \setminus \Omega$ is approximated through the sum of the integrals over the areas $A(el)$ enclosed between the boundary element and the half-lines connecting $\boldsymbol{x}_g$ with the extremes of the boundary element. We have

$$\int_{\mathbb{R}^n \setminus \Omega} \frac{1}{|\boldsymbol{x}_g - \boldsymbol{y}|^{n+2s}}\, d\boldsymbol{y} = \sum_{el=1}^{N_{bd}} \int_{A(el)} \frac{1}{|\boldsymbol{x}_g - \boldsymbol{y}|^{n+2s}}\, d\boldsymbol{y}\;, \tag{7}$$

where N_{bd} is the total number of the boundary elements. Each subdivision χ_i represents an approximation of the area $A(el)$, such that $\bigcup_{el} A(el) = \{\mathbb{R}^n \setminus \Omega\}$.

We implement an algorithm for the numerical resolution of the Riesz fractional Laplacian through (7) and all the above mentioned approximations introduced for the assembly of the stiffness matrix of the double integral over Ω.

3. The surface quasi-geostrophic (SQG) equations

In the framework of the Navier-Stokes equations for an incompressible and inviscid fluid under gravitational and Coriolis body force, it is possible to derive the SQG equation. The quasi-geostrophic assumption implies that the z component of the velocity vector w is negligible with respect to the components u and v along x and y, respectively. Thus, the SQG mathematical problem consists in finding $\boldsymbol{u} = \{u, v, 0\}$ and p such that

$$\begin{aligned} & \frac{\partial u}{\partial t} + u\frac{\partial u}{\partial x} + v\frac{\partial u}{\partial y} = g_x + 2\omega v \sin\varphi - \frac{1}{\rho}\frac{\partial p}{\partial x}\,, \\ & \frac{\partial v}{\partial t} + u\frac{\partial v}{\partial x} + v\frac{\partial v}{\partial y} = g_y - 2\omega u \sin\varphi - \frac{1}{\rho}\frac{\partial p}{\partial y}\,, \\ & g_z - \frac{1}{\rho}\frac{\partial p}{\partial z} = 0\,, \\ & \frac{\partial u}{\partial x} + \frac{\partial u}{\partial y} = 0\,, \end{aligned} \tag{8}$$

where φ is the latitude, ω is the angular velocity (earth rotation $7.272 \cdot 10^{-5} rad/s$) and $\boldsymbol{g} = \{g_x, g_y, g_z\}$ is the gravitational field.

We introduce the geopotential $\Psi(x,y,z,t)$ and the potential temperature $\theta(x,y,z,t)$. The geopotential Ψ is defined as the work done by the gravitational force required to lift a unit air mass at a given height h. The potential temperature θ is the temperature of a unit air mass brought adiabatically and reversibly from its initial state to a standard pressure p_0. These two quantities satisfy the hydrostatic relation $\partial\Psi/\partial z = g\theta/\theta_0$, where θ_0 is taken at the reference configuration. Usually, the potential temperature θ satisfies the energy balance equation and Ψ is expressed in terms of spherical harmonics functions [13].

The system (8) becomes

$$\begin{aligned} &\frac{\partial u}{\partial t} + u\frac{\partial u}{\partial x} + v\frac{\partial u}{\partial y} - f_0 v = -\frac{\partial \Psi}{\partial x}\,, \\ &\frac{\partial v}{\partial t} + u\frac{\partial v}{\partial x} + v\frac{\partial v}{\partial y} + f_0 u = -\frac{\partial \Psi}{\partial y}\,, \\ &\frac{\partial \Psi}{\partial z} = \frac{g\theta}{\theta_0} \qquad \frac{\partial \theta}{\partial t} + \boldsymbol{v}\cdot\nabla\theta = 0\,, \\ &\frac{\partial u}{\partial x} + \frac{\partial v}{\partial y} = 0\,, \end{aligned} \tag{9}$$

with $f_0 = 2\omega\sin\varphi$. We can write $\boldsymbol{u}$ in terms of the geopotential Ψ after the adimensionalization of (9) and by considering small variations around an equilibrium point in the z direction as

$$\boldsymbol{u} = \nabla^{\perp}\Psi, \qquad \nabla^{\perp}\Psi = \Big(-\frac{\partial\Psi}{\partial y}, \frac{\partial\Psi}{\partial x}\Big)\,. \tag{10}$$

Therefore, SQG systems can be studied by using only the geopotential and the potential temperature. Considering now the boundary restriction over the surface Ω of the quantities, as a result of the trace theorem, the ordinary Laplace operator is replaced by the fractional Laplace operator [14]

$$\begin{aligned} &\frac{\partial\theta}{\partial t} + \boldsymbol{u}\cdot\nabla\theta + \kappa(-\Delta)^s\theta = 0\,, \\ &\boldsymbol{u} = \nabla^{\perp}\Psi\,, \\ &(-\Delta)^{\frac{1}{2}}\Psi = -\theta\,, \end{aligned} \tag{11}$$

where $s \in \big[\frac{1}{2}, 1\big)$ and $0 < \kappa \ll 1$.

4. Numerical results

In this section, we first present some numerical results to validate the algorithm used for the numerical simulation of the Riesz fractional Laplacian. Numerical results of the SQG equations discretized in a finite element framework are then presented.

4.1. Validation of the fractional Laplacian algorithm

To test the presented numerical Riesz model, we compare our results with the two-dimensional reference values presented in [11]. Thus, we consider the two-dimensional domain $\Omega = \{x \in [-1,1], y \in [-1,1]\}$, meshed with 16×16 square cells. We solve the simple scalar equation $(-\Delta)^s\theta = f$, considering both a constant $f = 1$ and a variable $f = \sin(\pi x)\sin(\pi y)$ source terms.

In Figure 1, we compare the potential temperature θ_R obtained with the implementation of the Riesz model with the reference results. In particular, in Figure on the left we report the solution of the equation $(-\Delta)^{0.25}\theta = 1$ with homogeneous Dirichlet boundary conditions along the $y = 0$ axis of the Ω domain. On the right, we report the comparison between the reference and the obtained solution of $(-\Delta)^{0.25}\theta = \sin(\pi x)\sin(\pi y)$ with homogeneous boundary

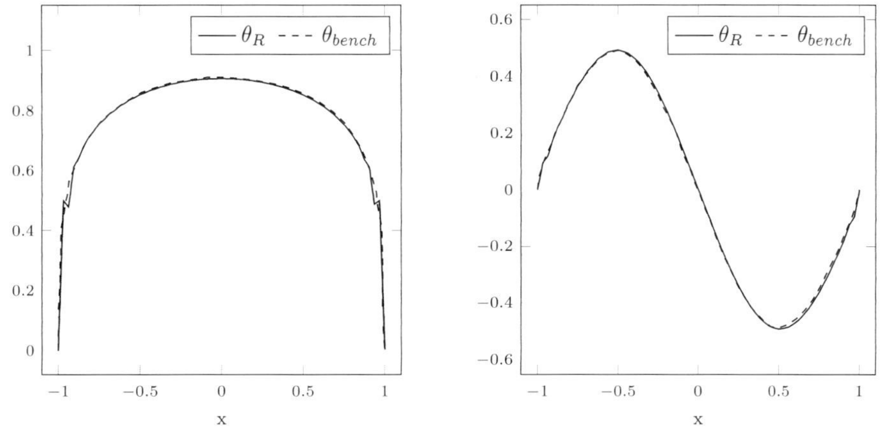

Figure 1: Comparison between the fractional Laplacian test case solutions for $s = 0.25$, with $f = 1$ (left) and $f = \sin(\pi x)\sin(\pi y)$ (right).

conditions. Our results obtained with the implemented algorithm are similar to the reference solutions in both cases, with the only exception of some wiggles near the boundaries of the domain. This effect is due to the numerical error arising from the numerical implementation of the external integration reported in equation (7). The presented finite element method for the numerical modeling of the fractional Laplacian is now applied to develop numerical models for quasi-geostrophic flows.

4.2. Numerical results of quasi-geostrophic flows

To show the robustness of the implemented algorithm for quasi-geostrophic simulations, we present now some numerical simulations of the model (11) equipped with various initial conditions. For each of the following simulations, homogeneous Dirichlet boundary conditions are considered on all the boundaries. All the numerical simulations have been performed on a square domain $\Omega = \{x \in [-\pi, \pi], y \in [-\pi, \pi]\}$, with a mesh of 1325 quadratic elements (5401 dof). The values of κ and s are set to 0.001 and 0.5 respectively, and the time step is $0.05s$. Concerning the quasi-geostrophic flows, we introduce now the kinetic energy and the helicity physical quantities to evaluate the numerical simulations. In particular, we define the kinetic energy $\mathcal{K}(\theta)$ as

$$\mathcal{K}(\theta) = \frac{1}{2}\int_\Omega \theta^2(\boldsymbol{x}, t)\, d\boldsymbol{x}\,, \tag{12}$$

and the helicity $\mathcal{H}(\theta)$ as

$$\mathcal{H}(\theta) = -\int_\Omega \Psi(\boldsymbol{x}, t)\theta(\boldsymbol{x}, t)\, d\boldsymbol{x}\,. \tag{13}$$

For the numerical simulations, we first consider a single vortex, resembling cyclonic circulations within the atmosphere. In particular, we consider the initial condition $\theta_0(x, y) = e^{-4x^2-16y^2}$. In the following, we refer to this case as IC_1. With the previous initial condition, the time evolution of the field θ is represented in Figure 2, at times $t = 0s$, $t = 5s$, $t = 10s$ and $t = 15s$. It can be noticed that the lack of filaments or secondary vortices is a result of the small eccentricity imposed by the initial condition. For this reason, the single vortex tends to collapse into a circular configuration.

In our second case, we consider a single vortex with higher eccentricity, with the initial condition $\theta_0(x, y) = e^{-x^2-16y^2}$. From now on, we refer to this case as IC_2. With the same time

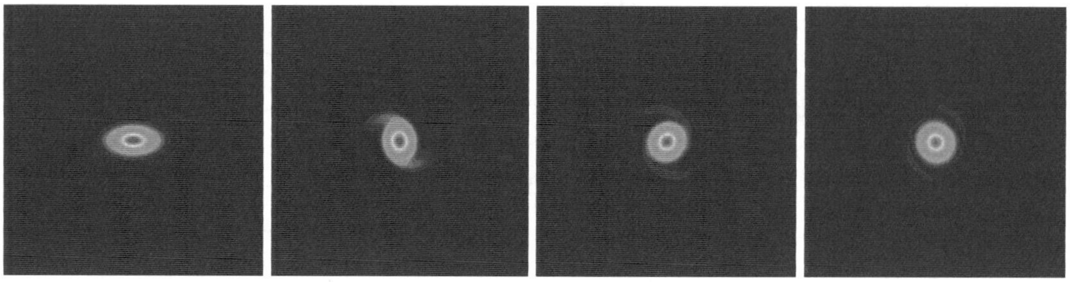

Figure 2: Evolution over time of the solution $\theta(x, y, t)$ with initial condition $\theta_0(x, y) = e^{-4x^2-16y^2}$ for (from left to right) $t = 0s$, $t = 5s$, $t = 10s$, $t = 15s$.

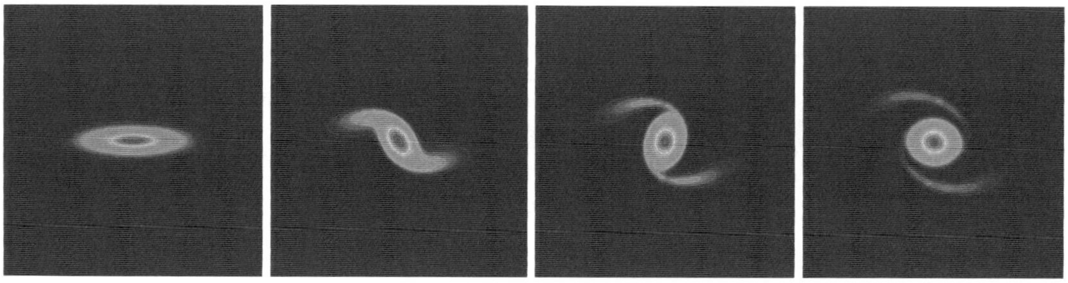

Figure 3: Evolution over time of the solution $\theta(x, y, t)$ with initial condition $\theta_0(x, y) = e^{-x^2-16y^2}$ for $t = 0s$, $t = 5s$, $t = 10s$, $t = 15s$.

steps reported above, in Figure 3, we present the time evolution of the imposed vortex. We can notice that the buoyancy turns into a spinning vortex characterized by the presence of thin filaments which tend to depart from the vortex center. This behavior is in accordance with the expectations.

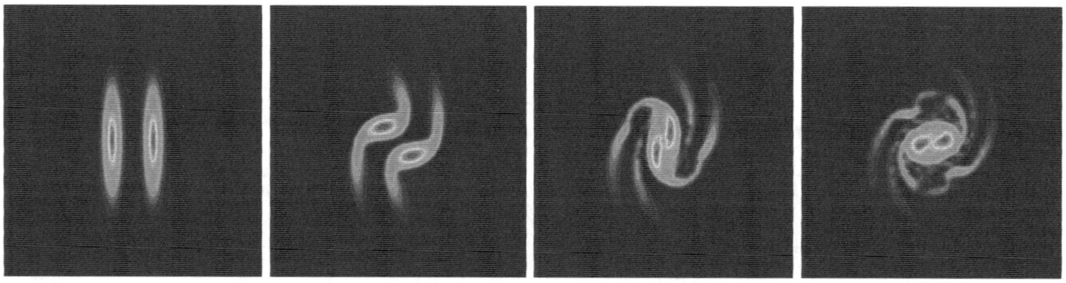

Figure 4: Evolution over time of the solution $\theta(x, y, t)$ with initial condition $\theta_0(x, y) = e^{-36(x-0.5)^2-y^2} + e^{-36(x+0.5)^2-y^2}$ for $t = 0s$, $t = 5s$, $t = 10s$, $t = 15s$.

Finally, we simulate the interaction between two different vortices, by considering the double Gaussian initial condition $\theta_0(x, y) = e^{-36(x-0.5)^2-y^2} + e^{-36(x+0.5)^2-y^2}$. We refer to this case as IC_3. In Figure 4, the interaction between the two vortices is shown at different times. At $t = 0s$, the two imposed gaussian fields do not influence each other. The two fields rotate as introduced in the cases IC_1 and IC_2, and after $t = 5s$ a first interaction between the two Gaussians can be observed. For $t \geq 10s$, the two vortices tend to merge and form a unique vortex. The development of a sharp layer between the two vortices can be noted. However, the vortices do not fully merge even though the intensity of the layer is reducing over time.

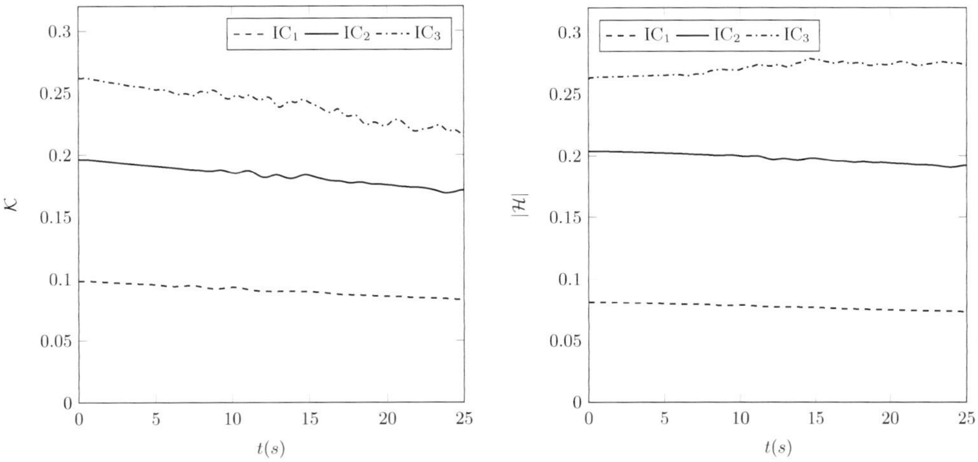

Figure 5: Kinetic energy $\mathcal{K}$ (left) and helicity $\mathcal{H}$ (right) in the three presented cases.

We propose now a comparison of the three presented initial conditions in terms of helicity and kinetic energy. In Figure 5 the kinetic energy and the helicity are plotted as a function of time t for each case. The cases IC_1 and IC_2 have similar linear trends. However, the kinetic energy $\mathcal{K}$ and the absolute value of the helicity $|\mathcal{H}|$ show larger values when the initial condition with the higher eccentricity is employed. This is an expected result since the rotation of the ellipse IC_2 generates a greater vortex. The interaction between the two vortices in the case IC_3 causes higher values and less regular evolution over time of $\mathcal{K}$ and $|\mathcal{H}|$.

The numerical simulation of the SQG problems requires high computational costs. Thus, it is important to develop a scalable parallel algorithm to reduce the time needed to have a solution. Since every processor needs information from all the other processors, the implementation of a parallel code for non-local assembly algorithms is not a straightforward task [12].

5. Conclusions

In this work, we have presented a technique for the numerical modeling of quasi-geostrophic flows based on a finite element discretization. Since the quasi-geostrophic models require the modeling of a fractional Laplacian operator, we have first implemented a parallel numerical algorithm based on the Riesz method to simulate such an operator. The implemented fractional Laplacian model has been tested through the comparison with benchmark results, showing consistent numerical results. By using this tool, numerical results of the implemented quasi-geostrophic model with homogeneous Dirichlet boundary conditions have been reported. Various initial conditions have been imposed, modeling the evolution of one or two vortices. The results are consistent with the expectations, and the developed algorithm can be considered a very interesting tool to simulate the quasi-geostrophic equations. However, further analysis on the stability and the convergence of the implemented model is necessary and it will be carried out in future works.

References

[1] Charney, Jule G *On the scale of atmospheric motions.* The Atmosphere — A Challenge. American Meteorological Society, Boston, MA (1990): 251-265

[2] Gill, Adrian E *Atmosphere—ocean dynamics.* Elsevier (2016).

[3] Lapeyre G, and Klein P *Dynamics of the upper oceanic layers in terms of surface quasi-geostrophy theory.* Journal of physical oceanography 36.2 (2006): 165-176.
[4] Charney, Jule G *Geostrophic turbulence.* Journal of the Atmospheric Sciences 28.6 (1971): 1087-1095.
[5] Pedlosky, J *Geophysical fluid dynamics.* Vol. 710. New York: Springer (1987).
[6] Constantin, P, Majda A J, and Tabak E *Formation of strong fronts in the 2-D quasigeostrophic thermal active scalar.* Nonlinearity 7.6 (1994): 1495.
[7] Di Nezza E, Palatucci G, and Valdinoci E. *Hitchhiker's guide to the fractional Sobolev spaces.* Bulletin des sciences mathématiques 136.5 (2012): 521-573.
[8] D'Elia M and Gunzburger M *The fractional Laplacian operator on bounded domains as a special case of the non-local diffusion operator.* Computers & Mathematics with Applications 66.7 (2013): 1245-1260.
[9] D'Elia M, Du Q, Glusa C, Gunzburger M, Tian X, and Zhou Z *Numerical methods for non-local and fractional models.* Acta Numerica, 29, 1-124.
[10] Bonito A, and Murtazo N. *Numerical Simulations of Surface Quasi-Geostrophic Flows on Periodic Domains.* SIAM Journal on Scientific Computing 43.2 (2021): B405-B430.
[11] Lischke A, et al. *What is the fractional Laplacian? A comparative review with new results.* Journal of Computational Physics 404 (2020): 109009.
[12] Aulisa E, Capodaglio G, Chierici A, and D'Elia M *Efficient quadrature rules for finite element discretizations of non-local equations*, Numer. Methods Partial Differ. Eq. (2021).
[13] Heiskanen, Weikko Aleksanteri. *Physical geodesy.* Determination of the Geoid from Ground Anomalies 8 (1967): 325-330.
[14] Ding, Zhonghai. *A proof of the trace theorem of Sobolev spaces on Lipschitz domains.* Proceedings of the American Mathematical Society 124.2 (1996): 591-600.

Journal of Physics: Conference Series 2177 (2022) 012014 doi:10.1088/1742-6596/2177/1/012014

NUMERICAL INVESTIGATION ON THERMAL BEHAVIOURS OF LATENT THERMAL ENERGY STORAGES WITH PCM PARTIALLY FILLED WITH ALUMINUM FOAM

Bernardo Buonomo, Oronzio Manca, Sergio Nardini, Renato Elpidio Plomitallo

Dipartimento di Ingegneria, Università degli Studi della Campania "Luigi Vanvitelli" via Roma 29, 81031 Aversa (Italy)

bernardo.buonomo@unicampania.it oronzio.manca@unicampania.it sergio.nardini@unicampania.it renatoelpidio.plomitallo@unicampania.it

Abstract. In this paper a numerical investigation on the effects of the metal foam into the Latent Heat Thermal Energy Storage System, based on a phase change material, is accomplished. The geometry of the system is a vertical shell and tube LHTESS made with two concentric aluminum tubes. The internal surface of the hollow cylinder is assumed at a constant temperature above the melting temperature of the PCM to simulate the heat transfer from a hot fluid. The other external surfaces are assumed adiabatic. The PCM is completely embedded in the volume between the two coaxial cylinders, it is pure paraffin wax with a low value of thermal conductivity and it melts in a range of temperatures. The metal foam is made of aluminum and it partially fills the volume starting from the internal cylinder with a thickness lower the differences between the external cylinder radius and internal one. The enthalpy-porosity theory is employed to simulate the phase change of the PCM and the metal foam is modelled as a porous media that follow the Brinkman-extended Darcy model. The governing equations are solved employing the Ansys-Fluent code. The results are shown in term of melting time, temperature and energy at varying of time and at varying the filling ratio.

1. Introduction

The environmental problems, the high cost of energy and the continuous and growing energy demand are important issues occupying a wide part of the scientific research. Nowadays the energy thermal demand is very variable both in temporal term and in the methods of use. This variation makes difficult to match the energy demand with the energy supply, so it is important to improve the energy storage systems to avoid wasting the excess energy. There are many types of thermal storage system (TES) among which the latent heat thermal energy storage systems (LHTESS) with phase change material (PCM) that represent the best choice because they work at quasi-constant temperature, they are very stable, and they have high energy density. They have a wide application field, both to high-temperature or low-temperature applications and they are indicated for solar thermal energy storage. Nevertheless, the main drawback is the low value of thermal conductivity of the PCM making the system useless for the thermal engineering applications. A way to resolve this problem is to combine the PCM with metal foam. Moreover, it can be interesting to realize a thermal storage with a fast-absorbing energy zone with a low thermal resistance and the external zone with higher thermal resistance to reduce the heat transfer losses toward the external ambient. This may be obtained opportunely combining the PCM and the metallic foam.

The latent heat thermal energy storage system (LHTESS) is based on the phase change materials (PCMs) [1,2] where the phase-change process can be utilized to absorb and release thermal energy related to high latent heat value of the material. However, the main weakness of Paraffin wax as PCM is the low thermal conductivity value and this leads the whole thermal storage system to be ineffective, from an engineering standpoint, because of the long and not-controlled time for melting. Such problem can be overcome with the use of the metal foams that seems to be a good solution to improve the thermal conductivity of the whole system. So, many authors have investigated the use of a metal foam with a PCM as Caliano et al. [3] that evaluated by numerical simulations the cooling energy charging and discharging processes of two different PCM-based cold thermal energy storage units. The first unit consisted of a cylindrical aluminum container partially filled with a biological PCM. The second one was similar, but it also included an aluminum metal foam to enhance the heat transfer inside the PCM. Also, Zhu et al. [4] studied the effect of the height ratio between the copper foam and the heat sink with PCM. The height of copper foam varies between 0 (no metal foam) and 20mm (the height of the heat sink). The bottom surface was heated, the other surfaces were thermally insulated. The results showed that higher height ratios improve the thermal performance of the heat sink but increase the weight and the costs of the whole system. An experimental investigation was accomplished by Zhao et al. [5], where the PCM in metal foam is heated for charging process and then it is cooled for discharging process. The metal foam is a copper foam and the PCM is the RT58 paraffin wax, employed for low-temperature storage application. By the results, the presence of metal foam is very significant because the heat transfer rate could be augmented between 5 and 20 times. In this paper, a partially filled LHTESS system is numerically studied. Specifically, the study is focused on a vertical hollow cylinder in five configurations below explained and an assigned temperature on the surface of the internal cylinder. Also, assuming that the heat transfer towards the external ambient is negligible. The results are shown in terms of melting time, average temperature of the system, and a performance parameter.

2. Physical model

The LHTESS system and its geometrical configuration are reported in Fig. 1. The LHTESS system is a hollow cylinder. The cylindrical domain is given in Fig. 1a. The axial symmetry is assumed and the problem under investigation is two-dimensional, as shown in Fig. 1b. The system, with the height equal to H, is enclosed between an internal cylinder with a radius equals to r_i and an external cylinder with a radius equals to r_e, in Fig. 1b. H is 100 mm, r_i is 6 mm and r_e is 38 mm. The gravitational acceleration g is directed along the z-axis. A temperature equal to 350 K on the internal cylindrical surface is assumed as boundary condition. The external surfaces are assumed adiabatic, as reported in Fig. 1c. The initial temperature is 300 K. Five ways to fill up the internal domain are considered.

In this paper, metal foam, where the pores are filled by PCM, is indicate as MF. The MF could completely occupy the domain, or a part of domain, therefore a filling ratio s is defined, and the MF radius (r_{mf}) is the radius of the MF inside the domain. In this work, s is equal to 0 for a domain filled up by pure PCM, so r_{mf} is 0. The filling ratio s is equal to 1/4, 2/4 or 3/4 to indicate that the domain is partially filled by pure PCM and MF, so r_{mf} is equal to 8, 16, 24 mm plus r_i. When s is equal to 1 the domain is totally filled by MF. The melting phenomenon of the PCM can be modelled with the enthalpy-porosity method [6], in which the solid-liquid interface is not explicitly traced but the method defines a mixed solid-liquid region that is described as a “pseudo” porous zone where the porosity is the liquid fraction. The Brinkman–Forchheimer-extended Darcy model and the local thermal equilibrium (LTE) assumption are assumed.

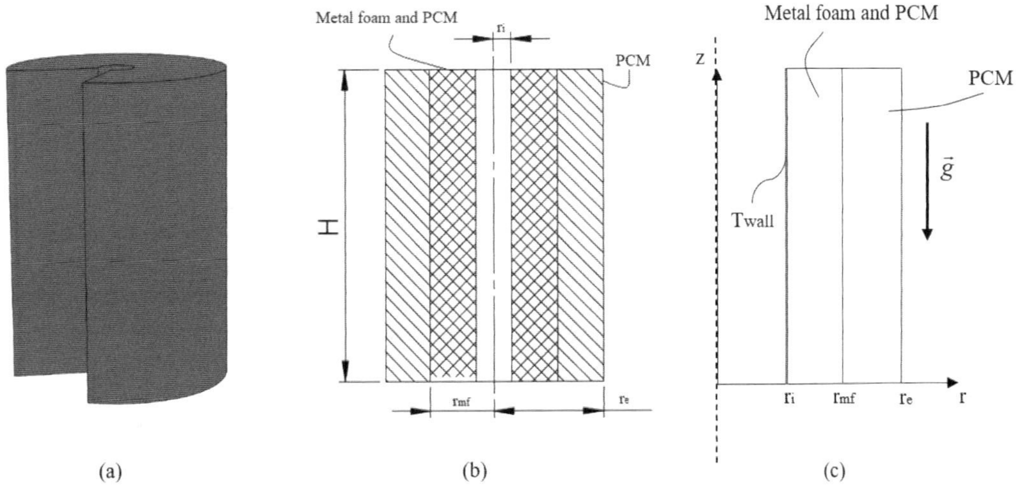

Figure 1:Latent heat thermal energy storage system geometry: (a) sketch of the thermal energy storage, (b) geometrical characteristics and (c) boundary conditions and numerical domain.

The following hypotheses are assumed in the present study:

1) A two-dimensional physical model is considered enclosed in two vertical coaxial cylinders.
2) The internal cylindrical surface is assumed isothermal, the external surface and the upper and lower bases of the cylinder are assumed adiabatic.
3) The Boussinesq approximation has been employed to model buoyancy effect in the momentum equation and the viscous effects in the energy equation were neglected.
4) The PCM thermal properties are assumed constant. The melting range is between 326-332 K.
5) The metal foam is isotropic and homogenous and it is simulated as a porous media.

According to the hypothesis the governing equations are the following:

- Continuity equation

$$\frac{1}{r}\frac{\partial(ru)}{\partial r}+\frac{\partial w}{\partial z}=0 \tag{1}$$

where u and w are the radial and axial velocity of the PCM in liquid phase, respectively, and r, z are the cylindrical coordinates.

- Momentum equations:

$$\begin{aligned}&\frac{\rho_p}{\varepsilon}\left(\frac{\partial u}{\partial t}+\frac{u}{\varepsilon}\frac{\partial u}{\partial r}+\frac{w}{\varepsilon}\frac{\partial u}{\partial z}\right)u=-\frac{\partial p}{\partial r}+\frac{\mu_p}{\varepsilon}\left(\frac{\partial^2 u}{\partial r^2}+\frac{1}{r}\frac{\partial u}{\partial r}-\frac{u}{r^2}+\frac{\partial^2 u}{\partial z^2}\right)+\\&-\frac{(1-\beta)^2}{(\beta^3+0.001)}A_{mush}u-\eta\left(\frac{\mu_p}{K}u+\frac{C_F}{\sqrt{K}}\rho_p u\sqrt{u^2+w^2}\right)\end{aligned} \tag{2}$$

$$\begin{aligned}&\frac{\rho_p}{\varepsilon}\left(\frac{\partial w}{\partial t}+\frac{u}{\varepsilon}\frac{\partial w}{\partial r}+\frac{w}{\varepsilon}\frac{\partial w}{\partial z}\right)u=-\frac{\partial p}{\partial r}+\frac{\mu_p}{\varepsilon}\left(\frac{\partial^2 w}{\partial r^2}+\frac{1}{r}\frac{\partial w}{\partial r}-\frac{u}{r^2}+\frac{\partial^2 w}{\partial z^2}\right)+\\&-\frac{(1-\beta)^2}{(\beta^3+0.001)}A_{mush}w-\eta\left(\frac{\mu_p}{K}w+\frac{C_F}{\sqrt{K}}\rho_p w\sqrt{u^2+w^2}\right)+\rho_p g\gamma(T-T_0)\end{aligned} \tag{3}$$

where ρ_p is the density of the PCM and it depends on temperature assuming the Boussinesq approximation, μ_p is the viscosity of the liquid PCM, p is the pressure.
In this investigation, the value of mushy constant (A_{mush}) is assumed equals to $5x10^5$ kg m^{-3} s^{-1}.

- The energy equation with local thermal equilibrium assumption is:

$$(\rho c)_{eff}\left(\frac{\partial T}{\partial t}+u\frac{\partial T}{\partial r}+w\frac{\partial T}{\partial z}\right)=k_{eff}\left[\frac{\partial}{\partial r}\left(\frac{1}{r}\frac{\partial}{\partial r}(rT)\right)+\frac{\partial^2 T}{\partial z^2}\right]-\varepsilon\rho_P H_L\frac{\partial\beta}{\partial t} \tag{4}$$

$(\rho c)_{eff}$ is a weighted average value:

$$(\rho c)_{eff}=(1-\varepsilon)\rho_{mf}c_{mf}+\varepsilon\rho_P c_P \tag{5}$$

In Eq. (5), ρ_{mf} and c_{mf} represent the metal foam density and specific heat, the metal foam porosity is ε and the density and specific heat of PCM are indicated by ρ_p and c_p. The effective thermal conductivity k_{eff} is evaluated based on the concept of series and parallel to heat transfer between solid matrix and fluid, the effective conductivity of the MF is expressed by [7]:

$$k_{eff}=A\left[\varepsilon k_f+(1-\varepsilon)k_s\right]+\frac{1-A}{\left[\frac{\varepsilon}{k_f}+\frac{(1-\varepsilon)}{k_s}\right]} \tag{6}$$

Where A is equal to 0.35 and in which k_f and k_s are the thermal conductivities of fluid phase and aluminum foam.

In Eq. (4) H_L represents the latent heat of the PCM whereas t is the time.

In the momentum and energy Eq. (2)-(4) are applied to both the clean region (PCM) and porous medium region (MF) using the following binary parameter:

$$\eta(x,y)=\begin{cases}0 & \text{clean region } (\varepsilon=1)\\ 1 & \text{metal foam region } (0<\varepsilon<1)\end{cases} \tag{7}$$

In Eq. (2)-(3) the last term simulates the presence of metal foam as a porous medium. In fact, in the last term there are: the Darcy law and the Forchheimer law. The permeability of the porous media is indicated by K and inertial drag factor is C_F. These terms are evaluated following the work of Battacharya et al. [7]. In fact, Aluminum metal foam properties were choice using sample 6 of their paper. The aluminum metal foam properties are: pores per inch (PPI) equal to 20; porosity equal to 0.9546; fiber diameter (d_f) equal to 0.0003 m; pore diameter (dp) equal to 0.0027 m; C_F equal to 0.093; permeability (K) equal to $1.3x10^7$ m^2.
To describe the mixed region of the enthalpy-porosity model, a new parameter is defined. It is the liquid fraction, β. It expresses the ratio between the liquid volume and the total volume of a cell of the mixed region. It has a value between 0 and 1. The β value depends on the local temperature in the cell and in particular:

$$\begin{cases} \beta = 0 & for \quad T < T_{solidus} \\ \beta = \dfrac{T - T_{solidus}}{T_{liquidus} - T_{solidus}} & for \quad T_{solidus} < T < T_{liquidus} \\ \beta = 1 & for \quad T > T_{liquidus} \end{cases} \tag{8}$$

β is 0 where the zone is totally solid, it is 1 where the zone is totally liquid. In Eq. (8), T is the local temperature and $T_{solidus}$ and $T_{liquidus}$ are the solid and liquid phase temperatures, respectively. The part of the domain below the first temperature is fully solid, while the part of the domain above the latter temperature is fully liquid. The melting occurs when the temperature is included between $T_{liquidus}$ and $T_{solidus}$. The description of the enthalpy-porosity method, during the melting or solidification, is completed considering a source term in the momentum equation which takes account of the solid part of the mixed zone. In fact, the mixed zone is modelled as a pseudo porous zone where the liquid fraction represents the porosity.
In the present work, the paraffin wax RT58 represent the PCM. The metal foam is made of aluminum. The choice of aluminum foam is dictated by its relative low cost, and its lower density and weight, in relation to other materials. Here are considered the following properties for RT58 [8]: density is 825 kg/m^3; thermal conductivity is 0.2 W/mK; specific heat is 2000 J/kg K; dynamic viscosity is 0.0269 kg/m s; thermal expansion coefficient is 0.00011 K^{-1}; latent heat is 160,000 J/kg; the melting range is 326-332 K. For aluminum [9] the following properties are considered: density is 2719 kg/m^3; thermal conductivity is 202.4 W/mK; specific heat is 871 J/kg K.

3. Numerical model

The governing Eqs. (1) – (4) are solved by means of Ansys-Fluent commercial code [10] which uses the finite volume method to discretize the problem. The SIMPLE algorithm is employed for the pressure-velocity coupling; PRESTO algorithm is used for the pressure calculation. The high order term relaxation is set equal to 0.75 for all variables, to avoid the numerical instability of the system. A transient analysis is performed with a time step size equal to 1 s. The convergence errors are assumed equal to 10^{-5}, for continuity and momentum equation, and 10^{-8}, for energy equation. The numerical grid (Fig. 2) is set of quadrilateral cells and it is uniform.

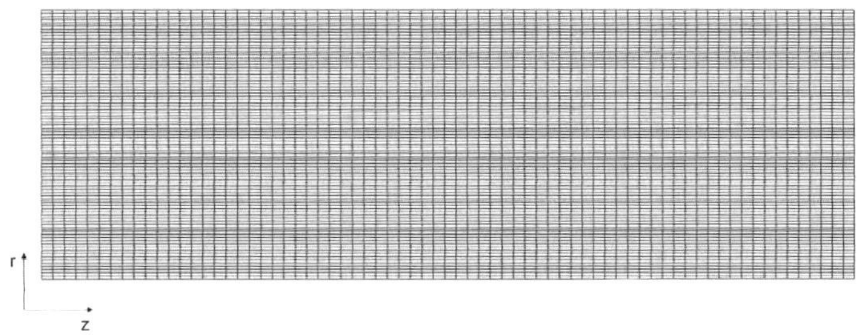

Figure 2: Computational domain

To check the accuracy of the results, a grid independency analysis is performed with three discretization: 1540, 5964, 23124 cells. In the analyzed case *s* is equal to 1/4. The liquid fraction and the average temperature as time function are considered to assess the grid independence comparing the values of the liquid fraction and the average temperature for each mesh and the results are given in Fig. 3. A grid with 1540 cells is employed to perform the study and it represents the best compromise between the solution accuracy and computing time. In fact, the error made by using the lower mesh is neglectable in this case (less than 1%).

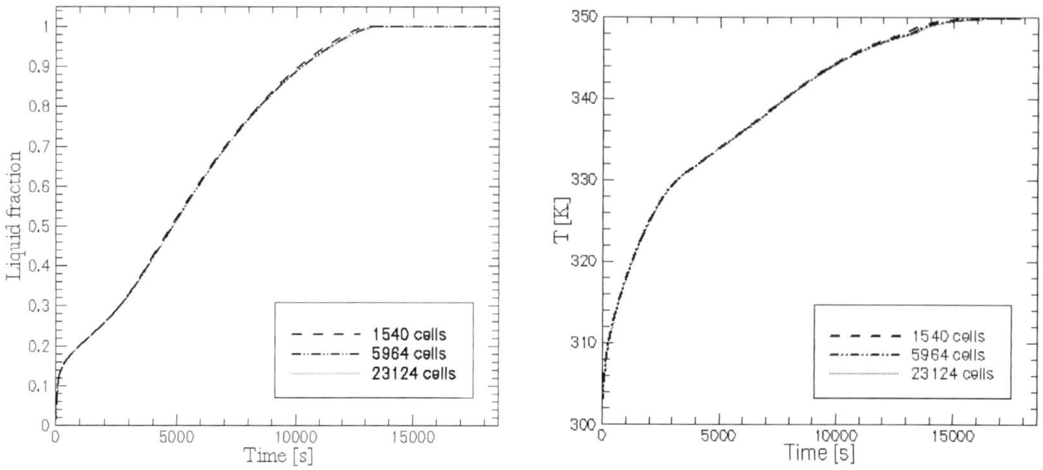

(a) (b)

Figure 3: grid independency analysis: (a) Liquid fraction as time function for all mesh size; (b) Average Temperature as time function for all mesh size.

The numerical simulations were performed on a workstation with two CPU Intel(R) Xeon(R) Silver 4108 1.80 GHz and with 128 Gb of installed ram. The maximum number of iterations was set equal to 200 for each time step. The total simulation time is 18600 s, and the average number of iterations is about 150 per each time step.

4. Results and discussion

In Fig. 4 there is the average evolution of liquid fraction for each value of filling ratio variable (*s*). As it can be seen the liquid fraction evolution is faster than the other one when *s* increase, so when the MF radius increase. The difference between the melting time of the case where the LHTESS is fille by only PCM and the case where the system is filled by only MF is about 1 order of magnitude.

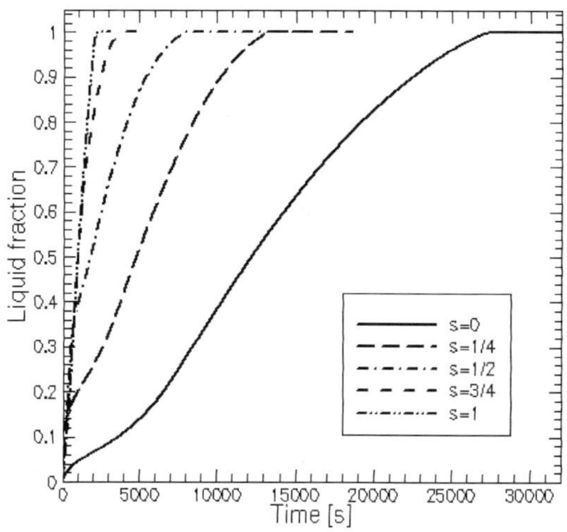

Figure 4: Liquid fraction as time function for all filling ratio

In Fig. 5 is reported the average temperature evolution as time function. The simulations were stopped when the average temperature reaches 350 K. Also, in this case the average temperature evolution is faster than the other one when *s* increase.

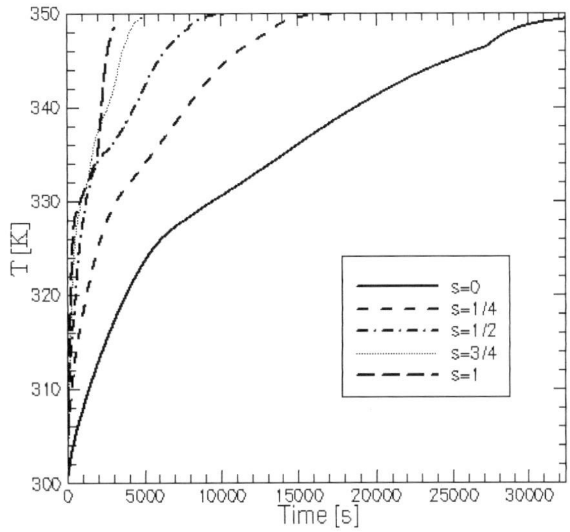

Figure 5: Average temperature as time function for all filling ratio

In Fig. 6 is shown the energy, as enthalpy, stored by the LHTESS. As it can be seen, the stored energy is the same for all filling ratio, but the velocity of storing change. In fact, the velocity of storing is highest when the system is filled up by only MF. It is slowest when the system is filled up by PCM. The enthalpy variable is the sum between sensible energy and latent one. The maximum stored energy is about 96,000 J.

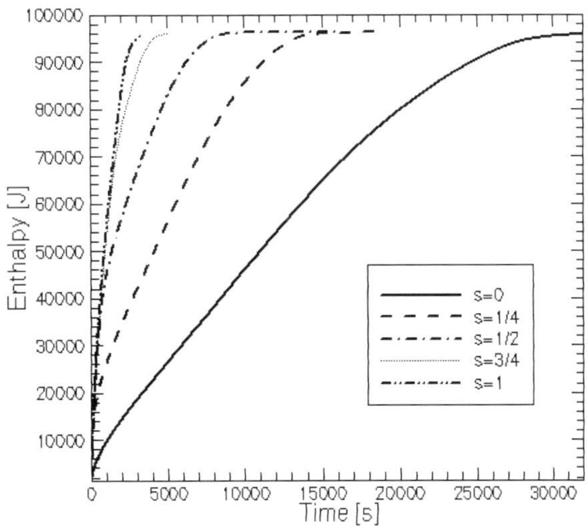

Figure 6: Enthalpy as time function for all filling ratio

5. Conclusions

A numerical investigation on the effects of PCM partially filled by metal foam into the Latent Heat Thermal Energy Storage System is accomplished. The geometry of the system is a hollow cylinder. The PCM is pure paraffin wax with a low value of thermal conductivity. It melts in a range of temperatures. The metal foam is made of aluminium and its pores are filled by PCM. It is denoted by MF. MF partially fills the cylinder. The aim is to understand the optimal filling ratio of the foam to have the better thermal performance. The results had shown that the melting time and the average temperature changing with the filling ratio. Also, the energy storing is faster while the filling ratio of MF increases.

Acknowledgements
This research was partially funded by MIUR (Ministero dell'Istruzione, dell'Università e della Ricerca), grant number PRIN- 2017F7KZWS and by Università degli Studi della Campania "Luigi Vanvitelli" with the grant number D.R. n. 138 under NanoTES project - V:ALERE program 2020.

References

[1] Y. Tian, C.Y. Zhao, A review of solar collectors and thermal energy storage in solar thermal applications, Appl. Energy 104 (2013) 538–553, https://doi.org/10.1016/j.apenergy.2012.11.051.
[2] S. Sabihuddin, A.E. Kiprakis, M. Mueller, A numerical and graphical review of energy storage technologies, Energies 8 (2015) 172–216, https://doi.org/10.3390/en8010172.
[3] M. Caliano, N. Bianco, G. Graditi, L. Mongibello, Analysis of a phase change material-based unit and of an aluminum foam/phase change material composite-based unit for cold thermal energy storage by numerical simulation, App. En. 256 (2019) 113921, https://doi.org/10.1016/j.apenergy.2019.113921.
[4] Z. Zhu, Y. Huang, N. Hua, Y. Zeng, L. Fan, Transient performance of a PCM-based heat sink with a partially filled metal foam: effects of the filling height ratio, Appl. Therm. Eng. 128 (2018) 966–972, https://doi.org/10.1016/j.applthermaleng.2017.09.047.
[5] C.Y. Zhao, W. Lu, Y. Tian, Heat transfer enhancement for thermal energy storage using metal foams embedded within phase change materials (PCMs), Sol. Energy 84 (2010) 1402–1412, https://doi.org/10.1016/j.solener.2010.04.022.
[6] V.R. Voller, C. Prakash, A fixed grid numerical modelling methodology for convection-diffusion mushy region phase-change problems, Int. J. Heat Mass Transf. 30 (1987) 1709–1719, https://doi.org/10.1016/0017-9310(87)90317-6.
[7] A. Battacharya, V.V. Calmidi, R.L. Mahajan, Thermophysical properties of high porosity metal foams, Int. J. Heat Mass Transf. 45 (2002) 1017-1031.
[8] N. S. Dhaidan, A. F. Khalaf, Experimental evaluation of the melting behaviours of paraffin within a hemicylindrical storage cell, Int. Comm. in Heat and Mass Transf. 111 (2020) 104476. https://doi.org/10.1016/j.icheatmasstransfer.2020.104476.
[9] C. Ji, Z. Qin, S. Dubey, F. H. Choo, F. Duan, Simulation on PCM melting enhancement with double-fin length arrangements in a rectangular enclosure induced by natural convection, Int. J. of Heat and Mass Transf. 127 (2018) 255–265. https://doi.org/10.1016/j.ijheatmasstransfer.2018.07.118.
[10] ANSYS-FLUENT Computational Fluid Dynamic Code Version 19.2 User Guide, Fluent, Inc., www.ansys.com; 2018.

A Eulerian-Lagrangian approach for the CFD analysis of airborne disease transmission in a car cabin

G. Grossi[1], F. Arpino[1], G. Buonanno[1], G. Cortellessa[1], L. Moretti[1], H. Nagano[2], L. Stabile[1]

[1]Department of Civil and Mechanical Engineering, University of Cassino and Southern Lazio, Cassino, FR, Italy

[2]Department of Mechanical Systems Engineering, Tokyo City University, 1-28-1 Tamazutsumi, Setagaya-ku, Tokyo, 158-8557, Japan

giorgio.grossi@unicas.it

Abstract. In indoor environments such as private cars, taxis and public transport, social distance cannot always be applied and virus concentrations can reach high levels. In this paper, a Eulerian-Lagrangian approach is adopted for the transient simulation of aerosol airborne dispersion in a passenger car cabin. The numerical tool, validated by PIV analysis, is applied to study three different scenarios during the respiratory activity. The car ventilation system and the position of the passengers affect the mass of inhaled aerosol and consequently the quality of the passenger compartment air.

1. Introduction

The risk of virus infection can be reduced by adopting different strategies such as social distancing and the adoption of masks in order to avoid virus diffusion particularly in indoor environments [1,2]. The thermodynamics phenomena related to breathing, coughing, speaking and sneezing are complicated therefore it is impossible to measure virus transport and diffusion in indoor environments [3].
In the last century, the large droplet route has been wrongly considered to be the primary route for most respiratory infections [4,5], and the associated social distancing of 1-2 m has been adopted in all the country of the world. The respect of a social distance represents a good measure in open spaces but it cannot be adopted in small indoor environments like car cabins where the occupants are in a close range and, during a car trip, the virus concentrations can reach crucial levels [6].
In fact, droplets of different sizes, coming from different areas of the respiratory tract, go through an atomization process by the passage of a high-speed air stream. Wells et.al. [7] studied the phenomenon of droplet evaporation during breathing from the mouth of a subject and the particle diameters reduction during the air injection. Further investigations were performed by Papineni et al. [8] adopting an optical particle detection system. In particular, the authors found that most of the particles present micrometres diameters (<100 μm) and, after an evaporation process, can reach the susceptible suspended in air (airborne particles) [9]. As said before, in open spaces there is a low risk of contracting infections because virus concentrations are very low [10]. On the contrary, in indoor environments such as public transport, hospital rooms, schools, on which a lot of infected people are in close contact even when social distance is applied, the infection risk is very high. Suggestions to reduce the increasing in virus concentrations in indoor consists in the improvement of the ventilation system, the surfaces disinfection,

Journal of Physics: Conference Series 2177 (2022) 012015 doi:10.1088/1742-6596/2177/1/012015

the reduction of the number of occupants, etc. However, more effective forecasting and analysis tools are desirable to allow a more detailed understanding about measures to be adopted inside restricted environments like passenger cars. Computational Fluid Dynamics (CFD) allow to obtain detailed information about transport and diffusion of aerosol also for the case of private transport such as for a car cabin. This technique get detailed spatial fluid flow distributions and aerosol transmission by solving a 3D mixed convection model in transient conditions [11,12] and it also allows to assess the ventilation conditions of aerosol transmission by solving a 3D mixed convection model in transient conditions. A lot of research was conducted to improve the airflow distribution in passenger cabins.

PIV (Particle Image Velocimetry) analysis were conducted by Yang et al. [13,14] to investigate airflow distribution and turbulence intensity in selected car cabin sections.

You et al. [15] investigated the suitable turbulence model to simulate air distribution in aircraft cabins with gaspers turned on. They showed that the SST k–ω model was more accurate than the RNG k–ε model for predicting the airflow distribution in an aircraft cabin. Mathai et al. [16] analysed how the windows opening and closing can affect the thermo-hygrometric comfort and the aerosol spread between the passengers discovering that addressing the airflow from the ventilation system away from passengers reduces the risk of transmission.

The numerical studies available in the scientific literature present same limitations which can be highlighted such as a simplified numerical modelling of the breathing phenomena in isothermal conditions and the absence of buoyancy effects. Moreover, such models do not distinguish the droplets variation in diameter during the respiratory activity and the effective mass flow rate injected during breathing by an emitter subject. In other cases, a velocity boundary condition is not set on the mouth of the receiver to properly simulate the breathing phenomenon, also considering the aerosol inhalation.

In the present paper, the OpenFOAM open-source software is adopted to simulate the transient non-isothermal airborne aerosol dispersion in a passenger car cabin. The model is based on a Eulerian-Lagrangian approach: mass, momentum and energy conservation equations are solved for the description of the fluid flow distribution within the computational domain, while Newton's equation is adopted for the diffusion and transport of particles. The numerical model was validated by comparing the numerical results with PIV measurements, performed by one of the authors of the paper and then applied to study the aerosol transport, spatial distribution and inhalation during respiratory activity, from one specific person (contagious subject) to another passenger (receiver subject) inside a car cabin for different ventilation scenarios.

2. CFD numerical model

CFD simulations were performed by the OpenFOAM code which can solve the mass, momentum and energy conservation equations for 3D transient and turbulent compressible fluid flow. The numerical tool gives information about the pressure, temperature and velocity contours in the investigated computational domain. Turbulence was modelled using the Unsteady Reynolds Averaged Navier Stokes (URANS) approach, solving the Shear Stress Transport (SST) k–ω model since it is well suited for simulating induced jet flows in closed environments, as proven by previous studies of You et al. [15] and Ullrich et al. [17]. Further information about the solved Partial Differential Equations (PDEs) can be found in [18].

The aerosol transport and diffusion were obtained by adopting a Lagrangian Particle Tracking (LPT) approach, based on a dispersed dilute two-phase flow. This model can be adopted when droplets are distant one from each other and their volume is lower than 10^{-3}. In this way, the mass momentum and energy equations are solved for the air flow (continuous phase) and the Newton's equation of motion is solved for each droplet (discrete phase). LPT equations are obtained by solving Equation 1 and Equation 2.

$$m_d \frac{d\boldsymbol{u}_d}{dt} = \boldsymbol{F}_D + \boldsymbol{F}_g \tag{1}$$

$$\frac{d\boldsymbol{x}_d}{dt} = \boldsymbol{u}_d \tag{2}$$

where $m_d\ (kg)$ is the mass of the droplet; $\boldsymbol{u}_d\ \left(\frac{m}{s}\right)$ represents the droplet velocity; $t\ (s)$ is the time; $\boldsymbol{F}_D\ (N)$ and $\boldsymbol{F}_g(N)$ are, respectively, the drag and gravity forces acting on the droplet; $\boldsymbol{x}_d\ (m)$ represents the trajectory of the droplet. The drag force is given by Crowe [19]:

$$\boldsymbol{F}_D = m_d \frac{18}{\rho_d \cdot d_d^2} C_D \frac{Re_d(\boldsymbol{u} - \boldsymbol{u}_d)}{24} \tag{3}$$

In Equation 3, $\rho_d\ \left(\frac{kg}{m^3}\right)$, $d_d(m)$ and Re_d represent, respectively, the density, diameter and Reynolds number of the droplet; $\boldsymbol{u}\ \left(\frac{m}{s}\right)$ is the air velocity. The droplet density was considered constant and equal to 1200 kg·m^{-3}. The Re_d was calculated as:

$$Re_d = \frac{\rho(|\boldsymbol{u} - \boldsymbol{u}_d|)d_d}{\mu} \tag{4}$$

where $\rho\ \left(\frac{kg}{m^3}\right)$ is the air density.
The drag coefficient, C_D, in Equation 3 is evaluated as a function of the droplet Reynolds number:

$$C_D = \begin{cases} \frac{24}{Re_d} & \text{if } Re_d < 1 \\ \frac{24}{Re_d}\left(1 + 0.15 \cdot Re_d^{0.687}\right) & \text{if } 1 \le Re_d \le 1000 \\ 0.44 & \text{if } Re_d > 1000 \end{cases} \tag{5}$$

Droplet collisions are considered to be elastic and the equations of motion for the droplets are solved assuming a one-way coupling between the continuum phase and the discrete phase: the flow field affects the droplet motion whereas the effect of the particles on the airflow is negligible.

3. The three investigated scenarios

The mathematical-numerical model described in Section 2 was applied to the analysis of droplets spread in a passenger car cabin during a 15-min travel, from one contagious subject (i.e. the driver in our analysis) to a susceptible receiver (i.e. another passenger). Three different scenarios were analysed: a) air supplied from front seat vents and susceptible sitting in the back-right seat (Figure 1a); b) air supplied from the windshield defroster vent and susceptible sitting in the back-right seat (Figure 1b); c) air supplied from the windshield defroster vent and susceptible sitting in the front seat (Figure 1c).
The numerical model validation was performed by a scaled-down glass model by adopting PIV analysis (Figure 2) but to investigate a real scenario it was increased in size by a scale factor equal to 1.7 as indicated in the study of Chang in [20] by reducing the seats inclination. A typical car volumetric flow rate equal to 238 m^3/h was imposed as boundary condition in inlet from the ventilation system and the volume of the scaled-up car cabin model is equivalent to that of a typical real car [17]. Inlet boundary condition for air temperature was set to 293 K and the heat exchange with the environment was not modelled. The mouths of the passengers were modelled as a circle with a 2 cm radius and a temperature of 308.15 K was adopted as boundary conditions for the resolution of the energy equation conservation equation.
An averaged fixed velocity value of 0.61 m/s was employed at the mouths of the emitter and the receiver as mean values of sinusoidal breathing during exhalation and inhalation reported by Abkarian et al. [21]. This choice allows to reach a quasi-steady state condition: once the quasi-steady state condition is

achieved, the flow field is frozen and is used to transport the particles injected by the emitter over time while breathing.
Droplets were assumed as liquid spheres whose dimensions, distribution and injection time were adopted from existing literature [3,22]. Evaporation phenomenon occurs soon after the injection [9], so in the present study the number and volume distribution of droplet after evaporation were considered. The simulations were carried out assuming the subjects to be mouth-breather.

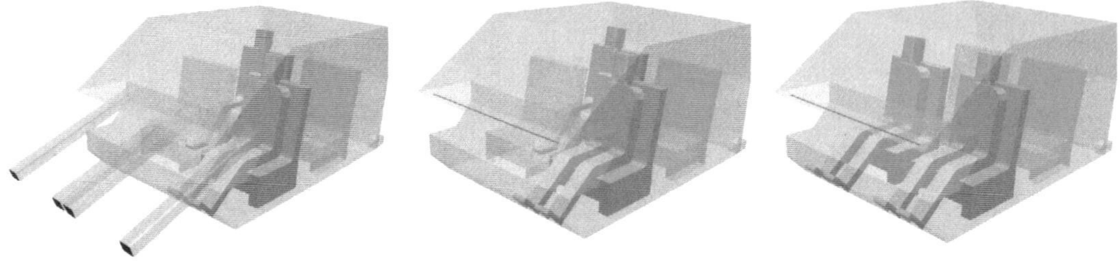

(a) front seat air vents supply mode – susceptible in the back-right seat

(b) windshield defroster air vent supply mode – susceptible in the back-right seat

(c) windshield defroster air vent supply mode – susceptible in the front seat

Figure 1 – The three analysed scenarios where the inlet sections are highlighted in red.

4. Validation of the numerical tool by PIV analysis

The flow field description within the passenger compartment was validated with the Particle Image Velocimetry (PIV) data which one of the authors obtained in a previous experimental campaign [14], for the scaled-down car cabin model shown in Figure 2a. Experimental analysis were performed along the longitudinal plane passing through the centre of the air vent in front of the driver seat (within the x-y plane at z=0.3945 m, as highlighted in Figure 4). The fluid flow is injected inside the cabin through four openings located on the dashboard, at a total flow rate equal to 100 m^3/h; in the experimental setup a rectangular duct is connected to each of the supply openings, causing a flow velocity profile to develop at the outlet section of the air vents (at the entry of the passenger compartment; for more details refer to [14]).
In the 3D model employed for numerical model validation, shown in Figure 2b, such ducts were considered to reproduce the velocity distribution at the inlet of the cabin and accurately replicate the experimental conditions. The adopted mesh was obtained by the tool *snappyHexMesh* and was determined on the basis of a proper grid sensitivity analysis. Four computational grids were realized and information about the grids parameters are illustrated in Table 1.

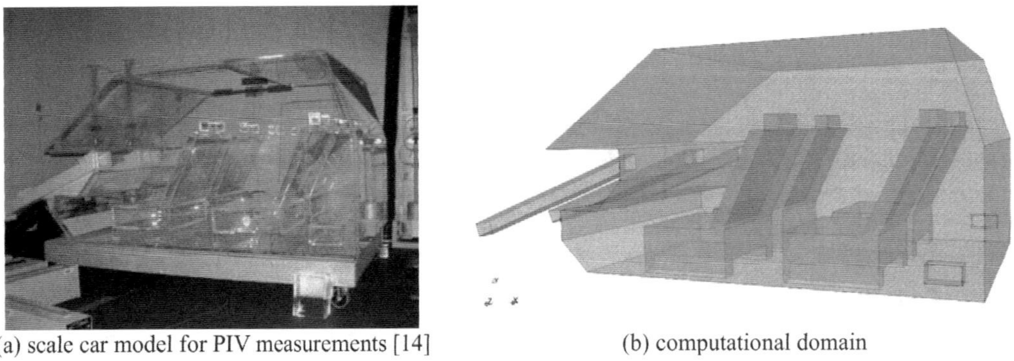

(a) scale car model for PIV measurements [14]

(b) computational domain

Figure 2 – Scaled glass model adopted for PIV analysis (a) and computational domain (b).

Table 1 – Computational grid information

Mesh #	Number of Cells	Skewness max	Non orthogonality max
1	355,534	1.44	50
2	743,232	3.45	50
3	1,559,007	2.19	50
4	2,905,119	3.78	50

The percent deviation amongst the velocity fields obtained with Mesh 3 and Mesh 4 is less than 5%, hence Mesh 3 has been used in the upcoming analysis. In Figure 3, a x-y slice of Mesh 3 for z=0.3945 m (measuring plane) is shown.

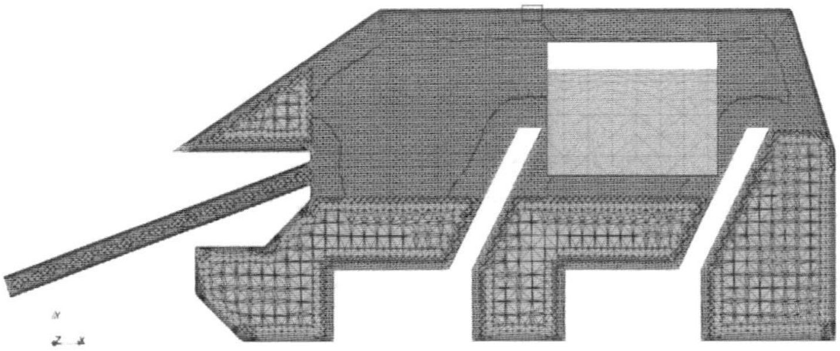

Figure 3 – Computational grid employed (Mesh 3, 1559007 cells) for velocity field validation.

The computational grid was refined near solid surfaces to properly capture the boundary layer; moreover, a refinement box was created in the upper zone of the car cabin to have a finer mesh in the area of most interest for the comparison with the experimental velocity profiles, without requiring unreasonable computational efforts. The minimum, maximum and average y^+ values obtained using Mesh 3 are 0.129, 41.484 and 5.472 respectively. PIV and CFD data were compared in terms of velocity profiles obtained in a a selected x-y plane at z=0.3945 m (see Figure 4), in four different sections: x=0.45 m, x=0.65 m, x=0.85 m and x=1.05 m.
On the whole, it was found that the airflow patterns predicted with CFD simulations are in a good agreement with the experimental results provided by Ozeki et al. [14], even when getting away from the inlet sections. Therefore, the mathematical-numerical model adopted in the present study well describes the flow field within the car cabin.

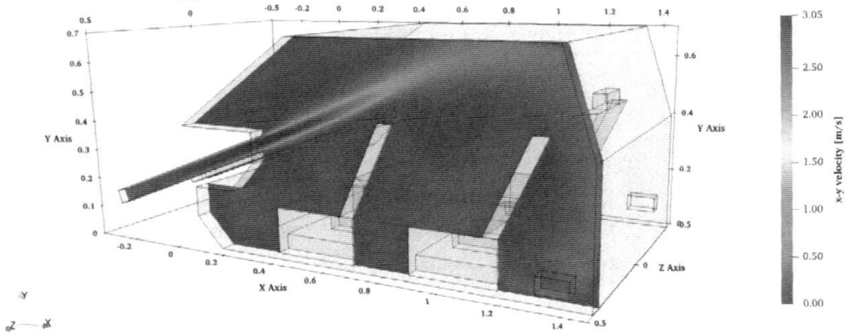

Figure 4 – Velocity contours for a slice located at z=0.3945 m.

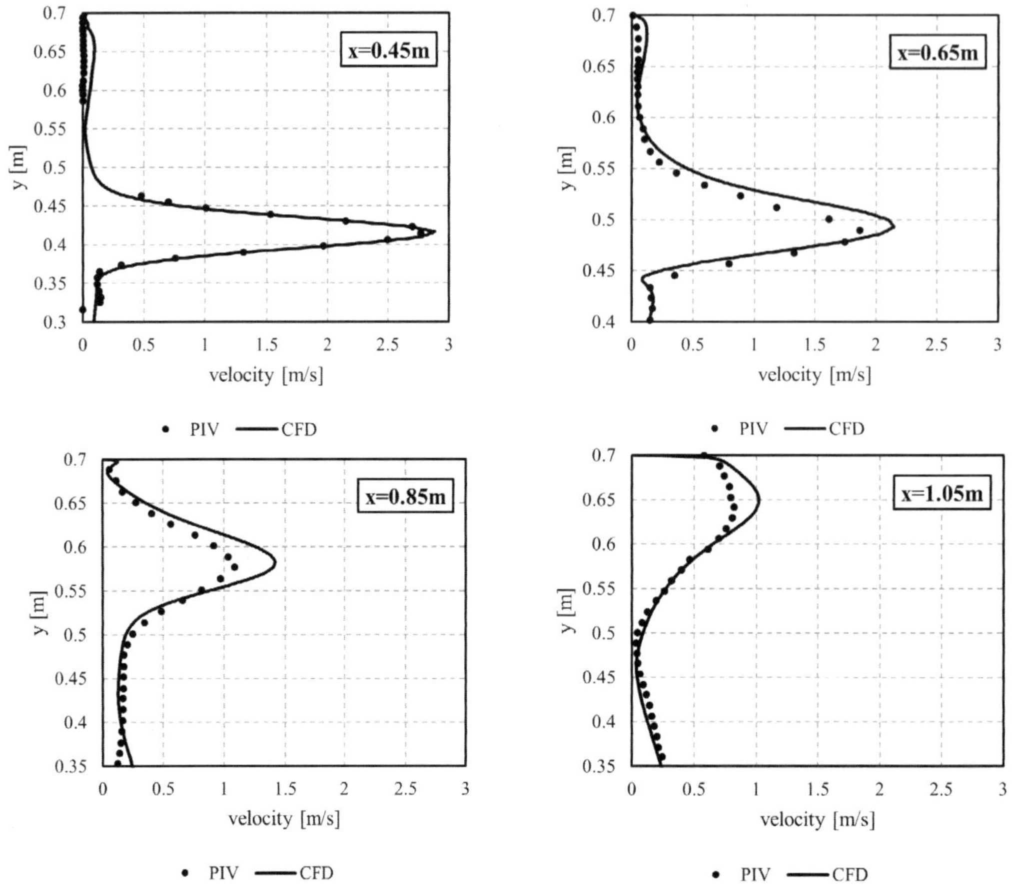

Figure 5 – Experimental and numerical velocity profiles comparison within a selected x-y plane at z=0.3945 m obtained in four different sections: x=0.45 m, x=0.65 m, x=0.85 m and x=1.05 m.

5. Results obtained by CFD analysis

In Figure 6 the particles position in the car cabin after 15 minutes, for the scenarios outlined in Section 3, is showed. Adopting the front seat vents (Figure 6a) the passengers do not receive particles because a recirculation area is formed above the dashboard and airborne droplets are not able to reach the subject position. On the contrary, turning on the windshield defroster (Figure 6b and Figure 6c), there is a quite uniform distribution of particles within the cabin. The passenger located in the back-right seat (Figure 6b), inhales about 20 particles (corresponding to a mass of $1.56 \cdot 10^{-15}$ kg), instead when he is placed in the front seat (Figure 6c), he receive about 15 particles (corresponding to a mass of $5.42 \cdot 10^{-16}$ kg). Although after 15 minutes the number of inhaled particles is not significant, by increasing the time spent in the car cabin the mass of inhaled aerosol could reach critical values. Therefore, a parametric analysis, varying the flowrate at the inlet and the time spent in the car, will be performed in later studies to assess how these aspects influence the motion, and thus the spread of droplets in the passenger compartment.

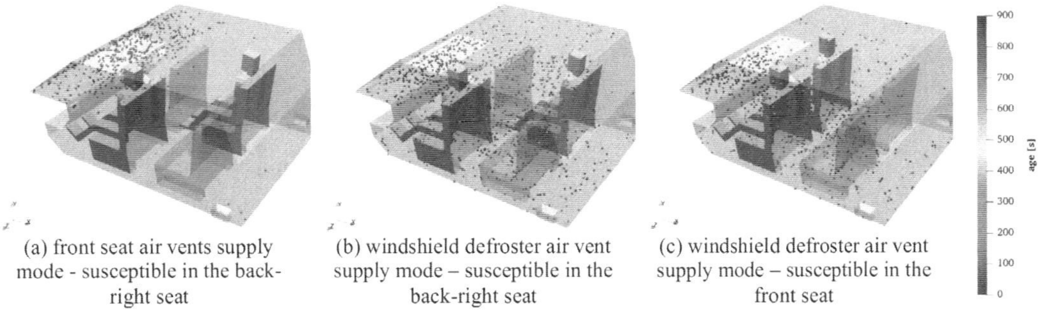

(a) front seat air vents supply mode - susceptible in the back-right seat; (b) windshield defroster air vent supply mode – susceptible in the back-right seat; (c) windshield defroster air vent supply mode – susceptible in the front seat

Figure 6 – Evolution of particles after 15 minutes.

6. Conclusions

In the present paper, the authors realized an innovative transient non-isothermal numerical model by using the open-source software OpenFOAM, able to simulate the transient airborne dispersion of aerosol in a passenger car cabin. The model is based on a Eulerian-Lagrangian approach: mass, momentum and energy conservation equations are solved for the description of the fluid flow distribution within the computational domain, while Newton's equation is adopted for diffusion and transport of particles. The numerical model was validated by comparing the numerical results with PIV measurements, performed by one of the authors of the paper. The validated numerical tool was applied to study the aerosol transport, spatial distribution and inhalation, during respiratory activity, from one specific person (contagious subject) to another passenger (receiver subject) inside a car cabin for different ventilation scenarios. The car ventilation system and the position of the passengers affect the mass of inhaled aerosol and consequently the quality of the passenger compartment air.

References

[1] Qian H, Miao T, Liu L, Zheng X, Luo D and Li Y 2021 Indoor transmission of SARS-CoV-2 *Indoor Air* **31** 639–45

[2] Blocken B, Malizia F, van Druenen T and Marchal T 2020 Towards aerodynamically equivalent COVID-19 1.5 m social distancing for walking and running

[3] Morawska L, Johnson G R, Ristovski Z D, Hargreaves M, Mengersen K, Corbett S, Chao C Y H, Li Y and Katoshevski D 2009 Size distribution and sites of origin of droplets expelled from the human respiratory tract during expiratory activities *Journal of Aerosol Science* **40** 256–69

[4] Chapin C V 1912 *The sources and modes of infection* (New York: J. Wiley & sons)

[5] Flügge C 1897 Ueber Luftinfection *Zeitschr. f. Hygiene.* **25** 179–224

[6] Buonanno G, Stabile L and Morawska L 2020 Estimation of airborne viral emission: Quanta emission rate of SARS-CoV-2 for infection risk assessment *Environment International* **141** 105794

[7] Wells W F 1934 ON AIR-BORNE INFECTION* *American Journal of Epidemiology* **20** 611–8

[8] Papineni R S and Rosenthal F S 1997 The Size Distribution of Droplets in the Exhaled Breath of Healthy Human Subjects *Journal of Aerosol Medicine* **10** 105–16

[9] Balachandar S, Zaleski S, Soldati A, Ahmadi G and Bourouiba L 2020 Host-to-host airborne transmission as a multiphase flow problem for science-based social distance guidelines *International Journal of Multiphase Flow* **132** 103439

[10] Coccia M 2020 Factors determining the diffusion of COVID-19 and suggested strategy to prevent future accelerated viral infectivity similar to COVID *Science of The Total Environment* **729** 138474

[11] Lewis R W, Nithiarasu P and Seetharamu K N 2004 *Fundamentals of the finite element method for heat and fluid flow* (Hoboken, NJ: Wiley)

[12] Scungio M, Arpino F, Cortellessa G and Buonanno G 2015 Detached eddy simulation of turbulent flow in isolated street canyons of different aspect ratios *Atmospheric Pollution Research* **6** 351–64

[13] Yang J H, Kato S and Nagano H 2009 Measurement of airflow of air-conditioning in a car with PIV *J Vis* **12** 119–30

[14] Ozeki Y, Yang J-H, Nagano H, Kato S, Nomura E, Inoue M and Kobayashi S 2008 Ventilation Characteristics of Modeled Compact Car Part 1 Airflow Velocity Measurement with PIV *SAE Int. J. Passeng. Cars – Mech. Syst.* **1** 631–9

[15] You R, Chen J, Shi Z, Liu W, Lin C-H, Wei D and Chen Q 2016 Experimental and numerical study of airflow distribution in an aircraft cabin mock-up with a gasper on *Journal of Building Performance Simulation* **9** 555–66

[16] Mathai V, Das A, Bailey J A and Breuer K 2021 Airflows inside passenger cars and implications for airborne disease transmission *Sci. Adv.* **7** eabe0166

[17] Ullrich S, Buder R, Boughanmi N, Friebe C and Wagner C 2020 Numerical Study of the Airflow Distribution in a Passenger Car Cabin Validated with PIV *New Results in Numerical and Experimental Fluid Mechanics XII* Notes on Numerical Fluid Mechanics and Multidisciplinary Design vol 142, ed A Dillmann, G Heller, E Krämer, C Wagner, C Tropea and S Jakirlić (Cham: Springer International Publishing) pp 457–67

[18] Versteeg H K and Malalasekera W 2007 *An introduction to computational fluid dynamics: the finite volume method* (Harlow, England ; New York: Pearson Education Ltd)

[19] Crowe C T 2012 *Multiphase flows with droplets and particles* (Boca Raton: CRC Press)

[20] Chang T-B, Sheu J-J, Huang J-W, Lin Y-S and Chang C-C 2018 Development of a CFD model for simulating vehicle cabin indoor air quality *Transportation Research Part D: Transport and Environment* **62** 433–40

[21] Abkarian M, Mendez S, Xue N, Yang F and Stone H A 2020 Speech can produce jet-like transport relevant to asymptomatic spreading of virus *Proc Natl Acad Sci USA* **117** 25237–45

[22] Johnson G R, Morawska L, Ristovski Z D, Hargreaves M, Mengersen K, Chao C Y H, Wan M P, Li Y, Xie X, Katoshevski D and Corbett S 2011 Modality of human expired aerosol size distributions *Journal of Aerosol Science* **42** 839–51

Journal of Physics: Conference Series 2177 (2022) 012016 doi:10.1088/1742-6596/2177/1/012016

Urban-Scale Energy Models: the relationship between cooling energy demand and urban form

G Mutani[1], V Todeschi[2] and S Santantonio[3]

[1] Politecnico di Torino, Department of Energy, Responsible Risk Resilience Centre
[2] Politecnico di Torino, Department of Energy, Future Urban Legacy Lab
[3] Politecnico di Torino, Department of Energy
Corresponding author e-mail: guglielmina.mutani@polito.it

Abstract. To enhance the quality of life in cities, it is necessary to improve the energy performance of buildings together with a sustainable urban planning especially in high-density contexts. Previous works investigated the building shape, the urban morphology, and the local climate conditions to optimize the energy performance for space heating of buildings. The aim of this study is to validate a GIS-based engineering model to simulate the hourly energy demand for space cooling in residential buildings at neighborhood scale and to assess the relationship between the urban form and the energy performance in terms of cooling energy demand. A place-based methodology was applied to six neighborhoods in the city of Turin (Italy), identified as homogeneous zones with different building characteristics and urban contexts. The hourly cooling demand of residential buildings was studied starting from the energy balance at building scale, and then was applied at block of buildings scale with the support of GIS. This model was validated with a comparison of the results using CitySim tool and ISO 52016 assessment. In order to investigate the relationship between cooling energy demand and urban form, the GIS-based engineering model was applied to five typical blocks of buildings with different construction periods. The results show how cooling energy demand varies according to building characteristics and urban morphology in a continental-temperate climate. By this analysis, it is possible to identify the optimal block of building shape in Turin ensuring lower energy consumptions during the cooling season with different types of buildings.

Key Words: urban-scale energy model, hourly energy simulation, engineering model, cooling energy demand, urban form, residential buildings.

1. Introduction

A growing number of countries and cities is setting targets for the reduction of greenhouse gas (GHG) emissions and for the mitigation of climate changes [1]. The Clean Energy for all European Package encourages all state members to increase energy efficiency and the use of renewable energy sources (RES) to promote sustainable development, especially for the buildings sector. Reducing the energy consumption of buildings is one of the main goals of European policies, in order to increase the livability and sustainability in urban areas [1], [2]. Especially in these contexts, an increase in energy demand for cooling is expected [3]. Analysis on the energy consumption of buildings at urban scale can support the identification of proper energy urban planning for both the design of new neighborhoods and the priority of retrofit interventions on existing building stock. The development of urban-scale energy models (USEMs) is an actual and very broad research field: a variety of energy models allows to assess the energy performance of buildings taking into account the relationships between the buildings and the surrounding environment [4]. The main characteristics of USEMs are: the approach used (top-down, bottom-up and engineering), the accuracy of input data, the time requested to process data and the possibility to integrate and manage variables at different scale [5]. USEMs are able to assess energy consumption at multiple temporal (hourly, daily, monthly and annual) and spatial scales (from building, neighborhood, district to urban scale). The GIS tools utilize existing databases (normally used for urban development plans), manage energy-related variables to identify GIS-based energy models, and provide the possibility to visualize spatiotemporal results [6].

1.1. Research background

Built environment and local climate conditions of urban areas influence energy performances of buildings [7], [8]. The high presence of anthropogenic heat sources and the scarcity of green areas increase the urban heat island (UHI) effects [9]. Urban morphology can modifies solar access, wind speed and direction, affecting buildings' energy performances [8]; it can occur in a negative or in a positive manner, according to heating or cooling season, generally compensating one the other. Studies have shown that more compact urban environment results in lower thermal losses, providing for a decrease in the heating energy demand suitable in cold climates. In contrast, in hot climate high built density has been associated with an increase in cooling energy consumption due to the reduction of wind speed and higher air temperatures. In the same way, wider street canyons provide better wind conditions at the expenses of an increase in solar radiation exposure of building façade that can affect the solar gains both on the opaque envelope and transparent components.

Tools and models have been developed to estimate building energy consumption at city scale, the most used are: CitySim, UMI, Simstadt, CityBES and open formats as CityGML. This one ensures the availability, quality and accessibility of spatial information, as expected by the INSPIRE EU Directive 2007/2/EC. Since the relationship between buildings and urban form influences the energy consumption, the urban environment needs to be described considering all energy-related parameters. The urban parameters that are mainly used to describe the urban morphology are: *building coverage ratio* (*BCR*, m^2/m^2), *building density* (*BD*, m^3/m^2), *main orientation of the streets and buildings* (*MOS*, -), *aspect ratio* or *height-to-width* ratio (*H/W*, m/m), *sky view factor* (*SVF*, -), *green area ratio (GAR, m^2/m^2)* and *normalized difference vegetation index* (*NDVI*, -) [5], [10], [11], [12].

1.2. Research objectives

The main purpose of this work is to validate a GIS-based engineering model to simulate the hourly energy demand for space cooling of residential buildings at neighborhood scale, as it was already validated for space heating energy demand [5]. Differently from models that need ad hoc databases, this simplified model uses existing open databases, normally utilized by urban planners. It elaborates these data with a GIS-based assessment to evaluate energy consumptions with spatial information and low times simulation, calculating and adapting some of the input variables from building to block of building scale. In addition, this work investigates the relationship between urban form and the energy performance of typical blocks of buildings during the summer season, to understand the effect of buildings typology and urban morphology on cooling energy demand.

2. Materials and Method

This section describes energy models and tools used to simulate cooling energy demand, the input data and the case study. The hourly cooling energy demand of residential buildings was simulated for six different neighborhoods in the city of Turin (Italy).

In the first part of this work, the GIS-based engineering model has been fixed up and then validated at urban scale by comparing the cooling demand simulated at two scales: (i) at building scale, considering five typical residential buildings for each of the six neighborhoods (30 buildings), and the comparison has been carried out with CitySim tool and ISO 52016 standard assessment; (ii) at block of buildings scale, five blocks of buildings with different urban shapes have been selected among the neighborhoods, and the simulations have been carried out with CitySim tool.

Journal of Physics: Conference Series 2177 (2022) 012016 doi:10.1088/1742-6596/2177/1/012016

In the second part, the GIS-based engineering model has been applied to the five different blocks of buildings in order to evaluate how the urban form influences the demand for space cooling considering also different types of buildings as a function of the relative construction periods (until 1918, 1961-70, 1981-90, and after 2006).

2.1. Urban-scale energy models and tools

According to previous works [2], [5], an existing GIS-based engineering model has been designed and validated for the space heating simulations at urban scale. In this study, the same model was improved to evaluate hourly cooling energy demand of residential buildings at building level and at block of buildings scale comparing its results with: CitySim tool [13] and the assessment of ISO 52016-1:2017. To design the GIS-based engineering model, the dynamic building thermal balance (according to ISO 52016-1:2017 and ISO 52017:2017 standards) was adapted to the neighborhood scale using GIS tools and the information of buildings from the municipal technical map. The GIS-based model is a lumped parameter model based on three thermodynamic systems (TSs): (i) the opaque envelope, composed by all opaque surfaces that separate the heated volume of the buildings from the external environment (or unheated zones); (ii) the glazing component, that separates the heated zone from the external environment; (iii) the inside part of the building which include internal partition and structures, air, occupants, and furniture. Equation 1 describes the dynamic heat balance for each TS ($\emptyset_T$ and $\emptyset_V$ are positive when the external temperature is minor than the internal one) and Figure 1 the application at block of building scale.

$$C_{TS}\frac{dT_{TS}}{dt} = \emptyset_{sol} + \emptyset_I - (\emptyset_C + \emptyset_T + \emptyset_V) \qquad (1)$$

where, for each TS: C is the heat capacity (JK^{-1}); T is the temperature of the TSs (K); t is the time (s); $\emptyset_{sol}$ is the heat flow rate from solar gains; $\emptyset_I$ is the heat flow rate from internal gains; $\emptyset_C$ is the heat flow rate from the cooling system; $\emptyset_T$ is the heat flow rate by transmission; $\emptyset_V$ is the heat flow rate by ventilation (usually in Italian residential buildings, the humidification and dehumidification of air is not controlled, then only sensible heat flow components were considered).

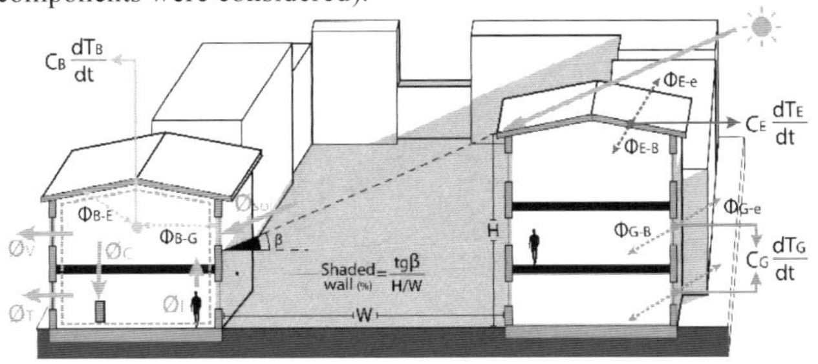

Figure 1. TSs of building (B), opaque envelope (E), and glazing (G) at block of buildings scale.

To describe the urban morphology and the buildings geometry, GIS tools have been applied to the built environment at district scale; in particular ArcGIS 10.7 and SOLWEIG 4.2 have been used to calculate the built areas, volumes, *SVF* at ground-roof level and *H/W* to characterize solar exposition and shadows. The novelty of this model is the application of the heat balance of buildings at the neighborhoods adapting some variables from the building scale to the district scale. In particular, two urban parameters (*SVF* and *H/W*) –that allow to include mutual shading and view factors of surrounding built-up context– were used to evaluate heat fluxes between the block of heated-cooled buildings and the external environment at a neighborhood scale [2]. Input data were processed and elaborated with the support of GIS tools [2,5]; the main information refer to: (i) geometrical characteristics of the buildings, such as the surface-to-volume ratio (*S/V*), the heat loss surfaces, the glazing area (that was assumed as 1/8 of the net floor area, Italian Decree 190/1975), and the heated net volume (75% of the gross volume); (ii) thermo-physical properties of the buildings according to the relative construction period: thermal transmittances (*U*) and thermal capacities (*C*) of opaque and transparent envelope, and solar energy transmittance of the glass ($g_{\perp}$); (iii) local climate conditions refer to climate data recorded by the nearest weather station.

Journal of Physics: Conference Series 2177 (2022) 012016 doi:10.1088/1742-6596/2177/1/012016

For the cooling demand simulation, firstly, the urban building thermal balance was applied at building scale in order to calibrate the model; secondly the input data were aggregated at block of buildings scale and the GIS-based engineering model was validated at this scale. Thanks to the flexibility of the methodology, the model can be quickly applied to different scales. The presented model is able to guarantee an accurate urban-scale results with less input data and the model simulation times are significantly lower than existing tools [5].

2.2. Case study

A place-based methodology was applied to six neighborhoods in the city of Turin (Italy), identified as homogeneous zones with different building characteristics and urban contexts [14]. In the city of Turin, the quota of residential buildings is around the 75% of the heated building stock, and the 80% of them was built before the 1970 (before the first Italian Law on energy savings in buildings). In this work the model was applied taking into account the construction period as a reference to identify different thermo-physical properties of the buildings, since the real energy consumption data for space cooling were not available. Table 1 shows the thermal transmittances (W/m²/K) and the thermal capacities (kJ/m²/K) of the opaque and transparent building components for nine construction periods used as input data.

Table 1. Thermo-physical properties of the buildings

Period	U_g	U_{wall}	U_{roof}	U_{floor}	$g_{\perp}$	$C_{envelope}$
	W/m²/K				-	*kJ/m²/K*
< 1918	5.9	1.45	1.8	1.75	0.82	504
1919-45	5.9	1.35	1.8	1.58	0.82	504
1946-60	5.9	1.18	1.8	1.23	0.82	283
1961-70	5.9	1.13	2.2	1.3	0.82	283
1971-80	5.9	1.04	2.2	1.21	0.82	257
1981-90	3.3	0.78	1.18	1.95	0.70	264
1991-00	2.7	0.7	0.68	0.8	0.70	274
2001-05	2.7	0.7	0.68	0.8	0.70	274
> 2006	1.8	0.46*	0.43*	0.43*	0.62	267

*Legislative Decree 311, 29 December 2006

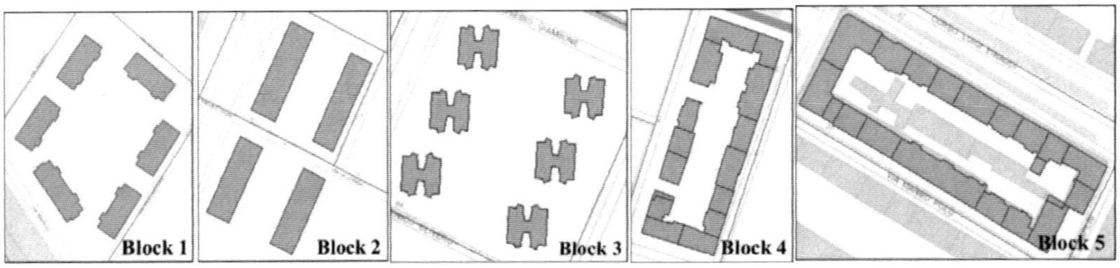

Figure 2. Blocks of buildings.

Table 2. Blocks of buildings' characteristics

Block of Buildings		Block 1	Block 2	Block 3	Block 4	Block 5
Neighborhood (name)		Arquata	Villaggio Olimpico		Mediterraneo	Crocetta
N. of buildings	-	6	4	6	15	20
Surface of flat	m²/flat	62	114	90	79	117
Components per family	Inh/fam	1.63	2.16	2.09	1.85	2.05
Prevalent Period	-	1961 - 1970	1971 - 1980	1961 - 1970	1946 - 1960	1961 - 1970
U_{wall}	W/m²/K	1.13	0.93	1.13	1.16	1.2
U_{roof}		2.2	1.73	2.2	1.73	2.02
U_{floor}		1.3	1.55	1.3	1.20	1.40
$U_{glazing}$		5.9	4.72	5.9	5.74	5.9
$g_{\perp}$	-	0.82	0.77	0.82	0.81	0.82
$C_{envelope}$	J/m²/K	282,518	260,199	282,518	282,094	345,277
Window-to-wall ratio	%	14	20	12	21	23
S/V	m²/m³	0.41	0.29	0.38	0.28	0.27
Prevalent Azimut	°	N/W = +125	N/W = +155	N/W = +165	N/W = +110	N/E = -150
		S/E = -55	S/E = -65	S/E = -85	S/E = -70	S/W = +30
BCR	m²/m²	0.16	0.31	0.12	0.31	0.38
BD	m³/m²	2.58	7.99	3.88	8.63	7.78
H/W	m²/m²	0.22	0.35	0.37	0.73	0.58
SVF	-	0.77	0.74	0.82	0.68	0.67

Journal of Physics: Conference Series 2177 (2022) 012016 doi:10.1088/1742-6596/2177/1/012016

The GIS-based engineering model has been calibrated and validated at urban scale by comparing the cooling demand simulated at a block of buildings scale. Figure 2 shows the five blocks of buildings selected. It is possible to observe different urban form: open-court buildings-Block 1, row-Block 2, tower-Block 3, courtyard-Block 4-5); blocks 4 and 5 represent the typical courtyards in the city of Turin with high-building density and different orientations. The main input data at block of buildings scale are reported in Table 2; the variables in the last rows well describe the different urban forms.
Energy simulations have been done according to the cooling season 2014 (same year used for the previous work [2]). Since in Turin the cooling energy consumptions is quite low, to analyze the results of the GIS-based model in the summer season, the warmer day of each month has been chosen: May 30th, June 12th, July 18th, August 5th and September 1st. In Table 3 the average ($T_{ae,avg}$) and maximum daily ($T_{ae,max}$) air temperature (°C), and the global horizontal irradiation (*GHI*, kWh/m^2) have been indicated.

Table 3. The warmer day of each month for the cooling season 2014.

	May 30th	June 12th	July 18th	August 5th	September 1st
$T_{ae,avg}$, °C	21.4	28.7	27.6	24.3	23.0
$T_{ae,max}$, °C (hour)	25.9 (4 p.m.)	33.6 (5 p.m.)	32.1 (7 p.m.)	29.6 (6 p.m.)	28.1 (5 p.m.)
GHI, kWh/m^2/day	6.91	7.36	6.98	7.08	6.23

3. Results and Discussion

This section presents the GIS-based engineering model calibration and validation by analyzing the monthly, daily and hourly results. The cooling energy demand of 30 typical buildings for each neighborhood has been simulated for the summer season 2014 comparing the GIS-based engineering model outputs to CitySim tool and ISO 52016 standard. Afterwards, the effect of urban forms on cooling energy performance has been investigated using the GIS-based engineering model according to four different periods of construction.

3.1. GIS-based engineering model validation

The results of the hourly simulation of three residential buildings are shown in Figure 3: 'ED-5' built in 1919-46 in Crocetta neighborhood; 'ED-2' built in 1946-60 in Mediterraneo neighborhood; and 'ED-3' built in 1981-90 in Villaggio Olimpico neighborhood. The hourly cooling energy profiles assessed with the GIS-based model (in red), the CitySim tool (in blue) and the ISO 52016 standard (in green) are compared.

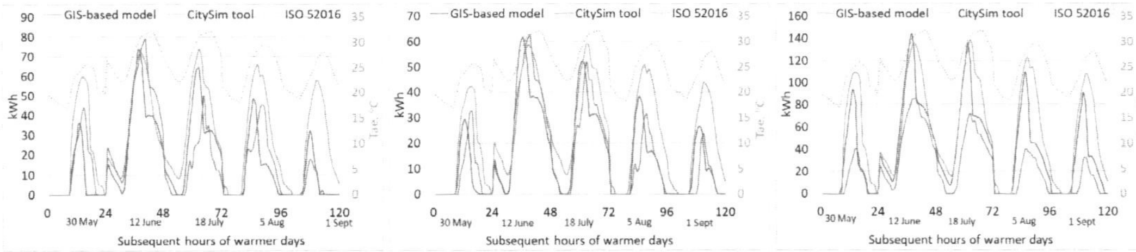

(a) 'ED-5' (1919-46, NW-SE) (b) 'ED-2' (1946-60, E-W) (b) 'ED-3' (1981-90, NE-SW)

Figure 3. Crocetta neighborhood: comparison of hourly cooling demand for the five warmer days: GIS-based model, CitySim tool and hourly method ISO 52016.

It can be observed that the hourly energy profiles have a typical trend related to the external air temperature T_{ae} (and to the solar irradiation). The GIS-based model and the CitySim tool results are very similar: the hourly energy peaks for the five days, and for both buildings are very close. The ISO 52016 standard simulates higher consumption than the other two tools, especially for month with lower external air temperature (May and September). These differences are mainly due to the fact that local climate conditions refer to the typical meteorological year (not to 2014); in this case, correlations between air

temperature/global horizontal irradiance and cooling demand have been used to compare the results according to the weather data used in the CitySim tool and the GIS-based engineering model.
The annual absolute relative error for the cooling season (April 15th - October 15th) calculated at building level is on average 30% (median 26%). There were minimum values of 21 and 22%, respectively in Raffaello and Crocetta neighbourhoods, and a maximum value of 70% in Villaggio Olimpico neighbourhoods. These results are compatible with the application of the GIS-based model which is on an urban scale and not on a building scale. The precision of the GIS-based model depends on the urban form, and it has been designed for the typical district of the city of Turin, the courtyard. Raffaello and Crocetta represent the typical neighbourhood of the city with compact condominiums built between 1946 and 1980, while Villaggio Olimpico was built more recently and it is characterized by towers and big isolated condominiums. It is necessary to consider that the relative error has less meaning when it refers to very low energy demands such as 2 to 6 kWh/m^3/year.

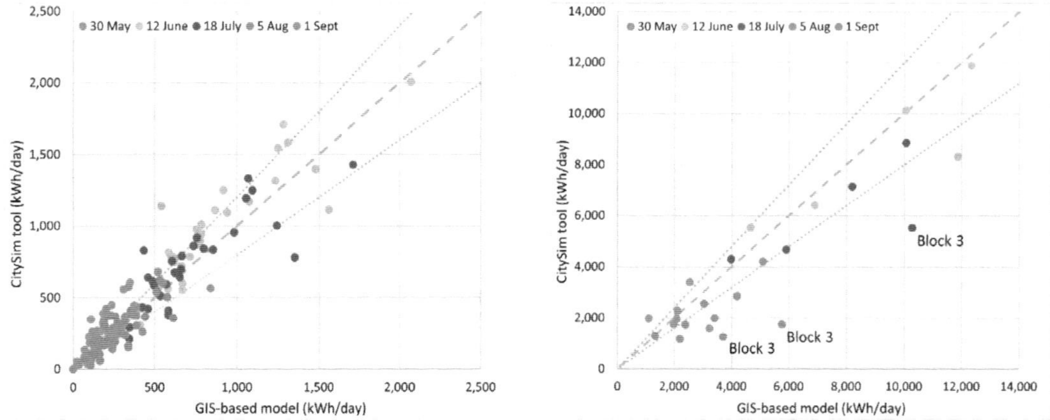

(a) Building scale: 30 buildings (b) Block of buildings scale: 5 blocks

Figure 4. Comparison of daily cooling demand for the five warmer days: GIS-based model and CitySim tool at building (a) and block of building scale (b).

Figure 4 shows the comparison of the daily cooling demand between the GIS-based model and the CitySim tool calculated for the five warmer days. In Figure 4a, the results refer to the energy simulations of the 30 residential buildings analyzed, while Figure 4b reports the cooling energy demand calculated for the five blocks of buildings. The GIS-based model shows a good accuracy in both cases, especially in June 12th and July 18th thanks to the highest daily temperatures, 33.6 and 32.1 °C, respectively.

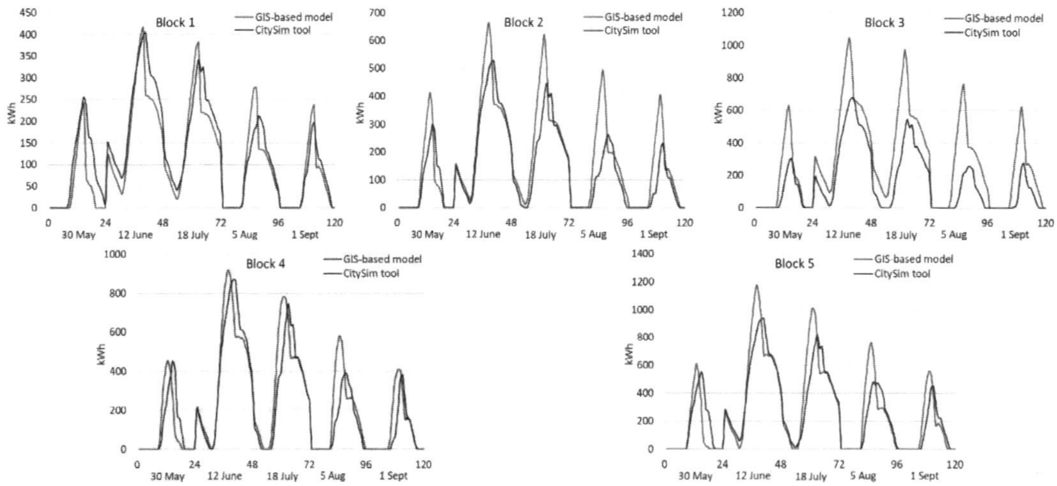

Figure 5. Comparison of hourly cooling demand at block of buildings scale for the five warmer days: GIS-based model and CitySim tool.

Figure 5 shows the comparison of the hourly space cooling demand between the GIS-based model (in red) and the CitySim tool (in blue) calculated for the five warmer days at block of buildings scale. As already observed in Figures 3 and 4, the model is less accurate for Villaggio Olimpico block 2 and especially for block 3. The precision also depends on the number of buildings in each blocks and on the scale of application [15], this is another reason why the GIS model is more accurate in blocks 4 and 5 with 15-20 buildings, compared to blocks 2 and 3 with 4-6 isolated buildings.

3.2. The effect of urban form on cooling demand

In this section the effect of urban form on cooling energy demand has been analysed by investigating the energy performance of five blocks of buildings (in Figure 2). This analysis takes into account different thermo-physical properties according to four construction periods: until 1918, 1961-70, 1980-90, and after 2006. The four construction periods have been selected considering a consistency variation of the thermal transmittance values of the opaque and transparent components and of the thermal capacities of the envelope that characterizes the buildings.

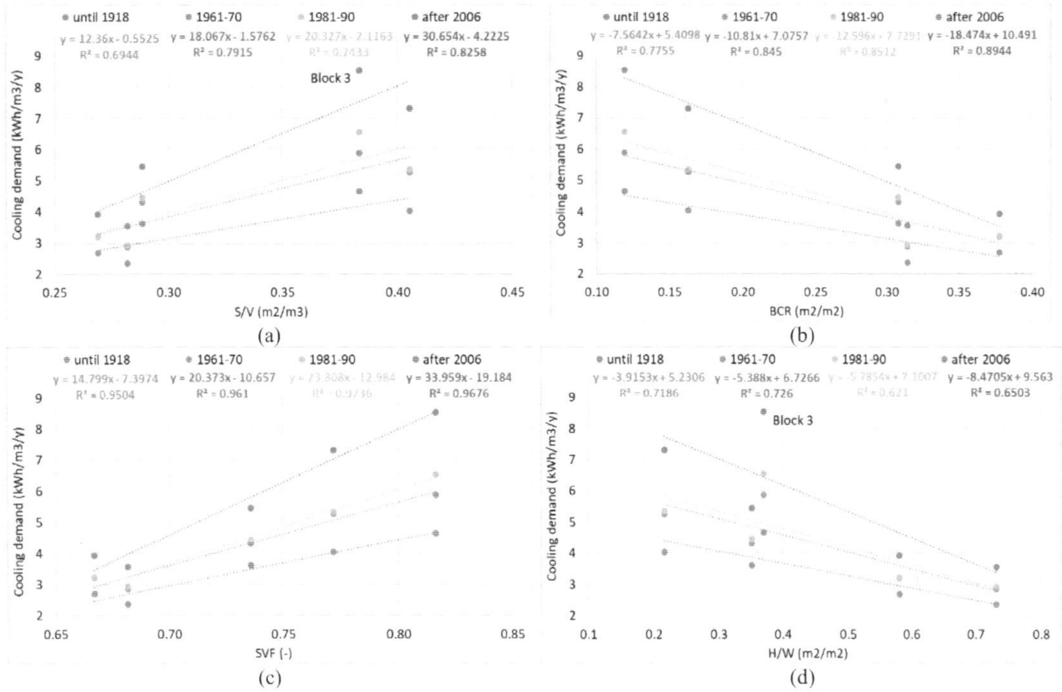

Figure 6. Simulation at block of buildings scale using the GIS-based engineering model for different urban forms and construction periods: correlation between the cooling demand and (a) the surface-to-volume ratio (*S/V*), (b) the building coverage ratio (*BCR*), (c) the sky view factor (*SVF*), and (d) the height-to-width ratio (*H/W*).

The main findings (in Figure 6) can be summarised as follows: (i) in all blocks of buildings investigated, older buildings have lower cooling energy demand, probably due to the higher thermal capacity values; (ii) cooling energy demand increases as *S/V* and *SVF* increase, and decreases with high values of *H/W* and *BCR*; (iii) the courtyard block (typical urban form in historical Turin districts), and the South-North orientation (block 4 in Mediterraneo) has the lowest cooling demand; (iv) the towers, behave abnormally and the GIS model is not accurate with this urban shape (block 3 in Villaggio Olimpico).

4. Conclusion

This work investigates how the urban form affects the energy performance of residential buildings during the cooling season. A GIS-based engineering model have been firstly validated for district-scale

applications, and then used to simulate the cooling energy demand in different neighborhoods with various morphologies.
This model uses urban variables mainly to evaluate the solar fluxes and the extra flux to the sky at block-scale. Thus, at the building scale the accuracy is low, but at the block scale it works with the exception of the isolated towers in block 3; the model has an accuracy that seems proportional to the number of buildings in the district. From the results it has emerged that the urban form and the shape of the built environment significantly influence the cooling energy demand of residential neighborhoods. In particular, courtyard with East-West orientation is a compact urban form with high building density, allowing a good shading on the internal facades of the courtyard and low cooling demand; in fact, it is a typical urban form of the historical districts in Turin.
Further developments of the GIS-based model are needed: (i) a sensibility analysis will be performed to evaluate the weight of the heat fluxes (*H/W* has a too high weight on solar gains); (ii) the thermodynamic system of glazing could be removed as it has a minimal impact on the energy demand of buildings; (iii) a more accurate description of the shaded environment exploiting the Solar Energy on Building Envelopes (SEBE) GIS-plugin.
Finally, the effect of the urban form to optimize the energy productivity of buildings will be investigated in order to improve self-sufficiency and self-consumption in densely built-up contexts. This GIS-based model could support policy makers by identifying energy efficiency measures and low carbon strategies that best suit each urban context, considering local constraints, and building codes: both in the design phase of new neighborhoods, and in the retrofit of existing buildings.

References

[1] J. Grove-Smith, V. Aydin, W. Feist, J. Schnieders, and S. Thomas, «Standards and policies for very high energy efficiency in the urban building sector towards reaching the 1.5°C target», *Current Opinion in Environmental Sustainability*, Feb. 2018, doi: 10.1016/j.cosust.2018.04.006.
[2] G. Mutani, V. Todeschi, and S. Beltramino, «Energy Consumption Models at Urban Scale to Measure Energy Resilience», *Sustainability*, 2020, 12(14), 5678, doi: 10.3390/su12145678.
[3] K. Lundgren and T. Kjellstrom, «Sustainability Challenges from Climate Change and Air Conditioning Use in Urban Areas», *Sustainability*, Jul. 2013, doi: 10.3390/su5073116.
[4] C. Cerezo, J. Sokol, S. AlKhaled, C. Reinhart, A. Al-Mumin, and A. Hajiah, «Comparison of four building archetype characterization methods in urban building energy modeling (UBEM): A residential case study in Kuwait City», *Energy and Buildings*, Nov. 2017, doi: 10.1016/j.enbuild.2017.08.029.
[5] V. Todeschi, R. Boghetti, J. H. Kämpf, and G. Mutani, «Evaluation of Urban-Scale Building Energy-Use Models and Tools—Application for the City of Fribourg, Switzerland», *Sustainability*, 2021, 13(4), 1595, doi: 10.3390/su13041595.
[6] P. Caputo and G. Pasetti, «GIS tools towards a renovation of the building heritage», *Energy Procedia*, Oct. 2017, doi: 10.1016/j.egypro.2017.09.388.
[7] M. Silva, V. Oliveira, and V. Leal, «Urban Form and Energy Demand: A Review of Energy-relevant Urban Attributes», *Journal of Planning Literature*, Nov. 2017, doi: 10.1177/0885412217706900.
[8] C. Chatzipoulka and M. Nikolopoulou, «Urban geometry, SVF and insolation of open spaces: London and Paris», *Building Research & Information*, Nov. 2018, doi: 10.1080/09613218.2018.1463015.
[9] A. Salvati, M. Palme, e L. Inostroza, «Key Parameters for Urban Heat Island Assessment in A Mediterranean Context: A Sensitivity Analysis Using the Urban Weather Generator Model», *IOP Conf. Ser.: Mater. Sci. Eng.*, Oct. 2017, doi: 10.1088/1757-899X/245/8/082055.
[10] Y. Ahn and D.-W. Sohn, «The effect of neighbourhood-level urban form on residential building energy use: A GIS-based model using building energy benchmarking data in Seattle», *Energy and Buildings*, Aug. 2019, doi: 10.1016/j.enbuild.2019.05.018.
[11] A. Salvati, M. Palme, G. Chiesa, and M. Kolokotroni, «Built form, urban climate and building energy modelling: case-studies in Rome and Antofagasta», *Journal of Building Performance Simulation*, Mar. 2020, doi: 10.1080/19401493.2019.1707876.
[12] E. Ahmadian, B. Sodagar, C. Bingham, A. Elnokaly, and G. Mills, «Effect of urban built form and density on building energy performance in temperate climates», *Energy and Buildings*, Apr. 2021, doi: 10.1016/j.enbuild.2021.110762.
[13] E. Walter and J. H. Kämpf, «A verification of CitySim results using the BESTEST and monitored consumption values», in *Proceedings of the 2nd Building Simulation Applications conference*, 2015, ISBN 9788860460745.
[14] G. Mutani, M. Carozza, V. Todeschi, and A. Rolando, «Urban-Scale Energy Models: relationship between urban form and energy performance», in *2020 IEEE 3rd International Conference and Workshop in Óbuda on Electrical and Power Engineering (CANDO-EPE)*, Nov. 2020, doi: 10.1109/CANDO-EPE51100.2020.9337760.
[15] G. Mutani and V. Todeschi, «An Urban Energy Atlas and Engineering Model for Resilient Cities», *International Journal of Heat and Technology*, Vol. 37(4), 2019, pp. 936-947, doi: 10.18280/ijht.370402.

Evaluation of the thermal-hydraulic performances of a once-through steam generator in nuclear fusion applications

Federico Galli, Cristiano Ciurluini*, Vincenzo Narcisi, Fabio Giannetti, Gianfranco Caruso

DIAEE – Nuclear Section, "Sapienza" University of Rome, Corso Vittorio Emanuele II, 244, 00186, Rome, Italy
* Corresponding author e-mail: cristiano.ciurluini@uniroma1.it

Abstract. After decades of operation in nuclear power plants, Once-Through Steam Generators (OTSGs) were recently proposed for nuclear fusion applications. In particular, they are supposed to be installed in the primary cooling systems of the European Union Demonstration fusion power plant (EU-DEMO). One of the key reactor components is the Breeding blanket (BB). Among the BB concepts that are currently under study, Water-Cooled Lithium-Lead (WCLL) option was considered for this work. The WCLL blanket is divided in two main subsystems, the breeder zone (BZ) and the first wall (FW), each one provided with an independent cooling circuit, named Primary Heat Transfer System (PHTS). Thermal power removed from BB by BZ and FW PHTS is driven to the Power Conversion System (PCS) to be converted into electricity. The thermal coupling is ensured by two OTSGs per system. At the Department of Astronautical, Electrical and Energy Engineering (DIAEE) of Sapienza University of Rome, a simulation activity was carried out to understand the component thermal-hydraulic behavior during DEMO normal operations. For calculation purposes, a full model of the steam generator was prepared by using a modified version of RELAP5/MOD3.3 system code. The computational activity performed allows to preliminary characterize the OTSG thermal-hydraulic performances during both pulse and dwell phases.

Keywords: DEMO, Primary Heat Transfer System, Balance of Plant, RELAP5, pulse/dwell transition.

1. Introduction

As an important step in the Roadmap to Fusion Electricity, European Union (EU) is performing a pre-conceptual design study of a Demonstration fusion power plant (DEMO), [1]. The reactor should demonstrate the capability of producing few hundred MWs of net electricity while operating with a closed-tritium fuel cycle. The Breeding blanket (BB) is one of the key reactor components. It accomplishes several functions such as cooling device, tritium breeder (ensuring reactor self-sufficiency) and neutron shield. Different BB concepts were selected to be investigated in the DEMO R&D strategy. Among them, there is the Water-Cooled Lithium-Lead (WCLL), [1], that is the option under study in the current work. The WCLL blanket is constituted by two main subsystems: the breeder zone (BZ) and the first wall (FW). Each one is provided with an independent cooling circuit, called Primary Heat Transfer System (PHTS). The former removes the thermal power generated in the BZ by the interactions between the lead-lithium (liquid breeder) and the neutrons coming from the plasma. The latter cools the FW component, heated up by the incident heat flux and by the neutron wall load. The thermal power removed by BZ and FW PHTS is directly driven to the Power Conversion System (PCS) for its conversion into electricity, [2]. Heat transfer between PHTSs and PCS takes place within two steam generators (per circuit). Once-Through Steam Generators (OTSG) were selected for this application, [2]. Even if this is the first time such technology was proposed for nuclear fusion applications, OTSGs has been used and operated for decades in the field of nuclear fission power plants, in particular in Pressurized Water Reactors (PWR) [3]. In such environment, a full thermal-hydraulic characterization of the steam generator was already available, [3][4][5]. In the framework of EUROfusion Work Package Balance of Plant (WPBOP), at the Department of Astronautical, Electrical and Energy Engineering (DIAEE) of Sapienza University of Rome, a simulation activity was performed to evaluate the thermal-hydraulic behavior of this fundamental component during DEMO normal operations, considering that the pulsed plasma regime significantly jeopardizes the OTSG performances. To reach this goal, a full model of the component was prepared by using a modified version of RELAP5/MOD3.3 system code. This new extended version was developed at DIAEE, in collaboration with ENEA, to enhance the code capability in simulating fusion reactors, [6]. It is actually considered the reference code version to carry out the future simulation activity involving DEMO reactor.

2. Once-Through Steam Generators for DEMO application

2.1. General description of the component

The OTSG design foresees a straight-tube, straight-shell layout, with flat tubesheets and hemispherical primary heads. [3]. An overview of the OTSG design is shown in Figure 1a. Primary system is bounded by hemispherical heads, tubesheets and tube bundle. Primary coolant enters from the OTSG top and flows downwards, exiting from the component bottom. Secondary side is the steam-producing section. It is bounded by the shell, named vessel, the tube outer surface and the tubesheets. A cylindrical shroud, called riser, surrounds the tube bundle and channels the secondary flow along the thermal height. Subcooled feedwater enters the steam generator laterally, in the lower vessel (downcomer) section. Firstly, it is preheated by aspirating steam coming from the tube bundle region (recirculated flow). Then, secondary water moves through the annular downcomer. It reaches the vessel bottom in nearly saturated conditions. Later, it rises in the central shroud where it boils to dry steam and then it is superheated. Once reached the top, steam is turned by the tubesheet and directed to the annulus between riser and vessel, in the OTSG upper section. Here, it flows downwards to the two outlet nozzles, connected laterally. Starting from the shroud bottom, as feedwater is converted to superheated steam, three heat transfer regions can be identified: Nucleate Boiling Region (NBR), Film Boiling Region (FBR) and SuperHeat Region (SHR), [3]. The former is where saturated feedwater begins to boil. Tube outer surface remains wetted while small bubbles rapidly form and break away from it. Thanks to the turbulence due to bubble formation, this heat transfer mode ensures a high heat transfer coefficient (HTC). For this, most of the primary-to-secondary thermal exchange occurs in the NBR. The nucleate and forced convective boiling continue until enough water is vaporized and the liquid layer is replaced by steam on the surface of the tubes. Therefore, film boiling occurs at high qualities after the dry-out point and fully develops within a very short axial distance. In the film boiling heat transfer, the heat flux is sharply reduced and heat transfer occurs by convection through the steam and evaporation of entrained liquid droplets in the saturated core. At FBR top, only dry steam is present. In the final SHR, thermal power transferred from primary fluid is used to produce superheated steam. The evolution of OTSG technology consists in the Integral Economizer Once-Through Steam Generators (IEOTSG), [3]. The IEOTSG is a true once-through steam generator, without recirculation in the secondary side. Feedwater is admitted near the bottom of the component. In the central riser, it is quickly preheated to saturation while dry steam is obtained at about the same elevation as in the OTSG. Once reached the upper tubesheet, superheated steam is diverted

Journal of Physics: Conference Series 2177 (2022) 012017 doi:10.1088/1742-6596/2177/1/012017

downwards through the annulus between riser and vessel. Differently from OTSG, outlet nozzles are installed in the lower part (slightly above the inlet ones) to maintain a high temperature of the shell. In the IEOTSG, a further region can be identified in addition to the ones already described for OTSG. It is located at the shroud bottom and called economize region. Here, the subcooled liquid is rapidly heated and brought to saturation temperature. Its extension is reduced since it is characterized by high HTCs in both primary and secondary sides. The IEOTSG design is reported in Figure 1b.

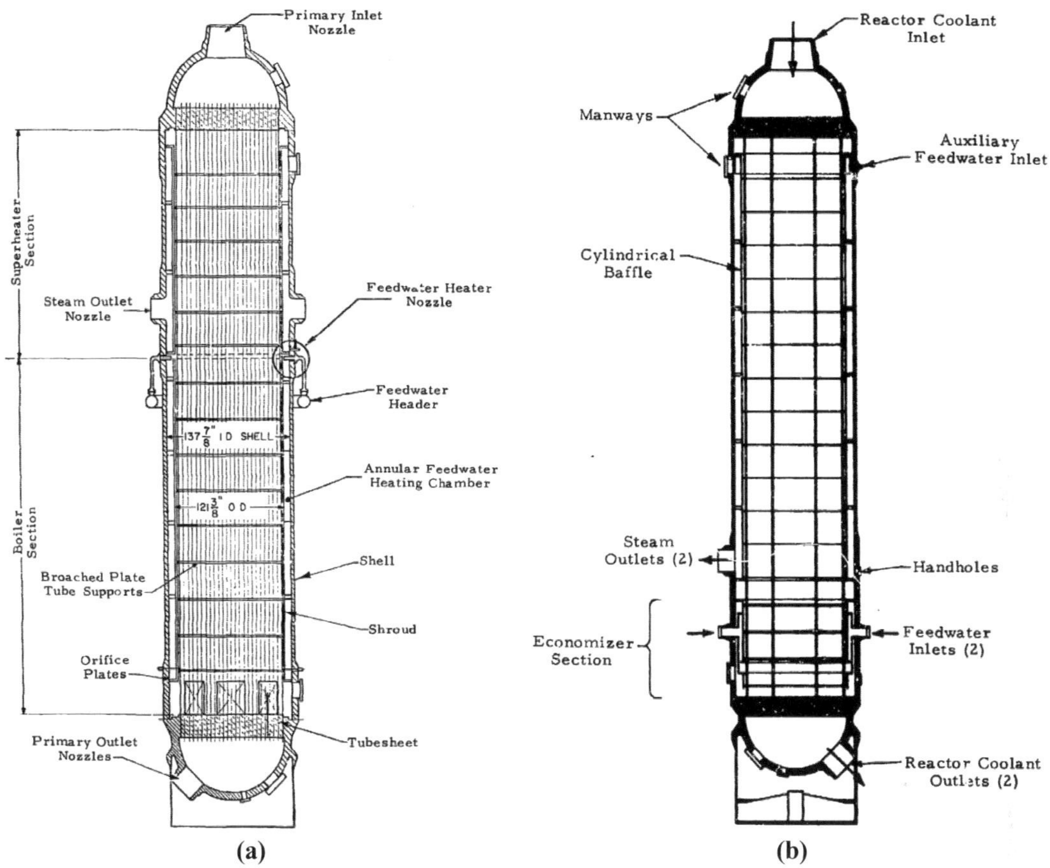

Figure 1. Design overview of OTSG **(a)** and IEOTSG **(b)** technologies, [3].

2.2. PHTS OTSGs pre-conceptual design

Regarding the OTSGs to be installed in DEMO BB PHTS, their design is still at a pre-conceptual stage, [2]. Such design was scaled from existent units still operating in nuclear fission power plants. Steam generator rated power was used as scaling factor. OTSG technology was considered to be appropriate to be installed in PHTS circuits since the primary (PHTS itself) and secondary (PCS) sides water thermodynamic conditions are comparable with respect to the ones of a PWR. Indeed, PHTS water enters the OTSGs (i.e. exits from the BB) at 601 K and it is cooled down to 568 K. Primary pressure is 15.5 MPa. On the secondary side, feedwater is admitted at 511 K and PCS reference pressure is set to 6.41 MPa. The main design data are reported in Table 1, for both BZ and FW steam generators. The connection between riser and vessel, allowing the recirculation, is located at nearly 60% of the thermal height.

Table 1. Pre-conceptual design parameters for BZ and FW OTSGs, [2].

Parameter	Unit	BZ OTSG	FW OTSG
Number of units (per PHTS)	-	2	2
Rated Power	MW	742	220
Number of tubes	-	7569	2197
Thermal height	m	12.99	12.99
Tube outside diameter	mm	15.88	15.88
Tube thickness	mm	0.89	0.89
Pitch to diameter ratio	-	1.28	1.28
Lattice	-	square	square
Vessel external diameter	m	2.9	1.5

2.3. RELAP5 model

As stated in section 2.2, even if rated power is quite different, BZ and FW OTSGs have the same pre-conceptual design rationale. Thus, steam generators of both primary cooling systems are supposed to have similar thermal-hydraulic (TH) performances during DEMO normal operations. A preliminary analysis on the BZ OTSG was already performed, [6][7]. For this reason, the current simulation activity focuses on the FW OTSG. A TH model of the component was prepared by using RELAP5/MOD3.3 system code. Its schematic view is shown in Figure 2. The same vertical mesh was adopted for the control volumes of all the RELAP5 components simulating the OTSG primary and secondary sides. Primary side was modelled with two branches, representing the inlet and outlet hemispherical heads, and an equivalent pipe component simulating the tube bundle. The PHTS hot leg was also included in the input deck. Inlet PHTS water thermodynamic conditions (temperature and mass flow, section 2.2) were set as boundary conditions (BCs). OTSG secondary side was simulated with four equivalent pipes, corresponding to lower/upper annular downcomer sections and to lower/upper central shroud sections. Three branches were used to link these pipe components, to manage the connections with the feedwater and steam lines and to simulate the recirculated flow. Feedwater line was added to the input deck by means of a dedicated pipe component. Steam lines were modelled up to the Turbine Stop Valves (TSV in Figure 2) and equipped with Safety Relief Valves (SRV in Figure 2). PHTS water temperature at blanket inlet (i.e., OTSG outlet) is a strict DEMO requirement. As a preliminary tentative, a temperature control system was associated to PCS feedwater. A Proportional-Integral (PI) controller was implemented to regulate the heat transfer inside the steam generator by tuning the secondary flow. In this way, the right PHTS temperature is obtained at OTSG outlet and the compliance to DEMO requirement is ensured. Control strategy for the steam generator will be refined and adapted once defined in the framework of the WPBoP research activity. Instead, feedwater inlet temperature (see section 2.2) was set as boundary condition. RELAP5 heat structures were used to simulate the thermal transfer taking place within steam generator, as well as the component heat losses. They also allow to account for the OTSG steel inventory (i.e., thermal inertia). Finally, what is worth to be emphasized are the correlations used by RELAP5 code to evaluate the heat transfer coefficient in both primary and secondary sides, [8]. If single-phase fluid is present, i.e. PHTS side and PCS side when feedwater is subcooled and steam is superheated, Dittus-Boelter correlation is adopted, [9]. Chen correlation is used for the nucleate boiling, [10], while Bromley model is implemented for the film boiling, [11]. The dryout quality is calculated by using the Groeneveld look-up tables, [12].

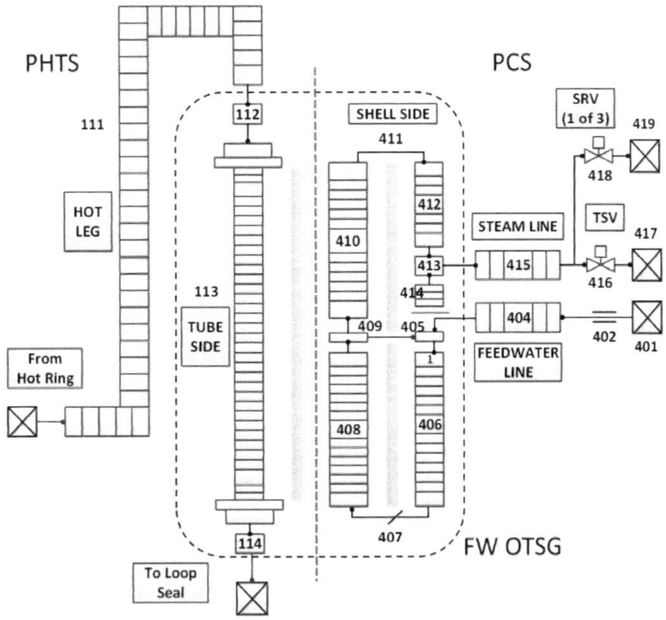

Figure 2. Schematic view of RELAP5 model developed for FW OTSG.

3. Results

3.1. Boundary conditions

Initially, the RELAP5 model discussed so far was used to preliminary assess the FW OTSG TH performances during pulse phase of DEMO normal operations. The current steam generator design, described in section 2.2, was tested in order to demonstrate its cooling capability with respect to BB and PHTS circuit. Standard procedure foresees that primary and secondary inlet conditions are imposed and the thermal power exchanged is checked. This procedure cannot be applied in the current simulation activity. In fact, during DEMO pulse, BB power and inlet/outlet temperatures (so also the mass flow) are strict requirements. For this, in the calculations performed, primary side inlet conditions are set while the PHTS water outlet temperature is monitored by the PI controller acting on the OTSG secondary flow (see section 2.3). Being all the primary side parameters set or controlled, also the steam generator exchanged power is imposed. On the secondary side, feedwater inlet temperature is a boundary condition since it depends on the feedwater preheaters train installed in the PCS upstream the OTSG. The parameters to be checked as simulation outcomes are the secondary and recirculation mass flows, as well as the steam outlet temperature.

3.2. Sensitivity analysis to assess the OTSG Model

Firstly, mesh and time step sensitivities were performed on the TH model presented in section 2.3. Thermal height (see Table 1) was divided in 20, 30 and 50 control volumes. A time step of 10^{-3} s was adopted. Incrementing the mesh number, the code predicts a higher steam outlet temperature (587, 589.3 and 590.6 K respectively) and, consequently, a lower secondary flow (116.2, 115.7, 115 kg/s). The mesh refinement also produces a decrease of recirculation inside the component (9.3, 8.9, 8.1 kg/s). These minor differences in the results are due to the precision in the assessment of the heat transfer regions within the steam generator (discussed in section 2.1). With a coarser mesh (20 control volumes), the NBR length is slightly overestimated. This increases the average HTC on the secondary side, that is the one limiting the overall steam generator heat transfer coefficient. To clarify this point, Figure 3 reports the primary/secondary side HTCs and the wall conductivity (computed as tube thermal conductivity, k, on tube thickness, s) against the normalized thermal height for the 30-mesh model. The normalized thermal height is considered going from the riser bottom to top. Primary side HTC and wall conductivity are nearly constant along the steam generator. Instead, secondary side HTC is of the same order of magnitude of the other two only up to the dryout occurring. In the residual length, corresponding to FBR plus SHR, the heat transfer between primary and secondary sides is strongly reduced. By increasing the mesh number, the model accuracy in predicting the effective height where dryout occurs is enhanced, even if the benefit is modest, as demonstrated by the small differences in the calculation outcomes. In conclusion, the 30-mesh model was selected to simulate the steam generator thermal height. With this nodalization, a time step sensitivity was carried out by varying this parameter from 1×10^{-4} s to 1×10^{-3} s. The results obtained by simulations do not change, either globally or locally. For this reason, they were not included in the paper and 10^{-3} s was the time step chosen to continue the computational activity. A further simulation was run by taking into account the RELAP5 additional model named Reflood, [8]. It was used to evaluate the 2D-conduction effect inside the tube wall. Without it, the system code performs 1D computations considering the wall conduction in the radial direction only. Instead, activating the Reflood, also axial conduction is taken into account. Secondary side HTC is the one mostly affected by the model activation. As shown in Figure 4a, the main effect to be detected is the increase of the peak before the dryout occurrence. Locally, the exchanged power rises significantly. Globally, the Reflood activation provokes a slight heat transfer increment, consisting in a higher steam outlet temperature (589.3 to 591 K) and a lower secondary flow (115.7 to 115.3 kg/s). In the following, to conservatively evaluate the steam generator TH performances, it was chosen to not consider this model.

3.3. Scoping calculations on DEMO pulse phase

Once selected the best settings for the FW OTSG model, several calculations were run to investigate the component design in different operative conditions, such as Beginning of Life (BoL) and End of Life (EoL), as well as to study alternative solutions. Simulations performed are collected in Table 2.

Table 2. DEMO pulse phase: main features related to calculations performed.

	Design Features				Simulation Outcomes		
ID	OTSG/ IEOTSG	BOL/ EOL	PCS pressure [MPa]	Feedwater Inlet Temp. [K]	Feedwater Mass Flow [kg/s]	Rec. Mass Flow [kg/s]	Steam Out Temp. [K]
C1	OTSG	BOL	6.41	511	115.7	8.9	589.3
C2	OTSG	EOL	6.41	511	121.4	6	565.9
C3	IEOTSG	BOL	6.41	511	115.8	-	589.1
C4	OTSG	BOL	6.00	507	113	9.6	593.1
C5	OTSG	BOL	6.80	515	120.1	5.4	579.6

To simulate the OTSG operative conditions at EoL, both tube fouling and plugging must be considered. The first reduces the overall heat transfer coefficient by adding two thermal resistances, related to deposits affecting tube internal and external surfaces; the second diminishes the heat transfer surface by postulating that a certain fraction of the total tube number is plugged and not available for thermal exchange. Primary/secondary deposit thermal resistances ($0.5/1 \times 10^{-5}$ m^2K/W) were derived from literature, [13][14], and set according to engineering judgment, while the target value for tube plugging (10%) was a design data for the OTSG, [2]. The heat exchange degradation, affecting the OTSG when passing from BoL (C1) to EoL (C2), is clearly visible in Figure 4b, comparing the secondary side HTC in both conditions. The main detectable effects are the appearance of an economize region at the bottom of the riser and the increase of the height where dryout takes place. To remove the steam generator rated power, at EoL, the control system strongly increments the PCS feedwater flow, provoking a significant reduction of the steam outlet temperature (compare cases C1 and C2 in Table 2). This is done to augment the secondary side velocity field and partially supply to the HTC drop due to the additional thermal resistances. At the elevation where the recirculation occurs, the steam quality at EoL is lower with respect to BoL and this reduces the recirculated contribution (see Table 2).

As stated before, IEOTSG design represents the evolution of the OTSG technology, [3]. The TH model described in section 2.3 was rearranged in order to match the peculiarities of the IEOTSG layout (see section 2.1). The main differences between models consist in the elimination of the junction allowing the secondary side recirculation flow (situated at 60% of the thermal height) and the location of the feedwater/steam inlet/outlet nozzles near the bottom of the component. The two different solutions were preliminary compared by considering only the thermal-hydraulic performances. The aim was to evaluate if IEOTSG technology could enhance the heat transfer between primary and secondary sides and so represent an interesting alternative to be studied for DEMO applications. Simulation outcomes are collected in Table 2, where case C1 refers to OTSG while C3 for IEOTSG. From the TH point of view, only small discrepancies can be observed between the two options. The main difference concerns the presence of an additional heat transfer region, the economize one, at the bottom of IEOTSG (see Figure 4c). Its length is quite reduced due to the high HTCs and the height corresponding to the dryout occurrence is nearly the same for both steam generator solutions. Figure 4c confirms the IEOTSG design features discussed in section 2.1. Thus, during DEMO pulse phase, the two technologies have a quite similar thermal-hydraulic behavior. However, further analyses must be performed to assess other technical aspects, for example the thermomechanical response.

Focusing again on the reference OTSG design, what is worth to be further investigated is the steam generator TH response to a modification in the PCS thermodynamic conditions. Secondary side pressure was increased and decreased with respect to the reference value (see Table 2). In every calculation, the feedwater inlet subcooling was kept, i.e. inlet temperature was modified proportionally to the saturation temperature. As shown by results contained in Table 2 (compare cases C1 and C4) and the secondary side HTCs reported in Figure 4d, decreasing PCS pressure produces a heat transfer increment within the component. In fact, steam exits the OTSG at higher temperature, lower secondary flow is required to remove the rated power and the height where dryout shows up is reduced. The prevalent effect influencing the steam generator performances is the increase of the primary/secondary side temperature difference in the NBR, where the majority of the thermal exchange takes place. The opposite OTSG behavior can be observed when PCS pressure is increased (compare cases C1 and C5 in Table 2 and secondary side HTC trends in Figure 4d).

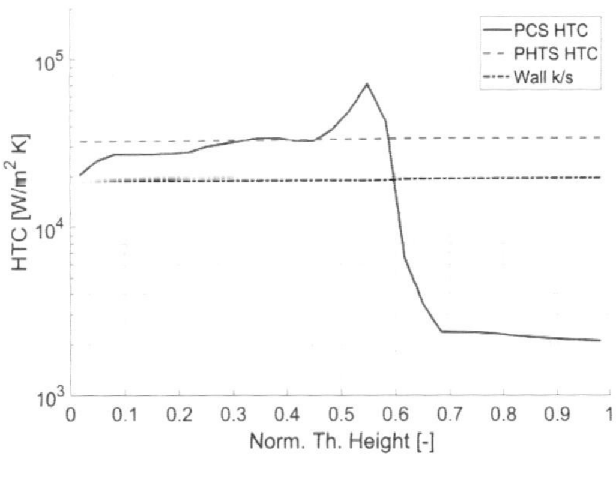

Figure 3. DEMO pulse phase, steam generator primary/secondary side HTCs and wall conductivity vs normalized thermal height (30-mesh model).

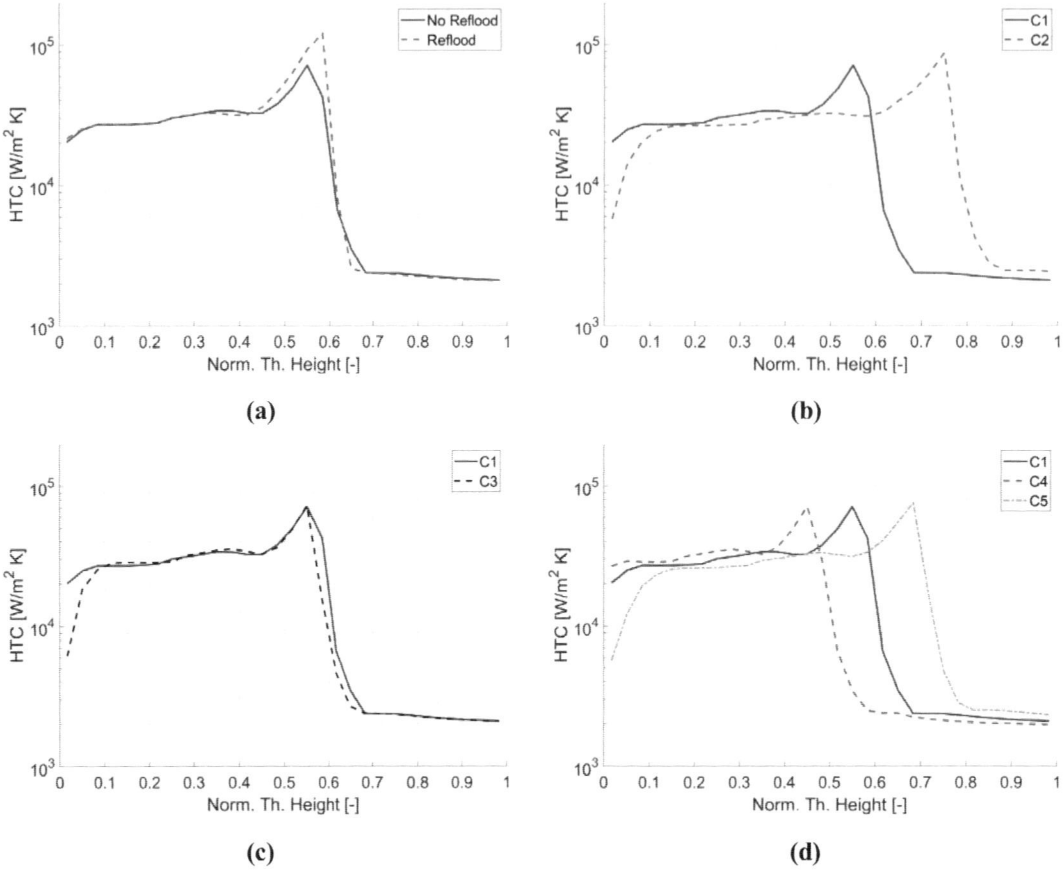

Figure 4. DEMO pulse phase, steam generator secondary side HTC vs normalized thermal height, comparison between simulation results: no Reflood vs Reflood **(a)**; OTSG at BoL (C1) vs EoL (C2) **(b)**; OTSG (C1) vs IEOTSG (C3) **(c)**; Different OTSG design for PCS pressure of 6.0 (C4), 6.41 (C1) and 6.8 MPa (C5) **(d)**.

3.4. Transient analysis on DEMO normal operations

Finally, FW OTSG behavior was evaluated during the overall DEMO normal operations, including both pulse and dwell phases. The reference plasma regime is composed by 2 hours of flat-top at full power, alternating with 10 minutes of dwell time. Power is ramp-down and ramp-up in nearly 100 s. During dwell time, only decay heat is still produced in the blanket, corresponding to nearly 2% of the reactor rated power. As for DEMO requirement, the BZ and FW primary pumps are always kept running at nominal velocity, in both pulse and dwell. Furthermore, during the latter, PHTS circuits must be operated at the system average temperature (nearly 583 K). A transient calculation was run by using case C1 (see Table 2) as initial condition. Plasma ramp down starts following 100 s of flat-top full power (grey background in Figure 5a to Figure 5d). Time was reset at the beginning of plasma ramp down, after which computation was run for 3000 s. According to DEMO requirement, a constant PHTS mass flow (equals to the one of pulse phase) was imposed for the overall simulation. Plasma ramp down and ramp-up were simulated by setting a variable BC for the PHTS temperature at OTSG inlet. To correctly simulate the feedback from BB, this boundary condition was derived from a previous analysis performed with a full model of the blanket and PHTS circuits, [7]. Although, in that case, OTSGs are present only in BZ primary circuit, as already discussed in section 2.3, the same TH behavior is expected for steam generators of both BZ and FW systems. For this reason, the BC derived from the study in [7] was considered appropriate for the current calculation. Finally, the PI controller acting on PCS feedwater was disabled. At its place, a variable trend was imposed (see Figure 5a). As preliminary tentative, a linear decrease and increase were adopted in correspondence of the plasma ramp down and ramp-up timing. During dwell, feedwater flow is kept at nearly 2% of the pulse value. In this way, the heat transfer within the steam generator is enough degraded to avoid PHTS overcooling and keep the primary system temperature at its average value, as for DEMO requirement. The appropriateness of this approach is demonstrated by PHTS outlet temperature trend shown in Figure 5b. Regarding the steam, two temperature peaks can be detected during pulse/dwell and dwell/pulse transitions (Figure 5b). These increases are rather limited (less than 5 °C in both cases) and do not cause concerns to the turbine operation. They are due to the temperature trend adopted as boundary condition for PHTS water at OTSG inlet (red line in Figure 5b, [7]). During ramp-down, the latter decreases slower with respect to the feedwater flow trend (blue line in Figure 5a). This delay is caused by the blanket thermal inertia, properly simulated in [7] and reported in the current work thanks to the specific boundary condition adopted. Instead, during ramp-up, PHTS inlet temperature rises faster with respect to feedwater flow. This depends on the particular shape of the plasma ramp-up curve adopted in [7]. It foresees a first sharp power increase, concentrated in the first few dozens of seconds, followed by a time interval where the power level is stabilized around the nominal value. The effect is a sudden rise in the PHTS inlet temperature, faster than the linear trend used for feedwater flow. However, in the future developments of the simulation activity, feedwater flow trend will be tuned to smooth these spikes. For example, it can be a little postponed in the pulse/dwell transition and slightly anticipated for the plasma ramp-up. What is worth to be noticed is that the liquid water inventory in the steam generator runs out during dwell time, as witnessed by Figure 5c. It shows the collapsed levels in the secondary side riser and lower downcomer, normalized with respect to the height of the components. During dwell, the OTSG is full of steam and this leads the secondary HTC to drop (Figure 5d), strongly limiting the overall heat transfer. This allows to keep the primary system temperature stable at the average value, as required.

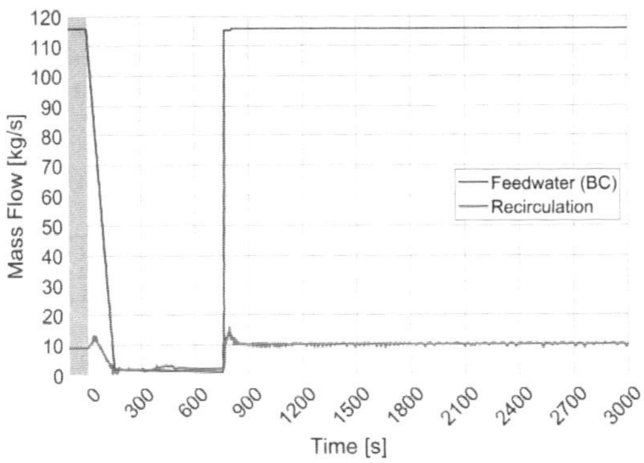

(a)

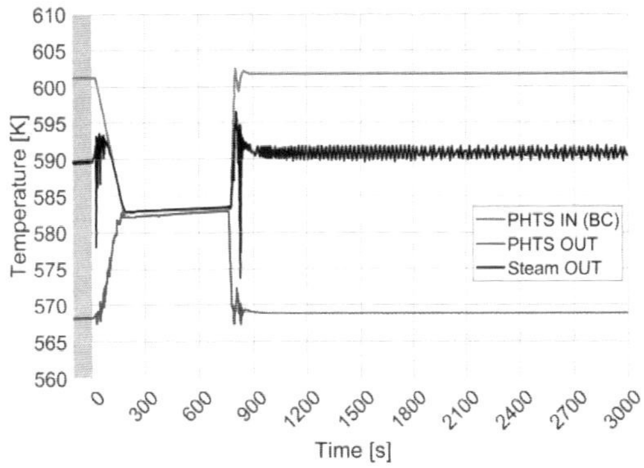

(b)

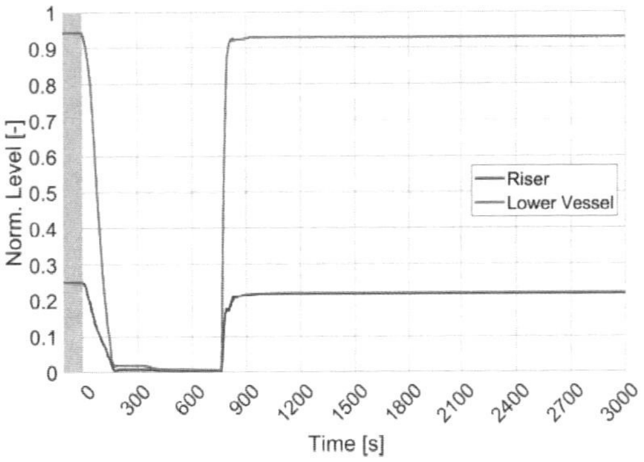

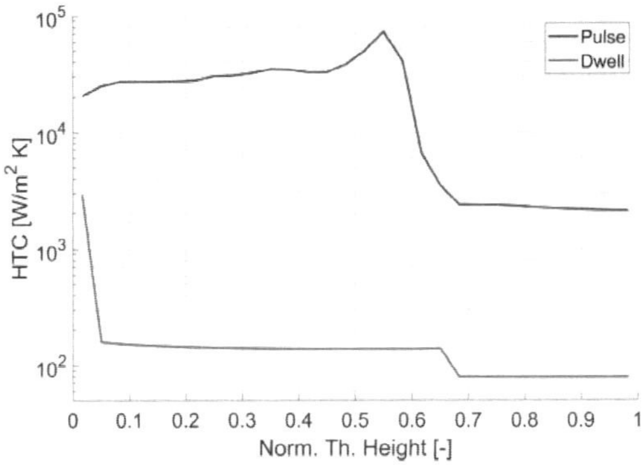

(c) (d)

Figure 5. FW OTSG main parameters during DEMO pulse and dwell phases: feedwater and recirculation flow rates **(a)**; main temperatures **(b)**; collapsed levels in the secondary side riser and lower downcomer **(c)**; secondary side HTC (comparison between pulse and dwell) **(d)**.

4. Conclusions

The pre-conceptual design of FW OTSG was assessed during DEMO normal operations by using a TH model developed with RELAP5/MOD3.3. Initially, the pulse phase was considered. The OTSG reference design was investigated in both BoL and EoL, individuating the main effects on the steam generator TH performances produced by the heat transfer degradation due to tube fouling and plugging. Then, OTSG and IEOTSG layouts were compared. Although, in IEOTSG, an additional heat transfer region (economize) is foreseen at the secondary side, no sensible differences in the global parameters were highlighted. Furthermore, it was studied the OTSG TH response to variations of the PCS thermodynamic conditions. A decrease/increase of the secondary system pressure produces an increment/reduction of the heat transfer within the component. Finally, a transient simulation was run to verify the OTSG capability to respect the DEMO requirements related to pulse/dwell transition and viceversa. A very preliminary management strategy (i.e. variable trend) for feedwater flow was proposed and analyzed. The simulation outcomes confirmed the capability of the FW steam generator to operate the primary system at a constant average temperature, as for DEMO requirement.

Acknowledgements

This work has been carried out within the framework of the EUROfusion Consortium and has received funding from the Euratom research and training programme 2014-2018 and 2019-2020 under grant agreement No 633053. The views and opinions expressed herein do not necessarily reflect those of the European Commission.

References

[1] G. Federici, et al., An overview of the EU breeding blanket design strategy as an integral part of the DEMO design effort. Fusion Eng. Des., 2019, 141, 30-42. https://doi.org/10.1016/j.fusengdes.2019.01.141.

[2] L. Barucca et al., *Pre-conceptual design of EU DEMO balance of plant systems: objectives and challenges*, Fusion Eng. Des., 2021, 169, 112504. https://doi.org/10.1016/j.fusengdes.2021.112504.

[3] The Babcock & Wilcox Company, *Steam, its generation and use*, 41st ed., Edited by John B. Kitto and Steven C. Stultz, 2005.

[4] The Babcock & Wilcox Company, *Thermal-hydraulic analysis of once-through steam generators*, EPRI project Final Report, EPRI NP-1431, Alliance, Ohio, USA, 1980. Available online at: https://www.osti.gov/servlets/purl/5262028.

[5] S. J. Green and G. Hetsroni., *PWR steam generators*, Int. J. Multiph. Flow, 1995, 21, 1-97. https://doi.org/10.1016/0301-9322(95)00016-Q.

[6] C. Ciurluini et al., *Study of the EU-DEMO WCLL Breeding Blanket Primary Cooling Circuits Thermal-Hydraulic Performances during Transients Belonging to LOFA Category*, Energies, 2021, 14 (6), 1541. https://doi.org/10.3390/en14061541.

[7] C. Ciurluini et al., *Analysis of the thermal-hydraulic behavior of the EU-DEMO WCLL Breeding Blanket cooling systems during a Loss Of Flow Accident*, Fusion Eng. Des. 2021, 164, 112206. https://doi.org/10.1016/j.fusengdes.2020.112206.

[8] The US Nuclear Regulatory Commission (USNRC), R*ELAP5/MOD3.3 code manual volume 4: models and correlations*, NUREG/CR-5535; USNRC: Washington, DC, USA, 1995. Available at: https://www.nrc.gov/docs/ML1103/ML110330271.pdf.

[9] F. W. Dittus and L. M. K. Boelter, *Heat Transfer in Automobile Radiators of the Tubular Type*, Publications in Engineering, University of California, Berkeley, 1930, 2, 443-461.

[10] J. C. Chen, *A Correlation for Boiling Heat Transfer to Saturated Fluids in Convective Flow*, Ind. Eng. Chem. Process Des. Dev., 1966, 5(3), 322-329. https://doi.org/ 10.1021/I260019A023.

[11] L. A. Bromley, *Heat Transfer in Stable Film Boiling*, Chem. Eng. Prog., 950, 46, 221-227. Available online at: https://escholarship.org/uc/item/0pj1211q.

[12] D. C. Groeneveld et al., *The 2006 CHF look-up table*, Nucl. Eng. Des., 2007, 237, 1909-1922, https://doi.org/10.1016/j.nucengdes.2007.02.014.

[13] M. A. Kreider et al., *A global fouling factor methodology for analyzing steam generator thermal performance degradation*, Proceedings of 3rd International Steam Generator and Heat Exchanger Conference, Toronto, Canada, June 21-24 1998. Available online at: https://inis.iaea.org/search/search.aspx?orig_q=RN:30031799.

[14] T. Prusek et al., *A methodology to simulate the impact of tube fouling on steam generator performance with a thermal-hydraulic code*, Proceedings of XI International Conference on Heat Exchanger Fouling and Cleaning, Enfield (Dublin), Ireland, 07-12 June 2015. Available online at: http://www.heatexchanger-fouling.com/papers/papers2015/17_Prusek_F.pdf.

Journal of Physics: Conference Series 2177 (2022) 012018 doi:10.1088/1742-6596/2177/1/012018

Transient analysis of OSU-MASLWR with RELAP5

Martina Molinari, Vincenzo Narcisi*, Cristiano Ciurluini, Fabio Giannetti

DIAEE - Nuclear Section, "Sapienza" University of Rome, Corso Vittorio Emanuele II, 244, 00186, Rome, Italy

* Corresponding author: vincenzo.narcisi@uniroma1.it

Abstract. The present paper deals with the assessment of the original and a modified version of RELAP5/MOD3.3 against the OSU Multi Application Small Light Water Reactor (OSU-MASLWR). The new implemented features regard suitable correlations for the heat transfer coefficient evaluation in helical geometry. Furthermore, two different modelling of the Helical Coil Steam Generator (HCSG) are assessed. In the first approach, HCSG's primary and secondary sides are collapsed in a single pipe component. In the second model, three equivalent pipes are conceived for the simulation of the three ranks composing the HCSG.

Two different power manoeuvring experiments are reproduced. The simulations highlight a satisfactory agreement in both the transients. Nevertheless, the modified code shows enhanced capabilities in the prediction of the HCSG operation. This is due to the improvements adopted in the modified version of RELAP5/MOD3.3 that allows a better modelling of the dryout phenomena occurring within helical tubes, as well as a better estimation of the primary side heat transfer coefficient. The better agreement of the heat exchange is propagated to the primary system, resulting in a more accurate prediction of the inlet and outlet core temperatures, and primary flow rate.

Key Words: *helical coil steam generator, natural circulation, transient analysis, heat transfer, small modular reactor*

1. Introduction

Best estimate System Thermal-Hydraulic (STH) codes (e.g., RELAP5/MOD3.3) are extensively used for licensing of light water reactors. To broaden code applicability to the domain of advanced nuclear power plants, it is needed to verify their modelling capabilities with respect to safety relevant thermal-hydraulic phenomena occurring in such systems. Comparison between experimental data and simulation outcomes constitutes a key point of the code validation.

In this framework, aiming at studying the operation of the Multi Application Small Light Water Reactor (MASLWR), the Oregon State University (OSU) designed and constructed the OSU-MASLWR non-nuclear test facility. The MASLWR is Small Modular Reactors (SMR), relying on natural circulation of the primary coolant in both normal operations and abnormal conditions [1]. According to the conceptual design, as it can be seen in Figure 1, the Reactor Pressure Vessel (RPV – item 5 in Figure 1) is surrounded by the High Pressure Containment (HPC – item 9 in Figure 1). The latter is partially filled with water providing pressure suppression and liquid makeup capability. The first containment vessel is hosted into a water pool, named Cooling Pool Vessel (CPV – item 10 in Figure 1), acting as

Journal of Physics: Conference Series 2177 (2022) 012018 doi:10.1088/1742-6596/2177/1/012018

the ultimate heat sink. The pressure suppression is actuated by the Automatic Depressurization System (ADS).

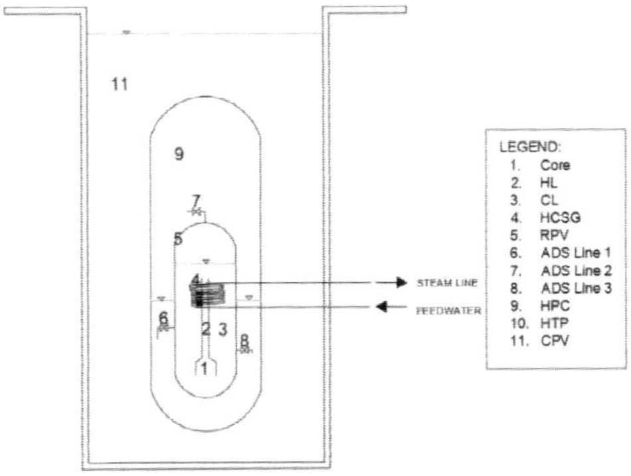

Figure 1: MASLWR conceptual design

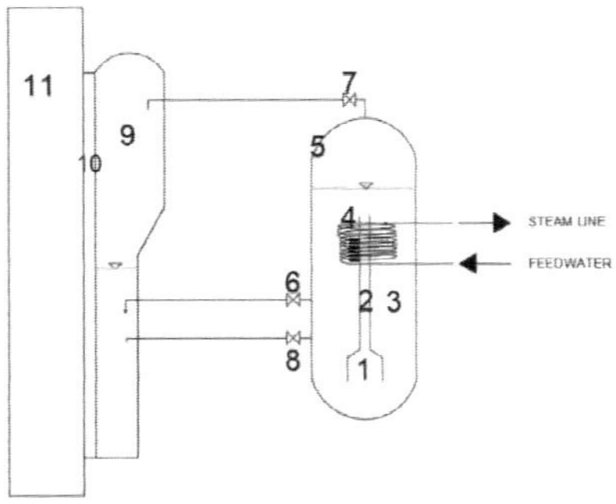

Figure 2: OSU MASLWR layout

One of the most relevant peculiarities of the MASLWR is the Helical Coil Steam Generator (HCSG – item 4 in Figure 1). The modelling of this component represents a crucial point in the safety and reliability analysis of such reactor. The aim of the present work is to assess the capabilities of RELAP5/MOD3.3, as well as of a modified version of the code, in the simulation of such component. This version has been developed at the Department of Astronautical, Electrical and Energy Engineering (DIAEE) of "Sapienza" University of Rome for advanced nuclear fission and fusion power plant applications [2], in collaboration with ENEA Brasimone Research Center. The main improvement for the present activity is to better simulate HCSG performances under different operative conditions. Some examples of the RELAP5 application for HCSG modelling were found in literature. To assess STH codes capabilities, two different validation benchmarks were carried out, using experimental data from OSU MASLWR experimental campaign. The first one was led by U.S. Department of Energy (DOE) with the OSU MASLWR 002 Test. This test has been simulated with different STH codes by several participants [3][4]. The RELAP5 models of the HCSG were developed following the modelling guidelines presented by Hoffer et al. [5]. The simulation results were in good agreement with experimental data and RELAP5/MOD3.3 had proven its capability to give a satisfactory qualitative prediction. The second benchmark, consisting in two exercises, was conducted by the International Atomic Energy Agency (IAEA): the first one regards the International Collaborative Standard Problem (ICSP) Test 1 and the ICSP Test 2, both reproduced through TRACE code [4], and the second one is related to the ICSP Test SP3 [6]. The latter was simulated by several participants, following the HCSG modelling guidelines presented by Hoffer et al. [5]. Although a general qualitative agreement with experimental data was observed, RELAP5/MOD3.3 showed some limitations in the prediction of the HCSG operation. In order to increase the simulation accuracy, the participants proposed an increase of the HCSG heat transfer area, or of the heat transfer coefficient (HTC), using a calibrated multiplicative factor greater than 1.0 [7].

To enlarge code applicability and improve its accuracy with respect to HCSG simulations, the DIAEE has developed an improved version of RELAP5/MOD3.3, that has been assessed reproducing the aforementioned tests.

2. Facility Overview

The OSU MASLWR test facility (Figure 2) reproduces the MASLWR conceptual design, presented in Figure 1, but RPV (item 5 in Figure 2), HPC (item 9 in Figure 2) and CPV (item 11 in Figure 2) are conceived as three separated vessels that accomplish the same safety operations proposed in the MASLWR design.

The facility is scaled at 1:3 length scale, 1:254 volume scale and 1:1 time scale with respect to MASLWR conceptual design [8].

The OSU MASLWR test facility consists of the RPV, the internals, and the Containment structures. The reactor core (item 1 in Figure 2) is reproduced with a Fuel Pin Simulator (FPS), consisting in 57 electrically heated cylindrical rods distributed in a 18.6 mm pitch square array with 1.33 pitch to diameter ratio. The FPS is located in the bottom region of the RPV. The HCSG (item 4 in Figure 2) is composed of vertical helical tubes and is located in the upper part. This configuration ensures the required thermal length between the power source and the heat sink, since core cooling is foreseen in natural circulation in both normal and abnormal conditions. The primary flow passes through the core and goes upstream within the Hot Leg (HL – item 2 in Figure 2) riser arriving to the Upper Plenum. Then it flows downstream through the SG and the Cold Leg (CL – item 3 in Figure 2), returning into the Lower Plenum. The secondary flow passes through HCSG, consisting of 14 helical tubes with two common headers, one for Feedwater (FW) inlet and one for the steam outlet.

Safety operations are ensured by the ADS, consisting of valves (item 6, 7, and 8 in Figure 2 and the final heat sink constituted by the HPC and the CPV, thermally coupled by a proper Heat Transfer Plate (HTP - item 10 in Figure 2).

3. Models

The nodalization of OSU MASLWR can be divided in five regions: primary and secondary systems, HPC, CPV and ADS.

The model consists of hydrodynamic components, that reproduce primary and secondary loops, and heat structures, that simulate heat exchange (where expected) and structural mass inventory. The modelling is developed with the sliced approach for the entire facility [9], keeping the length ratio of adjacent nodes into the range of 1 ÷ 1.25. The primary loop is simulated with pipe components and branch components, and nodalization features are defined respecting the actual geometry of the facility. The hydrodynamic characteristics of FPS are evaluated collapsing all the 57 subchannels in a single pipe component. In the input deck, its flow area is the total FPS flow area, while the hydraulic diameter is computed considering the geometry of the single subchannel. The heat structure of the FPS is in cylindrical geometry. In order to describe the thermal power supplied, a general table of the type power-to-time is used as boundary condition. Pressure drops through the core flow plate, FPS and HCSG primary side are set as Reynolds number dependent.

The ADS has been modelled with horizontal pipes and valves. In Figure 3 Vent Lines, System Blowdown Lines and Sump Return Lines are referred to components 43 & 45, 40 & 41, and 55 & 57, respectively. Both HPC and CPV have been modelled with two vertical parallel pipes connected with cross junctions, to better simulate buoyancy and mixing within the large volumes. In Figure 3, pipe components 61 and 62 stand for the HPC, while components 65 and 66 represent the CPV. It is worth emphasizing that such systems did not operate in the considered transient tests.

Regarding the secondary loop, boundary conditions are set with time dependent volumes (feedwater inlet temperature and steam outlet pressure) and a time dependent junction (feedwater inlet flow rate).

The focal point of the present analysis is the HCSG modelling. Two different nodalization schemes are developed, investigating the dependence of the simulation outcomes on the HCSG modelling choices. In the following, the two models are referred as Model 1 and Model 2. The geometrical scheme of the primary system is the same in both the models. The HCSG shell-side consists, in both the models, of a single equivalent pipe (component 22 in Figure 3) reproducing the free space in the SG shell-side. Concerning the secondary side, Model 1 is the simplest modelling approach. As a matter of fact, the HCSG tube-side is reproduced with a single equivalent pipe (component 26 in Figure 3), collapsing all the helical tubes. Such pipe component has the average tubes length and slope evaluated as weighted average, over the number of rotations of each coil. The total flow area considers all the 14 tubes. Components 27 and 32 in Figure 3, stand for the secondary side inlet and outlet headers.

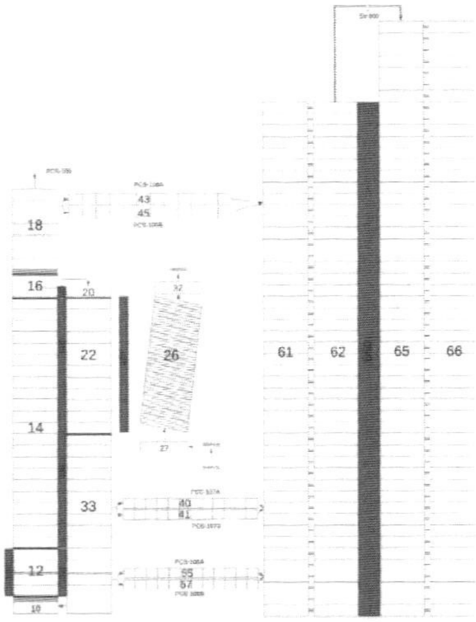

Figure 3: OSU MASLWR - RELAP5 Model 1

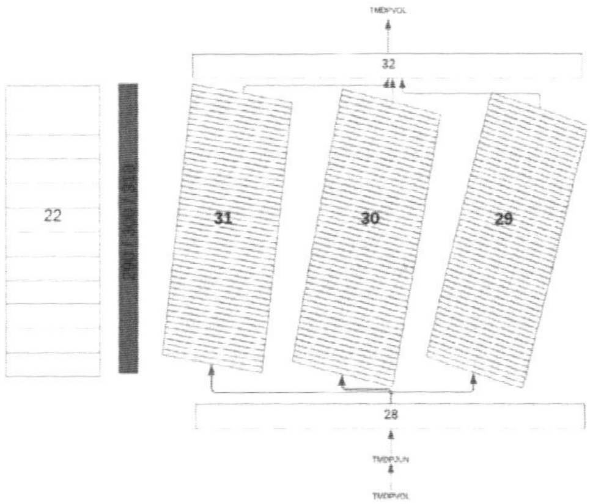

Figure 4: OSU MASLWR - RELAP5 Model 2 (zoom on HCSG)

The Model 2, shown in Figure 4 limited to the HCSG scheme, presents a greater level of detail on the secondary side: three different pipe components simulate the overall number of tubes, each representative of a single tube bank. Each pipe component of the secondary side has its own thermal coupling with the primary side, provided by heat structures 290, 300 and 310. Geometrical features of each component (e.g., length, inclination, and flow area) are evaluated as the average within the proper bank. The common inlet and outlet headers are simulated with component 28 and 32 (see Figure 4). A calibration procedure is required, aiming at the distribution of the flow rate through the three pipe components. Since any information about the orifices used at the tubes' inlet is not available, concentrated pressure drop coefficients were entered in the model in order to uniform the temperature at the outlet of components 29, 30 and 31 (respectively inner coil, mid coil ad outer coil). In particular, the temperature difference was kept below the value of the thermocouples' uncertainty.

All heat losses are simulated with heat structures, with a room temperature of 295 K and environmental HTC of 6.0 W/m^{2}K.

4. RELAP5/MOD3.3 modifications

RELAP5/MOD3.3 is a STH code for the best estimate analysis of nuclear systems. It is a code for the transient analysis, developed for the U.S. Nuclear Regulatory Commission (NRC). [10].

An improvement campaign is ongoing at the DIAEE to expand the applicability of RELAP5/MOD3.3 to advanced nuclear power plant applications [2]. Aiming at improving capability of the code to simulate a HCSG, some modifications have been added in the enhanced version of the code. Modified RELAP5/MOD3.3 includes new correlations for the evaluation of the HTC in helical geometry. For the shell-side (i.e., flow rate outside the tubes) Zukauskas correlations are added for the HTC evaluation. It takes into account the annuli created between the pipe banks, considering also the eventual variable pitches. Instead, for the in-tube flow, Mori-Nakayama correlation is implemented for the HTC estimation. In addition, specific two-phase flow maps are considered for this component, which take into account tube's geometrical characteristics.

5. OSU-MASLWR 002 Test

The OSU-MASLWR 002 Test investigated the primary system flow rate, driven by natural circulation, and the steam superheating in secondary system. There are 6 power levels, from 80 kW to 165 kW. According to the experiment, the primary flow rate increases as the power supplied to the FPS rises, allowing the FPS inlet and outlet temperature difference to be kept almost constant. Boundary and operating conditions are obtained from literature [1].

The stepwise trend, adopted for the heat source and for the feedwater flow rate, determines the primary flow rate presented in Figure 5a. The plot compares the experimental acquisition with the results obtained with default and modified versions of the code. Each version was used twice, considering both Model 1 and Model 2. It is worth emphasizing that the facility works in natural circulation, without active system involved in the movement of the primary coolant. Moreover, due to the small flow rate, experimental measurement was affected by visible fluctuations. Thus, Figure 5a presents the experimental primary flow rate with a colored band (please refer to online version of the paper) that does not stands for measurement uncertainty band but represents the oscillations of the acquisition.

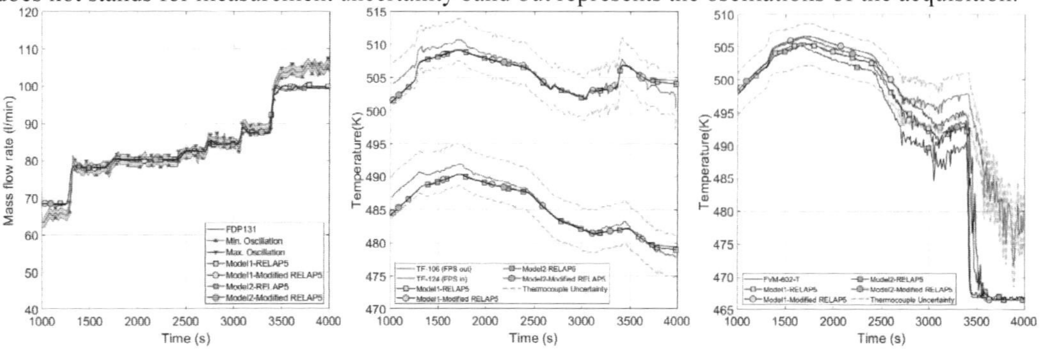

(a) (b) (c)

Figure 5: OSU MASLWR 002 - primary flow rate (5a), FPS inlet and outlet temperature (5b) and SG outlet temperature (c)

It is possible to notice that there are no appreciable differences between the trends predicted by the two code versions and between the two HCSG models. In the present test, both the codes and both the models provide a satisfactory prediction of the experimental data. In Figure 5a mainly two discrepancies can be seen: the first one at the beginning of the test and the second one in the highest power phase (after 3400 s). Regarding this latter, it must be underlined that initial conditions are poorly reported in literature [1], affecting the first phase of the simulations. In addition, there could be a certain difference between the estimated value of the pressure drops and the actual ones (no experimental data available), and these differences affect mostly the initial phase of the experiment Measurement uncertainties are not known for primary flow rate, but being flow rates very low, it is expected that uncertainties are larger than discrepancies between experiment and simulations. For this experiment, there are no considerable differences between the two code's version because the modifications do not have appreciable impact on natural circulation in that power range.

Figure 5b shows the FPS inlet and outlet temperature, respectively in purple line and red line for the experimental acquisitions. In this case the experimental data (solid lines) are reported with the error bands (dashed green lines). As for the primary flow rate, there are no appreciable disagreements between the trends predicted by the two code's versions and between the two models, all providing a satisfactory agreement with experimental data.

A slight difference between the two code versions is visible in the last phase of the experiment. As expected, with increasing HCSG power, the effects of the modification become visible. An overall good agreement between calculations and experimental data are observed. The minor discrepancy observed at the beginning of the test is mainly due to a lack of data concerning the phase preceding the experiment. The temperature difference along the FPS is kept almost constant throughout the test, confirming the good capability of the code to simulate system operation in the considered power range.

Figure 5c shows the HCSG steam outlet temperature, comparing the experimental acquisition (plus uncertainties) with computational predictions. For the whole experiment it is expected to obtain superheated steam. However, at the highest power level, the temperature tends to reduce towards the saturation temperature, while keeping superheated conditions. This phenomenon occurs because, in the final phase, the ratio between secondary mass flow rate and power to be removed is increasing, and for this reason HCSG outlet temperature is reducing. The simulations provide a satisfactory reproduction

Journal of Physics: Conference Series 2177 (2022) 012018 doi:10.1088/1742-6596/2177/1/012018

of the test, except in the final phase, where the code predicts saturated conditions at the HCSG outlet (at 3450 s, a steam void fraction of 0.8895 for RELAP5/MOD3.3 and 0.9987 for Modified RELAP5/MOD3.3). It is possible to quantify that discrepancy in 3.05 kW for RELAP5/MOD3.3, and about 1 kW for Modified RELAP5/MOD3.3 (i.e., 1.8% and 0.6% of the FPS power, respectively). One of the possible explanations for such deviation is the following. Being a one-dimensional code, RELAP5 is able to calculate only the average fluid thermodynamic properties related to a control volume. If thermocouple was installed at the tube center, it would measure the temperature of vapor phase, that in a two-phase condition here collects, and it could be slightly overheated. In this case, it would be possible that, on average, the fluid comes out from HCSG in saturated conditions, matching the code prediction.

Although the increasing of the ratio between secondary mass flow rate and power to be removed is well simulated, Figure 5c shows a sharper drop of the steam outlet temperature at the end of the test. It can be caused lack of information about feedwater inlet temperature throughout the test. Considering Model 1, the two versions of the code predict the dryout at the same level. On the other hand, adopting Model 2 a slight reduction of the dryout level is predicted by the modified code, even if irrelevant. This suggests that Model 2 is more affected by the improvement added in the modified RELAP5/MOD3.3, due to the diversification of heat exchange areas with the three pipe components.

One of the main outcomes of the present test is the overall agreement between the two versions of the code, caused by the low power levels assumed in this test. Thus, a similar experiment but with a most extended range of FPS power supplied is simulated (see section 6). Furthermore, both Model 1 and Model 2 proved capable of simulating the transient test performed on the facility. The difference between the two models is only quantitative. It can therefore be said that since the two models have the same boundary conditions, the second model is more affected by possible variations in secondary mass flow rate.

6. ICSP Test SP3

The ICSP Test SP3 [7] foresees seven power steps during the experiment in the range of 80 kW ÷ 320 kW. Boundary conditions for primary and secondary system are derived from [7], where a detailed description of the test is presented.

Figure 6a shows the primary flow rate, comparing the experimental acquisition of the natural circulation and the computational results. Experimental data are provided with a colored band (refer to online version of the paper) that accounts for the oscillations observed in the measurements (it is not representative of the measurement error band; information about uncertainties on the flow rate acquisition is not available in literature). An overall good agreement is observed over the whole test, although a slight overprediction is highlighted in the first phase of the test. The only difference between the two models, are the presence of fluctuations on Model 2. These fluctuations are caused by the secondary system and propagate to primary flow rate and FPS inlet and outlet temperatures. Such is widely investigated in the following.

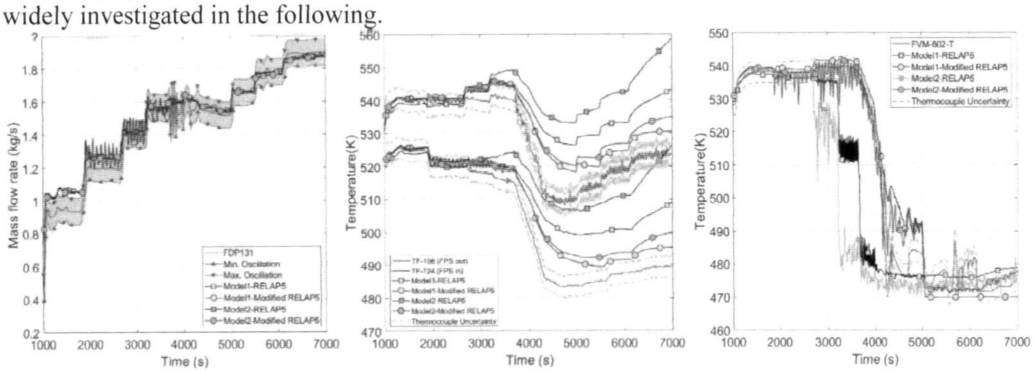

(a) (b) (c)

Figure 6: ICSP Test SP3 - primary flow rate (6a), FPS inlet and outlet temperature (6b) and SG outlet temperature (6c)

The FPS inlet and outlet temperatures are presented in Figure 6b, respectively in purple line and red line. Unlike the previous analysis, in the present test differences between the two versions of RELAP5/MOD3.3 are visible as the power increases above 200 kW (after 3500 s). The results of the Modified RELAP5/MOD3.3 are quantitatively closer to the experimental data for the whole experiment, especially at high power levels. These better results are due to improved prediction of HCSG heat exchange, improving cooling capabilities of the HCSG modelling. Following the satisfactory prediction of primary flow rate, the FPS inlet and outlet temperature difference is satisfactorily simulated by the modified code, for both Model 1 and Model 2.

Figure 6c presents the HCSG steam outlet temperature. An overall agreement between the two codes and the experimental data is observed up to 3000 s. After that, large qualitative and quantitative differences between the two versions of the code are observed. As power increases above 200 kW, the default version of the code shows limited capabilities to reproduce the experimental acquisition, determining a quick temperature drop at around 3000 s. This sharp temperature drop probably takes place because the dryout occurs very close to the HCSG outlet. According to RELAP5/MOD3.3 there is no dryout as opposed to Modified RELAP5/MOD3.3 which estimates the dryout in the last control volume of the component, and therefore there is a low overheating of the steam.

As already discussed for OSU-MASLWR 002 test, the default code is not able to simulate the correct heat exchange between primary and secondary system at high power levels. Modified RELAP5/MOD3.3, however, moves away from the trend of the default code not only maintaining the qualitative trend of the experimental data, but falling within the temperature uncertainty band.

Despite the overprediction between 4100 and 5000 s, the improvements introduced in the modified version of the code are evident. As expected, in the calculations performed with Modified RELAP5/MOD3.3, the height where the liquid water disappears is reduced. The differences between the two code versions are evaluated considering the control volumes where the single-phase steam occurs in HCSG secondary side. The major difference between the two versions occurs at high power levels and it is higher before 4000 s. Another effect of Modified RELAP5/MOD3.3 concerns the HTC: analyzing where saturated nucleate boiling occurs, as expected, there is an increasing number of control volumes where it shows up. This effect takes place because saturated nucleate boiling areas are the most affected by code modifications.

A substantial difference between the two models is the presence of fluctuations on the SG outlet temperature. Oscillations propagate from secondary side to primary side and are visible in both FPS inlet and outlet temperatures (Figure 6b) and primary flow rate (Figure 6a). Indeed, when HCSG secondary side parameters fluctuate, also the heat transfer within the steam generator is unstable. This results in oscillations in the primary side temperatures. Once the thermal field in the primary system is altered, also the natural circulation, whose driving force is the density difference between hot and cold leg water columns, starts to oscillate. In order to exclude the possibility that such fluctuations depend on the time step adopted for the simulation, a time step sensitivity has been carried out varying this parameter from $5 \cdot 10^{-4}$ s to $5 \cdot 10^{-3}$ s. Computational outcomes proved to be independent from the time step used for simulation purposes. Another explanation of this behavior could be the establishment of instabilities on the secondary system, typical of parallel channels connected to the boundaries via headers. Such fluctuations could appear for two-phase flow instabilities and/or for different distribution of the flow rate. It is worth noticing that experimental data on feedwater distribution through the parallel tubes are not available. A preliminary calibration of the pressure drops through the parallel channels has been carried out in Model 2, to obtain the same HCSG outlet temperature between pipe components. But there is no evidence that such control was performed in the experiment. Finally, it is not possible to exclude computational instabilities among the causes of the observed fluctuations.

Despite the presence of fluctuations, also in Model 2 the height where the single-phase steam appears with Modified RELAP5/MOD3.3 is reduced, and the number of control volumes where saturated nucleate boiling occurs is increased.

7. Conclusions

Both RELAP5/MOD3.3 and Modified RELAP5/MOD3.3 gave an overall satisfactory result at low power levels, showing the main differences at high power. At low power levels the modifications do not provide appreciable improvements. In fact, the default code is already able to predict the experimental data with satisfactory accuracy. It is essential that the modified code continues to predict in the same way what happens at low powers, because there must be no variations where the default version is already able to work. On the other hand, at high power the default code highlighted limits that were partially, but satisfactory, overcome by the modified version of the code which, therefore, proved to be more able to simulate the system operation in a wider operational range with respect to the default version.

From the point of view of the comparison between the two models, it must be taken into account that a more detailed model, based on parallel pipe components, may encounter instabilities beyond a certain power threshold. These instabilities of Model 2 can turn out to be more dampened in the case of using the Modified RELAP5/MOD3.3 and consequently the impact on the primary system is less important. With the same code version used, the two models have results that are consistent with each other. This evaluation is important from the point of view of the modification made to the code because it implies that it is probably not affected by the nodalization performed.

References

[1] Modro S M, Fisher J E, Weaver K D, Reyes J N, Groome J T, Babka P and Carlson T M 2003 *Multi-Application Small Light Water Reactor Final Report* (Idaho National Engineering and Environmental Laboratory Report) INEEL/EXT-04-01626

[2] Narcisi V, Melchiorri L and Giannetti F 2021 Improvements of RELAP5/MOD3.3 heat transfer capabilities for simulation in-pool passive power removal systems *Ann. Nucl. Energy* **160** 108436

[3] Mascari F, Vella G, Woods B G and D'Auria F 2012 Analyses of the OSU-MASLWR experiment test facility *Sci. Technol. Nucl. Ins.* **2012** 528241

[4] Mascari F, Woods B G, Welter K and D'Auria F 2019 *Validation of the TRACE code against small modular integral reactor natural circulation phenomena (Portland)* 18th Int. Topical Meeting on Nuclear Reactor Thermal Hydraulics NURETH-18 pp 6701-14

[5] Hoffer N V, Sabharwall P and Anderson N A 2011 Modeling a helical-coil steam generator for the next generation nuclear plant (Idaho National Laboratory Report) INL/EXT-10-19621

[6] Mascari F, De Rosa F, Woods B G, Weller K, Vella G and D'Auria F 2016 *Analysis of the OSU-MASLWR 001 and 002 tests by using the TRACE code* (Washington DC: U.S. Nuclear regulatory Commission) NUREG/IA-0466

[7] International Atomic Energy Agency 2014 *Evaluation of Advanced Thermohydraulic System Codes for Design and Safety Analysis of Integral Type Reactor* (Vienna: IAEA) IAEA-TECDOC-1733

[8] Demick N T, Galvin M R, Groome J T and Woods B G 2007 *OSU MASLWR Test Facility Description Report* (Vienna: IAEA) OSU-MASLWR-07001

[9] Mascari F, Vella G, Woods B G, Welter K, Pottorf J, Young E, Adorni M and D'Auria F 2010 Sensitivity analysis of the MASLWR helical coil steam generator using TRACE *Nucl. Eng. Des.* **241** 1137-44

[10] The RELAP5/Mod3.3 Code Development Team 2003 *RELAP5/Mod3.3 Code Manual Volume I: code structure, system models, and solution methods* (Washington DC: U.S. Nuclear Regulatory Commission) NUREG/CR-5535/Rev P3-Vol I

Effect of non-condensable gas in steam condensation at sub-atmospheric pressure condition

G. Giambartolomei[1*], A. Pesetti[1], M. Raucci[1], D. Aquaro[1]

[1] Department of Civil and Industrial Engineering (DICI), University of Pisa, Largo L. Lazzarino 2, 56122 Pisa (Italy)

*Corresponding author e-mail: guglielmo.giambartolomei@ing.unipi.it

Abstract. In the International Thermonuclear Experimental Reactor (ITER), a postulated Loss Of Coolant Accident (LOCA) in the Vacuum Vessel (VV) has to be managed with a pressure suppression system working at sub-atmospheric pressure. The operating conditions considerably differ from those experienced in the fission nuclear power plants such as BWR, since the ITER Tokamak works at very low pressure conditions and can withstand a maximum pressure of 0.15 MPa. For this reason, the pressure value must not exceed 10 kPa for a water temperature of 30 °C inside the Vapour Suppression Tanks (VSTs) that are the fundamental components of the Vacuum Vessel Pressure Suppression System (VVPSS). During a LOCA some non-condensable gases (mainly hydrogen and oxygen gases due to the water radiolysis or thermolysis) may be mixed in the steam and this could impair the condensation efficiency. In order to investigate the effects of non-condensable gas on DCC, we conducted a research program funded by ITER Organization at the laboratory of the University of Pisa: we designed and built a small-scale experimental rig to study the steam Direct Contact Condensation (DCC) with the presence of non-condensable gas and simulate the behaviour of a VST. Since DCC can occur with different characteristics, we ran 12 closed mode tests exploring all condensation regimes injecting a certain mass of air with the steam discharged in the subcooled water. The tests started at the saturation pressures corresponding to water temperatures ranging from 40 °C to 80 °C and ended when the free space volume reached the atmospheric pressure. From the analysis of the data acquired during the tests we observed that the condensation efficiency remained higher than 95%. We observed that despite this, the presence of a certain quantity of non-condensable gas has negative aspects on condensation: the condensation regime never reaches stability (the regimes quickly shift towards instability). Furthermore, the presence of air triggers a turbulence of the flow which interferes with the transfer of heat from the steam to the water. Not only the introduction of air into the flow increases linearly the pressure above the water head but its high temperature further contributes to the pressure increase. The air mixed with the flow also forms eddies that can trap the steam and transport it out of the water, preventing it from condensing. Apart from condensation, the most noticeable problem is the rapid increase in pressure inside the condensation tank.

1 Introduction

The main component of the International Thermonuclear Experimental Reactor (ITER) is the Vacuum Vessel (VV) in which nuclear fusion phenomena occur inside the plasma. To be able to form the latter, the pressure of the environment must be very low (high vacuum): the VV is therefore designed to withstand a maximum internal pressure of 0.15 MPa in a Loss of Coolant Event (LOCA). This accidental event, also called Ingress of Coolant Event (ICE) deriving from a plasma disruption that would cause damage to the hot plasma facing components, is postulated to occur during the normal operating activities of the reactor. In order to protect the VV from an over pressurization caused by these in-vessel coolant postulated leakages, the Vacuum Vessel Pressure Suppression System (VVPSS) has been designed alongside the reactor. The VVPSS resembles the design of the suppression systems of Boiling Water Reactors (BWRs) with the substantial difference that in the former, the Vapour Suppression Tanks (VSTs) are required to operate at sub-atmospheric pressure since, as previously mentioned, the ITER VV operates at extremely low vacuum conditions and for its structural integrity, the internal pressure must be limited to a maximum value of 0.15 MPa (absolute) in the case of a postulated LOCA.

The VSTs are 4 tanks measuring 4.7 m in height, with a diameter of 6.2 m and an inner volume of 100 m^3 each partially filled with water, 3 with 60 m^3 and one with 40 m^3. The 3 VSTs containing 60 m^3 of water, also called Large LOCA Tanks (LLTs), are designed to manage bigger LOCA events (cat. III and IV), while the one containing 40 m^3, also called Small LOCA Tank (SLT), is designed to manage smaller LOCA events (cat. II). Each VST is equipped with a submerged sparger to directly condense the steam. Figure 1 illustrates the four tanks located at the underground level -2 of the ITER reactor building.

Figure 1: Section view of the VVPSS

The fundamental phenomenon that governs this security system is the Direct Contact Condensation (DCC) at sub-atmospheric conditions and which has not yet been sufficiently studied. For this reason, the reliability, and the efficiency of this kind of fusion reactor safety system must be experimentally investigated. In this new scenario the situation can be very different from the atmospheric one: for example, for a given superheated steam temperature, the subcooling temperature needed to condense it is always higher (the saturation temperature is lower).

Moreover, the steam can react with the hot surfaces [1] and some non-condensable gases (hydrogen and oxygen gases due to the water radiolysis or thermolysis) may be mixed in it. The presence of non-condensable gases can drastically modify the phenomenon of condensation: air mixed in the steam DCC has been only marginally studied and only in atmospheric conditions [2]. In particular, the presence of non-condensable gases in the steam mixture modifies the flow and therefore the condensation regimes impacting on the level of condensation efficiency and oscillations [3,4].

Experimental and analytical studies were conducted to investigate the effect of non-condensable gas on a water hammer by [5]. Their experimental results indicated that the amplitude of pressure oscillation could be reduced by one order of magnitude with an increase in air mass fraction of up to 10%. The experimental results of [6] indicated that even a small amount of non-condensable gas could greatly hinder the steam condensation and facilitate the flow pattern transition from chugging (C) to oscillation condensation (OC) or stable condensation (SC).

In some cases it has been observed that air content had a great influence on unstable flow patterns: the interface of the unstable flow pattern extremely fluctuated at a low air mass fraction but stabilized when the air mass fraction exceeded a critical value [7].

Since there is no trace in the literature of studies on the effects of non-condensable gases on DCC at sub-atmospheric pressure condition and the effects of large amounts of air in the steam mixture are not known, we decided to study them in a small-scale experimental plant that will be described in the next section. Further scientific investigations about DCC at sub-atmospheric conditions will be carried out in a large-scale experimental facility by our research group [8].

2 Experimental activity

2.1 Experimental apparatus

Starting from 2014 the Department of Civil and Industrial Engineering (DICI) of the University of Pisa has begun a close collaboration with ITER Organization to validate the VVPSS. At first, we built a small-scale experimental apparatus to conduct hundreds of tests for various experimental campaigns. The main component of this small-scale experimental rig is the Condensation Tank (CT) with a scaling factor of 1/22 compared to the ITER VST inner volume. The CT (foreground of Figure 2) is a stainless steel and thermal insulated vessel able to withstand pressure values up to 4 kPa. It is heavily equipped with 28 temperature sensors (± 0.1 °C of accuracy), 8 pressure sensors (± 0.1 kPa of accuracy) and one differential pressure water level sensor (± 1 mm of accuracy).

The CT is equipped with a sparger that can be changed allowing different number of holes and their pattern trough the extremity that release the: it presents holes configurations with a vertical distance between the upper and the lower hole row axis set equal to 266 mm, as actually foreseen in the full-scale sparger design of the ITER VST (it is possible to change the number of holes during the experimental campaign). The hole diameter was set and fixed to 10 mm. What happens inside the tank is visible and can be recorded by means of 4 video cameras (GOPRO Hero 4).

Other fundamental components of the experimental rig are the Steam Generator (SG) and the Auxiliary Tank (AT), both visible in the background of Figure 2.

Figure 2: CT, AT and SG

The SG has an electrical power of 137 kW and it is able to provide superheated steam at a temperature of 140 °C at a maximum pressure of 1.5 bar abs. The purposes of the AT were to adjust the parameter of each test before letting the steam enter the CT and to use the water inside it to fix both the temperature and the water head level.

Between the SG and the sparger there are two steam control lines equipped with control valves and steam mass flow meters: a Coriolis mass flow meter for low steam mass fluxes and a Vortex steam mass flow meter for higher steam mass fluxes (up to 45 g/s). Parallel to the lines dedicated to steam, there is an air line which provides precise amounts of dry air in terms of weight which joins the steam line just before entering the sparger tube. All the lines are equipped with electrical resistance trace heating systems and are thermally insulated.

The CT and the AT are linked to a vacuum pump which is able to remove air up to about 10 kPa from the tank's free spaces. With this set-up we were able to reproduce the operational conditions of the VVPSS in terms of initial water temperature, water head level, initial pressure above the water surface, steam mass flow rate per hole and hole's distribution.

All the physical parameters characterizing the DCC are recorded by means of a Data Acquisition System (DAS) controlled by LabVIEW© software with suitable frequency acquisition rates.

2.2 Experimental procedures

As part of an experimental campaign in collaboration with ITER Organization we performed 12 tests with the purpose of studying the effect of mixing non-condensable gas with the released steam flow originating from the VV. The test conditions are defined to study the change induced in the steam condensation regimes by non-condensable gas mixture fraction added to the pure steam flow. Moreover, these tests are aimed to experimentally verify the heat transfer performances (pressure suppression and heat removal capabilities) when a given fraction (by weight) of non-condensable gas is mixed in the steam flow.

In these tests we used dry air as a non-condensable gas component, which was injected at the same temperature of the steam (130 °C) and conditioned in a separate and parallel flow control line. During all tests, the air mixing fraction is kept constant at 1/6 by weight (16.7 $\%_w$) with respect to the steam mass (as required by ITER Organization) throughout the test duration and the initial water head level was 1 m inside the CT. Both the quantity of steam and that of air are measured by means of appropriate flow meters.

We considered six steam condensation regimes (2 different tests for each), such as Chugging, Transitional Chugging, Condensation Oscillation, Bubbling Condensation Oscillation, Stable Condensation and Interfacial Oscillation Condensation, and five initial water temperature (40, 50, 60, 70 and 80 °C) using three sparger systems of 10 mm diameter hole patterns (12 holes, 18 holes and 24 holes) to maintain, whenever possible, the geometry and similitude scales unchanged. This guarantees the same steam mass flux per hole, as in full scale, and the ratio 0.5 between the whole area of the exit holes and the sparger pipe cross section. The total number of holes is determined by the maximum steam mess flow rate (45 g/s) obtainable by the steam generator (SG) and the required value of the mean steam mass flux to downstream pressure ratio per hole, necessary to establish the desired steam condensation regime to explore. All the 12 tests have been performed exclusively in closed mode: the test run is stopped when the pressure achieved in free space volume is 100 kPa. As demonstrated in [9], in the case of DCC at sub-atmospheric pressure, the fundamental parameters that describe the phenomenon are the steam mass flux $G_S = Q_S/A$ (kg/m^{2}s), the initial water temperature T_W, and the downstream pressure (at the exit of the sparger hole) P_W: it is therefore possible to build a 2D condensation regimes map. The initial points of each test on the regime map (G_S/P_W,T_W) are shown in Figure 3. At least two tests were carried out for each water temperature value.

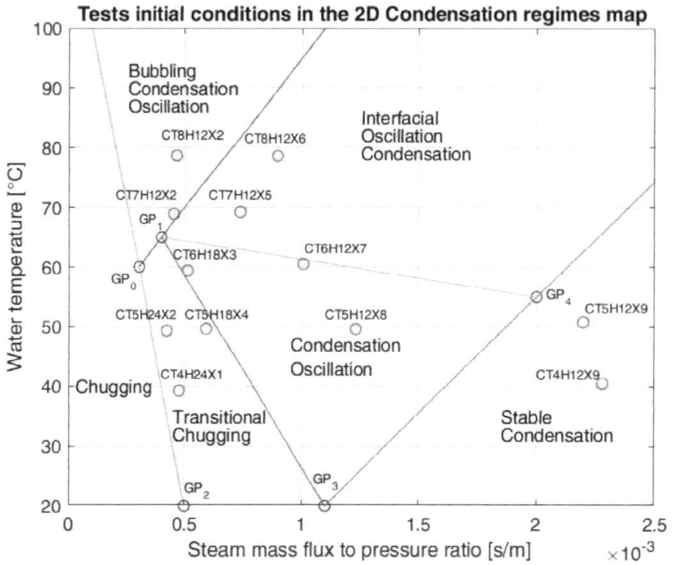

Figure 3: Distribution of the 12 tests (initial conditions) in the condensation regime map

3 Discussion of the results

As mentioned in 1, it has been observed that small amounts of air injected into the steam jet can have beneficial effects, but when the amount of non-condensable gas is as large as in this case, its effect can result in a reduction of the heat transfer coefficient, a reduction heat removal capability, and a reduced efficiency in the pressure suppression.

The fundamental parameter that represents the condensation efficiency is the Pressure Suppression Factor (PSF):

$$\mathrm{PSF(t)} = 100 \cdot \frac{\mathrm{M_S(t)} - \delta \mathrm{m_{ncs}(t)}}{\mathrm{M_S(t)}} = 100 \cdot \left(1 - \frac{\delta \mathrm{m_{ncs}(t)}}{\mathrm{M_S(t)}}\right) \qquad \text{Equation 1}$$

Where $\mathrm{M_S(t)} = \int_0^t \mathrm{Q_{Sp}(x)dx}$ (total steam mass discharged during a test) and $\delta \mathrm{m_{ncs}(t)} = \frac{18 \cdot \mathrm{V_{FS}}}{\mathrm{RT_{FS}(t)}} \delta \mathrm{P_{ncs}(t)}$ (mass accumulation of non-condensed steam in the free space volume). The partial pressure of non-condensed steam can be evaluated with three contributions: $\delta \mathrm{P_{ncs}(t)} = \Delta \mathrm{P_{FS}(t)} - \Delta \mathrm{P_{SAT}}\,(t) - \mathrm{P_{air}(t)}$ where $\Delta \mathrm{P_{FS}(t)}$ is the pressure change in the free space volume above the water head, $\Delta P_{SAT}(t)$ is the pressure saturation change in the free space volume and $\mathrm{P_{air}(t)}$ is the partial air pressure accumulated in the free space volume. The latter is calculated as $\mathrm{P_{air}(t)} = \frac{\mathrm{m_{air}(t)}}{29 \cdot \mathrm{V_{FS}}} \mathrm{RT_{FS}(t)}$ where $m_{air}(t)$ is the total air mass discharged in the tank ($\mathrm{m_{air}(t)} = \int_0^t \mathrm{q_{air}(x)dx}$) and $\mathrm{T_{FS}(t)}$ is the measured temperature in the free space volume.

From Equation 1 is clear that the higher the mass accumulation of non-condensed steam, the lower the PSF. On the other hand, if $\delta \mathrm{m_{ncs}(t)}$ is 0, the PSF is 100, i.e., all the discharged steam condenses in the pool.

For each test we calculated the PSF as a function of the variables included in Equation 1. Figure 4 summarizes the values of steam condensation efficiency versus the water temperature for all the performed tests.

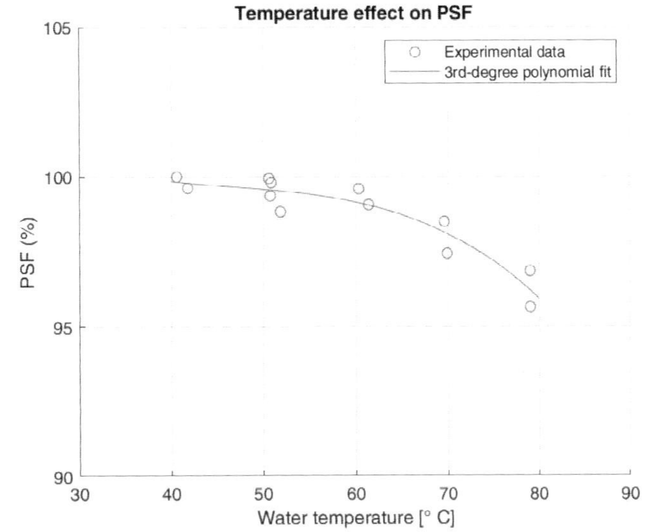

Figure 4: Effect of non-condensable gas (air, 16.7%) on the PSF as a function of the water temperature

For water temperature lower than 50 °C there is no great effect of the non-condensable gas on the condensation efficiency. For greater temperature values, the condensation efficiency is reduced at about 96%. But in these cases, the test duration to reach 100 kPa is very short (about 2 min) and the steam mass condensed is small.

Figure 5 shows the transient of the 12 tests in the regime maps: a shift of the flow regimes towards unstable regimes is clearly visible. The pressure in the free space volume P_{FS}, and also the downstream pressure P_W, increase linearly with time, resulting in the decrease of the parameter $X = G_S/P_W$, the steam mass flux to downstream pressure ratio. In the same figure the corresponding PSFs deduced from the experimental data are also shown.

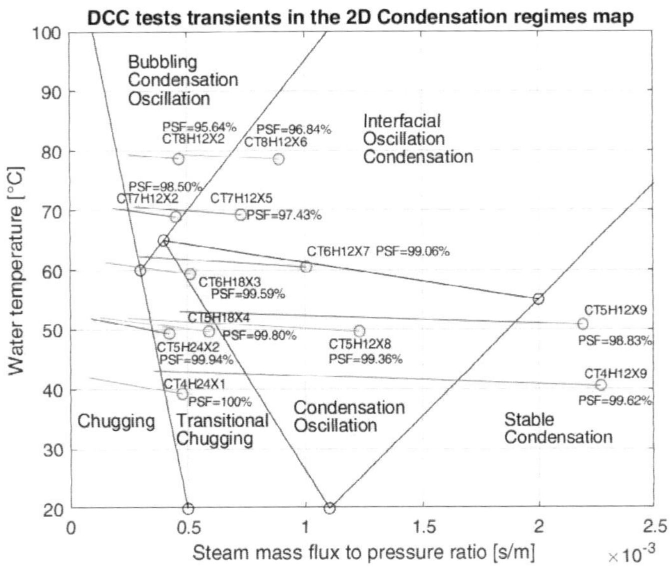

Figure 5: Measured PSF(%) during the test transients

When such a high fraction (about 16% by weight) of non-condensable gas is mixed with steam, the steam condensation is not able to reach the desired stability. Moreover, the disruption of the boundary along which the heat transfer occurs, caused by the onset of high turbulence of the mixed flow pattern, drastically decrease its effectiveness, resulting in the decrease of the steam condensation capabilities of the system. This adverse effect is accentuated with increasing temperature.

Another aspect to note is that, since the pressure in free space volume above the water head increases linearly, due to the accumulation of very hot non-condensable gas (about 130 ÷ 250 °C), the high temperature of this gas induces another contribution in pressure increase, which also reduce the pressure suppression factor. Moreover, the lower the heat transfer coefficient at the interface, the higher is the non-condensable gas temperature (which is released from high temperature conditions in the VV), the higher is also the pressure in free space volume where it accumulates, as long as the VVPSS system operates in closed mode. Furthermore, from image observation (Figure 6), a given amount of released steam surrounded by air whirlpools, appearing in the flow when air remains trapped, does not condense in the vertical path towards the water head and accumulates in the free space volume. From this figure it can also be deduced that the differentiation of the condensation regimes through the observation of the shape of the steam jet is no longer possible.

Figure 6: CT6H18X3 test: T_w = 60 °C, Q_{sp} = 18.8 g/s

However, the PSF remains greater than 95% for all the examined tests (water temperature ranging from 40 °C to 80 °C). It must be emphasized that the steam mass condensed during these tests is small because the tank quickly reaches the atmospheric pressure.

Considering the worst case (CT8H12X2 test), during the test, a total steam mass of M_S = 4.74 kg is injected into 1000 kg of water at a temperature T_w = 79 °C to reach 100 kPa. Given that the PSF for this test is 0.9564, the accumulation of non-condensed steam is m_{ncs} = (1-0.9564) · 4.74 kg = 207 g. In order to have the same situation in a Small Loca Tank (SLT) of ITER, the equivalent steam mass discharged should be about 40000 l · 4.74 kg / 1000 l = 190 kg, corresponding to about 30 kg of non-condensable gas (16% by weight). If this mass of non-condensable gas accumulated in the free space volume of the SLT (60 m^3), the pressure would rise to a value of: $P_{air} = \frac{m_{air}}{29 \cdot V_{FS}} RT_{FS} = 5$ bars $\gg$ 1.48 bars. For this reason, the steam-air mixture flow would stop very quickly to discharge in the VST because the pressure in the VV must remain low. This leads us to say that the non-condensable gas does not impair the steam condensation efficiency in a significant way, but it causes considerable issues from the point of view of the suppression tank pressure control or the accurate assessment of the mass of non-condensable gas.

4 Conclusions

As part of an experimental campaign, the effects on condensation of a certain amount of non-condensable gas injected together with steam were investigated. 12 tests exploring several condensation regimes have been conducted in a small-scale experimental apparatus discharging a mixture of superheated steam and hot air in a subcooled water pool. Our major conclusions are summarized as follows:

1. The condensation efficiency remains always greater than 95% (in all tests).
2. The injection of non-condensable gas leads to a quick shift towards unstable condensation regimes.
3. The presence of non-condensable gas brings to the onset of a high turbulent flow mix that leads to a disruption of the interfacial boundary along which the heat transfer occurs: this can impair the condensation capabilities of the system. The non-condensable gas can also organize in whirlpools that can trap the steam and carry it out of the water without it being able to condense.
4. The non-condensable gas accumulates in the free space volume above the water head and therefore the pressure increases linearly; the high temperature of this gas induces another contribution in the pressure increase.
5. Such a large amount of hot non-condensable gas causes considerable issue in terms of pressure control of the tank.

References

[1] Gaeta, M. J., Merrill, B. J., Bartels, H. W., Laval, C. R., and Topilski, L., 1997, "Short-Term Hydrogen Production Issues for ITER," Fusion Technol., 32(1), pp. 23–34.

[2] Kukita, Y., Namatame, K., Takeshita, I., and Shiba, M., 1987, "LOCA Steam Condensation Loads in BWR Mark II Pressure Suppression Containment System," Nucl. Eng. Des., 102(2), pp. 225–228.

[3] Zhao, Q., Cong, Y., Wang, Y., Chen, W., Chong, D., and Yan, J., 2016, "Effect of Non-Condensation Gas on Pressure Oscillation of Submerged Steam Jet Condensation," Nucl. Eng. Des., 305, pp. 110–120.

[4] Li, W., Meng, Z., Wang, J., and Sun, Z., 2018, "Effect of Non-Condensable Gas on Pressure Oscillation of Submerged Steam Jet Condensation in Condensation Oscillation Regime," Int. J. Heat Mass Transf., 124, pp. 141–149.

[5] Easom, B. H., 1992, "The Effect of Noncondensible Gas on Steam Bubble Collapse Induced Water Hammer", Doctoral Thesis, Massachussets Institute of Technology (MIT), June 1992.

[6] Liang, K. S., and Peter-Griffith, 1994, "Experimental and Analytical Study of Direct Contact Condensation of Steam in Water," Nucl. Eng. Des., 147(3), pp. 425–435.

[7] Yang, X., Liu, J., Fu, P., Chen, N., and Wei, J., 2019, "Experimental and Theoretical Study of Pressure Oscillation of Unstable Steam-Air Jet Condensation in Water in a Rectangular Channel," Int. J. Multiph. Flow, 119(July), pp. 14–27.

[8] Pesetti, A., Marini, A., Raucci, M., Giambartolomei, G., Olcese, M., Sarkar, B., and Aquaro, D., 2021, "Large Scale Experimental Facility for Performance Assessment of the Vacuum Vessel Pressure Suppression System of ITER," Fusion Eng. Des., 171, p. 112523.

[9] Lo Frano, R., Mazed, D., Aquaro, D., Del Serra, D., and Orlandi, F., 2017, "Experimental Investigation of Functional Performance of a Vacuum Vessel Pressure Suppression System of ITER," Fusion Eng. Des., 122, pp. 42–46.

Analysis of Fukushima Daiichi unit 4 spent fuel pool using MELCOR

M. D'Onorio, A. Maggiacomo, F. Giannetti, G. Caruso

Sapienza University of Rome, DIAEE, Corso Vittorio Emanuele II, 244, 00186 Rome, Italy

matteo.donorio@uniroma1.it

Abstract. A spent fuel pool is a temporary storage facility designed to store the nuclear fuel assemblies removed from the reactor core. Here, the depleted fuel is vertically arranged in racks structures, accurately spaced to ensure subcriticality safety margins. Until the events occurred at the Fukushima Daiichi power plants, severe accidents in spent fuel pools had never been considered as a major safety concern, mainly because of their slow progression. However, extremely unlikely accident events, aggravated by the loss of site power, highlighted the vulnerability of nuclear fuel stored within spent fuel pools. During these accident sequences, reactor fuel rods can generate enough decay heat to vaporize the coolant around them, causing, during unmitigated scenarios, the uncovering and melting of fuel bundles after several days. In past years, new features have been introduced in the MELCOR code for the evaluation of potential risks associated to accident progression in spent fuel pools. In this paper, the accidental sequence of a loss of cooling accident in a boiling water reactor spent fuel pool is analysed with version 2.2 of the MELCOR code. The spent fuel pool of the Fukushima Daiichi unit 4 has been selected as reference. Main accident events, such as fuel uncovering, claddings oxidation, hydrogen generation, and fission products release, have been analysed assuming the unavailability of safety systems.

1. Introduction

In nuclear power plants, during fuel replacement operations, the Fuel Assemblies (FAs) containing exhausted fuel are moved from the reactor core into a suitable storage facility known as Spent Fuel Pool (SFP). Such storage pools are also used to segregate fresh fuel assemblies before being moved into the core. Both irradiated and fresh fuel bundles are sorted in submerged steel racks to provide radiological protection from ionising radiations.

Spent fuel pools are equipped with auxiliary systems for the immediate cooling of the fuel rods. During normal operation, such cooling systems circulate subcooled water among the SFP racks ensuring the decay heat removal from depleted assemblies. However, if the main function of these systems is lost for a long time, fuel rods can generate enough decay heat to cause SFP's water evaporation and subsequent fuel assemblies uncovering and melting [1].

Prior to the event at the Fukushima Daiichi nuclear power plant, severe accidents in the spent fuel pool have never been considered a safety concern since their progression is slow enough to let operators begin mitigative actions. However, accident scenarios characterized by long-term loss of site power, such as the Fukushima one, highlighted a possible vulnerability of the nuclear fuel stored within SFPs.

After this accident which occurred in March 2011, international benchmark exercises were carried out to simulate the accident sequence and its consequences. Among them, one of the most important was in the framework of the NUGENIA+ project, carried out in 2015, in which participants investigated the Fukushima Daiichi occurrence of events by using different severe accident codes [2].

For the scope of this simulation, it has been decided to use the last available version of the MELCOR code, also focusing on phenomena related to radionuclide releases from fuel assemblies, which were not considered in the previous benchmark studies.

MELCOR is a computer code developed by SANDIA National Laboratories (SNL) for the US Nuclear Regulatory Commission (NRC) to simulate transients and the progression of severe accidents in nuclear power plants. MELCOR code comprises several major packages that let the user model the main systems of a reactor plant and the main phenomena occurring during a severe accident (e.g., zircalloy and steel oxidation, fuel degradation, core melting, and radionuclides release, transport, and deposition). These major packages include, for example, the CVH and the FL packages, which simulate the thermal hydraulics of the system; the COR package, which simulates the behaviour of structures and fuel inside the core; and the RN package, which simulates release, mobilization, and relocation of radionuclides.

Version 2.2.18019 of MELCOR has been used for this calculation [3][4]. The analysis of the progression and radionuclide release of the accident scenario is based on the layout of Fukushima Daiichi unit 4 SFP. The aim of this paper is to simulate a base case accident scenario to be used for futures uncertainty quantification analyses that will be carried out by Sapienza in the framework of the European H2020 MUSA Project [5].

2. Spent Fuel Pool modelling using MELCOR

According to TEPCO data [5], Fukushima Daiichi unit 4 Spent Fuel Pool is a rectangular pool of dimensions 12.2 × 9.9 × 11.8 m, located in the upper part of a Mark I reactor building with a free volume of 25800 m^3 [6]. During normal operation, the water level inside the SFP is 11.5 m. When the Fukushima accident occurred, inside the SFP of unit 4 reactor, there were 1535 fuel assemblies divided into hot (548 FAs), cold (783 FAs), and fresh fuel (204 FAs) [2]. As shown in Figure 1, the fuel assemblies were stored in 53 different racks, with the average assembly decay heat pattern.

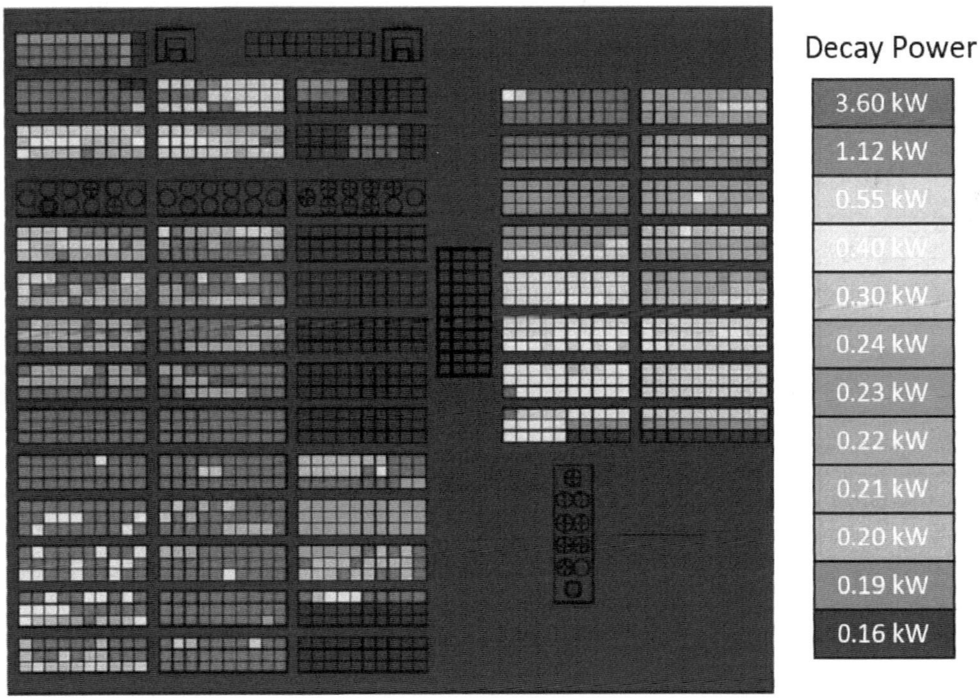

Figure 1 - Layout of spent fuel assemblies in the Fukushima unit 4 plant [2]

Journal of Physics: Conference Series 2177 (2022) 012020 doi:10.1088/1742-6596/2177/1/012020

The dimensional data used for this analysis are referred to the BWR STEP3 fuel assembly. Each fuel assembly is composed by 72 fuel rods in a 9x9 geometry and by a central water channel. Fuel bundles are encased in a channel box. There are also seven spacers between the rods in order to keep a constant distance between them. The fuel assemblies are stored inside rectangular racks, and each rack contains 30 FAs in a 3x10 geometry for a total of 53 racks. The steel walls of the racks are double walls with some space for water flow between them. The racks are also composed of 8 legs that keep the rack suspended into water to allow natural circulation and a baffle plate to sustain the fuel rods with holes to allow water flow [2].

Since the MELCOR COR package modelling is based on a cylindrical geometry, the fuel assemblies contained in the SFP have been modelled using concentric rings. As shown in Figure 3, the SFP is divided into 3 radial rings and 13 axial levels. The total number of core cells is 39. Hot fuel assemblies have been modelled in the central radial ring, which is surrounded by a second ring modelling cold FAs, and by a third ring modelling fresh FAs. With respect to the axial position of the rods, the active fuel is 3.71 m long, starting from a height of 0.3689 m to 4.0789 m. Fuel assemblies in each ring are surrounded by the stainless steel of the racks and by the water of the pool. All the masses for stainless steel, Zircalloy and fuel are evaluated from available TEPCO data.

Considering the low heat flux and the large amount of water, a detailed nodalization that considers each fuel assemblies' correct position is not required for our simulation purposes. Moreover, as a conservative assumption, the hot FAs are placed in the central ring of the COR model.

The cells modelled in the COR package are coupled with hydrodynamic package control volumes. In particular, COR cells containing nuclear fuel are coupled two by two with a control volume. In Figure 2, the thermal-hydraulic nodalization developed for this simulation is shown. It is based on the modelling of 26 control volumes and 49 flow paths. Each core channel (1 for each COR radial ring), modelling the fluid region inside the canister, has been axially split into 6 different control volumes. The three associated bypass control volumes, modelling the interstitial space between the channel box and the rack cell walls, have been simulated with three different control volumes (CV210, CV220, CV230 in Figure 2). The flow blockage model is activated for each core channel flow path.

Moreover, logically controlled flow paths connecting channels and bypass CVs are opened if the canister wall fails. The remaining water inside the pool is modeled with two large control volumes, one surrounding the core and one representing the upper part of the pool. Two time-independent control volumes have been used to model the reactor containment and the environment.

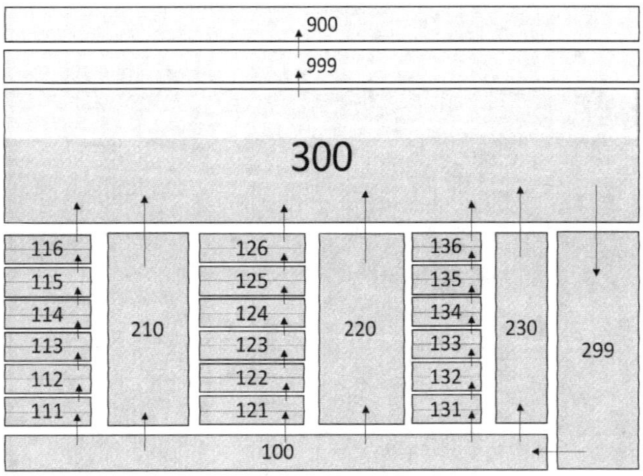

Figure 2. Fukushima Unit 4 SFP - Thermal-Hydraulic model

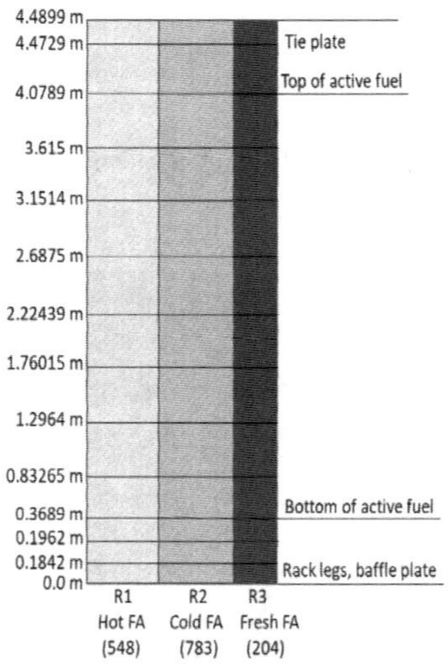

Figure 3. Axial COR nodalization

Structures surrounding the SFP have been modelled with 4 different 2-layer HSs composed of a steel liner of 0.635 cm and a 2 m thick layer of concrete. As a conservative assumption, to consider the pool as an adiabatic system, no heat exchange between the reactor building and the SFP wall structures has been assumed.

With the use of two different packages, namely the DCH (decay heat) and the RN (radionuclides) packages, MELCOR allows the user to input the decay heat power and the masses of radionuclides by defining radionuclide classes. MELCOR divides radionuclides in 17 chemical classes, indicating them with the most relevant element of the class itself. Through the DCH package, the user can define a decay power for each of these classes. Moreover, with the RN package, the user can define more specifics about radionuclides, and it is possible to define for each core cell the exact mass of present radionuclides for class. For the current analysis, radionuclides mass inventory calculated by ENEA and reported in [7] has been considered. Summing all the isotopes present for each class, an initial decay heat and radionuclides inventory have been calculated for both hot and cold fuel assemblies. As a conservative assumption, any radionuclide and decay power has been assumed for the third ring, containing only fresh fuel. The radionuclides masses for each class have been uniformly divided for the 8 cells (from axial level 4 to 11) containing active fuel, assuming a uniform distribution inside the rings.

The decay power for the 40 most relevant radionuclides is given in [7] in terms of kW/FA both for hot and cold fuel assemblies. By multiplying the given decay power for the FAs present in rings 1 and 2, a total decay power of 2.4 MW has been obtained. This has been divided into 1.9 MW and 0.5 MW for hot (ring 1) and cold (ring 2) fuel assemblies. As a conservative assumption, the decay power has been assumed to be constant throughout all the simulation time. The card for the combination of Iodine has been activated to simulate the formation of CsI and CsM in class 16 and 17, respectively.

3. Accident scenario and assumption

The postulated initiating event for the accident scenario is a loss of cooling accident occurring together with a total station blackout. The analysis has been performed assuming the unavailability of all emergency and mitigation systems. Because of the relatively low amount of decay power and the large water inventory, SFP postulated accident progression would result in a large time-consuming process. The accident sequence is assumed to start when the water level is just above the racks, in order to reduce the simulation time. For this initial condition, the total mass of water in the spent fuel pool has been evaluated to be 450 t (instead of the 1320 t of the full pool). A preliminary analysis highlighted that around 10 days are needed to reach this initial condition. Thus, for all the results described in the next session, it should be considered that for a complete accident sequence, including water evaporation from the top of SFP to rack levels, the beginning of simulation corresponds to 10 days. Since the top of the unit 4 reactor building of the Fukushima Daichi NPP was damaged by the hydrogen explosion, occurred around four days after the reactor SCRAM, its effect on mitigating radiological releases can be, conservatively, not considered.

4. Results

The predicted collapsed water level of the pool after the loss of cooling is shown in Figure 4. Since the initial bulk temperature of water has been set to saturation conditions, the water level decreases quite fast. The uncover of fuel starts after around 24 hours, reaching the end of active fuel after 6.6 days and the dry-out after around 9 days.

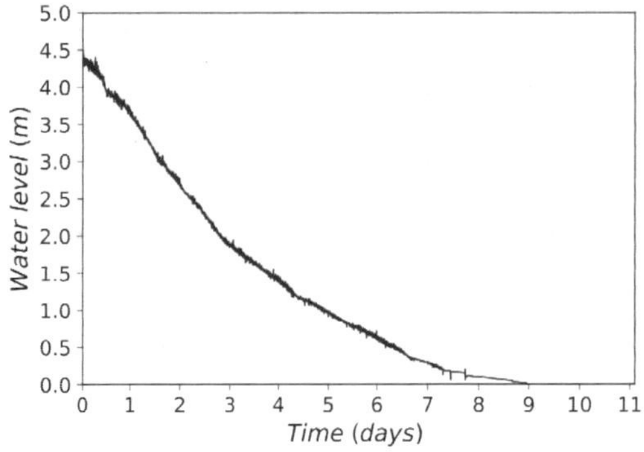

Figure 4. Water level

With the ongoing water level decrease and the consequent fuel uncovery, temperatures of core components start to increase. Figure 5 and Figure 6 illustrate the predicted cladding temperature and stainless-steel rack temperature at various axial levels for the central COR ring, respectively. The temperature of top-level cladding elements at the uppermost axial level begins to heat-up after around 36 h, when the active fuel is fully uncovered. The ability to remove decay heat generated by the spent fuel is reduced as the water level continuously drops, especially when most of the active fuels are exposed to air without any emergency mitigation available. From both Figure 5 and Figure 6, it is also possible to study the failure sequence and cladding, and rack supporting structure. The top-axial level rack cell failure occurs at 4.62 days when the temperature of the rack is 1700 K. Fuel clad failure occurs after around 6.8 days when a temperature of 2500 K is reached. The MELCOR interactive (INT) model was activated to simulate the eutectic temperature of the ZrO2–UO2 binary mixture. Two unique materials were introduced, UO2-INT and ZRO2-INT, with a eutectic temperature of 2500 K, as suggested by the Phébus experiment [8].

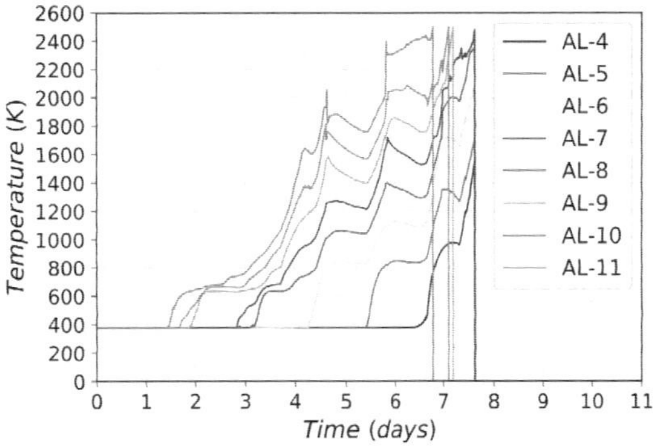

Figure 5. Fuel clad temperature in ring-1 for different axial levels

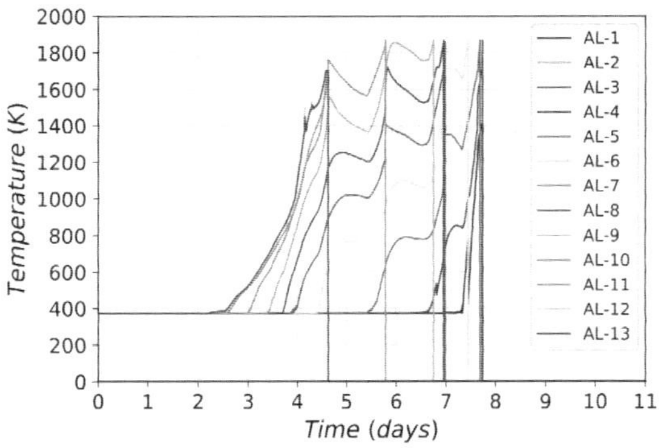

Figure 6. SFP rack temperature in ring-1 for different axial levels

The heat-up of cladding and stainless-steel structures led to the onset of oxidation reaction with steam. Zircalloy oxidation begins after a couple of days, while stainless-steel oxidation after around 4 days. In Figure 7, the waveform of oxide production is shown. After 9 days, the oxide masses are almost constant, reaching 28000 kg for ZrO2 and 26000 kg for SSOx. At the end of the simulation, around 16.0 % of Zr, and 13.7 % of SS, were oxidized, respectively. A large quantity of hydrogen gas will be generated as a product of the chemical reaction. As shown in Figure 8, hydrogen production follows the same trend of oxidation. At the same time, the hydrogen production is almost finished due to the complete evaporation of initial water, and the mass of hydrogen remains relatively constant at

1730 kg. In these conditions, it is possible to see that the progression of the accident is quite slow and let the operators have enough time to do mitigative actions. Hydrogen production results are in good agreement with the one obtained in past analyses related to loss of cooling accidents in SFP [9][10][11].

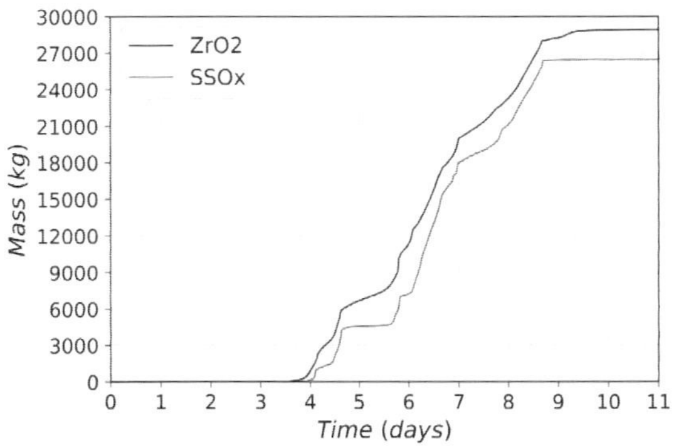

Figure 7. Zirconium and SS oxides production predicted by MELCOR

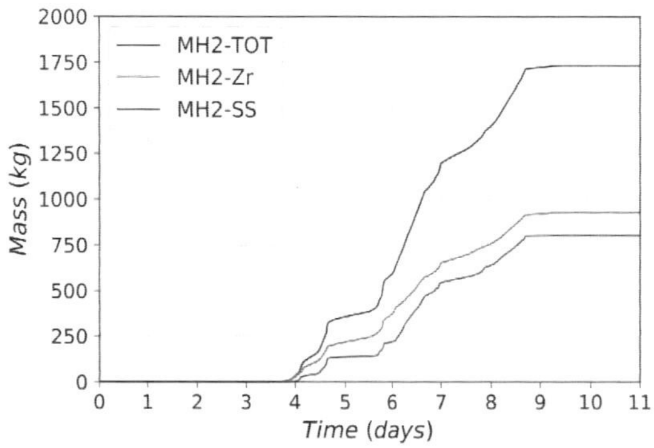

Figure 8. Mass of Hydrogen produced by oxidation reactions

When the cladding temperature reaches the oxidation temperature threshold (set to 600 K [12][13]) in the MELCOR oxidation model, the oxidation reaction in the presence of steam produces a strongly exothermic reaction that accelerates the rising of the cladding temperature. The oxidation heat generation rate in the spent fuel pool is shown in Figure 9. The heat generated is in good agreement with data obtained in the OECD/NEA Sandia Fuel Project program, conducted to study thermal-hydraulics and fire characteristics of spent fuel assemblies [14].

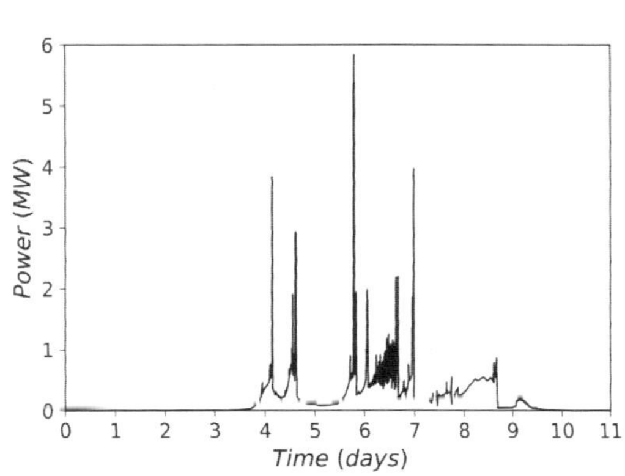

Figure 9. Oxidation heat generation rate

The volume fraction of damaged FAs is shown in Figure 10. The damage reaches 87 % since the third ring, not containing radionuclides and decay power, remains intact. Fuel rod collapse can occur through various means: loss of global support or local temperature exceeding the default temperature-based failure criterion. In the present simulation, rack baseplate failure occurs in radial rings 1 and 2. As shown in Figure 10, the ring 1 rack baseplate fails after around 6.3 days; before, the collapsed FAs are at axial levels 9 to 11. The rack baseplate of ring 2 fails after around 8.6 days, causing the failure of the cold fuel assemblies also if the threshold temperature of 2500 K is not reached in any axial node of ring two.

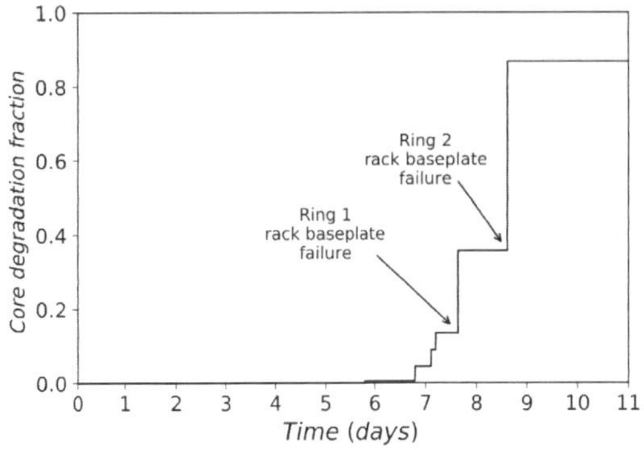

Figure 10. Core degradation fraction predicted by MELCOR

As degradation of core continues, radioactive fission gases contained in fuel assemblies are released and moved toward SPF and reactor building volumes. Source term estimation often focuses on the more volatile radionuclides such as caesium, iodine, and noble gases. Noble gasses are chemically not reactive and can contribute to the radiological doses by external exposure, generally with a small contribution on the total absorbed dose. Iodine and Caesium are volatile and chemically reactive elements and can contribute to internal exposure. Iodine may be present as the very hygroscopic molecules of Caesium Iodide (CsI), dangerous for its high solubility in water. Moreover, the Phebus FP program [8] indicated that Caesium-Molybdate (Cs2MoO4) is the dominant chemical form of the released Caesium, also if it is less volatile than CsOH and CsI. Analyzing MELCOR classes' contributions to released radionuclides, the most significant mass inventory is the chalcogens mass (including radioactive Tellurium), with about 16000 kg (Figure 11). Other relevant radiological MELCOR classes, shown in Figure 12, are: class 1 (grouping noble gases) with 933 kg; class 2 (containing alkali metals, such as CsOH) with 554 kg; class 16 (CsI) and class 17 (Caesium-Molybdate) with 83 kg and 20 kg, respectively.

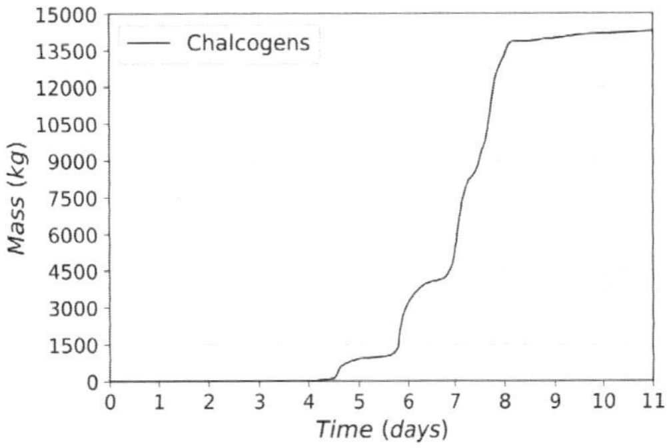

Figure 11. Mass of Chalcogens (Te) released from FAs predicted by MELCOR

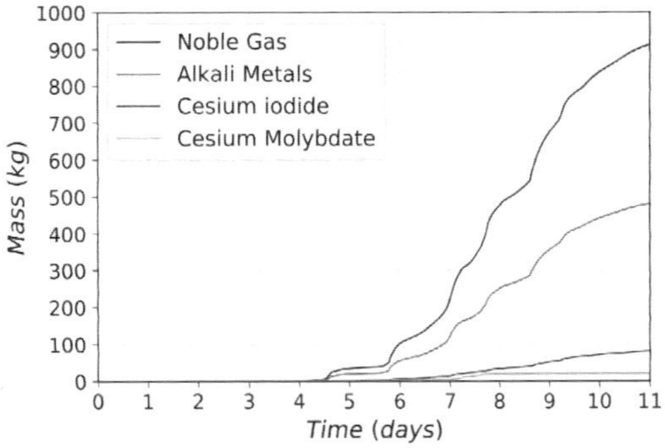

Figure 12. Mass of radionuclides released from FAs predicted by MELCOR

In Table 1, the comparison between initial and released mass is reported for the most relevant classes. Other classes have not been reported since radionuclides masses are very low (<5 kg) with respect to the ones shown in Table 1. As a final investigation, the release in different volumes was analyzed. Almost the totality of released radionuclides from the fuel assemblies has been deposited on the metal structures inside the pool and on the concrete structures of the pool. The released fraction outside the SFP is reported in Table 1.

Table 1. Radionuclides masses

Class	Initial masses	Mass released from FAs	Release fraction from FAs	Mass released outside the pool	Release fraction outside the pool
Class 1	1493.9 kg	933.15 kg	62.5%	871.27 kg	58.32%
Class 2	822.326 kg	554.47 kg	66.2%	66.24 kg	8.05%
Class 5	35018.0 kg	16094.3 kg	46%	1796. kg	5.12%

5. Conclusions

In the present work, the spent fuel pool of Fukushima Daiichi unit 4 has been modelled with MELCOR 2.2.18019 to evaluate the consequences of a loss of cooling accident in terms of fuel degradation and radionuclides release. Main results of the severe accident progression include fuel degradation, clad and racks structures heat-up and failure, hydrogen generation and radionuclide release. In particular, the large amount of water leads to the formation and production of large amounts of oxides and hydrogen. The studied accident, even starting from very conservative conditions such as the water in saturated conditions, let however a long time for mitigative actions to prevent fuel uncover and consequent overheating of the fuel rods since the progression of accident events is quite slow.

Release of radionuclides from cladding begins about 4 days after the initiating event and it strongly follows the progression of core damage and the ongoing oxidation. It was also shown that, in the long-term scenario, almost the whole core fails except for the ring containing fresh fuel that, having no decay power, does not heat up as fast as the other two rings. The failing of the core leads to the production of debris, but there is no damage to the pool walls, so the only concern for radionuclide release is the atmospheric one that is anyway a low part of radionuclides initial mass.

The analyses highlighted as the MELCOR could provide a valuable means for characterizing the consequences of severe accidents in spent fuel pools. However, as there are no corresponding experimental results, the validity of the simulation results cannot be assessed in detail, and uncertainty analyses are needed. As future work, this MELCOR model, developed in the framework of the MUSA project [5], will be used to perform sensitivity and uncertainty analyses to better explore the variability in the potential responses of SFPs systems undergoing severe accident conditions.

References

[1] OECD- Organisation for Economic Co-operation and Development, Phenomena Identification and Ranking Table, R&D Priorities for Loss-of-Cooling and Loss-of-Coolant Accidents in Spent Nuclear Fuel Pools, NEA/CSNI/R(2017)18

[2] Condreau O, Jäckel B, Rocchi F, Alcaro F, Angelova D, Bandini G, Barnak M, Behlerg M, Da Cruz DF, Dagan R, Drai P, Ederli S, Herranz LE, Hollands T, Horvath G, Kaliatka A, Kljenak I, Kotsuba O, Lind T, López C., Mancheva K, Matejovic P, Matkovič M, Steinbrückh M, Stempniewicz M, Thomas R, Vileiniskis V, Visser DC, Vokáč P, Vorobyov Y and Zhabin O 2018 *Severe accident code-to-code comparison for two accident scenarios in a spent fuel pool*, Annals of Nuclear Energy, 120, pp 880–887, https://doi.org/10.1016/j.anucene.2018.06.043

[3] Humphries LL, Beeny BA, Gelbard F, Louie DL and Phillips J 2015 *MELCOR Computer Code Manuals, Vol. 1: Primer and Users' Guide, Version 2.1.6840*, SANDIA REPORT SAND-2015-6691R, Sandia National Laboratories, Albuquerque, New Mexico. https://www.nrc.gov/docs/ML1704/ML17040A429.pdf

[4] Humphries LL, Beeny BA, Gelbard F, Louie DL and Phillips J 2017 *MELCOR Computer Code Manuals, Vol. 2: Reference Manual,* Version 2.1.6840, SANDIA REPORT SAND2017-0876O, Sandia National Laboratories, Albuquerque, New Mexico. https://www.nrc.gov/docs/ML1704/ML17040A420.pdf

[5] Luis E. Herranz, Sara Beck, Victor Hugo Sánchez-Espinoza, Fulvio Mascari, Stephan Brumm, Olivia Coindreau and Sandro Paci, The EC MUSA Project on Management and Uncertainty

of Severe Accidents: Main Pillars and Status, Energies 2021, 14(15), 4473, https://doi.org/10.3390/en14154473

[6] Gauntt RO, Kalinich DA, Cardoni JN, Phillips J, Goldmann AS, Pickering SY, Francis M, Robb K, Ott LJ, Wang D, Smith C, Germain S, Schwieder D and Phelan C 2012 *Fukushima Daiichi Accident Study (Status as of April 2012),* SAND2012-6173; Sandia National Laboratories, USA, https://doi.org/10.2172/1055601

[7] AIR-SFP Benchmark, Microsoft Excel file, Inventories_for_HF_and_CF_Revised_25_05_2020_V1, not publicly available

[8] Bernard C and Roland Z 2013 *The objectives of the Phébus FP experimental programme and main findings*, Annals of Nuclear Energy, 61, pp 4-10, https://doi.org/10.1016/j.anucene.2013.03.037

[9] Wang D, Gauld IC, Yoder GL, Ott LJ, Flanagan GF, Francis MW, Popov EL, Carbajo JJ, Jain PK, Wagner JC and Gehin JC 2012 *Study of Fukushima Daiichi nuclear power station unit 4 spent-fuel pool*, Nuclear Technology, 180, pp 205-215, https://doi.org/10.13182/NT12-A14634

[10] Carbajo JJ 2012 *MELCOR Model of the Spent Fuel Pool of Fukushima Dai-ichi Unit 4*, American Nuclear Society Meeting, Chicago, Illinois, USA

[11] Wu X, Li W, Zhang Y, Tian W, Su G and Qiu S 2014 *Analysis of the loss of pool cooling accident in a PWR spent fuel pool with MAAP5,* Annals of Nuclear Energy, 72, pp 198-213, https://doi.org/10.1016/j.anucene.2014.05.030

[12] Urbanic VF and Heidrick TR 1978. *High-temperature oxidation of zircaloy-2 and zircaloy-4 in steam*, Journal of Nuclear Materials, 75, pp 251–261, https://doi.org/10.1016/0022-3115(78)90006-5

[13] Coindreau O, Duriez C and Ederli S 2010 *Air oxidation of Zircaloy-4 in the 600-1000 °C temperature range: Modeling for ASTEC code application,* Journal of Nuclear Materials, 405, pp 207-2015, https://doi.org/10.1016/j.jnucmat.2010.07.038

[14] Martina Adorni, Luis E. Herranz, Thorsten Hollands, Kwang-Il Ahn, Christine Bals, Francesco D'Auria, Gabor L. Horvath, Bernd S. Jaeckel, Han-Chul Kim, Jung-Jae Lee, Masao Ogino, Zsolt Techy, Alexander Velazquez-Lozad, Abdelghani Zigh, Radomir Rehacek, 2016, *OECD/NEA Sandia Fuel Project phase I: Benchmark of the ignition testing,* Nuclear Engineering and Design, 307, pp 418-430, https://doi.org/10.1016/j.nucengdes.2016.07.016

Severe accident sensitivity and uncertainty estimation using MELCOR and RAVEN

M. D'Onorio[1], A. Giampaolo[1], F. Giannetti[1], F. Mascari[2], G. Caruso[1]

[1] Sapienza University of Rome, DIAEE, Corso Vittorio Emanuele II, 244, 00186 Rome, Italy

[2] ENEA, FSN-SICNUC-SIN, Via Martiri di Monte Sole, 4 - 40129 Bologna

matteo.donorio@uniroma1.it

Abstract. Today, using a best-estimate approach is a key factor in the simulation and prediction of thermal-hydraulic and other multi-physics phenomena occurring during nuclear severe accidents. The best-estimate approach requires the quantification of both epistemic and stochastic uncertainties of safety codes to be effective. This safety assessment approach, named Best Estimate Plus Uncertainty (BEPU), is being pursued as an alternative to traditional deterministic analyses that are intrinsically conservative with not clearly defined safety margins. While in a conservative approach, the results are expressed in terms of a set of calculated conservative values of parameters, in a best-estimate methodology, the results are expressed in terms of uncertainty ranges for the calculated parameters. The International Technical Nuclear Community has made great efforts to develop methods and tools for uncertainty and sensitivity analyses of severe accident codes. In this framework, Sapienza University of Rome has developed a new Python script to use RAVEN as a tool for the application of the BEPU methods within the safety analyses performed with MELCOR. The aim of this paper is to show the capabilities of the coupling between RAVEN and MELCOR by performing a statistical analysis to estimate the range of evolution of a severe accident scenario.

1. Introduction

In evaluating safety margins, the use of the Best Estimate Plus Uncertainty (BEPU) approach by coupling selected calculated parameters with the related uncertainty range is of great interest for the International Technical Nuclear Community. Considering the reached level of development and maturity of Severe Accident (SA) codes and their application on Severe Accident Management Guideline (SAMG) assessment, the discussion and application of SA progression analyses with uncertainty estimation is currently a key topic in the BEPU framework. Several SA codes, in fact, have been developed and validated against the available experimental data in the last decades to simulate the phenomena/processes taking place in an unmitigated accident determining potential core degradation.

In this framework, MELCOR [1][2][3] is a fully integrated severe accident code able to simulate the thermal-hydraulic phenomena in steady and transient condition and the main SA phenomena characterizing the Reactor Pressure Vessel (RPV), the reactor cavity, the containment, and the confinement buildings typical of Light Water Reactors (LWR). The estimation of the source term is obtained by the MELCOR code as well. MELCOR can be used with the Symbolic Nuclear Analysis

Package (SNAP) [4] to develop the nodalization and post-process data using its animation model capabilities. MELCOR has a modular structure. Each package simulates a different part of the transient phenomenology. In particular, the Control Volume Hydrodynamics (CVH) and FLow path (FL) packages simulate the mass and energy transfer between control volumes, the Heat Structure (HS) package simulates the thermal response of the heat structure, and the CORe (COR) module evaluates the behaviour of the fuel and structures contained in the core and lower plenum and their degradation phenomena. The CVH/FL packages' role that provides the boundary condition for other packages is to be underlined.

Sapienza University of Rome has developed a new Python [5] interface to couple the SA code MELCOR and Reactor Analysis and Virtual control ENviroment (RAVEN) software tool. RAVEN, developed at the Idaho National Laboratory (INL), acts as the control logic driver and post-processing tool for different applications [6]. RAVEN has been developed to perform parametric and probabilistic analysis based on the response of complex system codes to quantify the safety margins related to safety-related events [7]. Nowadays, RAVEN is a multi-purpose probabilistic and uncertainty quantification platform that can be coupled with any system code. For this purpose, a new code interface has been developed to couple MELCOR code with the RAVEN framework.

The aim of this paper is to show the capabilities of the coupling between RAVEN and MELCOR, performing statistical analysis to estimate the range of evolution of severe accident scenarios. For this purpose, the Fukushima Daiichi Unit 1 unmitigated accident sequence has been selected as the reference scenario. A set of seven parameters have been selected among the list of modelling coefficients investigated in the SOARCA (State-of-the-Art Reactor Consequence Analyses) project [8]. Transient results have been statistically analysed through the RAVEN post-processor.

2. Plant modelling using MELCOR

Fukushima Daiichi Unit 1 is BWR/3 of 1380 MW_{th} with a steam flow rate generation in nominal operation condition of 2400 t/h. The operation pressure in RPV is about 6.99 MPa considering an average core temperature of 558 K. Fukushima Daiichi Unit 1 employed a Mark I containment with a bulb-shaped drywell and a toroidal suppression chamber, connected with radial vents and downcomers immersed in the pool. There were 8 vent lines connecting through 80 downcomers the drywell volume with the suppression chamber [9][10].

In order to develop the Fukushima unit 1 MELCOR nodalization, following the SANDIA approach reported in the Fukushima Daiichi Accident Study [11], the nodalization has been based on the Peach Bottom reactor (different power but similar reactor layout). The references used to develop the BWR Peach Bottom nodalization are [12] and [13]. Starting from this model, the Fukushima Daiichi Unit 1 input was built with the data reported by TEPCO and applying a reasonable scaling factor for each component if the data were not available. The details of the developed MELCOR input deck (geometrical and steady-state operational condition comparison with plant data) are reported in [14] and [15].

The MELCOR nodalization was designed to have a reasonable computational time and a realistic prediction of the phenomena involved during the transient, assuring a reliable and accurate transient simulation. In MELCOR nodalization, developed by using SNAP [4], the core has been modelled by coupling the CVH package nodalization with the correspondent MELCOR code model of the COR package. In order to better simulate the local heat transfer process for a wide range of fluid conditions and structure surface temperatures, each core equivalent channel has been axially split into 5 control volumes. The reactor core is modelled in the COR package with 23 axial regions and 6 radial rings. The lower plenum has been modelled with the first 13 axial regions and the core with 10 axial regions. The hydraulics control volumes of the MELCOR nodalization are shown in Figure 1.

Journal of Physics: Conference Series 2177 (2022) 012021 doi:10.1088/1742-6596/2177/1/012021

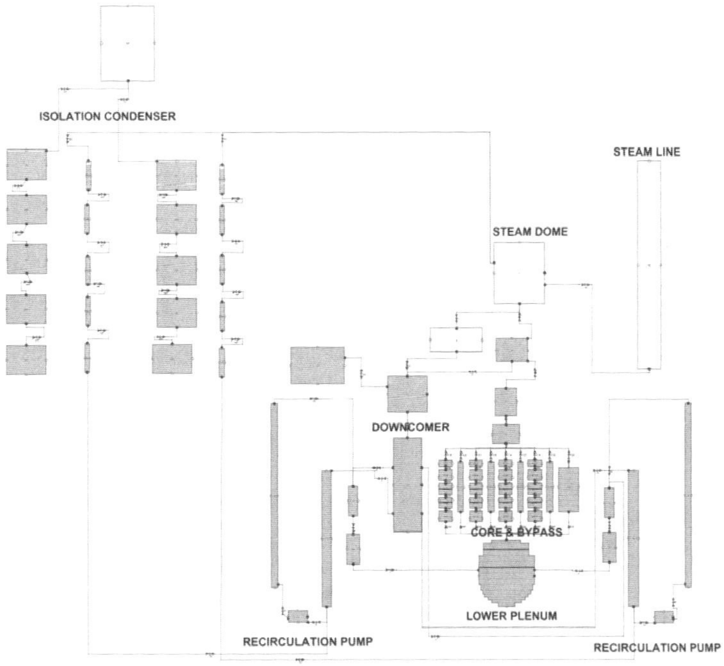

Figure 1. Thermal-hydraulic model of Fukushima Daiichi Unit 1 (made with SNAP).

3. RAVEN-MELCOR CODE COUPLING

A Python interface has been developed to allow the coupling between RAVEN and MELCOR. The interface has three manly functions, interpret the information coming from RAVEN, translate such information in the input of the driven code, and manipulate output data files to create a database. To deal with the first two functions, the "GenericCode interface" module, already implemented in RAVEN, has been coupled with a module that runs MELGEN and MELCOR executables. A Python output parser has been developed to allow RAVEN to store output data coming from MELCOR. The aim of this parser is to convert the PTF binary file generated by MELCOR into a single CSV file. The interface allows the creation of a CSV file with only plot variables required by the users to overcome the handling of large data files. So is possible to obtain a Hierarchical Data Format 5 (HDF5) database that comprises variables from all MELCOR packages (e.g., CVH, FL, COR, RN) and perform statistical analyses on the specified variables. The procedural framework used for the uncertainty quantification is shown in Figure 2.

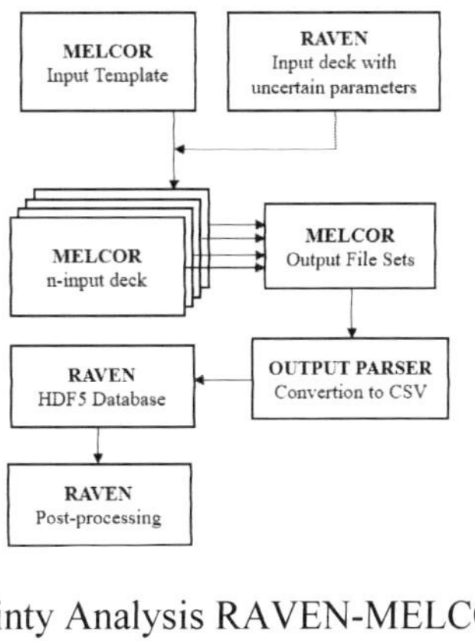

Figure 2. Uncertainty Analysis RAVEN-MELCOR flow diagram.

A MELCOR input deck is used as a template, and the chosen parameters are specified as a string with special characters. In such a way, RAVEN can identify such parameters and replace the string with values sampled from a specified distribution. Consequently, N MELCOR input-decks are generated. Once the HDF5 database has been generated, statistical analysis of the output sets can be performed.

3.1. Variables and sampling

A set of 7 parameters have been chosen to perform this study and to test the implemented code coupling between RAVEN and MELCOR. The probabilistic method to propagate input uncertainty has been selected to perform the analyses [17]. The parameters treated in this analysis are the recirculation pump seals leak flow area (plant parameter not available to the code user), and 6 MELCOR modelling parameters taken from the list of modelling parameters and sensitivity coefficient investigated in [16]. MELCOR code allows code-users to modify some important parameters model directly through code input as Sensitivity Coefficient (SC), to facilitate uncertainty analyses and sensitivity studies. The parameters selected are listed in Table 1. The Area_Seal parameter is related to the seal leak flow area. In typical BWR recirculation pumps, a combination of pressure unbalance and springs is used to keep the faces of the seal closed. If this combination is lost, a leakage from the pumps can occur. This parameter mainly affects the RPV blowdown phase and the pressurization of the DW containment. In MELCOR, the downward relocation of the core material is modeled as a constant velocity movement using a user-specified velocity of the falling debris (VFALL). Since core debris are cooled by surrounding water during the relocation phase, this velocity can affect the properties of debris relocated into the lower plenum, influencing lower head failure and melt release mode. Once debris relocates above the lower head, it increases its temperature based on debris-lower head heat transfer coefficient, input in MELCOR code as HDBLH. The default heat transfer coefficients are order-of-magnitude parameters that should be varied in sensitivity studies to determine their impact on lower head heat transfer and failure. After zirconium melting and candling by exceeding breakout temperature, UO_2/ZrO_2 eutectic reactions will form a complex mixture with a lower melting temperature than either ZrO_2 or UO_2. Temperature formation of eutectic could affect fuel failure and molten pools generation. In MELCOR the temperature at which eutectic will melt could be changed by varying the SC1132(1). The sensitivity coefficient 1141(2) governs the maximum melt Zr flow rate per unit width after breakthrough. A high flow rate corresponds to a higher mass of molten Zr relocating, thus lower local oxidation, temperature, and extended lifetime. Sensitivity coefficient 1502(2) specifies the minimum component mass below which will not be subject to the maximum temperature change criterion. In conclusion, the sensitivity coefficient 1250(1) is used to fix the temperature above which the conduction heat transfer model is enhanced where core debris is molten. This parameter is useful to capture the qualitative effects of convection in molten pools.

Once all the variables have been selected to perform an uncertainty quantification analysis, a sampling strategy needs to be employed. The sampling strategy is used to perturb the input space in relation to variable distributions. The sample size needed to obtain a significative output statistic was selected using the Wilks formula for two-sided statistical tolerance limits [18]. The required minimum number of computer code calculations becomes 93 for a 95% probability and 95% confidence level [17]. The Monte Carlo sampling strategy has been selected for this analysis, setting a limit of 250 calculations to consider possible code failures. Of the 250 MELCOR runs, 33 experienced numerical failures. The Monte Carlo method is one of the most-used sampling methodologies because it does not use a structured discretization of the input space but a cumulative probability and probability distribution function to compute the different values of the concerned variables. In table 1, the characterization of the input parameters is reported. However, this is a preliminary estimation based on engineering judgment, and it is the basis for showing the methodology used in the paper. Further studies are in progress to assess these PDFs and ranges.

Table 1. **MELCOR perturbed variables in input-deck.**

Variable	Description	Distribution	Unit	Data
Area_Seal	Recirculation pump seals leak flow area	Triangular	m^2	Mode: 0.12 E^{-3} Min: 0.06 E^{-3} Max: 0.4 E^{-3}
vfall	Velocity of falling debris	Triangular	m/s	Mode: 0.1 Min: 0.05 Max: 1.2
hdblh	Heat transfer coefficient from debris to lower head	Triangular	$\frac{W}{m^2\,K}$	Mode: 1000 Min: 50 Max: 1100
SC1132(1)	Core Component Failure Parameters - Temperature to which oxidized fuel rods can stand in the absence of unoxidized Zr in the cladding.	Normal	K	Mean: 2700 Sigma: 120
SC1141(2)	Core Melt Breakthrough Candling Parameters - Maximum melt flow rate per unit width after breakthrough	Triangular	Kg/s	Mode: 0.083 Min: 0.01 Max: 1.0
SC1502(2)	Minimum Component Masses - Minimum total mass of component subject to the maximum temperature change criterion for timestep control	Normal	Kg	Mean: 10. Sigma: 1.0
SC1250(1)	Conduction Enhancement for Molten Components - Temperature above which enhancement is employed	Normal	K	Mean: 2800 Sigma: 150.0

4. Results

Results have been statistically analysed through the RAVEN BasicStatistic post-processor. A dynamic statistical analysis has been performed, setting time as a pivot parameter. To better visualize uncertainty, the 0.05, 0.5 (median) and 0.95 percentiles have been selected, analysing different figures of merits. The Pearson's and Spearman's coefficients have been used to describe the correlation between the perturbed variables and the figure of merits. In particular, Pearson's coefficient has been analysed to measure the degree of linear correlation between a perturbed variables and the figure of merits. The Spearman coefficient has been used to measure the the degree of monotonous correlation between a perturbed variables and the figure of merits. Both coefficients span between -1 and +1.

Figure 3 and Figure 4 show the spectrum of predicted DryWell (DW) and WetWell (WW) pressure, respectively. In the early stage of the accident, the gradual increase of pressure is due to steam ejection from recirculation pumps seal leak. DW pressurization is also the result of hydrogen generated from Zircalloy oxidation reaction, which production starts about 2.5 hours after the reactor SCRAM.

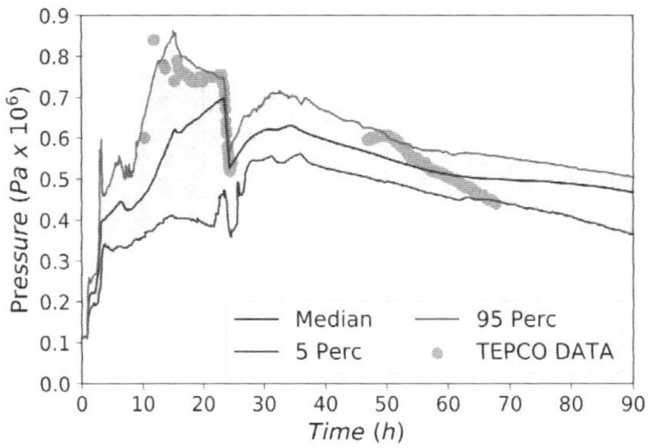

Figure 3. Variation range of Drywell pressure.

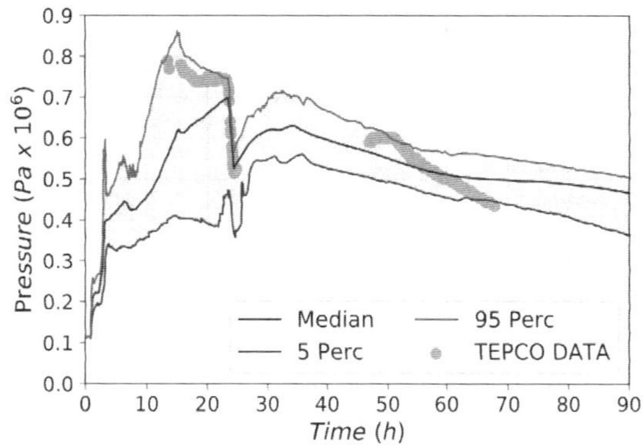

Figure 4. Variation range of Wetwell pressure.

The generated hydrogen reaches the DW and WW containment together with steam through leaks from RPV, causing a pressure increase in the containment. It is to underline that the results of the calculated data coupled with the related uncertainty band envelope mostly of the TEPCO plant data available related to both WW and DW pressure.

The pressure continued to increase until the operation of WW venting valves occurred 23 h 30 min after the earthquake. Before venting, the DW and WW pressure variation range is between 0.75 MPa and 0.44 MPa, with an expected value of around 0.7 MPa.

The venting was modelled by opening a flow path from the top of the WW to the environment. Venting activation is reported in TEPCO reference [9]. After, the venting pressure trend increases until around 33 h when the water injection with the fire engine was started.

The Pearson's and Spearman coefficient are defined between -1 and 1 and measure the linear and monotonic relationship strength between two variables. In Figure 5, the strong linear correlation between the recirculation pump seal leak area and the DW pressure is shown. The strong positive relationship highlights that as the leak area increases the pressure in containment increases, as expected. This correlation decreases with time since the long-term pressurization level of the DW depends mainly on the total inventory released and not from the leak area. Moreover, the Pearson and Spearman coefficients showed a weak correlation, lower than 0.2, between DW pressurization and other perturbed parameters.

Therefore, the lower head failure uncertainty is mainly related to the pump seal leak area, while the MELCOR modelling parameters and sensitivity coefficient, under investigation in this uncertainty application, have a low linear and ranked influence, as demonstrated by the Pearson and Spearman coefficients.

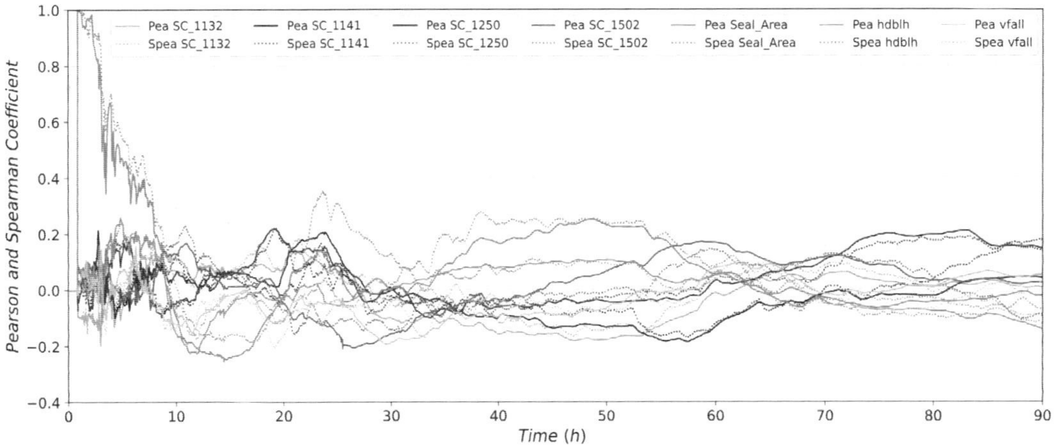

Figure 5. Pearson and Spearman coefficients for Drywell containment pressure.

The results of statistical analysis for pressure in RPV are shown in Figure 6. The first depressurization in the first hours after the SCRAM is caused by the Isolation Condenser (IC) operation. The pressure drop is predicted to start between 2 and 5.8 hours after the SCRAM and it is mainly associated with gas ejection from RPV, which can occur from the leak.

In Figure 7, the range of hydrogen mass that could be produced perturbing the input parameters is shown. At the end of calculations, a median value of 665 kg was produced, with a total mass ranging from 550 kg and 755 kg of quantiles. The onset of hydrogen production is not particularly affected by the perturbed parameters. From the output data analysis, SC1141(2) has a moderate linear correlation with hydrogen generation. Instead, the Area_Seal parameter has a weak negative correlation because its increment reduces the amount of steam available for the oxidation reaction.

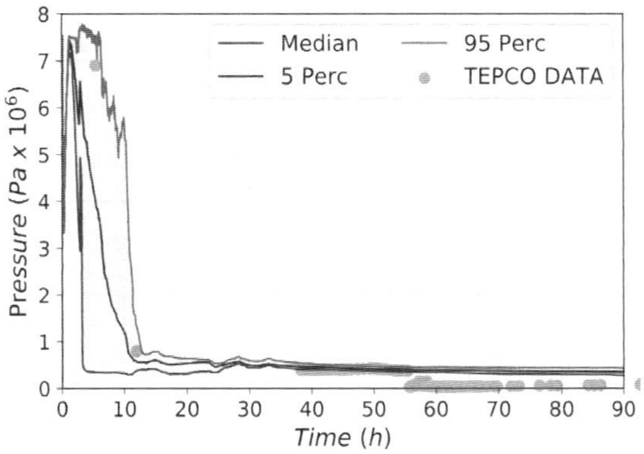

Figure 6. Variation range of RPV pressure.

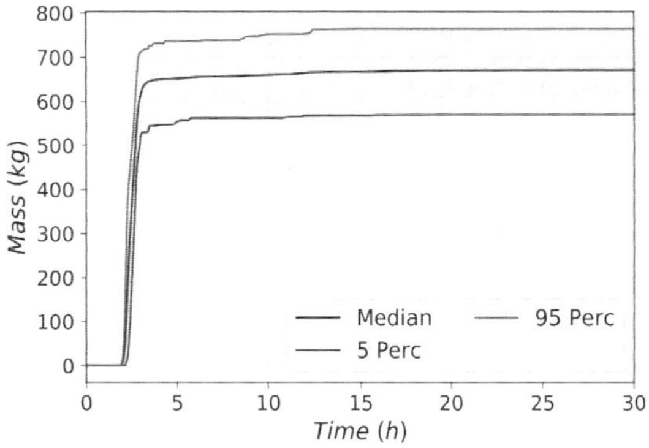

Figure 7. Variation range of mass of H_2 generated.

The Pearson and Spearman correlation coefficients have also been used for RPV pressure to evaluate the linearity and the correlation between the variables and the selected figures of merit during the transient, respectively. The time-dependent analysis is important to show how the correlation evolves during the phase of the transient. Figure 8 shows that the Area_seal is linear and strongly correlated with the RPV pressure trend in the first part of the transient. However, in this case, the trend is negative; in fact, an increase in leak area will cause a decrease in RPV pressure, as expected. Therefore, the RPV pressure is mainly related to the pump seal leak area uncertainty, while the MELCOR modelling parameters and sensitivity coefficient, under investigation in this uncertainty application, have a low linear and weak correlation, as demonstrated by Pearson and Spearman coefficient values, respectively. It is to be underlined that the results of the calculated data coupled with the related uncertainty band contain mostly available TEPCO plant data related to the RPV pressure, Figure 6.

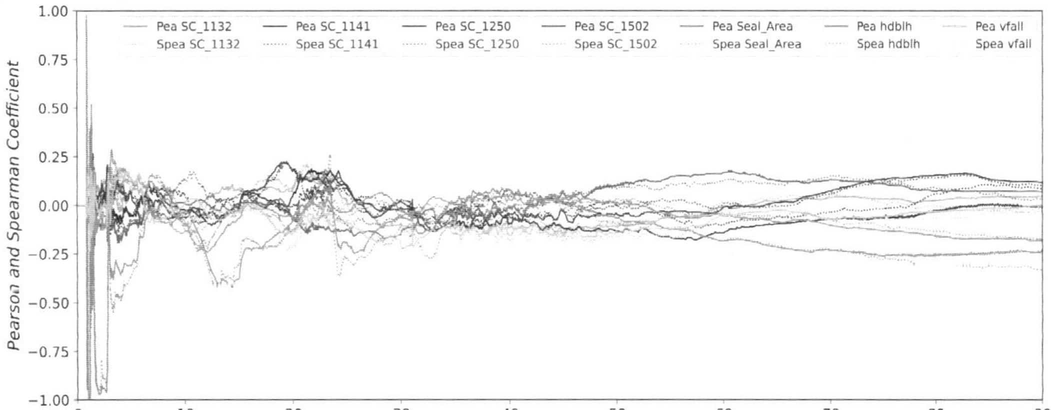

Figure 8. Pearson and Spearman coefficients for RPV pressure

5. Conclusions

The implemented code coupling between RAVEN and MELCOR has been successfully tested in the uncertainty analysis applied to the MELCOR simulation of the Fukushima Daiichi unit 1 severe accident progression. The present analysis highlights some uncertainty that produces differences in the timing of the major events, such as lower head failure and pressure behaviours. Considering the uncertainty band chosen (0.05, 0.5, and 0.95 percentiles), publicly available TEPCO plant data selected for this analysis are generally contained on it. Sensitivity analyses are in progress to study the effect of the nodalization details on the calculated data. Regarding the lower head failure time, it is to underline that there is a linear and strong correlation with the pump seal leak area. At the same time, the MELCOR modelling parameters and sensitivity coefficients, under investigation in this application, have a low linear and weak correlation, as demonstrated by Pearson and Spearman coefficients. Same considerations can be made for the pressure behaviours here investigated. In the future, a more detailed analysis will be performed, through the models implemented in RAVEN, by increasing the number of variables affected by uncertainty and including the uncertainty of the TEPCO plant data available in the literature.

References

[1] Humphries LL, Beeny BA, Gelbard F, Louie DL and Phillips J 2015 *MELCOR Computer Code Manuals, Vol. 1: Primer and Users' Guide, Version 2.1.6840*, SANDIA REPORT SAND-2015-6691R, Sandia National Laboratories, Albuquerque, New Mexico. https://www.nrc.gov/docs/ML1704/ML17040A429.pdf

[2] Humphries LL, Beeny BA, Gelbard F, Louie DL and Phillips J 2017 *MELCOR Computer Code Manuals, Vol. 2: Reference Manual,* Version 2.1.6840, SANDIA REPORT SAND2017-0876O, Sandia National Laboratories, Albuquerque, New Mexico. https://www.nrc.gov/docs/ML1704/ML17040A420.pdf

[3] Pescarini M, Mascari F, Mostacci D, De Rosa F, Lombardo C and Giannetti F 2017 *J. Phys.: Conf. Ser.* **923** 012009.

[4] Symbolic Nuclear Analysis Package (SNAP) 2021 Users Manual. Applied Programming Technology, Inc. Bloomsburg, PA

[5] G. van Rossum, Python tutorial, Technical Report CS-R9526, Centrum voor Wiskunde en Informatica (CWI), Amsterdam, May 1995

[6] Rabiti C, Alfonsi A, Cogliati J, Mandelli D, Kinoshita R, Sen S, Wang C, Talbot WP, Maljovec DP and Chen J 2017 *RAVEN User Manual,* INL/EXT-15-34123. https://inldigitallibrary.inl.gov/sites/sti/sti/Sort_4450.pdf

[7] Rabiti C, Alfonsi A, Cogliati J, Mandelli D and Martieneau R 2012 *RAVEN as Control Logic and Probabilistic Risk Assessment Driver for RELAP-7*, Proceeding of American Nuclear Society (ANS), 107, pp 333–335, https://www.osti.gov/servlets/purl/1060968

[8] Bixler NE, Ross K, Sallaberry C, Jones JA and Ghosh ST 2016. *State-of-the-Art Reactor Consequence Analyses Project: Uncertainty Analysis of the Unmitigated Short-Term Station Blackout of the Surry Power Station.* Available at https://www.osti.gov/servlets/purl/1380180

[9] TEPCO 2011 Fukushima Nuclear Accident Analysis Report. Tokyo: Tokyo Electric Power Company, Inc. December 2. Available at http://www.tepco.co.jp/en/press/corp-com/release/betu11_e/images/111202e14.pdf

[10] TEPCO, Plant Specifications of Unit 1, https://fdada.info/en/home2/accident2/specification2/

[11] Gauntt RO, Kalinich D, Cardoni J, Phillips J, Goldmann A, Pickering S, Francis M, Robb K, Ott L, Wang D, Smith C, Germain S, Schwieder D and Phelan Ch 2012. *FUKUSHIMA DAIICHI: ANS Committee Report, Fukushima Daiichi Accident Study (Status as of April 2012).* SANDIA REPORT SAND2012-6173 https://doi.org/10.2172/1055601

[12] Bixler N, Gauntt R, Jones J and Leonard M 2013 *State-of-the-Art Reactor Consequence Analyses Project Volume 1: Peach Bottom* Integrated Analysis. NUREG/CR-7110, Vol. 1, https://www.nrc.gov/docs/ML1315/ML13150A053.pdf

[13] Carbajo JJ 1993, *Severe Accident Source Term Characteristics for Selected Peach Bottom Sequences Predicted by the MELCOR Code*, NUREG/CR-5942, https://doi.org/10.2172/10182846

[14] Mascari F, Pescarini M, Giannetti F, Luciani I and Caruso G *Integral Calculations of Severe Accident Scenarios in PWR and BWRs*, ADPFISS-LP1-075, 2016

[15] Mascari F, D'Onorio M, Giannetti F, Caruso G and Naviglio A., *Analyses of unmitigated transients. Small break LOCA in PWR, Loss of Feed Water in PWR and SBO in BWR,* Report ADPFISS-LP1-103

[16] Ross K, Phillips J, Gauntt RO and Wagner KC 2014, *MELCOR Best Practices as Applied in the State-of-the-Art Reactor Consequence Analyses (SOARCA) Project*, NUREG/CR-7008, U.S. NRC. https://www.nrc.gov/docs/ML1423/ML14234A136.pdf

[17] Glaser H 2008 *GRS Method for Uncertainty and Sensitivity Evaluation of Code Results and Applications.* Science and Technology of Nuclear Installations, 798901, https://doi.org/10.1155/2008/798901

[18] Wilks SS 1941 *Determination of sample sizes for setting tolerance limits* Ann. Math. Stat., 12, pp. 91-96. https://doi.org/10.1214/aoms/1177731788

[19] ICANPS. (2012). Final Report Investigation Commitee on the Accident of Fukushima Nuclear Power Stations of Tokyo Electric Power Company. Available at https://www.cas.go.jp/jp/seisaku/icanps/eng/final-report.html

Optimisation strategies for solar assisted heat pumps coupled to traditional thermal fields

L.A. Tagliafico, A. Cavalletti, C. Marafioti, A. Marchitto*
DIME/TEC (University of Genoa), Via All'Opera Pia 5a, 16145 Genova Ge

annalisa.marchitto@unige.it

Abstract. The advantages of the solar assisted heat pumps are well-known but their diffusion on the market is still very limited. One of the main causes is associated to their complexity and almost always working under transient regime which can lead to frequent blocks especially when a low level of control and automatization is present. From the point of view of conceptual design, solar assisted heat pumps tend to be more subjected to blocks due to their serial asset. The present work illustrates the case study of the solar assisted heat pump pilot plant at Palacus, Genoa, equipped with a heat pump bypass which can directly connect the solar thermal field to either the domestic hot water or the space heating systems when the temperature in the circuit exceeds a fixed set point. This configuration allows therefore the use of the solar field either independently from the heat pump or interfaced with it, according to the external conditions. A numerical model within TRNSYS environment has been developed basing on the information collected on the pilot plant by the data acquisition system. A simulation of the working of the pilot plant along a typical year has been performed, evaluating its increased performance and optimisation thanks to the bypass.

1. Introduction

The earlier concept and the former applications of the Solar Assisted Heat Pumps (SAHPs) date back to the '70-'80s [1], especially for small scale application [2], [3]. Indeed, the theorical models, the numerical simulations and the small size prototypes carried out by researchers during these decades have clearly highlighted the potential of this technology as shown in the reviews such as [4], [5], [6].

In addition, SAHPs are not necessarily concerned with only industrial application, but they can be implemented even in civil buildings where the energy consumption for heating is significative. Many researches estimate that about 30% of the total energy supply is used for building space heating in Europe [7].

More in general, the reduction of fossil fuels consumption in such sectors has to be be faced with the use of renewable sources. Clearly, the interaction of different renewable sources is the key to succeed in this goal. Two main strategies can be identified:

- In parallel: this is the most typical configuration and it allows the working of the plant even if a part of it not active; nevertheless, the maximum exploitation of the renewable sources is not always granted. The classic example consists in a water storage which is heated by different sources, for instance a solar thermal circuit and then a heat pump which provides an additional heat flux to keep the water inside at the fixed set point. The maximum use of the solar thermal energy requires high temperatures inside the panels and therefore a lower efficiency is appreciated.

- Serial: one or more renewable sources are used in serial asset, to reach a higher performance taking advantage from the interaction of different renewable sources. Following the previous example about solar thermal energy, the panels are not directly interfaced with a water storage, but instead they can provide a heat flux to the heat pump, increasing its performance and COP. The heat pump then heats the water storage, like in the case of the SAHP. Clearly, the failure of one component would imply either the failure of the entire plant (due to the serial asset) or at least a significative reduction in its performance. In addition, the complexity of these system experiences an increase with respect to the parallel configuration and therefore a Data control and Acquisition System (DAS) becomes compulsory, to grant the full plant operativity under each environmental condition.

The case study presented below is concerned with an experimental SAHP installation which tries to match both strategies to achieve optimal performances. Within this context, the working of the plant cannot be considered under stationary regime and therefore a numerical simulation is required. Moreover, the COP of the plant has to be defined according to the transient regime and the added useful effect represented by the bypass. As a consequence, the definition for the Seasonal COP (SCOP) can be conceived as

$$SCOP = \frac{\sum_{i=1}^{365}\left(Q_{cond} + Q_{bypass}\right)}{\sum_{i=1}^{365} L_{HP}} \quad (1)$$

Where

Q_{cond} and Q_{bypass} represent respectively the heat associated to the condenser and the bypass, integrated over each day, [kWh]

L_{HP} is the electrical work of the compressor of the heat pump, [kWh]

2. Case study: the solar assisted heat pump at Palacus

The "Palacus" plant represents a real, full size pilot plant in which a hybrid solar assisted heat pump (SAHP-PVT) is interfaced with a traditional gas burner heating system. The heating system serves the sport palace (e.g. the reception, the lockers and the sports hall with a global heated volume of about 30,000 m3. As far as DHW consumption is concerned, the need is strongly variable and time dependent and its distribution tends not to be uniform over the day: in fact, the lockers are also used by the athletes using one of the six external tennis courts and one of the two football pitches. Two storage tanks of 1,500 liters each provide hot water to the utilities.

The main components of the SAHP-PVT (Figure 1) can be identified as follows:

- *Solar hybrid field*: the joint production of electricity (peak power production 20 kW_E) and heated water (peak power production 50 kW_T) has been designed to cover the electrical consumption of the elements of the circuit (e.g. valves, pumps and the heat pump itself) and increase the level of temperature on the cold side of the heat pump with consequent benefits on the mean seasonal COP. A bidirectional connection through the electricity meter to the national grid is available, in order to input any electric overproduction into the national grid and use it when needed following the criterion of the feed-in-premium strategy.
- *Heat pump system:* it is designed to cover the needs of the facility in terms of DHW and SH with a thermal nominal power of 62 kW_T and a related consumption of 12 kW_E. The connection of the cold side of the heat pump to the solar field increases the temperature at the evaporator and consequently the COP of the HP (up to a range of 5-5.5).
- *Supplementary gas burners and distribution system:* the hot water produced by the SAHP stored in the hot side storage tank is then sent to two supplementary gas burners which eventually heat the water to reach the chosen set points. The gas burners, respectively of 320 kW (used in winter when also SH needs are included) and 35 kW (used in summer when only DHW is required) represent the original heating plant of the structure. Their intervention should occur only in strict

environmental conditions when the renewable part of the plant cannot meet the thermal needs of the facility on its own.

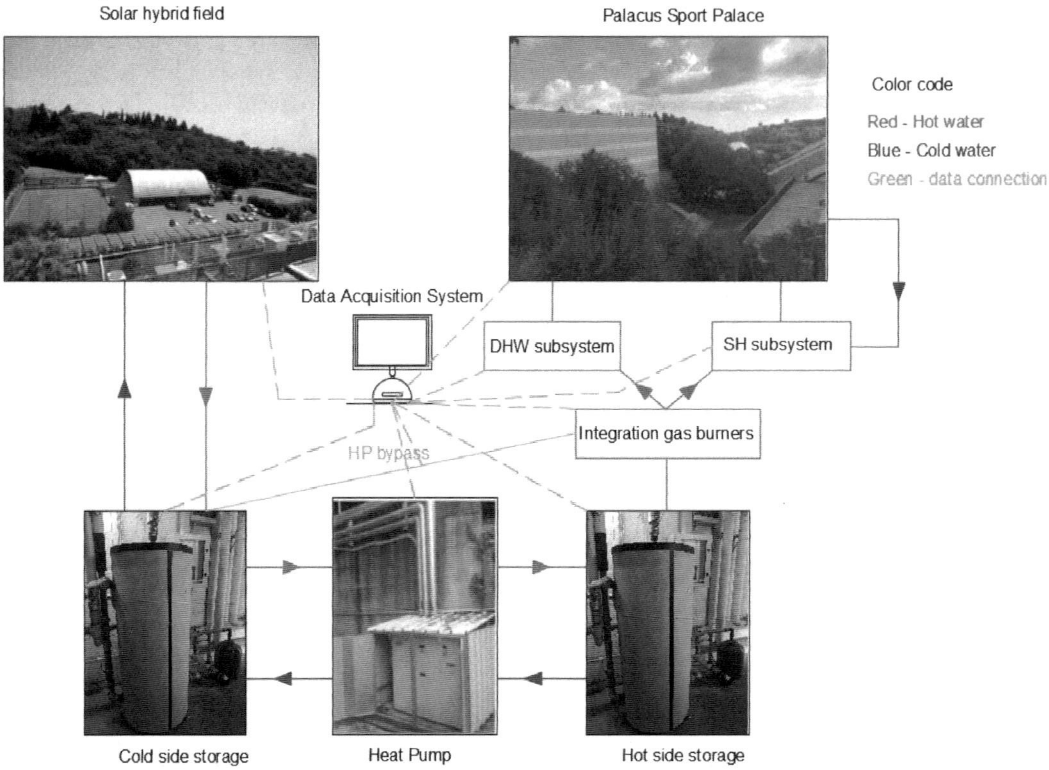

FIGURE 1. Block diagram of the pilot plant at Palacus Sports Centre, University of Genoa, with its Data Acquisition and control System (DAS).

- *Data Acquisition and control System (DAS):* its central position in Figure 1 is meant to remark both its importance to grant the correct working of the plant and the large set of measurement and control points (respectively about 50 measuring points and 15 actuators). An internet connected and remotely accessible PLC supervises and eventually operates on the plant, according to the strongly time-dependent environmental conditions and users' needs. The vast amount of data collected requires proper post-analysis and allows efficiency studies in order to build new regulation criteria and increase the plant efficiency also from the environmental point of view.
- *Heat pump bypass:* it allows a direct connection of the solar field (on the thermal side) to the hot side storage whenever the temperature of the water from the solar filed is high enough to directly support the needs (either DHW or SH). Actually, the condenser of the HP works at about 45-50 °C, so the control strategy activates the bypass each time similar temperatures are reached (About 45 °C). The working of the bypass depends on the installation site and on the performances of the solar field, although the bypass is most likely to be exploited during Summer. This particular issue allows an added degree of freedom in the plant working.

3. Numerical modelling

A simplified numerical model of the plant described has been implemented in TRNSYS environment (Figure 2). Despite of the data acquired by the DAS, a numerical modelling still plays a key role in plant optimisation since it allows the choice of the optimal configuration among the different virtual

ones. The added upgrade is represented by the acquired values of temperatures, mass flow rates and environmental conditions used to validate the model and the properties of the different elements employed.
The validation for most of components has been retrieved by the authors' previous work [8] while the missing information has been integrated with the technical datasheets and the measured values. In particular, the working of the numerical model for the solar panels has been validated under different measured climatic conditions (e.g. sunny day, day with clouds and variable air temperature) and comparing the simulated performance of panels with the measured one in that same period. Besides of the evaluation of the error committed (below about 5% which means about 1 °C for the working temperatures of 25-30 °C), two aspects deserve particular attention. Firstly, the measured information needs a processing where the datasets associated to error messages are discarded and the remaining part is associated to the different working of the panel. Secondly the validation of the element needs to be reproduced under different assumptions both on the environmental side and of the working of the entire plant, to have comparable results. For instance, a set of simulations was concerned with the panels with no water flow inside (the plant was off for maintenance).

Figure 2 provides an overview of the plant scheme about the components and the logical connections among them. The layers concerning the printing and integration of the results have been omitted to make the scheme clearer. According to the color code used, blue and red lines represent respectively the cold and hot water flow rates of each circuit. The dotted green connections provide information about weather (e.g. irradiation, wind speed, velocity, ambient temperature) while the dotted black lines are logical connections used to compute control values (e.g. for the controller of the solar circulation pump).

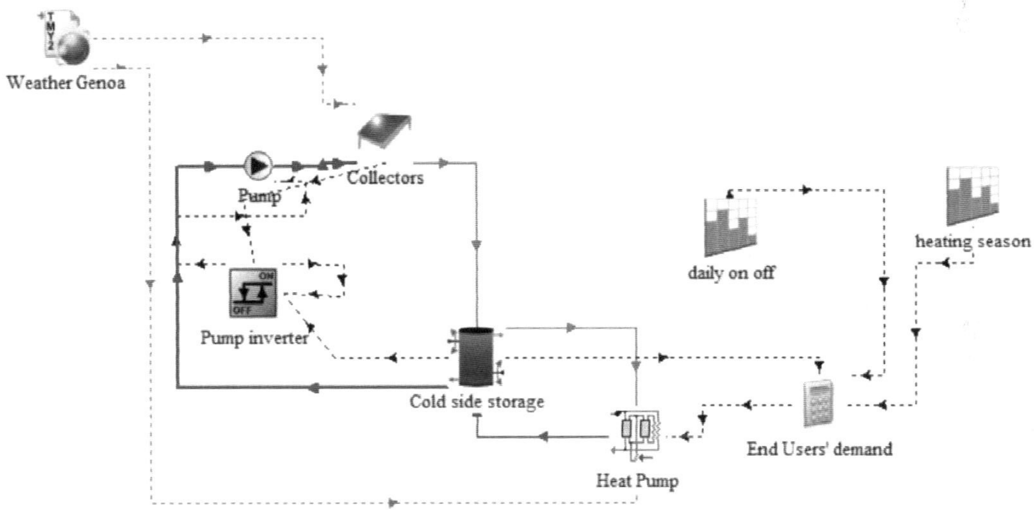

FIGURE 2. Simplified scheme of the solar thermal subsystem and of the heat pump installed at Palacus in TRNSYS environment. The red and blue lines are respectively the pipes with hot and cold water-flow rates. The dotted black lines represent logical connections used to control the system.

Briefly, the following components can be identified:

- *Solar thermal subsystem:* it is composed by the solar panels, the circulation pump and the cold side storage which is in turn interfaced with the built-in exchanger on the evaporator side of the heat pump. Actually, this model is concerned with the thermal analysis of the plant and therefore the electrical production of the hybrid panels has been neglected. The model for the solar collectors has been implemented with the nominal intercept efficiency, efficiency slope and curvature of the datasheet. A controller regulates the activation of the solar pumps only when the solar field can provide a significative heat flux to the storage. In particular an upper and a lower deadbands (5 °C and 1 °C) control the activation and the stop of the solar pump. In the end, the physical properties of the cold side storage have been chosen basing on the results of [8]. As a first approximation, no thermal losses between the solar field and the cold side storage have been assumed, since the gradient in temperature between the water inside the circuit and the external environment are on average small. Indeed, the thermal field temperature operating range when coupled to the heat pump can be considered within the interval of 5 °C – 18 °C which are comparable with the environmental temperatures for the installation site (Genoa).

- *Heat pump:* this component allows a dual working following a built-in criterion, either using the bypass, when the set point temperature of the outlet water from the storage is reached (the heat pump is off), or working in standard heating mode (the heat pump is on). Besides of this control, the end users' demand has been simulated with two forcing functions which simulate the period of the heating season and a daily simulation of the thermal need. In this way the heat pump is kept off independently from the external conditions during specific time intervals (e.g. during nights) while in the others it is free to switch to the best operating mode (either bypass or heating) according to the set points assigned.
- *Weather simulator:* the component available in TRNSYS reads the simulated information for the current site, Genoa, simulated by the program Meteonorm. This software is connected to the weather databases collected during the past decades by the weather stations set in Genoa, Polytechnic School, DICCA and in Genoa, Sestri Ponente.

4. Results of the numerical simulation

Figure 3 plots the useful effects represented by the heat flux at the condenser and the one associated to the bypass. The bypass activates only during the period in which the heat pump is off, since during the heating season the thermal load absorbed by the evaporator keeps the temperatures within the cold side tank far below the set point of 45 °C which activates the bypass.

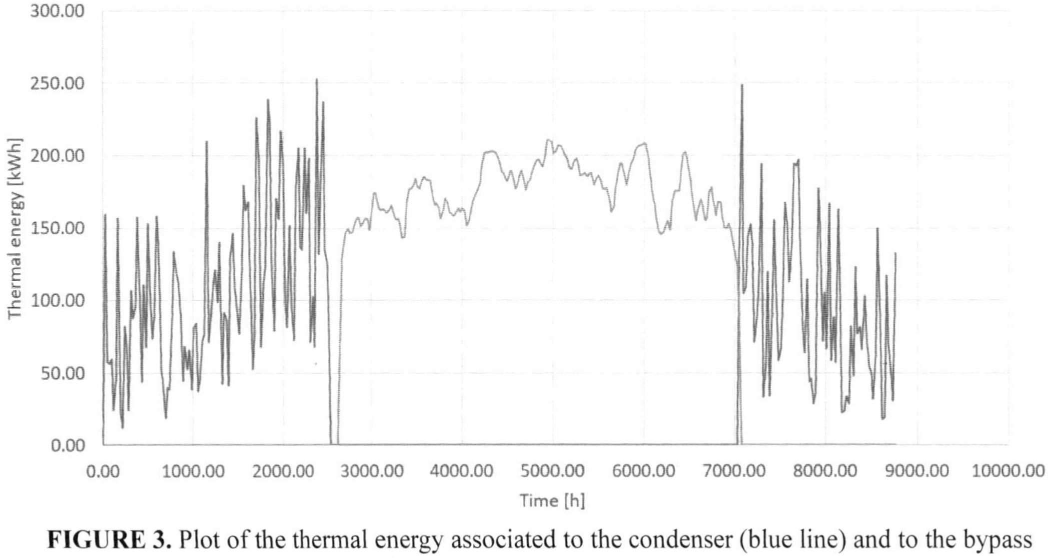

FIGURE 3. Plot of the thermal energy associated to the condenser (blue line) and to the bypass (orange line) during one year of simulation.

On the other hand, the summer period is characterized by high solar irradiation (Figure 4) and therefore the temperature in the cold side tank increases, also due to the absence of the thermal load absorbed by the evaporator (the HP is off).

As far as the heat pump performance is concerned, the average values for the COP during the heating seasons oscillate around about 3.8-3.9. Namely this condition considers the heat flux at the condenser as the unique useful effect (the bypass is not active) the heat pump benefits from the increase in temperature on the cold side.

Namely the environmental conditions of the installation site allow a good exploitation of the solar source as coupled to the heat pump. In other words, the sizing of the solar field and of the storage is adequate to the installed heat pump, since no overheating is appreciated in the cold side boiler when the HP is on. However, when the conditions don't allow the optimal use of the solar field, the bypass activates, providing a useful effect, even if the heat pump is not operating.

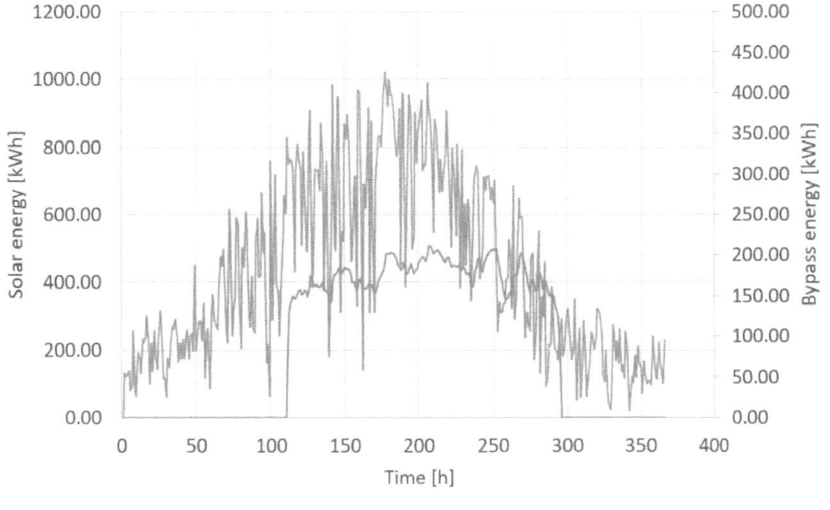

FIGURE 4. Plot of the solar thermal energy (left axis) and of the energy associated to the bypass (right axis)

In terms of seasonal COP this aspect plays a central role in increasing the performance of the plant. In fact, if the heat flux associated to the bypass is accounted as a useful effect (with almost zero external work required to the system), a seasonal COP of 6-6.4 can be reached, with a significative increase with respect to the COP obtained when referred only to the heating season (and with no bypass active).

With respect to the studies (experimental and theoretical) available in literature, the issue of the bypass has been seldom enquired. Actually, the average seasonal COP of similar experimental solar assisted heat pumps can reach values of 3.1 [9], 4.1 [10]. As far as theoretical models are concerned, values for seasonal COP are about 4.5 [11], 4 [12]. Anyhow, the obtained COP of 6 for the installation site considered (Genoa) is far above the limits available in literature.

More in general, the optimal application for the solar field is interfaced with the heat pump, especially during the periods in which solar radiation is low. On the other hand, the energy available during summer has to be used as much as possible, even if this means higher temperatures and lower performances of the panels when the bypass is on.

One last remark is concerned with the stability of the solution: the bypass has not to be used as an oscillating condition, but it can be used only when stable conditions have been reached. Otherwise, the solution would still be mathematically correct, but would lose any physical meaning since the plant cannot continuously switch from one asset to the other.

5. Conclusions and future developments

In the present work, the numerical model of a solar assisted heat pump, validated basing on the measurements acquired by the data acquisition system has been presented. While traditional solar assisted heat pumps are concerned with a serial asset in which the solar thermal field is constantly interfaced with the cold side of the heat pump, the present case study investigates the benefits associated to a bypass which allows to skip the heat pump whenever the temperature reached in the solar circuit is already high enough. With this simple variation of the plant scheme, a further degree of freedom is added since the plant can operate either directly using the solar field or using the solar assisted heat pump. In fact, in case of failure of the heat pump, the renewable part of the plant can still operate and be exploited.

The simulation performed in TRNSYS environment allows to analyse the plant under transient regime during a year simulated basing on the historical weather databases made available by the meteorological stations in Genoa. The results show the importance of the bypass and an estimation of the seasonal coefficient of performance of the plant is carried out, also accounting for the periods in which the heat pump is off and the solar field directly provides a useful heat flux either to the domestic hot water or the space heating system.

The presence of such bypass does not imply complex regulation criteria and it can be easily implemented in other solar assisted heat pumps at low cost. On the efficiency side, the seasonal COP of the plant experiences a significative increase, thanks to the direct exploitation of the solar thermal field during the summer and part of the spring/autumn. Results show how the SCOP experiences an increase of about 30%, thanks to the bypass which adds a useful effect with very low external work. In addition, the bypass does not work during the heating season, showing that the existing system has been correctly designed to provide a constant heat flux to the evaporator.

As far as the future developments are concerned, a wider analysis on different virtual installation sites has to be performed to carry out a comprehensive analysis on the advantages of this simple addition. As a consequence, the bypass is expected to result more effective in installations with low Degree Days, thanks to the availability of solar irradiation.

These simulations shall be performed adding the thermal losses due to the pipes connecting the solar panels to the cold side storage. Indeed, this term can be in first approximation negligible since the operating temperature range of the solar field coupled to the heat pump is almost comparable to the ambient temperatures. However, each time the bypass is activated (namely when high temperatures within the cold side storage are reached), the thermal losses associated to the connection pipes might not be negligible “a priori”.

References

[1] A. Cesarano, F. De Rossi, V. Naso, “La pompa di calore elioassistita: configurazioni d'impianto e loro caratteristiche. Una applicazione sperimentale”, La Termotecnica, Gennaio 1986.

[2] J. P. Chyng, C. P. Lee, B. J. Huang, “Performance analysis of a solar-assisted heat pump water heater” Solar Energy, vol. 74, pp. 33-44, 2003.

[3] S. K. Chaturvedi , D. T. Chen, A. Kheiredinne, “Thermal performance of a variable capacity direct expansion solar assisted heat pump” Energy Conversion and Management, vol. 39, pp. 181-191, 1998.

[4] A. Hepbasli, Y. Kalinci, “A review of heat pump water heating systems”, Renewable and Sustainable Energy Reviews, vol 13, pp. 1211-1229, 2009

[5] O. Kaara, K. Ulgena, A. Hepbasli, “Exergetic assessment of direct-expansion solar-assisted heat pump systems: review and modelling”, Renewable and Sustainable Energy Reviews, vol 12, pp. 1383-1401, 2008

[6] Kutbay Sezen, Azim Dogus Tuncer, Ali Ozhan Akyuz, Afsin Gungor, “Effects of ambient conditions on solar assisted heat pump systems: a review”, Science of the Total Environment, vol 778 pp. 194-212, 2021

[7] IEA. World Energy Outlook (Online). 2019 access date 01/05/2021 https://www.iea.org/fuels-and-technologies/heating

[8] L.A. Tagliafico, A. Cavalletti, C. Marafioti and A. Marchitto, “Solar assisted heat pump pilot plant management and troubleshooting by means of numerical modelling: a case study” Journal of Physics, Conference series, vol 1868, pp.12-29

[9] G. Besagnoi, L. Croci, R. Nesa and L. Molinaroli, "Field study of a novel solar-assisted dual-source multifunctional heat pump", Renewable Energy, vol 132, pp. 1185-1215, 2019
[10] J. Long, R. Zhang, J. Lu and F. Xu, "Heat transfer performance of an integrated solar-air source heat pump evaporator", Energy Conversion and Management, vol. 184, pp. 626-635, 2019
[11] J. Zhou, X. Zhao, X. Ma, Z.Qiu, J.Ji, Z.Du and M. Yu, "Experimental investigation of a solar driven direct-expansion heat pump system employing the novel PV/micro-channels-evaporator modules", Applied Energy, vol.78, pp. 484-495
[12] F. Calise, M. Dentice d'Accadia, R.D. Figaj and L. Vanoli "A novel solar-assisted heat pump driven by photovoltaic/thermal collectors: dynamic simulation and thermoeconomic optimization", Energy, vol. 95, pp. 346-366, 2016

Acknowledgments
The Palacus pilot plant, Un. of Genoa, has been developed with a co-funding of FSE Reg. Liguria, Bando Azione 2.1 "Efficienza Energetica e Produzione di Energia da Fonti Rinnovabili – Enti Pubblici" – Pos. n.2. This paper was included in the research activities of PRIN 2015 MIUR grant 2015M8S2PA "Clean Heating and Cooling Technologies for energy efficient smart grid". The control and monitoring systems have been completed with the co-funding of the University of Genoa within the project "Grandi Attrezzature" 2018, DR n. 3404, 19/07/2018).

Journal of Physics: Conference Series 2177 (2022) 012023 doi:10.1088/1742-6596/2177/1/012023

Cold Leg LBLOCA uncertainty analysis using TRACE/DAKOTA coupling

G Agnello[1], P A Di Maio[1], A Bersano[2], F Mascari[2]

[1] Università degli Studi di Palermo, Dipartimento di Ingegneria, Viale delle Scienze, Edificio 6, 90128, Palermo, Italy.
[2] ENEA, FSN-SICNUC, Via Martiri di Monte Sole 4, 40129, Bologna, Italy.

Corresponding author's email: giuseppe.agnello04@unipa.it

Abstract. Safety analyses for nuclear power plants were carried out in the past using a conservative approach. With the increase of the phenomenological knowledge, through experimental data, and computational power, it became possible to adopt best estimate thermal-hydraulic system codes to perform deterministic safety analyses. However, some uncertainties are still present in the models, correlations, initial and boundary conditions, etc. Therefore, it is fundamental to quantify the uncertainty of calculation. This approach is called "Best Estimate Plus Uncertainty" (BEPU). Among the available uncertainty analysis methodologies, the probabilistic method to propagate input uncertainty is widely adopted. In the present study, an uncertainty analysis of a cold leg large break loss of coolant accident in a generic PWR-900 MWe has been developed and it has been carried out coupling the best estimate thermal-hydraulic system code TRACE and the uncertainty quantification tool DAKOTA in the SNAP environment/architecture.

1. Introduction

In the safety analysis of Nuclear Power Plants (NPP), deterministic safety analyses play a fundamental role. Their goal is to demonstrate the safety of systems both during nominal and accidental transient conditions. In this framework, the computational tools, also called codes, permit to analyze the behavior of NPPs and assess their safety. These codes are subjected to a process of Verification and Validation (V&V). It is part of this process the evaluation of the qualitative and quantitative code accuracy in the prediction of the plant phenomena by comparing the code results against experimental data from Separate Effect Test Facilities (SETF) and Integral Test Facilities (ITF) [1,2]. Initially, computational tools used for deterministic analyses in NPP were developed adopting simplified models and the safety analyses were carried out using a conservative approach. With the increase of the phenomenological knowledge and computational power, it has been possible to develop computational tools, called "Best Estimate" (BE) codes, which adopt more realistic information concerning phenomena and plant behavior. However, in the application of BE codes there are still some sources of uncertainty affecting the calculation results (e.g. code uncertainty, representation uncertainty or nodalization effect, scaling issue, plant uncertainty, user effect, etc.) [3]. Therefore, to use a BE approach for the deterministic safety analysis, the Uncertainty Quantification (UQ) analysis of the calculation is also recommended.

Within this framework, the target of the present study is to use the probabilistic method to propagate input uncertainty to perform the UQ analysis of a postulated Large Break Loss of Coolant Accident

(LBLOCA) initiated by a double-ended guillotine break in the Cold Leg (CL) of a generic western type three-loops PWR-900 MWe, with the availability of active and passive Emergency Core Cooling Systems (ECCS). The UQ application has been developed coupling the BE thermal-hydraulic system code TRAC/RELAP Advanced Computational Engine (TRACE) v5 patch 6 [4,5,6] and the Design Analysis Kit for Optimization and Terascale Application (DAKOTA) toolkit [7,8] in the Symbolic Nuclear Analysis Package (SNAP) [9,10] environment/architecture.

2. Uncertainty quantification methodology description

2.1. Probabilistic method to propagate input uncertainty

Several methodologies have been developed in the past to perform UQ analyses. In general, these methodologies can be grouped in methods to propagate input uncertainty (probabilistic and deterministic methods) and the method to extrapolate output uncertainty. In general, the probabilistic method to propagate input uncertainty [11] is particularly suitable to be coupled with codes since it is based on the creation of a number of code runs with different uncertain input parameters to characterize the uncertainty of the output Figure Of Merits (FOMs), target of the analysis. The uncertain input parameters are characterized by a range of variation and a Probability Density Function (PDF) [12]. A random sampling (e.g. Monte Carlo or Latin Hypercube methods) of the selected uncertain input parameters is performed in order to define N sets of input parameters and code runs. The "Wilks Method" can be used to determine the minimum number of code runs N based on the number of FOMs investigated and on the requested probability content α and confidence level β [13,14]. The probability content α measures the portion of the theoretical distribution, obtained with all the possible results, included in the tolerance interval (one-side or two-side tolerance interval, considering respectively or only the 0.05 or 0.95 quantile or both) of our calculation and the confidence level is the probability that it holds. Based on Wilks, in case only one FOM is investigated, for the one-sided tolerance interval, N can be found by solving the following equation with respect to N:

$$1 - \alpha^N = \beta \tag{1}$$

If more than one FOM is investigated, for the one-sided tolerance interval, the minimum required number of code runs ca be found by solving the following equation with respect to N:

$$\beta = \sum_{j=0}^{N-p} \frac{N!}{(N-j)!\,j!} \alpha^j (1-\alpha)^{N-j} \tag{2}$$

where p is the number of FOMs investigated [15]. More information on statistical aspects of best estimate analyses can be found in [16]. The probabilistic method to propagate input uncertainty permits also to evaluate the statistical correlation (e.g. linear or monotonous) between the FOMs and the input uncertain parameters using correlation coefficients (e.g. Pearson's simple and Spearman's simple rank coefficient) [17, 18].

2.2. The DAKOTA toolkit in the SNAP environment/architecture

DAKOTA, developed by Sandia National Laboratories, is an open-source software written in C++ designed to perform parametric and uncertainty analysis in a fast and automatic way. The aim of this toolkit is to bridge simulation codes and analysis methods for parametric evaluation, uncertainty quantification and system optimization [19]. The DAKOTA toolkit is also provided as a plug-in for SNAP, which is a graphical interface designed to support the use of United States Nuclear Regulatory Commission (USNRC) codes (e.g. TRACE, RELAP5, MELCOR, PARCS, etc.). Using SNAP, it is possible to build the input-deck in a graphical user interface and to have a direct visualization of the code calculated data by using its animation capability. Through SNAP it is possible to set up the DAKOTA uncertainty analysis and to perform automatically all the steps needed for the UQ. Figure 1 shows a schematic representation of DAKOTA uncertainty analysis workflow in the SNAP environment/architecture [20].

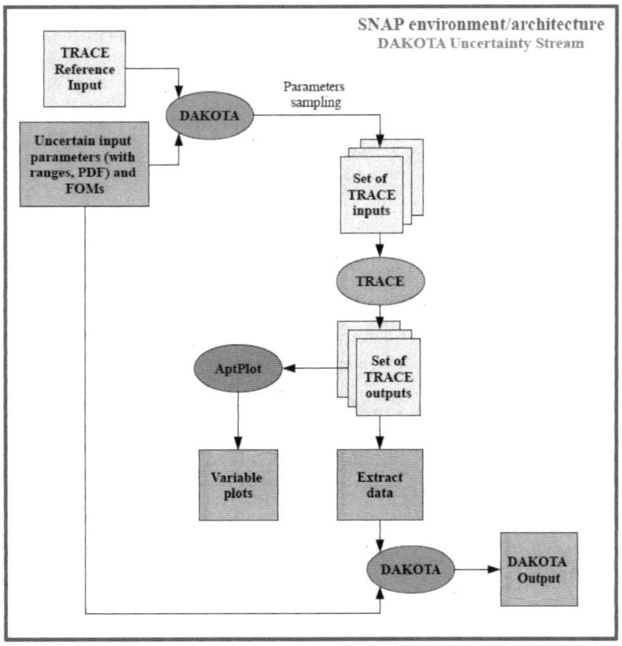

Figure 1. DAKOTA uncertainty analysis workflow for TRACE code in a SNAP environment/architecture [20].

DAKOTA is used at the beginning of the analysis to sample the uncertain input parameters and to generate the set of code inputs. Then, after the solution of the set of code inputs and the extraction of the desired data, DAKOTA performs the uncertainty analysis and applies statistical techniques to evaluate the statistical correlation between the selected uncertain input parameters and the FOMs considering four correlation coefficients: simple, partial, simple rank and partial rank. The simple coefficient, computed through the Pearson's correlation, is a measure of the degree of linear correlation between an input variable and an output variable and its value is comprised between -1 and 1. The partial coefficient also considers the effects of the other variables. The rank correlation coefficient is a measure of the degree of monotonous correlation. If two variables are monotonically related, the rank coefficient is -1 or +1. To compute the rank correlation, DAKOTA uses the Spearman's correlation [19].

3. Description of the generic PWR-900 MWe TRACE model and UQ application hypotheses

3.1. Description of the generic PWR-900 MWe TRACE nodalization

The BE code TRACE is developed by USNRC to perform thermal-hydraulic analysis of transient scenarios in light water reactors (BWR and PWR) and experimental facilities. It is a finite volume code with 3D capability and it is based on the two-phase fluid field equations. The adopted TRACE nodalization, shown in Figure 2, of the generic three loops western type PWR-900 MWe has been originally developed in [20] and further refined in [21]. The model is composed of 87 Hydraulic Components (HC) and 49 Heat Structures (HS). The three loops are modelled separately: one simulates the broken one (Loop A) and two simulate the intact loops (Loops B and C). The break has been modelled with a set of three valves; at the Start Of the Transient (SOT) one valve interrupts the connection between the two sections of the CL of Loop A and simultaneously the other two valves connect the two closed sections of the CL to the pipe component simulating the containment. A supplemental rod that simulates the hot rod in the reactor with a total peaking factor of 2.278 has been considered [22].

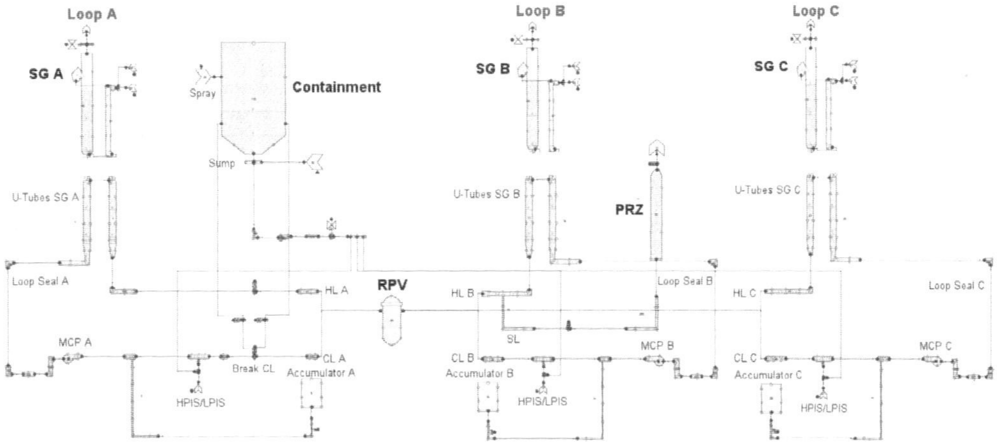

Figure 2. TRACE nodalization of the generic three-loops PWR-900 MWe developed with SNAP.

3.2. Uncertainty Quantification application hypotheses

The main goal of this study is not to be a detailed uncertainty study in terms of uncertain input parameters selected and FOMs limits identification but to develop a full UQ application with the coupling of TRACE and DAKOTA toolkit in the SNAP environment/architecture and to have some insights about the statistical correlation between the uncertain input parameters and the FOMs selected along the transient. As shown in Table 1, ten uncertain input parameters have been selected for this application: Safety Injection System (SIS) temperature, SIS characteristic, accumulator initial temperature and pressure, initial core power (multiplier influencing bot the nominal power and the decay power), the initial containment pressure, the peaking factor of the hot rod, the fuel rod gap size, the speed of the broken loop Main Coolant Pump (MCP) after the break and the speed of the intact loops MCPs after the break (multiplier influencing the speed of MCPs during the coast-down). All uncertain input parameters are based on the international program "BEMUSE" (Best-Estimate Methods-Uncertainty and Sensitivity Evaluation) promoted by the OECD "Working Group on Accident Management and Analysis" [23].

Table 1. Input uncertain parameters selected for the present analysis.

Uncertain input parameter	Reference value	Range of variation	PDF type
SIS temperature [K]	285	[275, 295]	Normal
SIS characteristic [-]	1	[0.95, 1.05] (multiplier)	Normal
Accumulator initial temperature [K]	325	[315, 335]	Normal
Accumulator initial pressure [bar]	40.8	[38.8, 42.8]	Normal
Initial core power [-]	1	[0.98, 1.02] (multiplier)	Normal
Initial containment pressure [bar]	1.013	[0.85, 1.15]	Uniform
Peaking factor [-]	2.275	[2.16, 2.39]	Normal
Gap size [m]	9.95E-05	[7.95E-05, 1.19E-04]	Normal
Speed MCP broken loop [-]	1	[0.98, 1.02] (multiplier)	Normal
Speed MCPs intact loops [-]	1	[0.9, 1.1] (multiplier)	Normal

Three FOMs have been selected: the hot rod cladding temperature, the Reactor Pressure Vessel (RPV) collapsed coolant level and the containment pressure. With a probability and confidence level of 95%, for the one-sided tolerance interval, a minimum of 124 runs is required as previously discussed. The Monte Carlo sampling method has been chosen for this UQ application. All 124 runs were correctly executed without any failure.

4. LBLOCA transient uncertainty analysis

4.1. FOMs transient results

At the SOT the core power decreases due to the activation of the SCRAM. The high-pressure difference between the Primary Cooling System (PCS) and the containment causes a huge loss of primary coolant through the break determining a rapid PCS depressurization (blowdown phase). The coolant flow rate in the core drops and it flows downwards in the core region and then upward in the downcomer to reach the break position in the CL. When the flow condition changes from single phase to two-phase conditions this determines a drastic reduction of heat removal in the core. The heat stored in the fuel is redistributed, leading to the first cladding peak temperature, as shown in Figure 3. The cladding temperature passes from the nominal value of around 617 K to a value of around 950 K around 6 s after the SOT in the reference calculation.

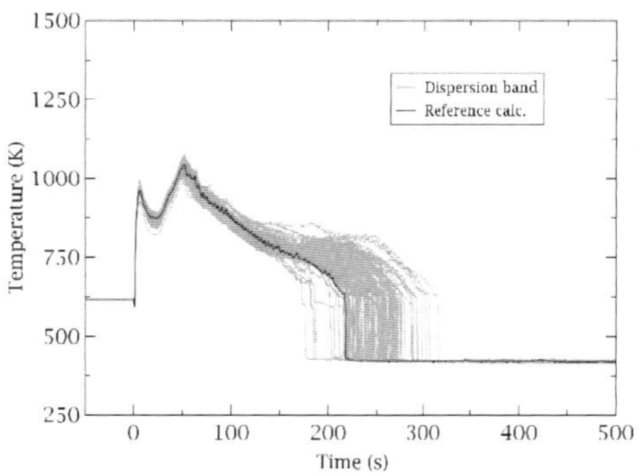

Figure 3. Hot rod cladding temperature reference calculation and dispersion band predicted by TRACE.

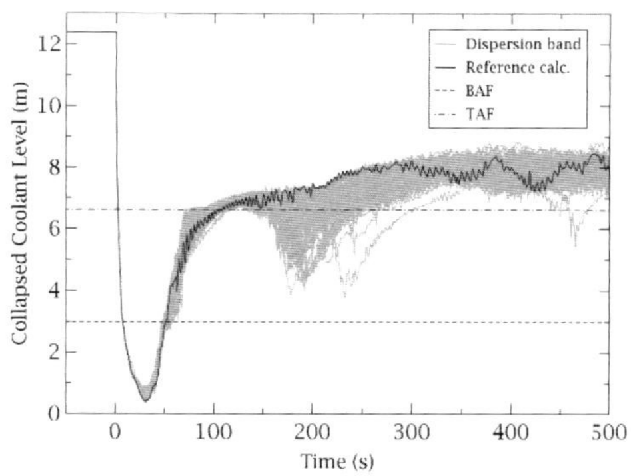

Figure 4. RPV collapsed coolant level reference calculation and dispersion band predicted by TRACE.

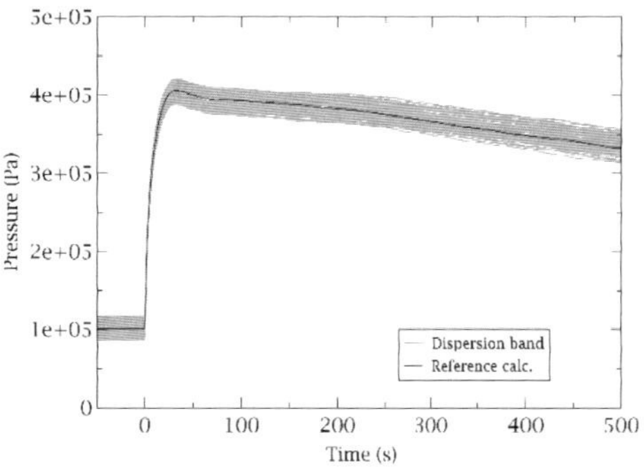

Figure 5. Containment pressure reference calculation and dispersion band predicted by TRACE.

The first cladding temperature peak presents a dispersion band of around 70 K. The expulsion of the primary coolant through the break also determines the RPV collapsed coolant level drop, as shown in Figure 4. In this first part of the transient, the RPV collapsed coolant level does not present any relevant dispersion band. The coolant discharge through the break determines the increase of the containment pressure, as shown in Figure 5, which presents a peak with a dispersion band of about 4.0E4 Pa. When the PCS pressure drops, the HPIS/LPIS and the accumulators are activated and discharge water in the PCS. This determines the increase of RPV collapsed coolant level and the start of the refill phase. As shown in Figure 4, the RPV collapsed coolant level increases and reach the BAF (Bottom of Active Fuel) after about 50 s after the SOT in the reference calculation, with a time dispersion of around 10 s. During the refill phase the core is mainly uncovered, the heat is not removed from the fuel rod; the cladding temperature rises and reaches the second cladding temperature peak of about 1075 K at around

50 s after the SOT in the reference calculation, with a dispersion of about 85 K. The RPV collapsed coolant level continues to increase during the reflood phase and reaches the TAF at about 120 s after the SOT in the reference calculation, with a time dispersion band of about 45 s. The complete rewetting of the cladding surfaces by the coolant determines the "core quenching" with a strong temperature decrease. This phenomenon is predicted by TRACE and it happens around 220 s after the SOT in the reference calculation, with a time dispersion band of 150 s.

4.2. Correlation analysis of the LBLOCA transient

A computation of the correlation coefficients between the selected uncertain input parameters and the FOMs has been performed at selected time values during the transient. From Figure 6 to 8 the Pearson's simple correlation coefficients and Spearman's simple rank correlation coefficients for the three FOMs are shown.

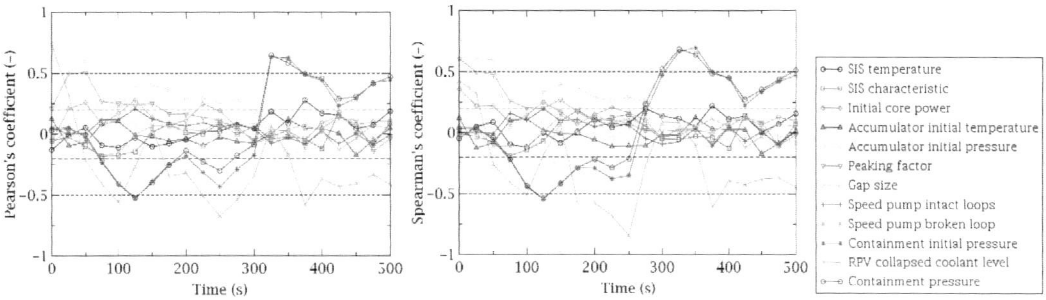

Figure 6. Pearson and Spearman's coefficients for the hot rod cladding temperature.

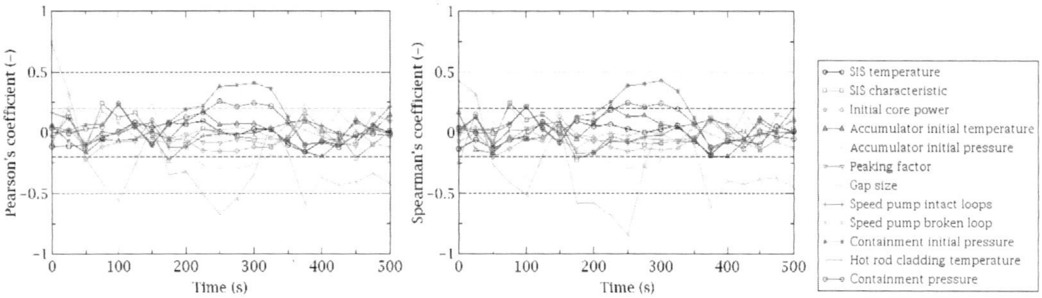

Figure 7. Pearson and Spearman's coefficients for the RPV collapsed coolant level.

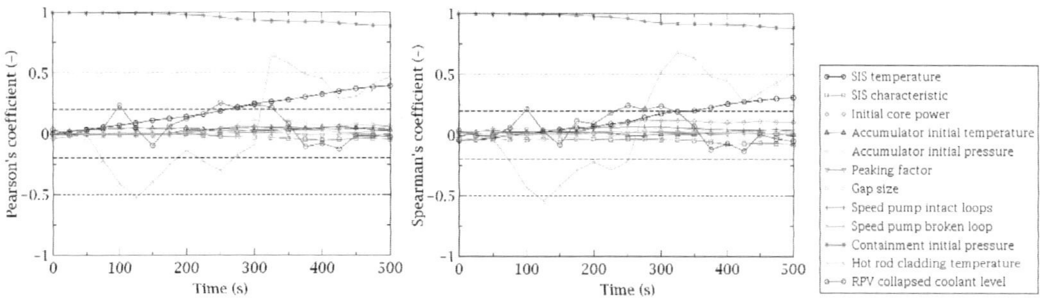

Figure 8. Pearson and Spearman's coefficients for the containment pressure.

As indicated in [17], for the Spearman coefficient, if the coefficient is higher than 0.5 (or lower than -0.5) the correlation is significant, if it is between 0.2 and 0.5 (or -0.2 and -0.5) the correlation is moderate, otherwise it is low. In this study the same threshold values have been adopted for the Pearson coefficient, as done in [24]. In relation to the hot rod cladding temperature, the peaking factor presents a significant monotonous correlation after the SOT and a moderate monotonous and linear correlation between 25 and 50 s after the SOT. The gap size presents a significant monotonous and linear correlation between 25 s and 50 s after the SOT. The initial core power presents a moderate monotonous correlation at the SOT. The hot rod cladding temperature presents, also, a significant linear and monotonous correlation with the initial containment pressure during the reflood phase of the transient (at about 300 s after the SOT). The containment initial pressure presents also a moderate linear and monotonous correlation with the RPV collapsed coolant level. The containment pressure, as expected, presents a high significant monotonous and linear correlation with the initial containment pressure along the transient. The SIS temperature presents a moderate monotonous and linear correlations during the long-term cooling phase. In relation to the correlations between the FOMs, it is to underline a significant positive linear correlation in the initial phase of the transient and a moderate negative correlation during the reflood and the long-term cooling phases of the transient between the hot rod cladding temperature and the RPV collapsed coolant level. The containment pressure and the hot rod cladding temperature present a significant linear and monotonous correlation in the reflood phase of the transient. The RPV collapsed coolant level presents a moderate positive correlation with the containment pressure at the end of the reflood phase and the begin of the long-term cooling phase.

5. Conclusions

In this activity an UQ analysis has been conducted using the probabilistic method to propagate input uncertainty. The postulated transient analysed has been a LBLOCA initiated by a double-ended guillotine break on the CL of a generic three-loops PWR-900 MWe. The main goal of this study is not to be a detailed uncertainty study but to develop a full UQ application with the coupling of TRACE and DAKOTA toolkit in the SNAP environment/architecture and to have some insights about the statistical correlation between the input parameters and the FOMs selected along the transient. The study is conducted by coupling the thermal-hydraulic BE code TRACE, developed by USNRC, and the uncertainty quantification tool DAKOTA, developed by Sandia National Laboratories, in the SNAP environment/architecture. Ten uncertain input parameters have been selected, based on the BEMUSE program. Three FOMs have been considered: the hot rod cladding temperature, the RPV collapsed coolant level and the containment pressure. The minimum number of code run for the selected probability and confidence level is obtained based on the Wilks method. The transient behaviour and the dispersion band of the selected FOMs have been analysed. In particular, the two cladding temperature peaks present a dispersion band of 70 K and 85 K respectively. The end of the quenching phenomenon presents a time dispersion of about 150 s. The end of the refill and of the reflood phases present a dispersion time of about 10 s and 45 s respectively. The containment pressure presents a peak with dispersion band of about 4.0E4 Pa. A statistical correlation analysis has been conducted, considering the Pearson's simple and the Spearman's simple rank coefficients computed by DAKOTA. A significant linear and monotonous correlation has been identified between the hot rod cladding temperature and the gap size in the initial phase of the transient. As expected, a significant linear and monotonous correlation has been underlined between the initial pressure of the containment before the beginning of the transient and the pressure of the containment along the transient. It has been underlined a moderate linear and monotonous correlation between the SIS temperature and the containment pressure. In a future study the nodalization can be further improved, new uncertainty input parameters and FOMs can be added in order to have a more complete evaluation of the uncertainty quantification results.

References

[1] Mascari F, Nakamura H, Umminger K, De Rosa F and D'Auria F, Scaling Issues For The Experimental Characterization Of Reactor Coolant System In Integral Test Facilities And Role Of System Code As Extrapolation Tool, *Proceedings Of International Topical Meeting on Nuclear Reactor Thermal Hydraulics*, NURETH 2015 **6**, pp. 4921-4934.

[2] NEA/CSNI *A state-of-the-art report on scaling in system themal hydraulics applications to*

nuclear reactor safety and design NEA/CSNI/R(2016)14(2017).
[3] IAEA International Atomic Energy Agency 2008 *Best Estimate Safety Analysis for Nuclear Power Plants: Uncertainty Evaluation*, Safety Reports Series.
[4] U.S. Nuclear Regulatory Commission 2020 TRACE v5.0 Patch 6 Theory Manual.
[5] U.S. Nuclear Regulatory Commission 2020 TRACE v5.0 Patch 6 User's Manual Volume 1: Input Specification.
[6] U.S. Nuclear Regulatory Commission 2020 TRACE v5.0 Patch 6 User's Manual Volume 2: Modeling Guidelines.
[7] Dalbey K R, Eldred M S, Geraci G, Jakeman J D, Maupin K A, Monschke J A, Seidl D T, Swiler L P, Tran A , Menhorn F and Zeng X 2020 Dakota, A Multilevel Parallel Object-Oriented Framework for Design Optimization, Parameter Estimation, Uncertainty Quantification, and Sensitivity Analysis: Version 6.13 Theory Manual.
[8] Adams B M, Bohnhoff W J, Dalbey K R, Ebeida M S, Eddy J P, Eldred M S, Hooper R W, Hough P D, Hu K T, Jakeman J D, Khalil M, Maupin K A, Monschke J A, Ridgway E M, Rushdi A A, Seidl D T, Stephens J A, Swiler L P and Winokur J G 2020 Dakota, A Multilevel Parallel Object-Oriented Framework for Design Optimization, Parameter Estimation, Uncertainty Quantification, and Sensitivity Analysis: Version 6.13 User's Manual.
[9] Applied Programming Technology, Inc. 2021 Symbolic Nuclear Analysis Package (SNAP) User's Manual.
[10] Applied Programming Technology, Inc. 2020 Uncertainty analysis User manual, Symbolic Nuclear Analysis Package (SNAP).
[11] Glaeser H 2008 *Science and Technology of Nuclear Installation* Vol 2008, 798901. DOI: https://doi.org/10.1155/2008/798901
[12] Stephens M E, Goodwin B W and Andres T H 1993 *Reliability Engineering & System Safety* **42**, pp. 271-291. DOI: https://doi.org/10.1016/0951-8320(93)90094-F
[13] Wilks S S 1941 *The Annals of Mathematical Statistics* **12**(1), pp. 91-96. DOI: https://doi.org/10.1214/aoms/1177731788
[14] Wilks S S 1942 *The Annals of Mathematical Statistics* **13**(4), pp. 400-409. DOI: https://doi.org/10.1214/aoms/1177731537
[15] Bajorek S M and Gingrich C 2013 *Uncertainty Methods Framework Development for the TRACE Thermal-Hydraulics Code by the U.S.NRC*, OECD/CSNI Workshop on Best Estimate Methods and Uncertainty Evaluations, NEA/CSNI/R(2013)8/PART2.
[16] Guba A, Makai M and Pál 2003 *Reliab. Eng. Syst. Safe.* **80**, pp. 217-232. DOI: https://doi.org/10.1016/S0951-8320(03)00022-X
[17] Gamble K A and Swiler L P 2016 *Uncertainty Quantification and Sensitivity Analysis Applications to Fuel Performance Modeling*, SAND2016-4597C.
[18] Rodgers J L and Nicewander W A 1988 *The American Statistician* **42**, pp. 59-66. DOI: https://doi.org/10.1080/00031305.1988.10475524
[19] https://dakota.sandia.gov/
[20] Bersano A and Mascari F 2019 *atw* **64**, pp. 526-532
[21] Agnello G 2020 *Analisi di incidenti di tipo LOCA con codici di sistema e relativa valutazione dell'incertezza dei risultati*, Master Thesis, Università degli Studi di Palermo.
[22] Iguchi T, Okubo T and Murao Y 1988 *Journal of Nuclear Science and Technology* **25**(6), pp. 520-527. DOI: https://doi.org/10.1080/18811248.1990.9731149
[23] Perez M, Reventos F, Batet L, Guba A, Toth I, Mieusset T, Bazin P, De Crécy A, Borisov S, Skorek T, Glaeser H, Joucla J, Probst P, Ui A, Chung B D, Oh D Y, Pernica R, Kyncl M, Macek J, Manera A, Freixa J, Petruzzi A, D'Auria F and Del Nevo A 2011 *Nuclear Engineering and Design* **241**, pp. 4206-4222. DOI: https://doi.org/10.1016/j.nucengdes.2011.08.019
[24] Bersano A, Mascari F, Porfiri M T, Maccari P, Bertani C 2020 *Fusion Engineering and Design* **159**, 111944. DOI: https://doi.org/10.1016/j.fusengdes.2020.111944

Journal of Physics: Conference Series 2177 (2022) 012024 doi:10.1088/1742-6596/2177/1/012024

Analysis of an unmitigated 2-inch cold leg LOCA transient with ASTEC and MELCOR codes

G Agnello[1], S Ederli[2], P Maccari[3], F Mascari[4]

[1] Università degli Studi di Palermo, Dipartimento di Ingegneria, Viale delle Scienze, Edificio 6, 90128, Palermo, Italy
[2] ENEA – Casaccia, Via Anguillarese 301, 00123, Roma, Italy
[3] UNIBO, DIN, Via dei colli 16, 40136, Bologna, Italy
[4] ENEA – Bologna, Via Martiri di Monte Sole 4, 40123, Bologna, Italy

Corresponding author's email: giuseppe.agnello04@unipa.it

Abstract. The analyses of postulated severe accident sequences play a key role for the international nuclear technical scientific community for the study of the effect of possible actions to prevent significant core degradation and mitigate source term release. To simulate the complexity of phenomena involved in a severe accident, computational tools, known as severe accident codes, have been developed in the last decades. In the framework of NUGENIA TA-2 ASCOM project, the analysis of an unmitigated 2-inch cold leg LOCA transient, occurring in a generic western three-loops PWR-900 MWe, has been carried out with the aim to give some insights on the modelling capabilities of these tools and to characterize the differences in the calculations results. The ASTEC V2.2b code (study carried out with ASTEC V2, IRSN all rights reserved, [2021]), and MELCOR 2.2 code have been used in this code-to-code benchmark exercise. In the postulated transient, the unavailability of all active injection coolant systems has been considered and only the injection of accumulators has been assumed as accident mitigation strategy.

1. Introduction

The interest of international nuclear scientific community, after the Chernobyl and, more recently, Fukushima Dai-ichi accident, has focused on the field of Severe Accident (SA) and on the development of new strategies to mitigate core damages and possible releases of Fission Products (FPs) to the environment. To develop and consolidate Severe Accident Management (SAM) measures, the study on SA sequences led to the development of simulation tools, known as SA codes, for simulating the main phenomena occurring in a Nuclear Power Plant (NPP) during postulated accidents. Several experimental programs in the field of SA phenomena have been conducted to validate the models and correlations implemented in the codes and provided a valuable "assessment database" to assess SA simulation tools [1]. However, the analyses of the current state-of-art shows that there is a need to reduce some uncertainties still present [2] and a consequent investigation of phenomena/processes not investigated in detail in geometric prototypical facility with prototypical material should be addressed. In relation to this, for example, the complex multi-physics phenomena taking place during the degradation and relocation of the core material determines uncertainties on the zircaloy area exposed to steam for oxidation process and limited data from full-scale experimental facility are available for code validation. Therefore, discrepancies in some core degradation phenomena can be still observed when comparing the results as predicted by different simulation tools considering the different core degradation models implemented in the codes [3]. Code-to-code benchmark exercise [4] involving code developers could contribute to identify code modelling differences affecting the codes [5]. Code-to-code exercise involving code users could characterize the user effect on code prediction [3,6]. The present code-to-code benchmark exercise, developed in the framework of NUGENIA TA-2 ASCOM (ASTEC COMmunity) collaborative project [7], coordinated by IRSN, has been conducted to investigate the main discrepancy in predicting thermal-hydraulic and core degradation in-vessel phenomena of ASTEC [8] and MELCOR [9] codes during an unmitigated 2-inch Cold Leg (CL) Loss Of Coolant Accident (LOCA) in a generic PWR 900 MWe three-loops western type.

2. Transient characteristics and hypothesis of the study

The NPP considered in this analysis is a generic PWR-900 MWe three-loops western type and the SA sequence selected for the presented study is an unmitigated Small Break LOCA [4,10] in which only the accumulators are considered available. The accident initiator event considered is a 2-inch break of the CL of loop 1, where the pressurizer (PRZ) is located. The initial plant operation point, before the Start of the Transient (SOT), is the nominal full-power condition. The break event is coupled with the loss of offsite Alternating Current (AC) power and the failure of all the diesel generators determining the unavailability of the following systems: primary pressure control systems (heaters and PRZ spry), Chemical and Volume Control systems (CVCS), Reactor Coolant Pump (RCP) seal injection, Active safety injection systems (High Pressure Injection System, HPIS, and Low Pressure Injection System, LPIS), Motor-driven Auxiliary Feedwater (MDAFW) system. The following hypotheses are also considered: reactor SCRAM and Steam Generators (SGs) isolation at SOT, independent failure of the Turbine Driven Auxiliary Feedwater (TDAFW) pump, no primary boundary structures thermal induced degradation phenomena, primary and secondary side Steam Relief Valves (SRVs) availability throughout the accident evolution, Safety Valves of Pressure Compensator (SEBIM) manually stuck open when the core exit temperature reaches 650°C. The transient has been analyzed until the vessel Lower Head (LH) failure.

3. ASTEC code description and reactor input-deck main characteristics

The ASTEC code (Accident Source Term Evaluation Code) [8], developed by the French "Institut de Radioprotection et de Sûreté Nucléaire" (IRSN), aims at simulating an entire SA sequence in nuclear water-cooled reactors. ASTEC includes several coupled modules that can deal with the different SA phenomena. The code version used in this activity is ASTEC V2.2b. The ASTEC reactor input-deck used is the generic PWR-900 MWe input-deck delivered by IRSN with the ASTEC V2.2b code.

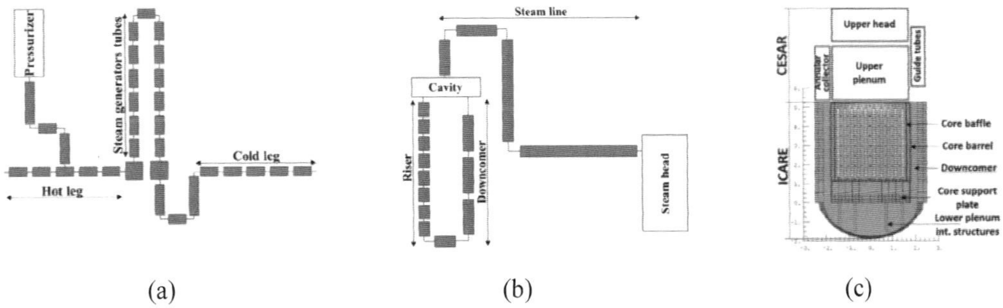

Figure 1. ASTEC Loop 1 (a), SCS (one loop) (b) (CESAR) and RVP (c) (CESAR and ICARE) nodalization [11].

The Primary Cooling System (PCS), part of the Reactor Pressure Vessel (RPV) (Upper Head -UH-, Upper Plenum -UP-, annular collector and the guide tubes) and Secondary Cooling System (SCS) are modelled with the volumes of CESAR module [12], the module of ASTEC dedicated to the simulation of thermal-hydraulic phenomena. The three primary loops are modelled separately: in Figure 1a the loop 1 containing the PRZ is shown. The accumulators have been modelled using special volumes defined as accumulators, with the availability of multiple dedicated settings (e.g. gas expansion model, etc.). The accumulator opening has been set with a control logic that permits the emergency coolant discharge when the primary pressure falls below the pressure set-point value. The SCS, shown in Figure 1b, includes the SGs and the steam lines. The RPV structures, shown in Figure 1c, from the LH to the upper core plate, are modeled with ICARE module [13], the ASTEC module dedicated to the simulation of the in-vessel core degradation phenomena. The core is divided into 5 radial fluid channels (containing 5 groups of fuel rods) and 16 axial segments and two other fluid channels simulate respectively the core bypass (BP) and the downcomer (DC).

4. MELCOR code description and reactor input-deck main characteristics

MELCOR [9] is a fully integrated SA code, developed at Sandia National Laboratories (SNL) for the US Nuclear Regulatory Commission (USNRC), able to simulate the thermal-hydraulic phenomena in steady-state and transient condition and the main phenomena occurring in a SA. MELCOR can be used with Symbolic Nuclear Analysis Package (SNAP) [14] to develop the nodalization and for the post processing of data. MELCOR has a modular structure and is based on packages simulating a different part of the transient phenomenology. The MELCOR code version used is MELCOR 2.2 18019. The MELCOR nodalization [3,6,10,15,16] of the generic PWR-900 western type three-loops, shown in Figure 2a, developed by using SNAP, was designed to have a reasonable computational time and realistic prediction of the thermal hydraulics and core degradation phenomena.

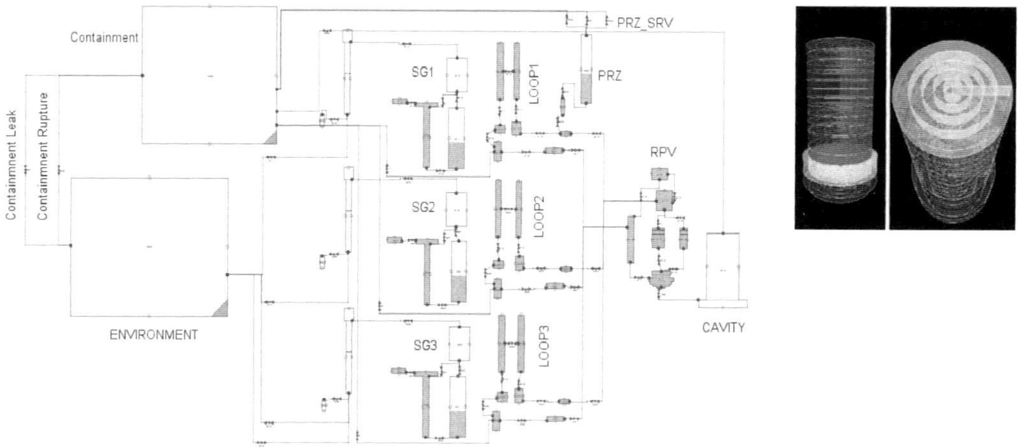

(a) (b)

Figure 2. MELCOR PCS, SCS (a) and core (b) nodalization developed using SNAP.

It has been developed considering the plant information's reported in the 2013 Foucher reports [17,18] and the ASTEC input files provided by IRSN during the EU-CESAM project [19]. The prediction capability of this MELCOR input-deck has been already revised along the EU-FASTNET project [20,21] and EU-IVMR project [22,23]. The three loops have been modelled separately. The SG side is composed of three different volumes representing the SG downcomer, the SG riser and the SG cavity. The accumulators have been modelled by one CVH volume for each loop, in adiabatic condition, and by control functions to set the accumulator opening/closing. The accumulators opening begin when the primary pressure falls below the pressure set-point. The thermal-hydraulic CVH nodalization of the RPV features different hydraulic regions simulating Lower Plenum (LP), core, core BP, UP, UH and DC. The core is modelled by a single hydraulic region, CVH package, coupled with the correspondent model of the COR package. A temperature of 2450 K is considered for modelling the melting temperature of the mixture UO_2/ZrO_2. The core, within the COR package, is modelled with 17 axial regions and 6 radial regions; 5 radial regions are used to model the core region, as shown in Figure 2b.

Journal of Physics: Conference Series 2177 (2022) 012024 doi:10.1088/1742-6596/2177/1/012024

5. Transient evaluation and codes results analysis

5.1. Thermal-hydraulic evolution of the transient

Before the transient simulation, a steady state analysis has been evaluated for both the two codes and the comparison have been already reported in [6]. At the SOT (t = 0 s), as hypothesized in both the codes, reactor SCRAM and the SGs isolation take place. It follows the drop of the core steady fission power to decay power. The break opening determines a primary coolant leakage to the containment with a consequent RPV blowdown and fast PCS pressure decrease. As shown in Figure 3, the primary coolant mass flow rate through the break is greater in ASTEC than in MELCOR in the first 200 s of the transient and, subsequently, it is greater in MELCOR till 1500 s. The total amount of water mass discharged through the break till 1500 s is greater in MELCOR calculation.

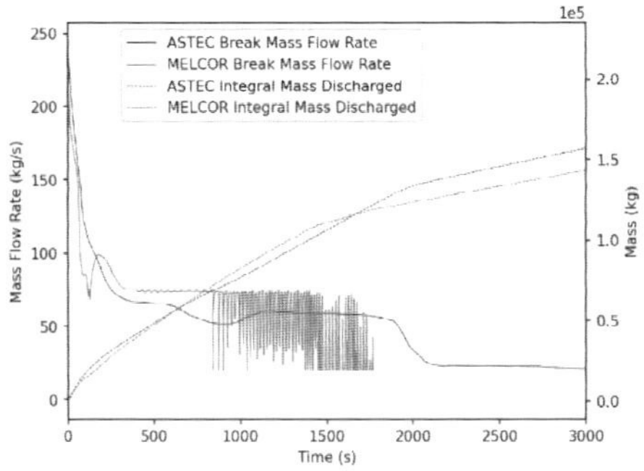

Figure 3. Break mass flow rate and primary coolant mass discharged from the break.

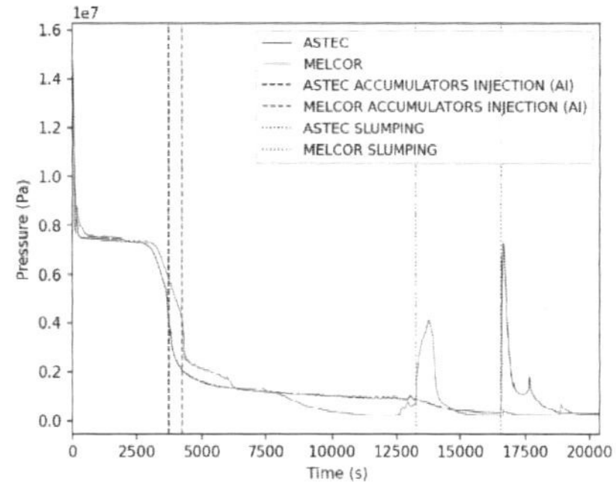

Figure 4. Primary pressure (PRZ pressure).

As shown in Figure 4, along this first blowdown phase, the primary pressure behavior for ASTEC and MELCOR is similar from both a qualitative and quantitative point of view. The isolation of the SGs takes place leading to a SGs secondary side pressure temporary increase: when the pressure set point of SGs SRVs is reached, the SGs start the cycling phase with the release of steam, at about 38 s and 30 s after the SOT in the SG1 respectively in ASTEC and MELCOR. After a rapid depressurization, the primary pressure stabilizes at the saturation pressure. During this phase of the transient, considering the dimension of the break, part of the core decay heat is removed by the coolant discharged through the break and part by the SGs. The primary coolant boil-off determines the decrease of the core collapsed coolant level, as shown in Figure 5 where the relative core collapsed coolant level is reported for the two codes (Bottom of Active Fuel, BAF, is the value 0 and the Top of Active Fuel, TAF, is the value 1). Some discrepancies between the two codes calculations in the core level decrease for the first 2500 s of the transient are underlined: MELCOR simulates a faster decrease of the core collapsed coolant level than ASTEC due to the major quantity of coolant discharged through the break in the first 1500 s of the transient, as shown previously in Figure 3. In term of pressure, after the pressure plateau, this determines a faster pressure decrease in ASTEC than in MELCOR with the consequent start of the accumulator injection at about 3713 s in ASTEC and 4236 s in MELCOR. A different accumulators behavior, in terms of initial water mass, starting time, injection rate and total amount of emergency coolant injected, is also highlighted in Figure 6 where the water mass of accumulator of loop 1 is shown. The accumulators start the injection first in ASTEC than in MELCOR, and the major injection rate of ASTEC, for a longer period, determines a more rapid rise of the core collapsed coolant level than in MELCOR and allows to reach a higher level in the core. The boil-off of the primary fluid continues determining the fluid level decrease which is faster in MELCOR calculation due to the earlier stop of the first accumulators injection at about 6150 s. In ASTEC the accumulators injection continue until about 12900 s with the total accumulators discharge. The core water level reaches the BAF in the ASTEC case at about 3000 s and at about 13000 s after the accumulators injection. In MELCOR calculation, characterized by an initial slower depressurization, the BAF is reached only after the accumulator injection at about 10000 s.

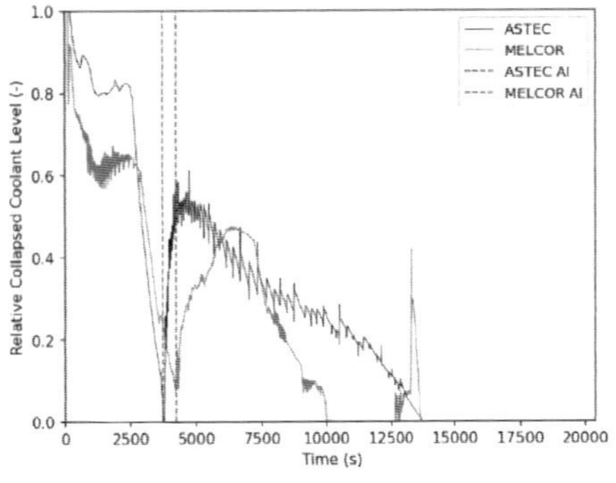

Figure 5. Relative core collapsed coolant level.

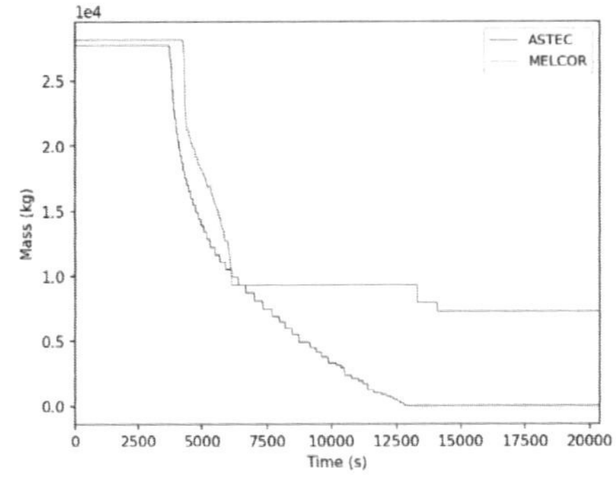

Figure 6. Loop 1 accumulator water mass.

5.2. Core degradation phase

The core uncovery determines the core heat-up and the onset of chemical reaction between zirconium and steam in the RPV. The energy released by the exothermic oxidation is added to the core decay power causing an acceleration of the heat-up rate with a consequent temperature escalation and hydrogen generation. At about 4000 s, the differences in terms of core uncovery between the two codes determine some discrepancies in terms of cladding temperature, as shown in Figure 7 and Figure 8 where the maximum intact cladding temperature for each fuel ring are represented for ASTEC and MELCOR calculations. When the accumulators start the injection in ASTEC, the maximum cladding temperature presents a value of around 1100 K; in MELCOR, instead, the maximum cladding temperature presents a higher value of around 1600 K; this difference influences the H_2 generation along this phase of the transient. In the MELCOR case the H_2 generation starts and around 100 kg of H_2 are produced before the accumulator injection.

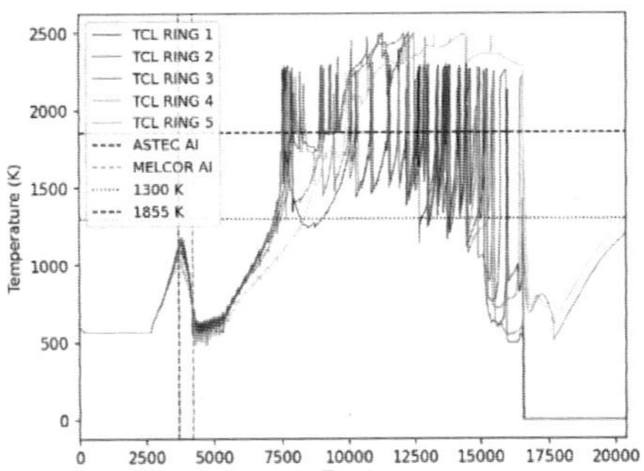

Figure 7. ASTEC maximum intact cladding temperature for each fuel ring.

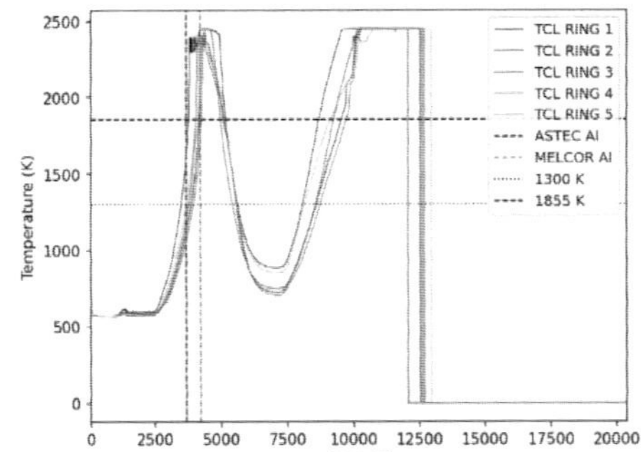

Figure 8. MELCOR maximum intact cladding temperature for each fuel ring.

In ASTEC calculation no significant oxidation and consequently a H_2 production, before the accumulator injection, is predicted. ASTEC calculation presents a first cladding temperature peak of about 1250 K at about 3760 s and MELCOR calculation presents a first cladding temperature peak of about 2500 K at about 4300 s. After the partial reflooding of the core, the maximum intact cladding temperature decreases at about 600 K in both the codes predictions. With the cladding temperature decrease, the oxidation rate of Zr and consequently the H_2 and ZrO_2 generation rate stop in MELCOR calculation as shown in Figure 9, 10 and 11 where the H_2 cumulative generation, the Zr, ZrO (only for ASTEC code) and ZrO_2 masses along the transient evolution are reported. At about 5500 s in ASTEC and 7500 s in MELCOR, the core water level returns to decrease and the cladding temperature returns to growing. Then, the H_2 generation starts considerably in ASTEC (at about 7100 s) and starts again in MELCOR (at about 9500 s) after the plateau of about 4200 s.

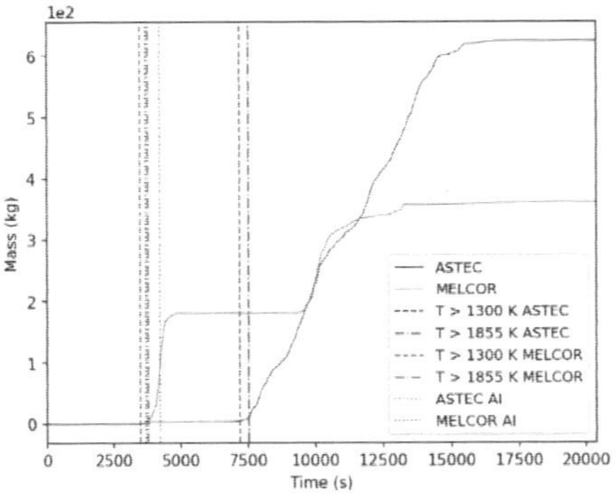

Figure 9. H_2 cumulative generation.

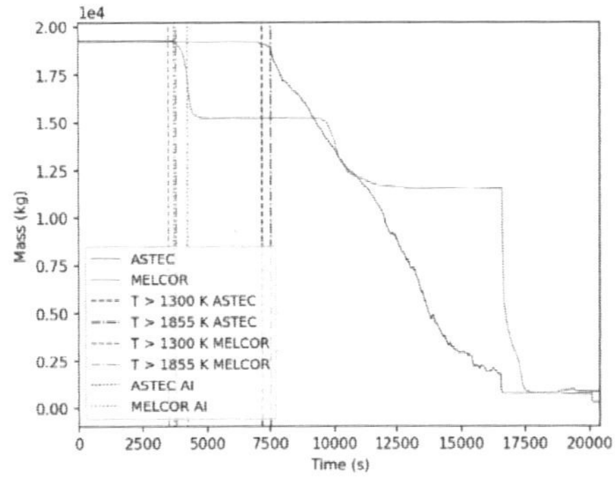

Figure 10. Zr core mass.

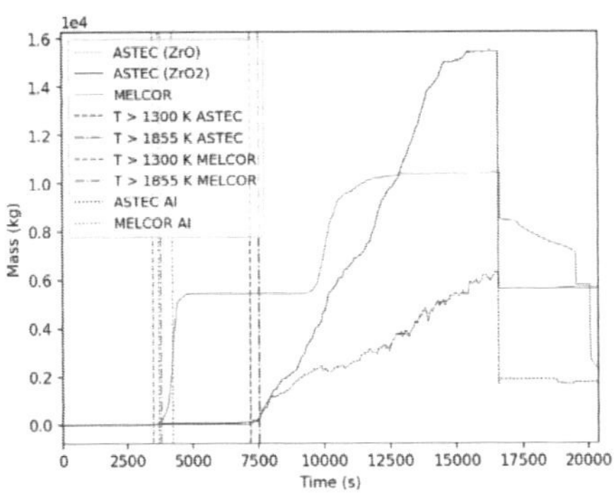

Figure 11. ZrO (only for ASTEC case) and ZrO_2 core mass.

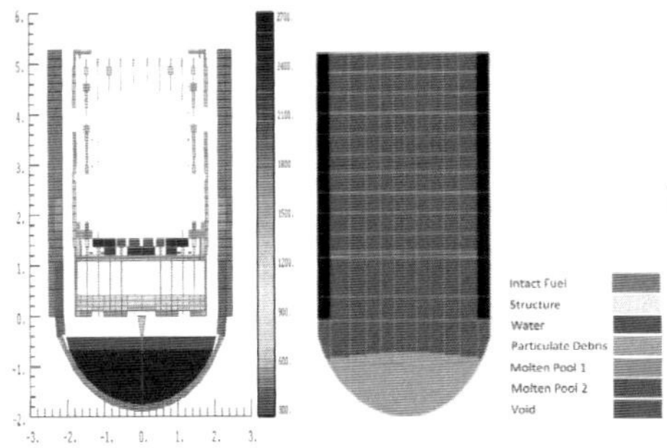

Figure 12. ASTEC (left) and MELCOR (right) core composition before lower head failure.

The core collapsed level continues to decrease faster in MELCOR than in ASTEC. Along this phase the core degradation and relocation phenomena progress in both the simulations. The loss of core geometry determines the relocation of core material in the LP (slumping), predicted both in the two codes. The slumping occurs first, at about 13253 s, in MELCOR and later in ASTEC at about 16564 s. The corium relocates to the LP where it remains for a time of about 3830 s in ASTEC and about 6907 s in MELCOR before the LH failure which occurs at about 20160 s in MELCOR and at about 20394 s in ASTEC. These discrepancies could be caused by the difference in the corium behaviour predicted by the two codes. In fact, in ASTEC the corium in the LP is composed of molten corium layers (oxide at the bottom and metal at the top) and the presence of debris bed particles is negligee. In MELCOR the corium in the LP is composed predominantly by particulate debris at the bottom and a small quantity of molten metal at the top, as shown in Figure 12. In Table 1 are presented the relevant phenomenological aspect predicted by ASTEC and MELCOR codes with the specific timings and discrepancies. The total mass of H_2 produced by Zr oxidation process is about 600 kg in ASTEC and 350 kg in MELCOR. One of the reasons of this difference could be the MELCOR prediction of a first generation of H_2 before the accumulator injection; therefore, a first limited degradation of the core takes place. The second accumulator injection determines a further source of steam. Since the H_2 generation depends on the fuel cladding surface available to the reaction with the steam, a less damage core produces more H_2. Furthermore, the accumulators inject more water in ASTEC and more steam is available. Therefore, in ASTEC along the second peak cladding temperature more H_2 is produced.

Table 1. Relevant phenomena aspects predicted by ASTEC and MELCOR codes.

Relevant Phenomenological Aspects [s]	ASTEC	MELCOR	DISCR [%]*
SOT, SCRAM, SGs isolation	0	0	0
SG 1,2,3 cycling inception	38,30,30	30,25,25	21.05,16.67,16.67
Core TAF uncovered	204	126	38.24
H_2 generation start	2923	3336	14.13
T > 1300 K (before accumulators injection)	-	3505	-
T > 1855 K (before accumulators injection)	-	3798	-
Start of accumulators 1,2,3 discharge	3713,3713,3713	4236,4238,4238	14.09,14.14,14.14
Core BAF uncover	3773	-	-
T > 1300 K (after accumulators injection)	7208	8086	12.18
T > 1855 K (after accumulators injection)	7528	8707	15.66
Core BAF uncover	13702	10006	26.97
Slumping	16564	13253	19.99
LH failure	20394	20160	1.15

*discrepancy in the Table 1 has been calculated as the percentage difference between the parameters evaluated by the two codes.

6. Conclusions

In this present work, developed in the framework of NUGENIA TA-2 ASCOM project, a code-to-code benchmark activity with ASTEC and MELCOR codes has been conducted considering a Small Break LOCA initiated by a 2-inch break on the CL of a generic PWR-900 MWe three loops western type. The unavailability of all active injection coolant systems has been considered and the injection of accumulators has been assumed as only accident mitigation strategy. In relation to the phase where thermal-hydraulic phenomena are dominant, the two codes predict the same qualitative behaviour while some quantitative differences are related to break flow rate and the accumulator behaviour: both affect the core level behaviour and the consequent partial or total core uncovery phenomena. This determines the cladding temperature increase and the consequent H_2 generation before and after the accumulator injection. Before the accumulator injection a more evident core uncovery, as in the case of MELCOR calculation, can determine a H_2 generation and, therefore, limited degraded core. After the accumulator injection, since the H_2 generation depends on the fuel cladding surface available to the reaction with the steam, a less damage core can produce more H_2, as in the case of ASTEC simulation (350 kg of H_2 are produced by MELCOR, 600 kg of H_2 are produced by ASTEC). Also, accumulators injection behaviour can have an influence on the H_2 generation due to the injection rate and total mass injected. In the case of ASTEC accumulators inject more water and, therefore, more steam is available. Once the accumulators injection ends, the core will start to heat-up again and degradation phenomena start to be dominant. This phase is still characterized by some uncertainties on phenomena such as core melt progression, core relocation to LP, LP debris bed behaviour, and LH failure. Even though MELCOR predicts a faster degradation than ASTEC (also related to the different accumulator behaviour) and a consequent earlier corium relocation to the LP, the different retention time of the corium in the LP leads to limit the code-to-code discrepancy to about 1% for the LH failure event. The observed discrepancies underline some modelling differences between the two codes and details studies are in progress to characterize them as well as the user effect in view of uncertainty estimation. In relation to the user effect, in the case of MELCOR code the effect of the accumulator modelling is under investigation. In fact, in this study a CVH model is used to model the accumulator, but, currently, an accumulator packageis available and an adiabatic and isothermal approximations, used for modelling the accumulators gas expansion on the transient progression prediction, are available for the user [4,9].

References

[1] Mascari F, Nakamura H, Ummlnger K, De Rosa F and D'Auria F 2015 *Scaling Issues for the Experimental Characterization of Reactor Coolant System in Integral Test Facilities and Role of System Code as Extrapolation Tool*, Proceedings of NURETH 2015 **6**.

[2] Van Dorsselaere J P, Mustoe J, Power S, Adorni M, Schaffrath A and Nieminen 2016 *Kerntechnik* **81**, Issue 5, pp. 527-534. DOI: https://doi.org/10.3139/124.110734

[3] Mascari F, De La Rosa Blul J C, Sangiorgi M and Bandini G 2019 *Analyses of an Unmitigated*

Station Blackout Transient in a Generic PWR-900 with ASTEC, MAAP and MELCOR Codes, NUREG/IA-0515.

[4] De la Rosa Blul, Brumm S, Mascari F, Lee S J and Carenini L 2018 *ASTEC-MAAP Comparison of a 2 Inch Cold Leg LOCA until RPV Failure*, Science and Technology of Nuclear Installations, Vol 2018. DOI: https://doi.org/10.1155/2018/9189010

[5] Belon S, Bouillet C, Topin V, Bonneville H and Andrews N 2017 *Insight of core degradation simulation in integral codes throughout ASTEC/MELCOR crosswalk comparisons and ASTEC sensitivity studies*, Proceedings of the 8th European Review Meeting on Severe Accident Research ERMSAR-2017, Warsaw, Poland, May 16-18, 2017.

[6] Maccari P, Mascari F, Ederli S and Manservisi A 2021 *J.Phys.: Conf. Ser.* **1868** 012019. DOI: https://doi.org/10.1088/1742-6596/1868/1/012019

[7] Chatelard P 2018 *ASCOM project overview and description of work*, SARNET-ASCOM-COORDI-P1.

[8] Chatelard P, Belon S, Bosland L, Carénini L, Coindreau O, Cousin F, Marchetto C, Nowack H, Piar L and Chailan L 2016 *Annals of Nuclear Energy* **93**, pp. 83-93. DOI: https://doi.org/10.1016/j.anucene.2015.12.026

[9] MELCOR Computer Code Manuals, Vol.1: Primer and Users' Guide, SAND 2015-6691 R; Vol. 2: Reference Manual, SAND 2015-6692 R; Vol. 3: MELCOR Assessment Problems, SAND 2015-6693 R; Sandia National Laboratories, USA.

[10] Mascari F, Guglielmelli A and De la Rosa Blul J C 2018 *Analysis of a postulated 2-inch Cold Leg LOCA severe accident in a PWR-like 900 MWe with MELCOR code*, Proceedings of the International Conference Nuclear Energy for New Europe, NENE 2018, Portorož, Slovenia, September 10-13.

[11] Ederli S, Mascari F and Carenini L 2019 *ASTEC simulation of In Vessel Retetion Strategy applied to a generic PWR 900 MWe*, ERMSAR2019, Prague, March 18-20.

[12] Piar L 2015 *CESAR physical and numerical modelling*, IRSN Report n° PSN-RES/SAG/2015-00332.

[13] Coindreau O 2016 *ASTEC V2.1: Physical modelling of the ICARE module*.

[14] Symbolic Nuclear Analysis Package (SNAP) 2021 Users Manual. Applied Programming Technology, Inc. Bloomsburg, PA.

[15] Mascari F and Ederli S 2019 *Analyses with MELCOR code of an unmitigated SBO scenario with in vessel retention strategy applied to a generic PWR-900 MWe*, 18th International Topical Meeting on Nuclear Reactor Thermal Hydraulics, NURETH 2019, pp. 2497-2510.

[16] Pescarini M, Mascari F, Mostacci D, De Rosa F, Lombardo C and Giannetti F 2017 *J. Phys.: Conf. Ser.* **923** 012009. DOI: https://doi.org/10.1088/1742-6596/923/1/012009

[17] Foucher L, ASTEC V20R3 PWR900 Like ASTEC Input Deck, Rapport n° PSNRES/SAG/2013-451.

[18] Foucher L, ASTEC V20R3, PWR900 Like ASTEC Steady State Calculation, Rapport n° PSN-RES/SAG/2013-466.

[19] Nowack H, Chatelard P, Chailan L, Hermsmeyer St, Sanchez V and Herranz L 2018 *Annals of Nuclear Energy* **116**, pp. 128-136. DOI: https://doi.org/10.1016/j.anucene.2018.02.021

[20] https://cordis.europa.eu/project/id/662284/it

[21] Mascari F et al. 2019 *FASTNET scenarios database development and status*, Proceedings of the 9TH European Review Meeting on Severe Accident Research (ERMSAR2019), Prague, Czech Republic, March 18-20.

[22] Fichot F, Carénini L, Sangiorgi M, Hermsmeyer S, Miassoedov A, Bechta S, Zdarek J and Guenadou D 2017 *Status of the IVMR Project: First steps Towards a New Methodology to Assess In-Vessel Retention Strategy for High-power Reactors*, Proceedings of the 8th European Review Meeting on Severe Accident Research - ERMSAR-2017 Warsaw, Poland, 16-18 May 2017.

[23] https://cordis.europa.eu/project/id/662157/it

Journal of Physics: Conference Series 2177 (2022) 012025 doi:10.1088/1742-6596/2177/1/012025

EFFECT OF REAL TEMPERATURE DATA ON THE SEASONAL COEFFICIENT OF PERFORMANCE OF AIR SOURCE HEAT PUMPS

Vincenzo Ballerini, Matteo Dongellini, Eugenia Rossi di Schio, Paolo Valdiserri

Department of Industrial Engineering
Alma Mater Studiorum – University of Bologna, Bologna, Italy
eugenia.rossidischio@unibo.it

Abstract. In this paper, a transient analysis is performed in order to evaluate how effective climate data affect the determination of the seasonal coefficient of performance of ASHPs. Three Italian cities, characterized by different climates, are considered as reference case studies and the influence of real meteorological data, collected from different years, on the heat pump SCOP is performed. The analysis is carried out by employing the dynamic software TRNSYS. Numerical results show small variations in SCOP, while a significant influence on annual thermal energy demand can be observed. The analysis also underlines the absence of a general correlation between annual thermal energy supplied by the heat pump and SCOP and between HDD and SCOP.

1. Introduction

Currently, the energy demand for heating buildings in the European Union (EU) overlies an important part of the overall energy demand. The important share of energy requested for space heating and air conditioning has direct consequences on greenhouse gas emissions, thus displaying a clear environmental impact. To achieve the goal of reducing CO_2 emissions, a possible solution is the use of heating systems powered by electricity (as heat pumps), combined with a decarbonization of the power grid, using renewable energy sources [1]. The installation of heating systems based on heat pumps is a growing market in the EU, with some differences between countries especially related to the heat pump type [2]. Currently, the majority of installed units are air-source heat pumps (ASHPs) because these systems have a lower installation costs compared with other typologies, such as ground-source heat pumps (GSHPs).

ASHPs are indeed especially effective for space heating in temperate and mild climates, and therefore could be particularly effective in Italy. In the literature, the Seasonal Coefficient of Performance (SCOP) of ASHPs is usually determined by using standard weather data of cities/regions taken from a Test Reference Year (TRY). Reference years can be constructed according to the procedure reported in the standard ISO 15927-4 [3], starting from hourly meteorological data taken from a long time series (10 years or more). Nonetheless, effective ambient temperature may significantly vary every year and, for this reason, the effective energy performance of ASHP systems might underperform (or outperform, of course) the calculated one. In the literature there are a lot of works about dynamic simulation concerning heat pump heating systems coupled to residential and non-residential buildings. Conversely, there are few papers discussing on the influence of real weather condition (i.e. outdoor air temperature, which can significantly vary from a year to another for the same place) on the ASHPs performances [4]-[7]. This kind of analysis could be relevant to prevent the underestimation or overestimation of benefits using a heat pump in terms of affordability and environmental impact of that technology.

In this paper, the results of dynamic simulations will be presented considering three different heat pump systems coupled with the same building, located in three different Italian cities: S. Benedetto del Tronto (Central Italy, 42°58' North, 13° 53' East), Milan (North of Italy, 45°28' North, 9° 10' East) and Livigno (North of Italy, 46°28' North, 10° 8' East). Dynamic simulations have been carried out considering real weather data (i.e. outdoor air temperature and solar radiation) referred to the three abovementioned cities for a period of 8 years (2013-2020). The influence of the weather compensation of supply water temperature set-point on the heat pump seasonal performance will be also analyzed.

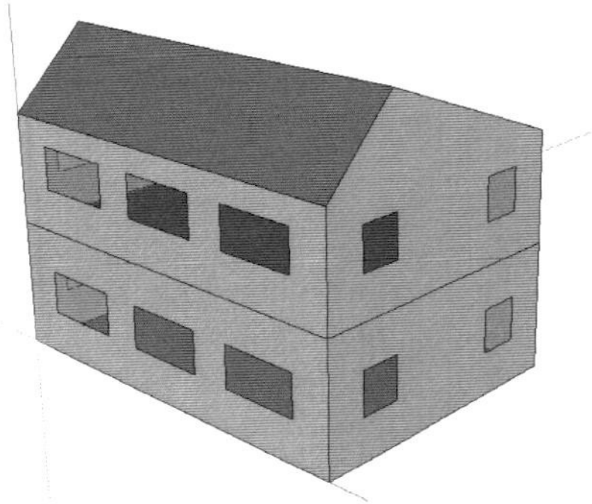

Figure 1. 3D model of the building.

2. Data and setting of the analysis

2.1. Building

The building used for simulations derives from IEA SHC Task 44 [8]. The building considered is SFH45, a single family-house building characterized by two floors with a total floor area of 360m^2 and by a thermal load of 45kWh/m^{2}y in the climate of Strasbourg. Some information about building envelope components is reported in Table 1 and a 3D representation of the building is showed in Figure 1.
The building was modeled through TRNSYS type 56, the multizone building type implemented in TRNSYS [9]-[10]. Analysis will focus only on the heating period since the largest share of buildings' energy demand is still related to space heating.

Table 1. Geometrical and thermo-physical data of the building envelope components. [11]

Assembly	Thickness (m)	U-Value (W/m^{2}K)
External wall	0.318	0.364
Ground floor	0.385	0.241
Floor	0.185	0.235
Roof	0.285	0.694
Internal wall	0.200	0.885

2.2. Heat pumps

Three inverter heat pumps are considered, as reported in Fig. 2: the first heat pump (Ariston Nimbus N50), namely Unit 1, was combined with the building located in S. Benedetto del Tronto, Unit 2 (Ariston Nimbus N70) with that located in Milan and Unit 3 (Galletti Hiwarm Compact 023) with the building located in Livigno. Figure 2 shows the thermal power supplied by the heat pumps, as well as the coefficient of performance (*COP*), as functions of the external air temperature, for different compressor frequency values. That Figure shows that Unit 1 and Unit 2 behave similarly referring both to the thermal power and the *COP*, while Unit 3 displays higher thermal power values. As in [12], the considered heat pumps are selected in order to totally satisfy the thermal energy demand of the building, i.e. to guarantee the set-point temperature for the indoor air of 20°C when the outdoor ambient temperature reaches the winter design temperature, reported in Table 2.

2.3. Heating system

Figure 3 shows the layout of the heating system, consisting of the heat pump, the terminal units, given by 2-pipe three-speeds fan-coils, a circulation pump and a storage of 0.2 m^3. Two heating zones are considered, corresponding to the two floors of the building, and each thermal zone presents a fan-coil. Each emitter is characterized by an on-board controller, which selects the fan speed on the basis of indoor air temperature. The storage is the same for all considered scenarios, while fan-coils have been sized to satisfy the thermal power required by the single floor of each building. More in detail, for the building in Milan the two fan-coils have a nominal heating capacity of 10kW and 7kW, for the building in S. Benedetto each fan-coil has a nominal heating capacity of 5.5kW, while for Livigno the size of

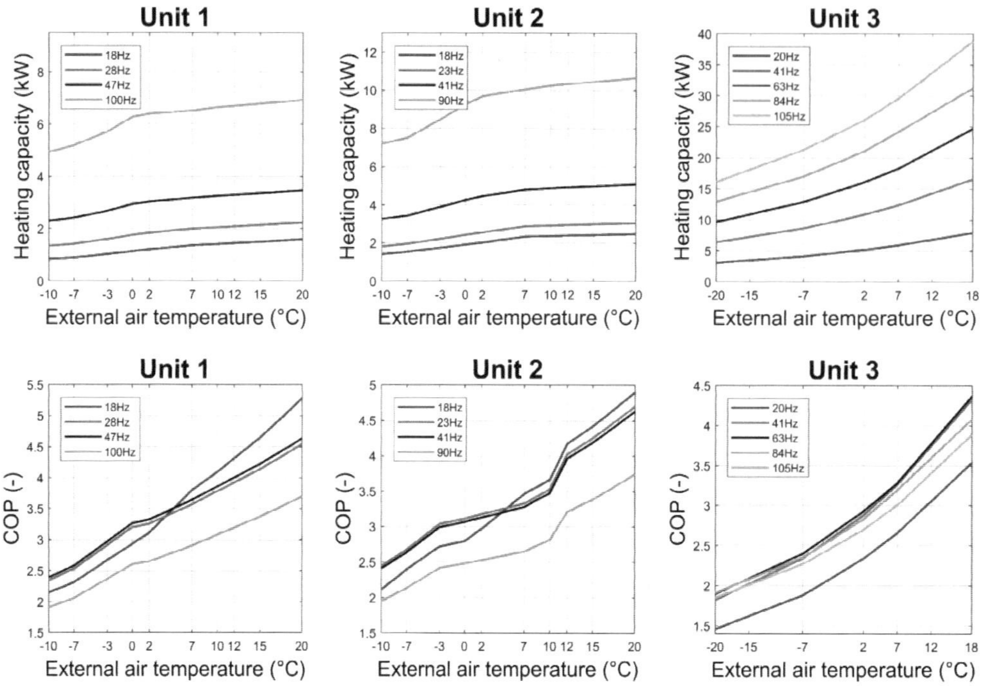

Figure 2. Heating capacity and *COP* of the three HPs used in simulations.

each unit increases due to the higher thermal load. In this location, the nominal thermal power of the fan-coils are equal to 12kW and 10kW, respectively.

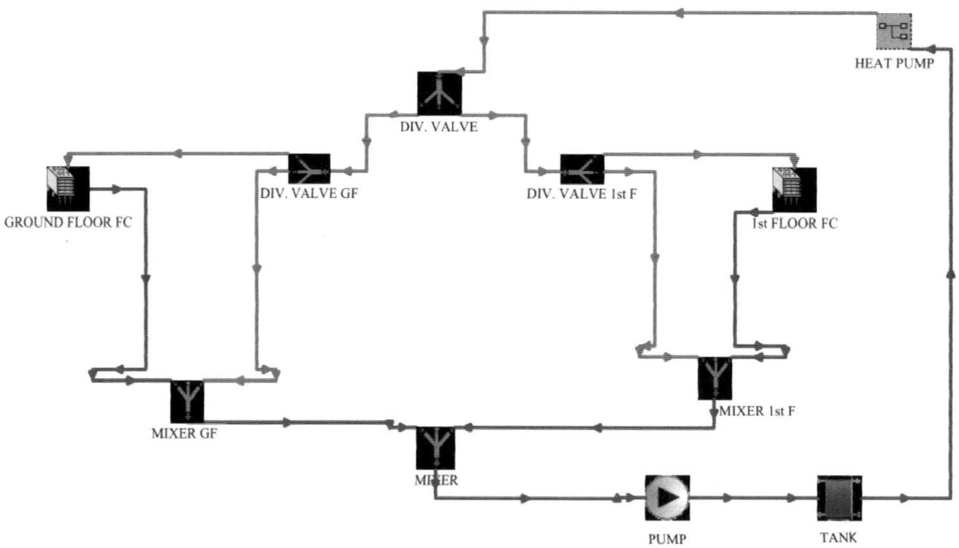

Figure 3. TRNSYS layout of the heating system.

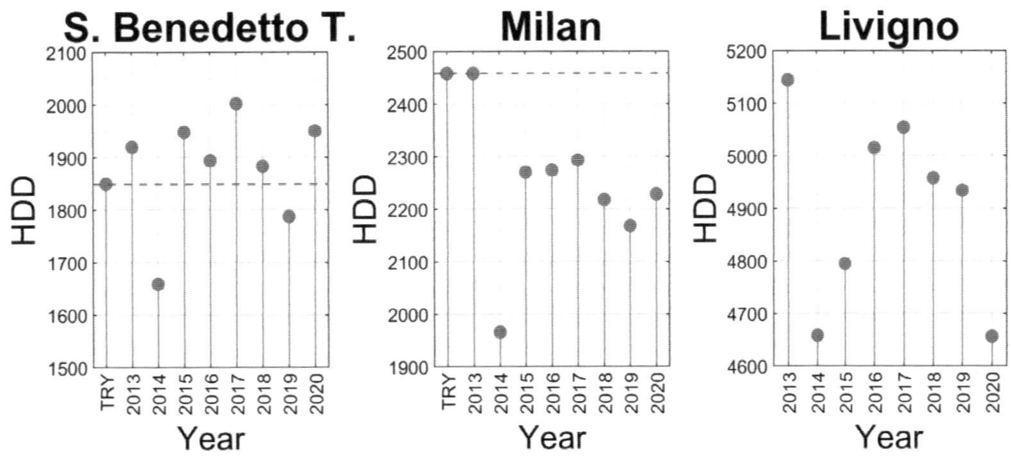

Figure 4. Heating Degree Days (HDD) of the three cities considered (period 10 October - 30 April), for years 2013-2020 and for TRY of S. Benedetto del Tronto and Milan.

2.4. Climatic data

The three locations considered in this work are classified, according to the current Italian law (i.e. DPR 412/1993), in three different climatic zones. Each zone is characterized by a different standard heating season. Table 2 shows the standard heating season, the external winter design temperature, the standard climatic zone and the degree days for the selected municipalities.

Table 2. Heating season, design temperature t_{des}, standard climatic zone and heating degree days (HDD) for the three considered municipalities.

Municipality	Heating season	t_{des} (°C)	Climatic zone	HDD
S. Benedetto del Tronto	1 November – 15 April	-1	D	1593
Milan	15 October – 15 April	-5	E	2404
Livigno	1 January – 31 December	-19	F	4648

The climate data for the three above mentioned municipalities are available in regional services websites: Arpa Lombardia[13] for Milan and Livigno and SIRMIP Marche [14] for San Benedetto del Tronto. In the present analysis, on account of the availability and quality of the data, attention is paid to

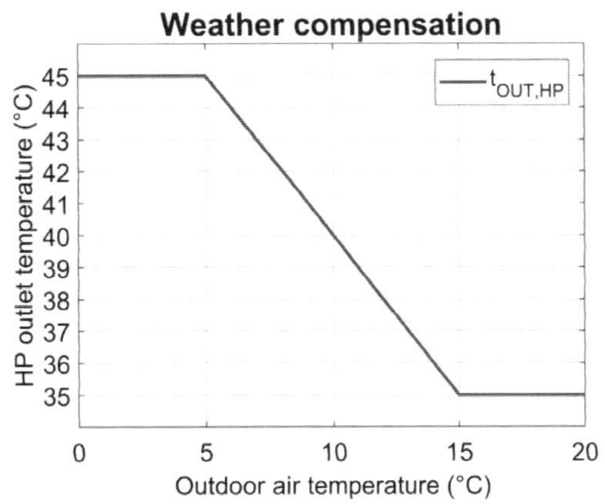

Figure 5. Weather compensation control used in simulations for HP outlet temperature.

8 years (2013-2020). For all locations, external temperature, global radiation on horizontal and relative humidity data are downloaded, as well as data on precipitation, wind direction and speed for Livigno and Milan. The quality of data is good, with less than 5% missing data for most years. The exception is the year 2018 for San Benedetto del Tronto, for which several values of radiation and relative humidity were not present. For this reason, this year was not considered due to a large lack of data.
The open data were used, through Meteonorm software [15], in order to develop the climate files to be used as input for TRNSYS. These climate files display real data, eventually integrated, in case of lack of data, with historical series related to the considered location.
From the open (and reworked) data, the value of the real heating degree days (HDDs) for each location and each year was obtained, considering a base temperature of 20°C, as is shown in Figure 4.

2.5. Setting of the analysis
Annual simulations are carried out, taking into account the same building and the different heat pumps presented in previous sections. The operating period of the heating system is assumed as explained in Table 2. The standard period corresponding to the climatic zone of each municipality is considered for simulations.
First, a simulation is carried out for each year of the considered period and for each location, considering a water supply temperature from the heat pump fixed at 45°C and a temperature difference with respect to the return of 5°C. Then, further simulations are also carried out, with reference to the same years, but assuming a climatic compensation, i.e. considering a decreasing water supply temperature from 45°C for an external temperature below 5°C up to 35°C for an external temperature above 15°C. The compensation curve is shown in Figure 5. The simulations were carried out considering a time step of 30 s.
Attention is focused on the thermal energy supplied by the heat pump and its electric energy use, both evaluated on a seasonal basis, and the seasonal performance factor of the heat pump (*SCOP*), defined as

$$SCOP = \frac{ET}{EE}$$

where *EE* (kWh) is the total annual electric energy supplied to the heat pump and *ET* (kWh) is the total annual thermal energy provided by the heat pump to the plant.
The influence of the heat pump defrost cycles on the seasonal energy performance of the unit has been neglected. The climatic data are, as already discussed, mostly complete and the integration with experimental radiation data, in addition to the ambient air temperature, helps to obtain a more accurate simulation of the behavior of the building.

3. Results and discussion
In Figure 6, the main results of the present analysis are shown for the fixed HP water outlet temperature; for Milano and San Benedetto del Tronto simulations considering the test reference year (TRY) are reported as well.
The figure shows an important variation along the years in SCOP and thermal energy supplied by the heat pump to the building, especially with reference to San Benedetto del Tronto and Milano.
In detail, let us discuss the case of S. Benedetto del Tronto. The simulations indicate an important variation in terms of thermal energy supplied from the heat pump to the heating system, with a value referred to TRY of 4800 kWh/year and a maximum value for the year 2013 (5674 kWh/year, increased by 18% referring to TRY) and a minimum value for year 2020 (3009 kWh/year decreased by 37% referring to TRY). Analysis also shows variations in obtained SCOP, with a value of 3.30 for simulation using the test reference year, a maximum value of 3.52 for year 2014 (+6.7% referring to TRY simulation) and a minimum value of 3.30 for year 2017 (equal to TRY simulation). Moreover, comparing SCOP values and thermal energy demand for San Benedetto del Tronto, Fig. 6 shows that a strict correlation between those two quantities cannot be established; in fact, for example, referring to years 2018 and 2019 SCOP values are similar but thermal energy supplied are very different.

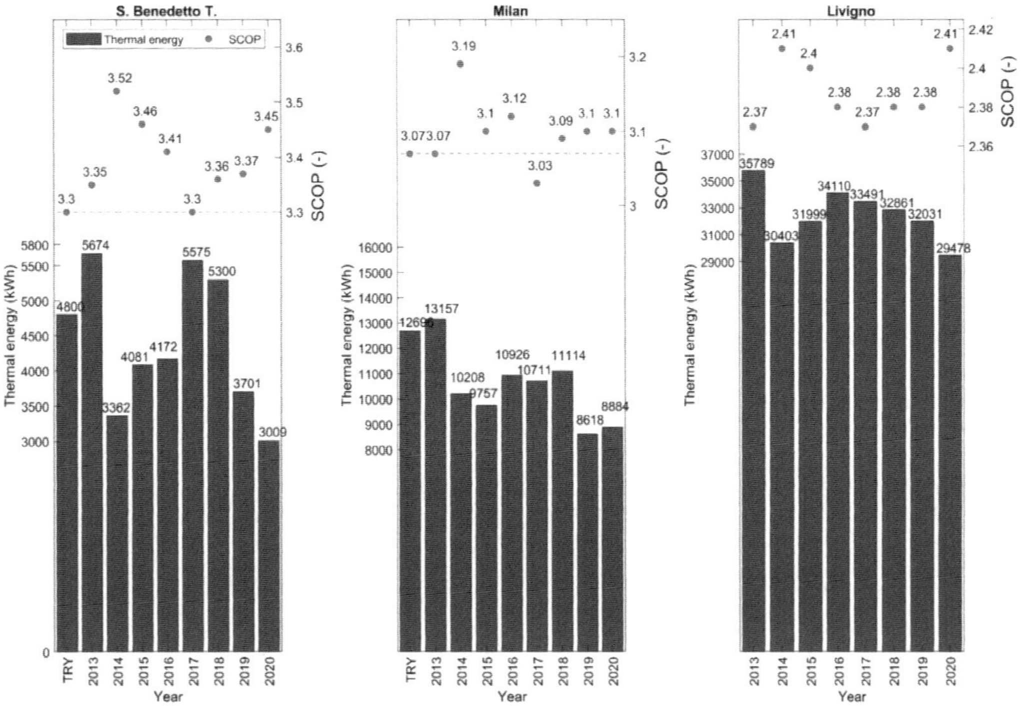

Figure 6. Results of simulations for selected municipalities (annual thermal energy and SCOP), considering an outlet temperature from the HP fixed at 45°C.

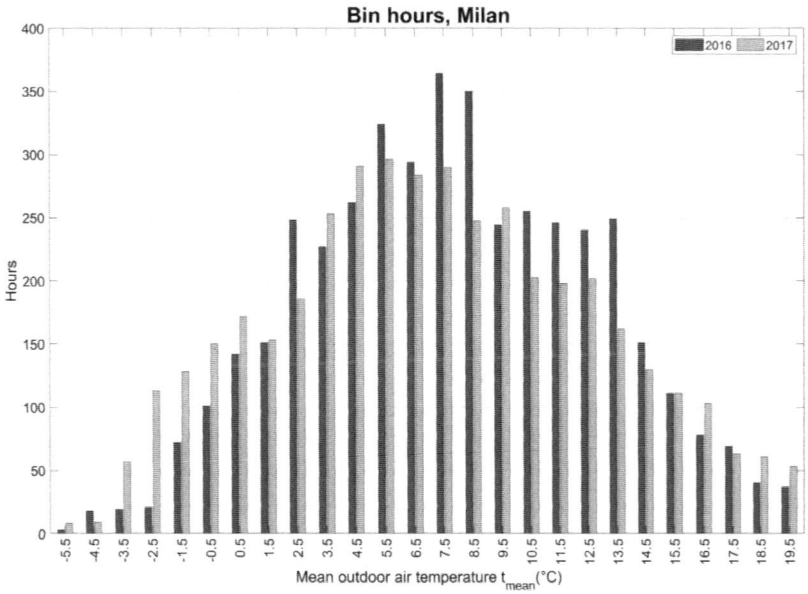

Figure 7. Bin hours for Milan, year 2016 and 2017.

Additionally, there is no correlation neither between HDD and SCOP, nor between HDD and thermal energy supplied.

Similar conclusions may be drawn with reference to Milan and for Livigno, but Livigno has the peculiarity that SCOP value show very little differences (range for Livigno 2.37-2.41, Milan 3.03-3.19,

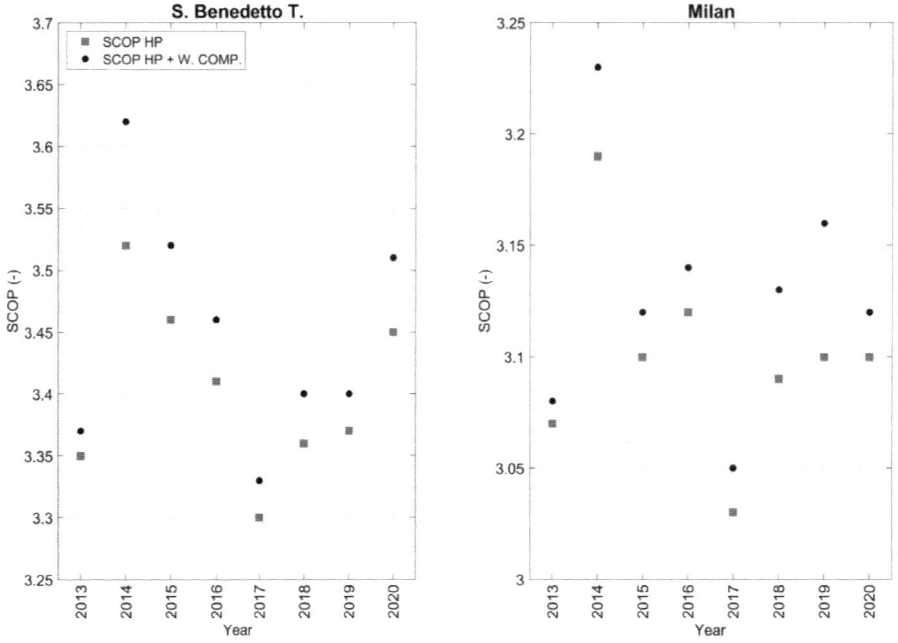

Figure 8. Comparison between HP with fixed outlet water temperature and weather compensation.

S. Benedetto del Tronto 3.30-3.52). In addition, results of simulation for Milan using the Test Reference Year show an annual thermal energy request of 12695kWh comparable to the thermal request of year 2013 (13157kWh).

An explanation of the absence of correlation between HDD and SCOP/thermal energy and between thermal energy/SCOP could be related to the number of hours in correspondence of which the heat pump is active at a given outdoor air temperature. In Figure 7, the number of hours in correspondence of which the outdoor air temperature is in a range of $(t_{mean} \pm 0.5)°C$ is reported, for year 2016 and 2017 and referring to Milano. It can be noticed that for several hours during 2017 the outdoor air temperature is lower than 1°C. For these values of outdoor air temperature, HP's *COP* are smaller than for higher outdoor air temperatures and this may display a relevant impact on SCOP.

Finally, simulations considering the weather compensation for S. Benedetto del T. and Milan are performed. The obtained results are shown in Figure 8. In the comparison with the case avoiding weather compensation, no substantial differences are noticed in terms of thermal energy supplied by the HP for heating; on the contrary, some little benefits may be noticed in terms of SCOP, both for Milan and S. Benedetto del Tronto, especially for years characterized by higher outdoor air temperature during heating season.

4. Conclusions

In the present paper, the effect of real weather data on the seasonal performance of HPs has been investigated. Simulations were done with TRNSYS and show little differences between different years, showing small variations in SCOP (less than 7%) but significant variations on annual thermal energy demand (up to 37%). The analysis also underlines the absence of a general correlation between annual thermal energy supplied by the heat pump and SCOP and between HDDs and SCOP.

Moreover, the use of weather compensation for the considered building does not offer significant benefits for the reduction of energy consumption, while is linked to only a slight increase of SCOP.

Further research might investigate the importance of real temperature data on the heat pump seasonal performance including the effect of defrost cycles. In fact, energy losses linked to defrost cycles could significantly affect the heat pump seasonal performance [16]. In future investigations we wish to double check how defrost energy losses are usually less relevant on seasonal energy performance for HPs installed in places characterized by cold and dry winters like Livigno and more relevant for HPs installed in places characterized by a humid climate.

References

[1] Thomas Nowak, Heat Pumps - Integrating Technologies to Decarbonise Heating and Cooling; European Heat Pump Association, 2018
[2] Thomas Nowak EHPA Market Report and Statistics Outlook 2019; European Heat Pump Association (EHPA), 2019
[3] UNI EN ISO 15927-4:2005, Prestazione termoigrometrica degli edifici – Calcolo e prestaione dei dati climatici – Parte 4: Dati orari per la valutazione del fabbisogno annuale di energia per il riscaldamento e il raffrescamento, 2005
[4] G. A. Mouzeviris, K. T. Papakostas, Air-to-water heat pumps: the impact of climate, compressor technology, water output temperature and sizing on the seasonal coefficient of performance for heating, IOP Conference Series: Materials Science and Engineering, Volume 997, 2020
[5] G. A.. Mouzeviris, K. T. Papakostas, Seasonal heat performances of air-to-water heat pumps in the Greek climate, IOP Conference Series: Earth and Environmental Science, Volume 410, 2019
[6] M. Dongellini, C. Naldi, GL. Morini, Seasonal performance evaluation of electric air-to-water heat pump systems, Applied Thermal Engineering, 2015
[7] C. Naldi, M. Dongellini, GL. Morini, Climate influence on seasonal performances of air-to-water heat pumps for heating, Energy Procedia, 2015
[8] Dott, R.; Haller, M.; Ruschenburg, J.; Ochs, F.; Bony, J. IEA-SHC Task 44 Subtask C Technical Report: The Reference Framework for System Simulations of the IEA SHC Task 44/HPP Annex 38: Part B: Buildings and Space Heat Load. IEA-SHC.; 2013
[9] Klein, S.A.; Duffie, A.J.; Mitchell, J.C.; Kummer, J.P.; Thornton, J.W.; Bradley, D.E.; Arias, D.A.; Beckman, W.A.;Braun, J.E.; et al.; TRNSYS 17: A Transient System Simulation Program; University of Wisconsin: Madison, WI, USA, 2010
[10] Klein, S.A.; Duffie, A.J.; Mitchell, J.C.; Kummer, J.P.; Thornton, J.W.; Bradley, D.E.; Arias, D.A.; Beckman, W.A.; TRNSYS 17—A TRaNsient SYstem Simulation Program, User Manual. Multizone Building Modeling with Type 56 and TRNBuild. Version 17.1; University of Wisconsin: Madison, WI, USA, 2010
[11] Dongellini, M.; Valdiserri, P.; Naldi, C.; Morini, G.L. The Role of Emitters, Heat Pump Size, and Building Massive Envelope Elements on the Seasonal Energy Performance of Heat Pump-Based Heating Systems. Energies 2020, 13, 5098, doi:10.3390/en13195098
[12] E. Rossi di Schio, V. Ballerini, M. Dongellini, P. Valdiserri, Defrosting of air-source heat pumps: Effect of real temperature data on seasonal energy performance for different locations in Italy, Applied Sciences 11, 8003, 2021
[13] Arpa Lombardia, https://www.arpalombardia.it/Pages/Meteorologia/Richiesta-dati-misurati.aspx
[14] Sirmip Marche, http://app.protezionecivile.marche.it/sol/indexjs.sol?lang=it
[15] Meteotest: J. Remund, S. Müller, M. Schmutz, D. Barsotti, C. Studer, R. Cattin. Meteonorm Handbook part I-II, 2020
[16] Vocale, P.; Morini, G.L.; Spiga, M. Influence of Outdoor Air Conditions on the Air Source Heat Pumps Performance. Energy Procedia 2014, 45, 653–662, doi:10.1016/j.egypro.2014.01.070

Journal of Physics: Conference Series 2177 (2022) 012026 doi:10.1088/1742-6596/2177/1/012026

Entraining of dust and steam flux in a water pool: experimental and numerical simulations of vapor condensation and dust deposition

Miriam Ibba[1*], Alessio Pesetti[1], Michele Raucci[1], Luca Berti[1], Guglielmo Giambartolomei[1], Donato Aquaro[1]

[1] Department of Civil and Industrial Engineering – University of Pisa
Largo Lucio Lazzarino 56122, Pisa, Italy

* Corresponding author's e-mail: m.ibba1@studenti.unipi.it

Abstract. This paper deals with experimental tests of steam condensation in a water pool at atmospheric pressure and temperature in the range 15-100 °C. The activity is performed in the frame of a research program, funded by the ITER organization, for the study of dust deposition, produced in the ITER Vacuum Vessel and entrained by the steam and non condensable gas into the Pressure Suppression Tanks in the case of a Loss of Coolant Accident. The steam condensation into a subcooled water pool has been investigated to characterise the condensation regimes occurring during dust deposition tests. The dust distribution on the tank walls strongly depends on the steam jet length and on the effective heat transfer coefficient. Few grams of dust reduce the water transparency, therefore separated tests without dust overcome this drawback. Measurements of the lengths and surfaces of the steam jets (which permit to calculate effective average heat transfer) have been performed by means of image analyses and have been compared with theoretical correlations. The comparison showed a good agreement between the experimental data and theoretical correlations. Useful data have been obtained for implementing numerical models of dust deposition and for understanding the dust distribution on the tank wall obtained in the experimental tests.

1. Introduction

Steam Direct Contact Condensation (DCC) in water pool has many applications in nuclear field, mainly the pressure suppression systems of Boiling Water Reactor (BWR). A similar application, but at sub-atmospheric pressure, is envisaged in the ITER nuclear fusion reactor, presently under construction. The main advantage of DCC is the great heat exchange between the steam and the subcooled water. The heat transfer coefficient is also enhanced by the turbulence at the steam-liquid interface.

Many researchers have studied the DCC at atmospheric pressure. Kim, H. Y. et al. [1] and Wu, X. Z. et al. [2] determined empirical correlations of the length of the steam cavity, also called "jet plume", which is directly related to the heat transfer coefficient. Condensation regime maps versus the steam mass flow rate and the water temperature have been reported by Cho, S. et al. [3].

Steam direct condensation at sub-atmospheric pressure was mainly studied at the Laboratory "Bruno Guerrini" of Department of Civil and Industrial Engineering (DICI)-University of Pisa [5, 10]. These studies, funded by ITER Organization, refer to the qualification of the ITER Vacuum Vessel Pressure Suppression System (VVPSS). This safety system has been designed to manage the Ingress of Coolant Event (ICE cat. IV) accident, postulated to occur in the case of a rupture of the cooling piping of shielding blanket or the divertor.

During this event, the dust produced and accumulated inside the Vacuum Vessel (VV), could be transported into the Pressure Suppression Tanks. The dust deposition and the necessary activities of decontamination are considered important issues from the safety point of view.

In the frame of the research activity on DCC, performed at the DICI of the University of Pisa, an experimental campaign is being carried out, funded by the ITER Organization, aiming to analyze the dust deposition and decontamination in a reduced scale pressure suppression system [11]. The experimental tests consist of introducing air and/or steam and dust into a condensation prismatic tank (Fig. 1 and Fig. 2) partially filled of softened water. The main objectives are to determine the amount of dust deposited on different surfaces of the tank, the amount of dust entrained during the water discharge as well as by the air flow during the scrubbing operation.

The size of the dust, accumulated in the Vacuum Vessel, varies in a very wide range from tens of nanometers to hundreds of micrometers and its shape is flake-like, irregular or, sometime, spherical. The composition of these particles reflects that of the Vacuum Vessel walls and that of the substances used for plasma conditioning. Tungsten and Beryllium produced by the erosion of the divertor and first wall are the main compound of the dust. In the case of accidental events in the VV, like Loss of Coolant Accident (LOCA), above all the beryllium dust (mainly as beryllium oxide) is transported to the pressure suppression system and to the draining system.

Figure 1 – Experimental set up used for the dust deposition by means of a steam or air flux

The beryllium or beryllium oxide, characterized by a high chemical reactivity and toxicity, have been simulated by means of Aluminum oxide (Al_2O_3) dust, having suitable granulometry.

This paper is focused on the characterisation of the condensation regimes occurring during the tests of the dust deposition. Experimental steam condensation tests, in a water pool at 15-100 °C of temperature and at atmospheric pressure, have been carried out without the dust injection. The condensation regimes, which occur during the dust deposition, have been analysed by video cameras. These separated tests overcame the poor water transparency drawback occurring during the alumina injection (Fig. 3). Measurements of the lengths and of the surfaces of the steam jets, performed by means of image analyses, have been compared with theoretical correlations and permitted to obtain useful data for the numerical simulations of dust deposition [11].

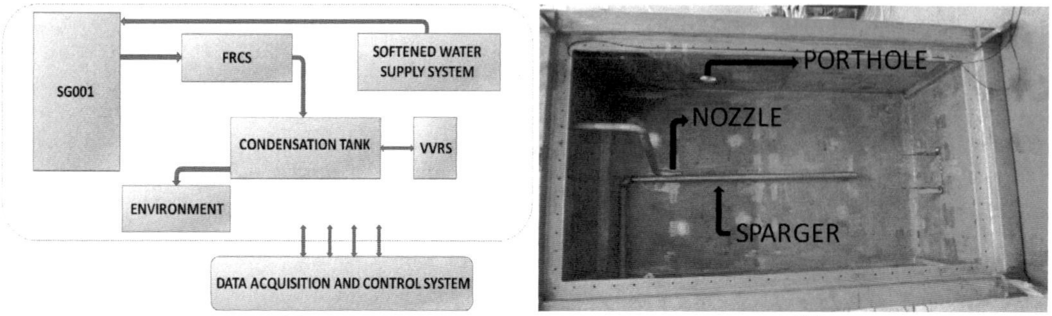

(a) (b)

Figure 2 – Block diagram of the experimental set up (a) and location of the nozzle and sparger in the condensation tank (b).

Figure 3 – Injection of steam and dust in the condensation tank

2. Experimental set up for dust deposition and decontamination tests

Figures 1 and 2 illustrate the experimental set up, located at the "B. Guerrini" Laboratory of the DICI-University of Pisa, used for the tests of deposition and decontamination of the dust. The main components are: a condensation prismatic tank (CT) (about 3.5 m^3 of volume); a superheated Steam electric Generator (SG001); a flow rate control system (FRCS); a dust auger doser (DAD), temperature and pressure sensors, a visualization video recording system (VVRS) and a data acquisition and control system (DACS).

The condensation tank (1.460 m wide, 2.3 m long and 0.99 m high), insulated by rock wool, is filled with 1.73 m^3 of softened water (about the 50% of the internal volume). The steam-air-dust flow (separately or mixed each other, depending on the test) enters in the condensation tank through a central steam nozzle (10 mm of diameter and located at 250 mm from the lower surface of the tank). Steam flow, for water heating purposes, or air flow in dust scrubbing tests are sent into the tank through a 40.9 mm of internal diameter horizontal holed sparger, located in the middle of the tank, at about 100 mm from the bottom surface. The horizontal sparger is provided by 92 holes located on two rows 25° apart in the circumferential direction. The holes diameter is 4 mm and the longitudinal pitch is 30 mm (Fig. 2). The design specifications of the superheated Steam electrical Generator (SG001) are: 130 kW rated power; 50 g/s of steam mass flow rate at a pressure of 1.5 bar absolute and 150 °C of temperature. SCRs allow to control the Steam Generator power.

The Flow Rate Control System (FRCS) is made of two pipe lines each designed for a different flow rate range:

- high flow rate range 5-50 g/s, controlled by a vortex Siemens flow rate sensor, with an accuracy of about 3%, and a Spirax-Sarco control valve;

- low flow rate range 0.1-7 g/s, controlled by an Endress-Hauser Coriolis flow rate sensor, with an accuracy of about 5%, and a Spirax-Sarco control valve.

The two control lines, thermally insulated with rock mineral wool, are also equipped by pressure and temperature sensors (uncertainties of measurements of ± 0.1 kPa and ± 0.1 °C respectively), duly calibrated according to the International standards and externally heated by means of electric resistances. The dust flow rate is controlled by means of an auger doser (0.1-1.6 g/s of range) and it is entrained in the tank by a flow of steam or air.

The visualization video recording system (VVRS) consists of two video cameras located outside the CT, one on the lateral side of the tank and the other on the upper cover, both in correspondence of glass portholes. Two screens allow to constantly monitor the test. The data acquisition and control system is implemented by the NI® Labview code.

3. Experimental tests

As previously mentioned, the deposition of the dust, entrained by a steam or a non-condensable gas flow, occurs in the actual safety system of ITER (VVPSS) over a wide range of temperature (15-100 °C) and pressure (10-150 kPa absolute). The experimental tests, performed with the experimental set-up described in the previous paragraph, have been performed at atmospheric pressure and in the range 15-100 °C. Dust was injected with steam in discontinuous steps through the nozzle, between which only steam was entrained in the condensation tank. Dust batches of about 200-300 g, until a total amount of 3 kg, were injected at different temperatures during a test, on the full temperature range 15-100 °C. Between two injections of dust batches, the steam was also flowed into the horizontal sparger to increase the water temperature more quickly. The dust deposition on the tank walls and on the internal components depends strongly on the steam thermodynamic conditions as well as the jet characteristics.

Few grams of dust reduce the transparency of the water and do not permit to visualize the steam jet (Fig. 3). For this reason, a complete test was performed without dust. In about 6 hours of test, the water temperature reached 100 °C starting from an initial temperature of 15 °C. As in the tests with dust, the steam flowed into the tank through the nozzle alone or with the sparger at interval of temperature. The complete test has been analysed considering 13 different water temperature ranges: 5 ranges refer to the steam flowing only through the nozzle and 8 temperature ranges refer to the steam injected at the same time from the nozzle and sparger. Table 1 reports the mass flow rates and the steam temperatures values corresponding to all the examined ranges.

Moreover, in the same Table, the condensation regimes have been identified by means of the map proposed by Cho, S. et al. [3], for DCC at atmospheric pressure versus the steam mass flow rate per unit of flux area, G, and the water temperature, T_w. This map has been derived [3] from horizontal steam jet, nozzles diameter from 5 to 20 mm and water temperature in the range 0–100 °C. The condensation regime is stable (SC) at high value of G and small value of T_w. Reducing G, no stable condensation regimes (Condensation Oscillation (CO), Transitional Chugging (TC), Chugging (C)) occur. At high water temperature, greater than 80 °C, the regime is called Interfacial Oscillation Condensation (IOC) at high G and Bubbling Condensation Oscillation (BCO) at small G. A non condensation area is located at a T_w greater than 90 °C.

Figure 4 illustrates the condensation regimes of the 13 analysed T_w ranges obtained plotting the experimental data on the Cho map. The steam mass flow rate of the nozzle is equal to the critical value. Therefore, the condensation regimes have the same G and move on vertical line remaining until 80 °C in the CO area. For T_w between 80 and 90 °C the regime is BCO. The steam injected through the sparger holes condenses in a more unstable area (Chugging and Bubbling Condensation Oscillation).

		Regime	T_w [°C]	T_s [°C]	q_s [g/s]	G [kg/(m²s)]	L/D	HTC [kW/(m²K)]
1	N	CO	25.3	108.2	13.3	169.5	6.5	986.5
2	N	CO	34.9	104.7	13.3	169.5	6.2	944.6
	S	C			18.6	16.1	/	/
3	N	CO	39.2	105.7	13.3	169.5	5.1	1186.8
	S	C			20.9	18.1	/	/
4	N	CO	45.8	108.4	13.3	169.9	5.8	874.9
5	N	CO	52.9	104.9	13.3	169.9	4.3	/
	S	C			12.4	10.7	/	/
6	N	CO	57.9	106.3	13.3	169.9	4	931.5
	S	C			13.8	11.9	/	/
7	N	CO	62.6	108.3	13.3	168.8	2.5	873.9
8	N	CO	70.3	106.0	13.3	168.8	2.7	/
	S	C*			13.3	11.6	/	/
9	N	CO	78.4	108.4	13.5	171.8	1.9	629.64
10	N	BCO*	82.7	106.4	13.3	168.7	2.3	/
	S				12.2	10.6	/	/
11	N	BCO	90.2	106.5	13.3	171.8	1.8	/
	S	BCO*			11.9	10.3	/	/
12	N	NC	96.1	108.5	13.3	169.2	2	671.6
13	N	NC	98.9	106.6	13.3	169.2	1.8	906.6
	S				11.1	21.1	/	/

Table 1 – Summary of main experimental results. (N=nozzle; S=sparger)

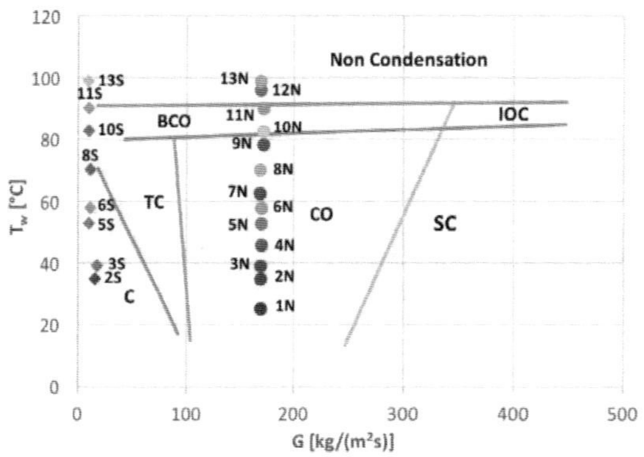

Figure 4 – Steam condensation regimes obtained during the tests

Figure 5 shows the steam jet shapes at the different condensation regimes (labelled versus the T_w values). As the temperature increases the steam plume gets greater and greater but it remains close to the nozzle.

4. Discussion of the results

The experimental results show that also with constant steam mass flow rate (critical value), the steam plume changes with the water temperature and the dust can settle at different distant from the nozzle.

The distribution of the dust on the tank walls depends strongly on the path of the steam in the water. A numerical model of the experimental rig is implemented and simulations are performed using the computer code ECART [12], validated as a best estimate source term code for safety analysis of nuclear fission and fusion reactors.

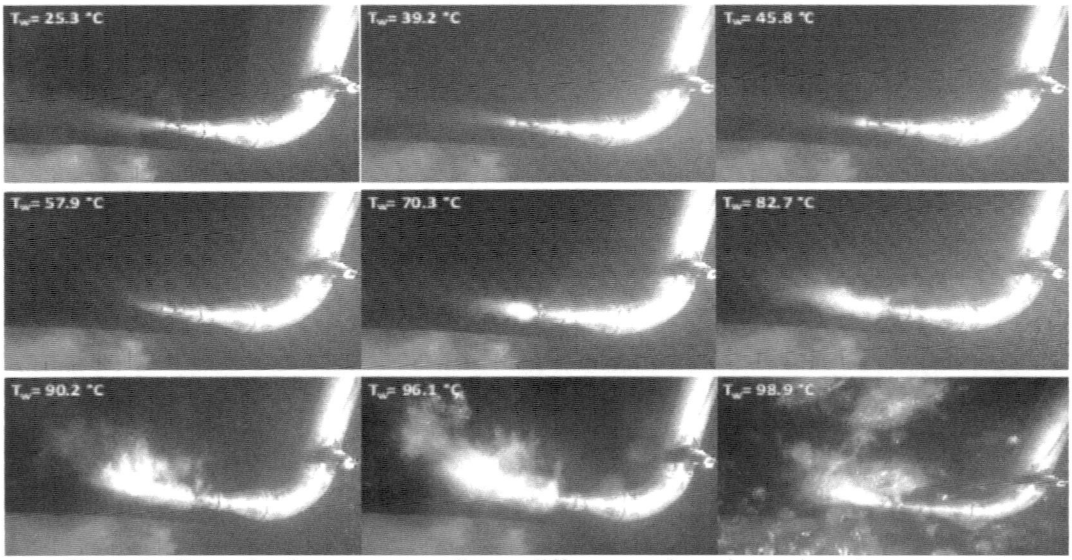

Figure 5 – Steam jet shape corresponding to different condensation regimes (temperatures)

Figure 6 shows a numerical simulation of the injection of air and Al_2O_3 dust in water with a fine grain size particles (lognormal distribution with mean=0.1 μm and standard deviation σ=2). The 92.42% of the dust is deposited in the water in the condensation tank (Vol-4 Tank), the 7.58% is retained in the sparger piping (Vol-2 and Vol-3 piping). The condensation tank (Vol. 4) is considered as a single volume and, therefore, they do not discriminate the walls at different distance from the nozzle [11]. Actually, the experimental results show a variable distribution of the dust (greater in front of the sparger at greater distance from it). Therefore, very important parameters for understanding the experimental results are the jet length and the effective heat transfer coefficient.

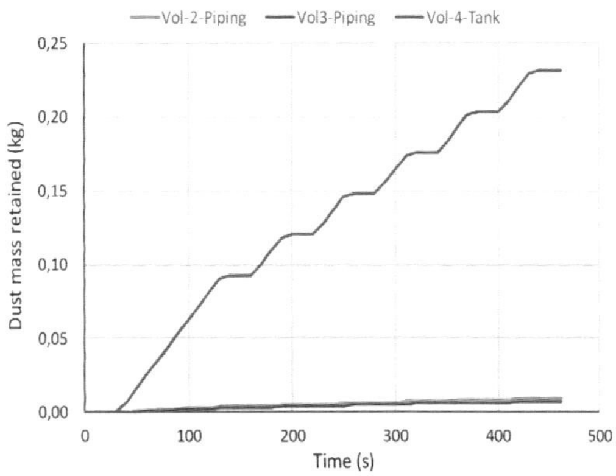

Figure 6 – Dust retained on the tank walls (numerical simulations [11])

An open-source software for image elaboration (ImageJ-National Institute Health-USA [13]) has been used to determine, at different water temperature, the steam jet profile (estimating the 3-D surface) and its penetration length L, referred to the nozzle internal diameter, D (Table 1).

Kim, H. Y. et al. [1] and Wu, X. Z. et al. [2] determined theoretically and experimentally the penetration length of steam jet flowing by submerged horizontal nozzle in quiescent subcooled water. The steam jet shape depends on steam mass flow rate, nozzle direction, water head at the nozzle level and the degree of subcooling of pool water.

Kim, H. Y. et al. [1] proposed (for nozzles having diameter in the range 5-20 mm; steam mass flux, G, in the range 250-1188 kg/m²s and pool water temperature in the range 30-80 °C) the following empirical equation:

$$\frac{L}{D} = 0.503B^{-0.70127}\left({G}/{G_m}\right)^{0.47688} \quad (1)$$

Wu, X. Z. et al. [2] proposed (for nozzles having diameter of 11 mm; steam mass flux, G, in the range 133-370 kg/m²s and pool water temperature in the range 20-50 °C) the following empirical equation:

$$\frac{L}{D} = 0.1B^{-1.37}\left({G}/{G_m}\right)^{1.86} \quad (2)$$

The above correlations are based on the dimensionless condensation driving potential, B, and on ratio between steam mass flow rate and its critical value, G/G_m. The condensation driving potential represents the condensation capacity defined by the equation:

$$B = C_p \frac{\Delta T_{sub}}{H_{fg}} \quad (3)$$

where C_p is the specific heat of water, ΔT_{sub} is the water subcooling and H_{fg} is the latent heat of vaporization. In our test, the ratio G/G_m is equal to 1, being the steam mass flow rate equal to the critical value.

Kim, H. Y. et al. [1] proposed for heat transfer coefficient (for horizontal and vertical nozzles having diameter in the range 5-20 mm; steam mass flux, G, in the range 250-1188 kg/m²s and pool water temperature in the range 35-80 °C) the following empirical equation:

$$h_{ave} = 1.4453C_pG_mB^{0.03587}\left({G}/{G_m}\right)^{0.13315} \quad (4)$$

Kim, Y. S. et al. [14] proposed a similar empirical equation with different coefficients (for horizontal nozzles having diameter in the range 1.35-7.65 mm, steam mass flux up to 1500 kg/m²s and pool water temperature in the range 13-87 °C):

$$h_{ave} = 1.3583C_pG_mB^{0.0405}\left({G}/{G_m}\right)^{0.3714} \quad (5)$$

where C_p is the isobaric heat capacity and G_m is the critical steam mass flux.

Figure 7-a shows the comparison of steam penetration length normalized to the nozzle diameter versus the parameter B, as determined experimentally and calculated by the correlations 1) and 2). The effective average heat transfer coefficient, h_{ave}, for different T_w ranges (Fig. 7-b) has been determined applying the following formula:

$$h_{ave} = \frac{G\ A_e(H_s-H_f)}{A_i(T_s-T_w)} \quad (6)$$

being

A_e: is the area of the nozzle hole;

A_i: is the jet surface area;

H_s, H_f: are the steam and water enthalpy, J/kg.

The comparison of the L/D values shows that both the correlations (1) and (2) agree very well to the experimental results. In particular, the correlation (1) fits the experimental results in the whole B range, while the correlation (2) is valid in the 20-50 °C water temperature range and, therefore, it does not give realistic values out this range (for B smaller than 0.032, corresponding to $T_w > 90$ °C).

The experimental average heat transfer coefficients (Fig.7-b) are in good agreement with the value determined by both the correlations (4) and (5). The maximum difference between experimental and theoretical values is smaller than 25%.

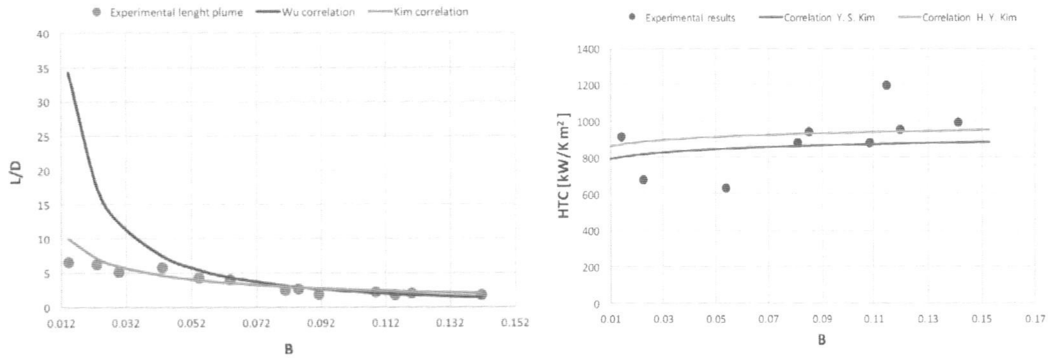

(a) (b)

Figure 7 – Comparison of experimental and predicted values of L/D (a) and HTC (b).

5. Conclusions

This paper deals with experimental tests of steam condensation in a water pool at atmospheric pressure and temperature varying between 15 and 100 °C. The activity is performed in the frame of a research program, funded by the ITER Organization, for the study of the deposition of the dust, produced in the ITER Vacuum Vessel, and entrained by the steam and non condensable gas into the Pressure Suppression Tanks in the case of a Loss of Coolant Accident.

The main aim of the tests described in the paper is to determine the steam condensation regimes during dust injection in the pool. In fact, the distribution of the dust on the wall tank depends strongly on the steam jet length and on the effective heat transfer coefficient. The experimental set up is provided by video cameras but few grams of dust reduce the water transparency. Therefore, separated tests without dust overcome this drawback. Measurements of the lengths and of the steam jets surfaces (which permit to calculate effective average heat transfer), by means of image analyses, have been compared with theoretical correlations.

The comparison showed a good agreement between the experimental data and theoretical correlations. Therefore, these correlations could be used also for determining these parameter (L/D and HTC) at different conditions in order to obtain useful data for implementing numerical models of dust deposition and for understanding the dust distribution on the wall tank obtained in the experimental tests.

References

[1] Kim, H. Y. et al., 2001, Int. J. Energy Res., 25(3), pp. 239–252.
[2] Wu, X. Z. et al., 2009, Experimental Thermal and Fluid Science, 34 (2010), 10-19.
[3] Cho, S. et al., 1998, First Korea-Japan Symposium on Nuclear Thermal Hydraulics and Safety, NTHAS98, Pusan, Korea.
[4] Du, Y. et al., 2020, Front. Energy Res., 8, p. 119.
[5] Pesetti, A. et al., Fusion Engineering and Design, Volume 168, Year 2021, Doi: 10.1016/j.fusengdes.2021.112639.
[6] Pesetti, A. et al., Fusion Engineering and Design, Volume 171, October 2021, Doi: 10.1016/j.fusengdes.2021.112523.
[7] Pesetti, A. et al., Journal of Physics: Conference Series, 1868 (2021) 012018. Doi:10.1088/1742-6596/1868/1/012018.
[8] Giambartolomei, G. et al., Journal of Physics: Conference Series, Volume 1599, Issue 1, 22 September 2020, Doi: 10.1088/1742-6596/1599/1/012024.

[9] Pesetti, A. et al., Fusion Engineering and Design, Volume 158, September 2020, Doi: 10.1016/j.fusengdes.2020.111746.
[10] Pesetti, A. et al., Proceedings of 27 the International Conference on Nuclear Engineering, ICONE, Code 150404, Tsukuba, Japan, 19-24 May 2019.
[11] Ibba, M. et al., Journal of Nuclear Engineering and Radiation Science, NERS-21-1025. Doi: 10.1115/1.4050768.
[12] Paci, S. et. al., "Validation of the ECART code for the safety analysis of fusion reactors", Fusion Engineering and Design, **75–79**, 2005, pp. 1243–1246.
[13] ImageJ, National Institute Health-USA, https://imagej.net/
[14] Kim, Y. S. et al, Journal of the Korean Nuclear Society Vol. 29, Number 1, pp. 45-57, (1997).

Journal of Physics: Conference Series 2177 (2022) 012027 doi:10.1088/1742-6596/2177/1/012027

elastoCaloric (eC) Cooling: recent studies and preliminary results on materials and devices

Luca Cirillo[1*], Adriana Rosaria Farina[1], Adriana Greco[1], Claudia Masselli[1]

[1]Department of Industrial Engineering, University of Naples Federico II, Piazzale Tecchio 80, 80125, Napoli, Italy
Corresponding author: luca.cirillo2@unina.it

Abstract. Refrigeration is a process aimed at lowering the temperature of an environment compared to the outside temperature. It has always played a fundamental role in modern industry: food, for the preservation of food; in the air conditioning of buildings, to make them habitable and even in the field of medicine and biology, for the preservation of samples and particular therapies. Today refrigeration systems, are responsible for about 17% of the world's electricity consumption. Therefore, the scientific community is researching solid-state refrigeration, which is 50% more efficient than vapour compression systems. Solid-state refrigeration is based on some materials' caloric effect, which can release or absorb latent heat when subjected to an external field (magnetic, electric, pressure or strain) release. Elastocaloric Refrigeration is one of the most minor explored frontiers, unlike magnetocaloric and electro-caloric systems. Some preliminary studies confirm that the eC systems have the advantage of being able to decrease the intensity of the external field (with a relative increase of the COP). In this work, a review on Ec systems is carried out, with particular attention to the description of the properties of the materials used, the various types of stress application. The results have been presented in terms of temperature, entropy, latent heat and COP.

1. Introduction

To date, a quarter of the total electricity consumption already goes to cooling devices in any form. With global warming underway, the use of cooling and refrigeration systems is expected to increase even more (could double by 2050, according to the United Nations Environment Program). In this regard, scientists and engineers are working on new cheaper, cleaner, and more efficient refrigeration technologies [1, 2]. The point is to abandon vapour compression technology for a new technology that should be completely new. Hydrochlorofluorocarbons (HCFCs) are gases used worldwide in refrigeration, air-conditioning, and foam applications, but they are being phased out under the Montreal Protocol, depleting the ozone layer. HCFCs are both ODS (Ozone Depletion Substances) and potent greenhouse gases: the most used HCFC is nearly 2000 times more potent than carbon dioxide in terms of its Global Warming Potential (GWP). HFCs that will suffer an 85% reduction by 2036 are currently the most widely used types of refrigerants [3-8]. The scientific community in recent decades has been strongly oriented to find new sustainable solutions in the field of refrigeration so that current systems based on vapour compression can be improved, as no types of refrigerants that are dangerous for the environment are used. From this point of view, solid-state refrigeration represents one of the most exciting alternatives and is gradually gaining enormous success worldwide [9]. Some studies show that solid-state refrigeration guarantees 50% higher energy efficiency than traditional systems. In addition, the low maintenance and high helpful life of these systems also guarantee excellent reliability. Among the advantages above, solid-state systems have a very minimal acoustic impact. This fact allows them to be used in healthcare facilities, for worship, and in general in all those places where the acoustic impact is quite significant. The determining factor of solid-state refrigeration is the caloric effect of the materials used: a thermal effect capable of supplying or releasing heat if subject to an external field (magnetic [10], electrical [11], pressure [12], or simply axial stress [9]). To these effects is coupled magnetocaloric [13–15], electrocaloric [16], elastocaloric and barocaloric refrigeration or heat pumping [17, 18]. Among the caloric materials, there are the elastocaloric, which subjected to stress cycles produce significant temperature variations; moreover, unlike magnetocaloric or electrocaloric materials, they do not need an applied field rather expensive from an energy point of view. They are also called SMA (Shape Memory Alloy): the elastic effect depends on the austenitic stability temperature (A_f). A SMA is superelastic when its martensite temperature must be below a specific M_f value (Figure 1a). These temperatures show a direct correspondence with the effort applied for phase change (austenite-martensite), making these materials enjoyable in various engineering sectors, from the automotive to the medical sector [13]. In this work, the aim is to collect and analyse all the recent research activities based on eC refrigeration by reference to engineering parameters such as COP, type of application stress (tensile, compressive or torsion). There are many numerical studies, but few research groups have created a prototype of an eC cooler. This study will try to define the pros and cons of this new solid-state technology to realise an experimental prototype for industrial and residential applications in the future.

2. Elastocaloric Cooling Cycle – AeR

The eC cycle is similar to that used for solid-state refrigeration through magnetocaloric [21] or electro-caloric materials. The reference model and the simplest of the caloric cycle is that of Brayton illustrated in figure 1. The Brayton thermodynamic cycle comprises four processes: loading, heat exchange with constant load; unloading; heat exchange with constant load.

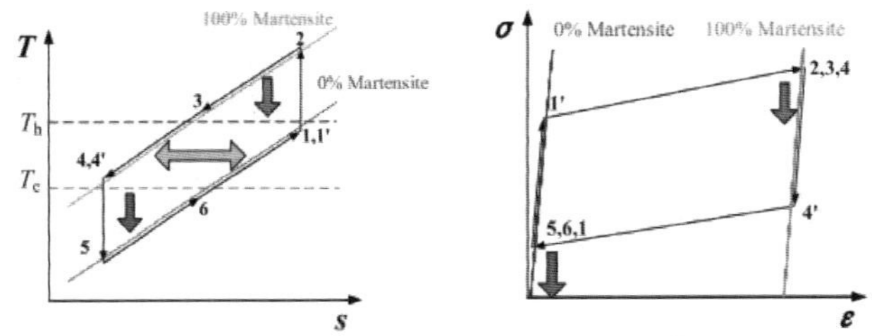

Figure 1: Typical Brayton cycle for elastocaloric cooling

- Loading: the eC material is subject to mechanical stress in adiabatic conditions. The transition from the austenite phase to the martensite phase takes place at a specific stress value. The generated effect is the increase in the temperature of the material by a certain amount ΔT_{ad};
- Heat Transfer at Constant Load: the auxiliary fluid (air or water) exchanges heat with the eC material. The generated effect is the enhancement of the fluid temperature and, at the same time, the decrease of the eC material temperature;
- Unloading: the applied stress is reduced to the eC material, making the transition from the martensite phase to the austenite phase. The generated effect is the decrease of the eC material temperature;
- Heat Transfer at Constant Load: the auxiliary fluid exchanges heat with the eC material. The generated effect is the decrease of the fluid temperature, which is lower than the temperature of the environment to be refrigerated.

By neglecting the irreversible contributions to the entropy change, the absorbed heat q_{ad} during the transition from phase four to phase one derived from the following equation:

$$q_{ad} = \int T\, ds \tag{1}$$

The COP of the eC cooling cycle defined as:

$$COP = \frac{q_{ad}}{w_0} \tag{2}$$

3. Recent studies on Shape Memory Alloys (SMAs) and their elastocaloric properties

The critical point of the eC cooling system is identifying the suitable material to be used. Therefore, the study of the characteristics of SMAs is one of the most intense research activities in the eC field. The most recent studies show that the chemical composition of the material plays a fundamental role in obtaining the eC effect, avoiding the phenomenon of thermal and mechanical hysteresis, and improving production processes. The elastocaloric effect (eCe) can be defined as the temperature change (ΔT_{ad}) in the eC material following the application of the uniaxial stress in an adiabatic transformation. The material increases its temperature during loading and decreases it during unloading. If the process is isothermal, an entropy variation in the material itself is detected.

Table 1 reports as comprehensive as possible of SMA materials together with the eCe shown, while they are arranged as tube, rod, wire, and bulk. The table reports only recent experimental results. Binary alloy of Ni-Ti, are the benchmark Shape Memory Alloys (SMA) with elastocaloric properties at room temperature. The elastocaloric materials can be classified in: Shape Memory Alloys (SMA) and in Shape Memory Polymers (SMP). The SMA class can be further divided in Ni-Ti-based (doped with Cu, Co, Pd, etc.), Cu-based (doped with Al, Ni, Zn, Mn, etc.) alloys. Nickel-Titanium alloys can be binary, ternary or quaternary. With respect to the binary, ternary alloys with the addition of vanadium or aluminum to the alloy helps to reduce the stress hysteresis and to enhance fatigue life. The further addendum of a fourth element, like copper, allows to shift the operative temperature range where the Austenite Martensite transformation occurs . Among the Cu-based elastocaloric alloys the attentions goes on the ternary Cu-Al-Mn. The presence of Aluminum ensures an enhancement of the heat exchange coefficients. The SMP is a class embracing the so-called soft materials like rubber or polymers. As general consideration, these materials are cheaper than Ni-Ti alloys as well as improved is the fatigue life; the major limit is represented by the much larger elongations (volume variations) needed to make the elastocaloric effect manifesting (up to 600%) that make them not to easy to be employed in experimental devices. PVDF-based polymer shows different caloric effects, including elastocaloric and barocaloric, due to their polymorph properties [36]. Additionally, PVDF has other advantages: ease of fabrication (shape), low cost, and low processing temperatures. Further, PVDF being a cross-linked polymer (with high crystallinity) promises the high potential for eC refrigeration applications. Moreover, as Chluba et al. in 2015 [21], the copper addition carries a remarkable enhancement of the fatigue-life allowing the material to experiment up to 106 loading-unloading cycles. On the other hand, the expense to be paid reduces latent heat concerning the binary Ni-Ti. Also, a ternary NiTi-based alloy where this element is Pd was studied. However, the alloys were considered favourable for energy application since the consequence observed was an augmentation of the stress-strain hysteresis [22]. Recently, the quaternary Ni-Ti alloys are under investigation. Frenzel et al. [23] noticed that the Co addition to the ternary Ni-Ti-Cu alloy results in a shifting possibility of the transformation temperatures is always observing high latent heat values. A strong dependence of transformation temperatures and latent heat on Ni concentration has been demonstrated in the study reported in Figure 2. Their study shows a correlation between transformation temperatures, A_f, and the latent heat ΔH (Figure 2a). Besides, their results show a linear behaviour by adding Ni, Cr, Co, Fe, and V to stoichiometric NiTi: a decrease in A_f is always associated with a decrease in latent heat. In binary Ni-Ti shape memory alloys, the latent heats ΔH of the transformation from austenite to martensite decrease when Ni is added to stoichiometric NiTi; this decrease in ΔH directly results in a decrease of A_f. Figure 2b shows the latent heat of transformation from austenite to martensite as a function of the width of the thermal hysteresis ΔT_H. Low thermal hysteresis widths and large latent heat in the eC cooling are fundamental to choose the SMA materials to obtain a high cooling efficiency because thermal hysteresis is directly proportional to the width of the mechanical hysteresis, as demonstrated in the experimental analysis of Jaeger et al. [24]. From the literature, it is possible to determine a good compromise of SMA eC material with a high coefficient of performance (COP): the quaternary SMA NiTiCuV. The quaternary SMA NiTiCuV provides promising eC properties and shows minor functional effects during thermomechanical training.

Table 1. SMA materials and their elastocaloric properties.

Material	Shape	σ_{max} [MPa]	T_{max} [K]	ΔT_{ad} [K]	Δs_{iso} [J kg^{-1} K^{-1}]	Reference
$Ni_{51.5}$ $Ti_{48.5}$	Rod Tube Honeycomb	1600	294	4.1	/	[25]
$Ni_{50.8}$ $Ti_{49.2}$	Wire encapsuled	10	35	14	/	[26]
Ni Ti	Wire	500	333	11.1	/	[27]
Ni Ti	Wire	500	343	27	/	[28]
Ni Ti	Wire	600	319	23.7	/	[28]
Ni Ti	Wire	790	342	24	/	[29]
Ni Ti Cu	Film	300	363	5	/	[21]
Ni Ti Cu Co	Bulk	420	315	14.4	/	[30]
Ni Co Mn Al	Rods	1228	/	6	/	[31]
Ni Ti Pd Hf	Ingot	-700	388	30	/	[32]
Cu Al Mn	Wire	150	340	3.9	16.1	[33]
$Cu_{71.1}$ $Al_{17.2}$ $Mn_{11.7}$	Microwire	400	316	12.6	21	[34]
PVDF	Bulk	15	298	1.8	/	[35]
PVDF-TrFE-CTFE	Bulk	200	368	18.6	/	[35]
P(VDF-TrFE) 65/35	Bulk	-100	376	32.2	/	[36]
P(VDF-TrFE)	Bulk	-100	354	65	/	[36]

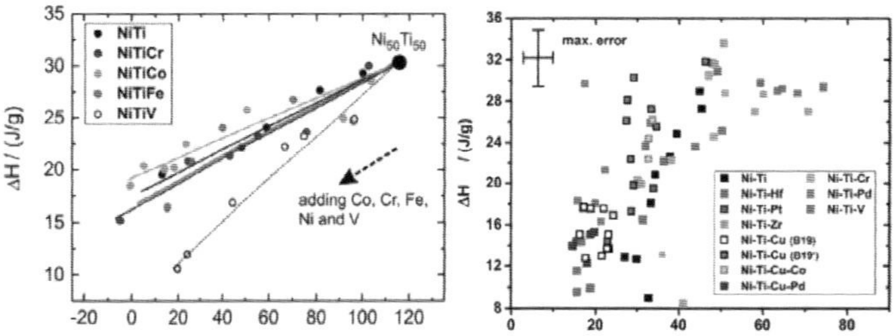

a b

Figure 2. a) Latent heats ΔH and transformation temperatures correlation for various NiTi-based SMAs; b) Key parameters: latent heat and width of thermal hysteresis

4. Recent studies on elastocaloric cooling device

The most used alloy is the NiTi-based one as they have more developed shape memory properties, readily available on the market, and have an excellent helpful life for eC applications. Other elements can be added to these alloys to increase the latent heat and increase the transformation temperature and the alloy's helpful life. The pioneers of elastocaloric device developing were Cui et al. [37] in 2012 introducing the very first elastocaloric proof-of-concept for cooling, mounting NiTi wires (with 0.5 mm diameter) placed between two rings (the upper one is inclined). The rotation of the rings and the convective heat exchange with air (the HTF) ensures the exploitation of the AeR cooling cycle. The device achieved 17 K as maximum adiabatic temperature change (under 8.5% strain applied for stretching). The main problem of this device was the low lifetime of the refrigerant mounted with a high percentage of cracking. One of the prototypes presented to the scientific community was that of Tusek et al. [38]. With their prototype of an eC regenerator, built using the NiTi binary alloy as SMA and water as the working fluid, they managed to obtain ΔT values of 15.3 K and a maximum COP value of 7. The regenerator is formed by SMA plates enclosed between two clamps on which the loading and unloading cycle is applied. Two circuits (hot and cold) are connected to the regenerator in which the working fluid flows, which transfers and absorbs heat with two exchangers placed outside, as shown in Figure 3a. This is achieved here by designing SMA foil deflection as a heat sink with tilted geometry. During unloading, the foil is brought in contact with a nearly flat heat source to allow heat absorption. In this way, only a single actuator is needed to strain the SMA foil and switch thermal contacts between source and sink to direct heat flow (as shown in figure 3b).

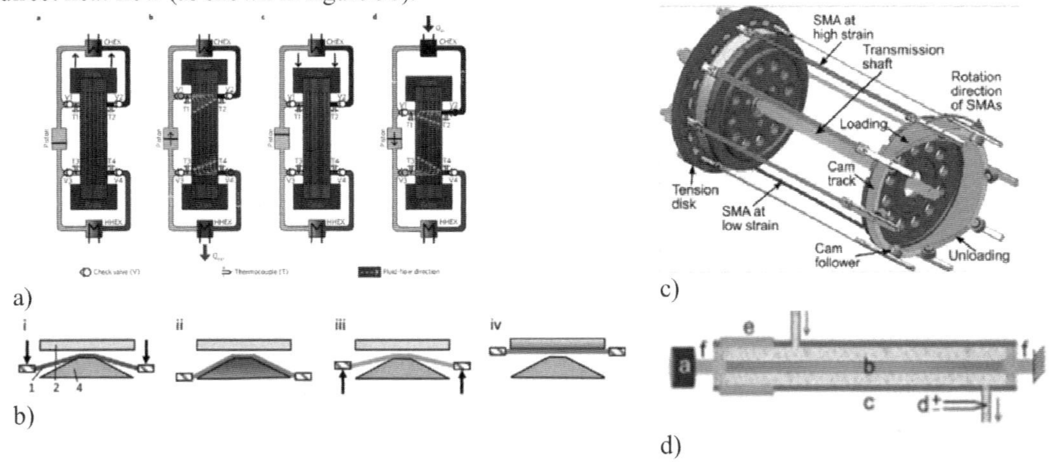

Figure 3. Elastocaloric Prototype: a) Tusek et al. [29]; b) Bruederlin et al. [39]; c) Kirsch et al. [37]; Wang et al. [40]

Table 2. Preliminary results of recent elastocaloric cooling devices.

Type	Material	Stress application	σ_{max} [MPa]	Working Fluid	ΔT_{ad} [K]	COP	ΔH [J g^{-1}]	Reference
SMA tube	NiTi	linear	1100	water	24.6	11	10	[41]
SMA plate	NiTi	linear	440	water	14	/	/	[38]
SMA plate	NiTi	linear	/[1]	water	15.3	7	/	[37]
SMA foil	NiTiFe	deflection of the bridge centre	/[2]	air	14	3.3	/	[39]
SMA wire	NiTi	rotary tensile	400	air	20	/	/	[40,29]
SMA wire	$Ni_{52.6}Ti_{47.4}$	torsional stress	/[3]	water	8	/	9.4	[43]
Ferroelectric polymer	PVDF	torsional stress	/[4]	water	-3.0	8.8	/	[45]

The need to find a model capable of predicting the quantity of SMA material to be inserted, the sizing of the prototype, in general, has led to the definition of rotary prototypes presented in the works [29, 38, 40, 33, 21]. Recently, the German group [37] of Saarland University, introduced a new rotary elastocaloric prototype mounting wires of $Ni_{45}Ti_{47.25}Cu_5V_{2.75}$ whose fatigue life is longer than the binary Ni-Ti alloys and exhibiting a ΔT_{ad} of 28 K if loaded with a strain rate of 0.05 s^{-1} up to a maximum strain of 8.5%. Preliminary results on the device guaranteed 8.4 as COP and 250 as cooling power with a total mass of 50 g of elastocaloric material. In Figure 3c is reported a schematic of the German rotative prototype. The figure clearly shows that by placing multiple strands of eC material on a concentric circumference to another (for example, the circumferences of two discs), the two discs' synchronous rotation allows the elements of eC material to experience different phases of the cycle at the same time.

[1] Maximum applied strain 4%

[2] The stress application is in terms of Δl. In this case is equal to 2 mm

[3] The stress application is in terms of strain/spring index. In this case is equal to 75%

[4] The stress application is in terms of strain/spring index. In this case is equal to 60%

During the rotation, some wires will be in traction, others in the relaxation phase allowing a continuous flow of cold air because each wire will experience during one rotation one complete cycle. The cyclical traction (loading) and relaxation (unloading) of the eC material induce a temperature profile along the device's circumference. In recent years Wang et al. [47] have proposed a new application of stress to the eC element SMA: torsion. Significant, reversible temperature changes result from twist insertion and removal from a single NiTi shape memory wire and from plying and replying to these wires. They demonstrated that cooling by a change in yarn or fibre twist, which they call twisto-caloric cooling. They named coolers that use twist changes for refrigeration "twist fridges". A schematic device of their work is shown in figure 3d. The most important results and characteristics of the previously mentioned eC chiller prototypes are shown in Table 2.

5. Conclusions

This work introduces a new frontier of refrigeration: elastocaloric cooling. A general overview of the thermodynamic cycle is described to define the characteristics of solid-state refrigeration. In elastocaloric refrigeration, the key parameters are the austenitic transformation temperature and the martensitic transformation temperature. To obtain satisfactory results and for the thermodynamic cycle to be efficient, the elastocaloric material plays a fundamental role. Many studies propose the NiTi alloy, as the construction cost, the ease of supply, an excellent elastic behavior make this material suitable for exploiting the caloric phenomenon. Recently, new NiTi-based alloys such as NiTiCoCu or NiTiV are being studied. It has been noted that subjected to cyclic loads it allows a longer useful life than binary NiTi alloys. Furthermore, these materials exhibit reduced hysterical thermal phenomena during the loading and unloading phase, which consequently allow having lower mechanical hysterical phenomena. This fact allows to have for having rather exciting coefficients of performance (COP). Finally, a general overview was carried out on the prototypes presented to the scientific community up to now. The number of prototypes is low, so there are high margins for developing new device ideas that optimize the caloric phenomenon and improve the device's efficiency. To date, there is no Italian prototype of an elastocaloric cooler. However, there is a collaboration between the University of Naples Federico II, the University of Genoa and the CNR of Naples to study and develop the first Italian elastocaloric prototype to be presented to the scientific community. One of the biggest challenges to be faced is that of improving the durability of the high-calorie material. To date, the most used material in prototypes is Nitinol (Ni-Ti), which, however, has been observed to fail due to fatigue in the most demanding applications. The Italian prototype will be rotary, but unlike Kirsch et al. [37], the application of the stress will take place using hydraulic pistons which will guarantee lower friction forces compared to the model proposed by the German group, with consequent improvement of the COP values.

Acknowledgements

This research was funded through the project "SUSSTAINEBLE" - FISR2019 04798 granted by FISR – Fondo Integrativo Speciale per la Ricerca (Italian special supplementary fund for research).

Reference

[1] F. Cascetta, R. Di Lorenzo, S. Nardini, and L. Cirillo, "A Trnsys Simulation of a Solar-Driven Air Refrigerating System for a Low-Temperature Room of an Agro-Industry site in the Southern part of Italy," *Energy Procedia*, vol. 126, pp. 329–336, 2017, doi: 10.1016/j.egypro.2017.08.259.

[2] O. Manca, L. Cirillo, S. Nardini, B. Buonomo, and D. Ercole, "Experimental Investigation on Fluid Dynamic and Thermal Behavior in Confined Impinging Round Jets in Aluminum Foam," *Energy Procedia*, vol. 101, no. September, pp. 1095–1102, 2016, doi: 10.1016/j.egypro.2016.11.149.

[3] A. Greco, R. Mastrullo, and A. Palombo, "R407C as an alternative to R22 in vapour compression plant: An experimental study," *Int. J. Energy Res.*, vol. 21, no. 12, pp. 1087–1098, 1997, doi: 10.1002/(SICI)1099-114X(19971010)21:12<1087::AID-ER330>3.0.CO;2-Y.

[4] C. Aprea, A. Greco, A. Maiorino, and C. Masselli, "The drop-in of HFC134a with HFO1234ze in a household refrigerator," *Int. J. Therm. Sci.*, vol. 127, no. January, pp. 117–125, 2018, doi: 10.1016/j.ijthermalsci.2018.01.026.

[5] C. Aprea, A. Greco, and A. Maiorino, "An experimental investigation of the energetic performances of HFO1234yf and its binary mixtures with HFC134a in a Étude expérimentale des performances énergétiques du HFO1234yf et de ses mélanges binaires avec du HFC134a dans un réfrigérateur domestique," *Int. J. Refrig.*, vol. 76, pp. 109–117, 2017, doi: 10.1016/j.ijrefrig.2017.02.005.
[6] C. Aprea, A. Greco, and A. Maiorino, "An experimental evaluation of the greenhouse effect in the substitution of R134a with CO2," *Energy*, vol. 45, no. 1, pp. 753–761, 2012, doi: 10.1016/j.energy.2012.07.015.
[7] A. Greco and G. P. Vanoli, "Experimental two-phase pressure gradients during evaporation of pure and mixed refrigerants in a smooth horizontal tube. Comparison with correlations," *Heat Mass Transf. und Stoffuebertragung*, vol. 42, no. 8, pp. 709–725, 2006, doi: 10.1007/s00231-005-0020-7.
[8] C. Aprea, A. Greco, and A. Maiorino, "The substitution of R134a with R744: An exergetic analysis based on experimental data," *Int. J. Refrig.*, vol. 36, no. 8, pp. 2148–2159, 2013, doi: 10.1016/j.ijrefrig.2013.06.012.
[9] J. Tušek *et al.*, "The Elastocaloric Effect: A Way to Cool Efficiently," *Adv. Energy Mater.*, vol. 5, no. 13, pp. 1–5, 2015, doi: 10.1002/aenm.201500361.
[10] C. Aprea, A. Greco, and A. Maiorino, "The use of the first and of the second order phase magnetic transition alloys for an AMR refrigerator at room temperature: A numerical analysis of the energy performances," *Energy Convers. Manag.*, vol. 70, pp. 40–55, 2013, doi: 10.1016/j.enconman.2013.02.006.
[11] C. Aprea, A. Greco, A. Maiorino, and C. Masselli, "Electrocaloric Refrigeration: an innovative, emerging, eco-friendly refrigeration," *J. Phys. Conf. Ser.*, vol. 755, no. 1, 2016, doi: 10.1088/1742-6596/755/1/011001.
[12] C. Aprea, A. Greco, A. Maiorino, and C. Masselli, "The use of barocaloric effect for energy saving in a domestic refrigerator with ethylene-glycol based nanofluids: A numerical analysis and a comparison with a vapor compression cooler," *Energy*, vol. 190, p. 116404, 2020, doi: 10.1016/j.energy.2019.116404.
[13] C. Aprea, A. Greco, and A. Maiorino, "Modelling an active magnetic refrigeration system: A comparison with different models of incompressible flow through a packed bed," *Appl. Therm. Eng.*, vol. 36, no. 1, pp. 296–306, 2012, doi: 10.1016/j.applthermaleng.2011.10.034.
[14] C. Aprea, A. Greco, A. Maiorino, and C. Masselli, "Energy performances and numerical investigation of solid-state magnetocaloric materials used as refrigerant in an active magnetic regenerator," *Therm. Sci. Eng. Prog.*, vol. 6, no. January, pp. 370–379, 2018, doi: 10.1016/j.tsep.2018.01.006.
[15] C. Aprea, A. Greco, and A. Maiorino, "A numerical analysis of an active magnetic regenerative refrigerant system with a multi-layer regenerator," *Energy Convers. Manag.*, vol. 52, no. 1, pp. 97–107, 2011, doi: 10.1016/j.enconman.2010.06.048.
[16] C. Aprea, A. Greco, A. Maiorino, and C. Masselli, "Une comparaison de différents matériaux dans un cycle actif électrocalorique régénératif avec un modèle numérique à 2D," *Int. J. Refrig.*, vol. 69, pp. 369–382, 2016, doi: 10.1016/j.ijrefrig.2016.06.016.
[17] C. Aprea, A. Greco, A. Maiorino, and C. Masselli, "The employment of caloric-effect materials for solid-state heat pumping," *Int. J. Refrig.*, vol. 109, pp. 1–11, 2020, doi: 10.1016/j.ijrefrig.2019.09.011.
[18] A. Greco, C. Aprea, A. Maiorino, and C. Masselli, "A review of the state of the art of solid-state caloric cooling processes at room-temperature before 2019," *Int. J. Refrig.*, vol. 106, pp. 66–88, 2019, doi: 10.1016/j.ijrefrig.2019.06.034.
[19] J. Mohd Jani, M. Leary, A. Subic, and M. A. Gibson, "A review of shape memory alloy research, applications and opportunities," *Mater. Des.*, vol. 56, pp. 1078–1113, 2014, doi: 10.1016/j.matdes.2013.11.084.
[20] C. Aprea, A. Greco, A. Maiorino, and C. Masselli, "Analyzing the energetic performances of AMR regenerator working with different magnetocaloric materials: Investigations and viewpoints," *Int. J. Heat Technol.*, vol. 35, no. Special Issue 1, pp. S383–S390, 2017, doi: 10.18280/ijht.35Sp0152.
[21] C. Chluba *et al.*, "Ultralow-fatigue shape memory alloy films," *Science (80-.).*, vol. 348, no. 6238, pp. 1004–1007, 2015, doi: 10.1126/science.1261164.
[22] J. Frenzel, G. Eggeler, E. Quandt, S. Seelecke, and M. Kohl, "High-performance elastocaloric materials for the engineering of bulk- and micro-cooling devices," *MRS Bull.*, vol. 43, no. 4, pp. 280–284, 2018, doi: 10.1557/mrs.2018.67.
[23] J. Frenzel, A. Wieczorek, I. Opahle, B. Maaß, R. Drautz, and G. Eggeler, "On the effect of alloy composition on martensite start temperatures and latent heats in Ni-Ti-based shape memory alloys," *Acta Mater.*, vol. 90, pp. 213–231, 2015, doi: 10.1016/j.actamat.2015.02.029.
[24] S. Jaeger *et al.*, "On the widths of the hysteresis of mechanically and thermally induced martensitic

transformations in Ni-Ti-based shape memory alloys," *Int. J. Mater. Res.*, vol. 106, no. 10, pp. 1029–1039, 2015, doi: 10.3139/146.111284.

[25] H. Hou *et al.*, "Fatigue-resistant high-performance elastocaloric materials made by additive manufacturing," vol. 1121, no. November, pp. 1116–1121, 2019.

[26] X. Zhao, "Elastocaloric Performance of Pseudoelastic NiTi Coiled Wires," *Shape Mem. Superelasticity*, vol. 7, no. 1, pp. 101–108, 2021, doi: 10.1007/s40830-021-00310-9.

[27] S. M. Kirsch *et al.*, "NiTi-Based Elastocaloric Cooling on the Macroscale: From Basic Concepts to Realization," *Energy Technol.*, vol. 6, no. 8, pp. 1567–1587, 2018, doi: 10.1002/ente.201800152.

[28] D. J. Sharar, J. Radice, R. Warzoha, B. Hanrahan, and B. Chang, "First Demonstration of a Bending-Mode Elastocaloric Cooling 'Loop,'" *Proc. 17th Intersoc. Conf. Therm. Thermomechanical Phenom. Electron. Syst. ITherm 2018*, pp. 218–226, 2018, doi: 10.1109/ITHERM.2018.8419513.

[29] J. Tušek, K. Engelbrecht, L. P. Mikkelsen, and N. Pryds, "Elastocaloric effect of Ni-Ti wire for application in a cooling device," *J. Appl. Phys.*, vol. 117, no. 12, 2015, doi: 10.1063/1.4913878.

[30] Z. Yang, D. Cong, Y. Yuan, R. Li, H. Zheng, and X. Sun, "Large room-temperature elastocaloric effect in a bulk polycrystalline Ni-Ti-Cu-Co alloy with low isothermal stress hysteresis," *Appl. Mater. Today*, vol. 21, p. 100844, 2020, doi: 10.1016/j.apmt.2020.100844.

[31] H. Xuan *et al.*, "The effect of Co on elastocaloric and mechanical properties of Ni-Co-Mn-Al alloys," *Solid State Commun.*, vol. 301, no. August, p. 113706, 2019, doi: 10.1016/j.ssc.2019.113706.

[32] E. Acar, G. P. Toker, S. Saedi, H. Tobe, and H. E. Karaca, "Thermo-mechanical Characterization of an Aged Ni 45 . 3 Ti 39 . 7 Hf 10 Pd 5 Shape Memory Alloy," *JOM*, vol. 73, no. 2, pp. 640–645, 2021, doi: 10.1007/s11837-020-04449-1.

[33] B. Yuan, "materials Elastocaloric effect with small hysteresis in bamboo- grained Cu – Al – Mn microwires," *J. Mater. Sci.*, vol. 54, no. 13, pp. 9613–9621, 2019, doi: 10.1007/s10853-019-03592-8.

[34] B. Yuan, S. Zhong, M. Qian, X. Zhang, and L. Geng, "Elastocaloric effect in bamboo-grained Cu71.1Al17.2Mn11.7 microwires," *J. Alloys Compd.*, vol. 850, p. 156612, 2021, doi: 10.1016/j.jallcom.2020.156612.

[35] S. Patel, A. Chauhan, R. Vaish, and P. Thomas, "Elastocaloric and barocaloric effects in polyvinylidene di-fluoride-based polymers," vol. 0641090641, 2016, doi: 10.1063/1.4942000.

[36] G. Bai, D. Liu, and C. Gao, "Phenomenological analysis of elastocaloric effect in ferroelectric poly (vinylidene fluoride-trifluoroethylene) copolymers Phenomenological analysis of elastocaloric effect in ferroelectric poly (vinylidene fl uoride-tri fl uoroethylene) copolymers," *J. Appl. Phys.*, vol. 164105, no. August, 2019, doi: 10.1063/1.5124901.

[37] S. M. Kirsch, F. Welsch, N. Michaelis, M. Schmidt, A. Schütze, and S. Seelecke, "Continuously Operating Elastocaloric Cooling Device Based on Shape Memory Alloys: Development and Realization," 2018.

[38] J. Tušek *et al.*, "The Elastocaloric Effect: A Way to Cool Efficiently," *Adv. Energy Mater.*, vol. 5, no. 13, 2015, doi: 10.1002/aenm.201500361.

[39] F. Bruederlin, H. Ossmer, F. Wendler, S. Miyazaki, and M. Kohl, "SMA foil-based elastocaloric cooling: From material behavior to device engineering," *J. Phys. D. Appl. Phys.*, vol. 50, no. 42, 2017, doi: 10.1088/1361-6463/aa87a2.

[40] S. Qian, Y. Wang, L. Yuan, and J. Yu, "A heat driven elastocaloric cooling system," *Energy*, vol. 182, pp. 881–899, 2019, doi: 10.1016/j.energy.2019.06.094.

[41] S. Qian *et al.*, "Design of a hydraulically driven compressive elastocaloric cooling system," *Sci. Technol. Built Environ.*, vol. 22, no. 5, pp. 500–506, 2016, doi: 10.1080/23744731.2016.1171630.

[42] J. Tušek, K. Engelbrecht, and N. Pryds, "Elastocaloric effect of a Ni-Ti plate to be applied in a regenerator-based cooling device," *Sci. Technol. Built Environ.*, vol. 22, no. 5, pp. 489–499, 2016, doi: 10.1080/23744731.2016.1176809.

[43] J. Tušek, K. Engelbrecht, D. Eiksen, S. Dall'Olio, J. Tusek, and N. Pryds, "A regenerative elastocaloric heat pump," *Nat. Energy*, p. 16134, 2016, doi: 10.1038/NENERGY.2016.134.

[44] F. Welsch *et al.*, "Elastocaloric Cooling: System Design, Simulation, and Realization," in *ASME 2018*, 2018, pp. 1–9.

[45] R. Wang *et al.*, "Torsional refrigeration by twisted, coiled, and supercoiled fibers," *arXiv*, vol. 221, no. October, pp. 216–221, 2019.

[46] J. Cui *et al.*, "Demonstration of high efficiency elastocaloric cooling with large Δt using NiTi wires," *Appl. Phys. Lett.*, vol. 101, no. 7, pp. 27–31, 2012, doi: 10.1063/1.4746257.

[47] R. Wang, X. Zhou, W. Wang, and Z. Liu, "Twist-based cooling of polyvinylidene difluoride for mechanothermochromic fibers," *Chem. Eng. J.*, no. November, p. 128060, 2020, doi:

10.1016/j.cej.2020.128060.

Analytical and numerical modelling of biomass gasification in downdraft gasifiers

L. Moretti[1], F. Arpino[1], G. Cortellessa[1], M. Dell'Isola[1], G. Ficco[1], G. Grossi[1], F. Zuena[1], M. Di Palma[2], L. Vanoli[2]

[1]Department of Civil and Mechanical Engineering, University of Cassino and Southern Lazio, via G. Di Biasio 43, Cassino, Italy
[2]Department of Engineering, University of Naples Parthenope, Naples, Italy;
Corresponding author e-mail: linda.moretti@unicas.it

Abstract. The ever-growing energy demand and environmental pollution level have pushed research interest toward the development of new promising technologies to utilize bioenergy sustainably. In this scenario, biomass gasification has been regarded as a promising renewable energy resource that, if efficiently exploited, can contribute to the reduction of dependency on fossil fuel and CO_2 emissions from the power sector. Nevertheless, exploitation of biomass gasification still requires to overcome technological and logistical issues. In this work, the authors propose a general thermodynamic model able to predict the steady operating conditions of biomass gasification. Assuming thermodynamic equilibrium, the model is able to predict temperature, mass flow rate and composition of produced syngas. Moreover, a numerical model of the biomass gasification system has been developed by using the commercial software Aspen Plus. It is a steady-state zero-dimensional equilibrium model, based on the mass and energy balances, assuming the Gibbs free energy minimization. Aspen Plus allows building models of customized operative units using the Fortran code. In order to assure the performance of both analytical and Aspen Plus models, numerical results are compared with experimental data available in the scientific literature for downdraft gasifiers.

1. Introduction

Depletion of fossil fuels, global warming effect and environmental pollution and deterioration significantly pushed the research and development of renewable energy technology during the past few decades. In this scenario, gasification of biomass for power production showed an increased interest worldwide, as it is considered particularly interesting for decentralized power plants, especially in developing countries.

Gasification is a thermochemical conversion process through which the carbonaceous solid fuels are converted into a gaseous fuel. This process occurs in high temperature (i.e. 500 – 1300 °C) and high temperature environment (i.e. > 1 MPa) [1] and under sub-stoichiometric conditions of the oxidant (e.g., air, vapor, oxygen, or carbon dioxide). The amount of oxidizing agent supplied is limited in order to avoid complete combustion and to transform part of the initial energy into chemical energy, in form of a gaseous fuel known as syngas or producer gas, mainly composed of carbon monoxide, hydrogen, carbon dioxide, nitrogen, methane, tar and ash. The physical and chemical properties of the syngas significantly depend on: i) the biomass properties, ii) the oxidant agent, iii) the operating conditions, iv) the heating rate and v) the type of gasifier used [2]. In particular, the Equivalence Ratio (ER) defined as the ratio of the actual fuel/air to the stoichiometric fuel/air ratio is one of the main operating parameters affecting the gasification process [1]. The typical ER values range for the biomass gasification process varies between 0.2 and 0.4. Temperature also affects the equilibrium reactions in the gasification process. It is noted in various works [3] that an increase in temperature leads to an increase in the production of syngas and the reduction of tar. The optimum combination of this parameters is crucial in developing a sustainable utilization of biomass resources through gasification process. In a recent work, it was found that the optimum temperature range is from 780 °C to 790 °C and the equivalence ratio is approximately 0.7 for biomass gasification using steam as oxidizing agent [4].

Even though biomass gasification is a mature technology, some technological barriers limit its large-scale commercial development [5]. A better insight of gasification phenomena and their dependence on operational parameters is desirable to support its commercialization. To this aim, a numerical model of the gasification process may represent a very useful tool to better understand the physical and chemical mechanisms that occur inside the gasifier. Mathematical models for the biomass gasification process can be categorized into: i) kinetic rate models, ii) thermodynamic equilibrium models, iii) computational fluid dynamics models (CFD) and iv) artificial neural networks (ANNs) models. Kinetic models are based on the mathematical description of the reaction kinetics of the main reactions and their application requires the knowledge of reaction rates, residence time, reactor hydrodynamics (e.g., superficial velocity, diffusion rate) and geometry of gasifier. Thermodynamic equilibrium models are based on the assumption that the system of chemical reactions involving in the process reaches the state of thermochemical equilibrium inside the gasifier reactor [6]. This approach does not require details of system geometry or estimate the necessary time to reach that equilibrium, being suitable to study the influence of operating parameters on the gaseous product composition of the process. The choice of the most suitable modelling approach is clearly determined by the scope of analysis and the experimental data available. Complex models provide detailed information on mechanisms, but the computational cost and the number of input data required are usually high. Although equilibrium models are valid under chemical equilibrium conditions, they are valuable because they can predict thermodynamic limits of a gasification system, leading the process design, evaluation and improvement [7].
In this paper, the performances of non-stoichiometric equilibrium models are investigated by comparing numerical results with experimental data available in the scientific literature for downdraft gasifiers. To this aims the authors present and discuss the results obtained by employing the commercial code Aspen Plus and a non-commercial analytical model implemented in Matlab.

2. Analytical model

The proposed analytical model describes the process of biomass gasification using a concentrated parameter approach, under the hypotheses of: i) average steady state; ii) uniform and constant properties of biomass inside the gasifier; iii) local thermodynamic equilibrium; iv) complete chemical reactions; v) uniform and constant temperature of biomass bed. Input data for calculation are: i) the mass flow rate of biomass feedstock and its temperature; ii) temperature and volumetric flow rate of gasification air; iii) biomass composition.
To develop the model, the chemical formula of biomass is defined as $CH_xO_yN_z$, where x, y and z are the number of hydrogen, oxygen and nitrogen atoms, respectively [8]. The latter is calculable knowing the elemental composition of biomass based on ultimate and proximate analysis. The biomass gasification process is modelled assuming thermodynamic equilibrium, according to the following global reaction:

$$CH_xO_yN_z + wH_2O + m(O_2 + 3.76N_2) = n_{H_2}H_2 + n_{CO}CO + n_{CO_2}CO_2 + n_{CH_4}CH4 + \left(\frac{z}{2} + 3.76m\right)N_2 \quad (1)$$

where w is the amount of moisture per mole of carbon and $m = O/C$ is the amount of oxygen per mole of carbon. The number of moles of the different species, n_i, on the right-hand side represent the unknown quantities of the problem. The moles produced of H_2, CO, CO_2, H_2O and CH_4 were determined solving the following simultaneous five equations [8–10]:

Carbon mass balance:

$$n_{CO} + n_{CO_2} + n_{CH_4} - 1 = 0 \quad (2)$$

Hydrogen mass balance:

$$2n_{H_2} + 2n_{H_2O} + 4n_{CH_4} - x - 2w = 0 \quad (3)$$

Oxygen mass balance:

$$n_{CO} + 2n_{CO_2} + n_{H_{2O}} - w - 2m = 0 \quad (4)$$

Methanation reaction:

$$C + 2H_2 = CH_4 \quad (5)$$

Water shift reaction:

$$CO + H_2O = CO_2 + H_2 \quad (6)$$

The equilibrium constants for the reactions (5) and (6) are available in the scientific literature [11–13] and are not reported here for brevity. The gasification temperature must be known to calculate the equilibrium constants and hence, the mass balances and chemical equilibrium relations have to be performed together with the energy conservation equation for the gasification process, written as:

$$h^0_{f,CH_xO_yN_z} + wh^0_{f,H_2O} + \frac{\dot{n}_{O_2}}{\dot{n}_{CH_xO_yN_z}} h_{T_{IN},O_2} + \frac{\dot{n}_{N_2}}{\dot{n}_{CH_xO_yN_z}} h_{T_{IN},N_2} = n_{H_2} h_{T_{OUT},H_2} + n_{CO} h_{T_{OUT},CO} + n_{CO_2} h_{T_{OUT},CO_2} + n_{CH_4} h_{T_{OUT},CH_4} + n_{H_2O} h_{T_{OUT},H_2O} \quad (7)$$

where $h^0_{f,CH_xO_yN_z}$ and h^0_{f,H_2O} are the formation enthalpy of organic matter and water in reference condition (298 K and 101325 Pa) in kJ/kmol, respectively; n_i represents the number of moles produced, for the same chemical species, per mole of consumed fuel; lastly, $h_{T,i}$ is the enthalpy associated with the i-th chemical species at the temperature T and it is the sum of the formation enthalpy and the difference of enthalpy between the state in which the substance is found and the reference status. The enthalpy of formation for organic matter, $h^0_{f,CH_xO_yN_z}$, in the reactant is calculated according to the relation suggested in [14]:

$$h^0_{f,CH_xO_yN_z} = LHV + \sum_{k=prod} \left[n_k (\bar{h}^0_f)_k\right] \quad (8)$$

where $(\bar{h}^0_f)_k$ is the enthalpy of formation of product k under complete combustion of the solid fuel and LHV is the lower heating value of the dry fraction of biomass calculated as function of biomass composition calculated conforming to the following equation [15]:

$$LHV = 0.3491 \cdot C + 1.1783 \cdot H + 0.1005 \cdot S - 0.1034 \cdot O - 0.0151 \cdot N - 0.0211 \cdot ash - 9 \cdot H \cdot h_{fg} \quad (9)$$

where C, H, S, O, N, ash are the mass fraction of carbon, hydrogen, oxygen, nitrogen, sulphur and ash in the biomass and h_{fg} is the enthalpy of vaporization of water at reference conditions.

3. Aspen Plus model

A zero-dimensional numerical model to simulate downdraft gasification is developed through the commercial software Aspen Plus based on the Gibbs free energy minimization formulated as follow:

$$min_{n_i}\, G/RT = \sum_{i=1}^{N} n_i \,(\Delta G^0_{f,i}/RT) + ln\, n_i \sum_{i=1}^{N} n_i \quad (10)$$

where $\Delta G°_{f,i}$ is the standard formation free Gibbs energy of species i and n_i the molar amounts of elements i
Aspen Plus involved the following main phases: i) specification of the stream class with a selection of the property method, specification of the system components from a database and identification of the conventional and non-conventional components; ii) setting of the flow chart in which the block units and their connections with the relative flows are indicated; iii) definition of the operating conditions of the individual blocks. The biomass is considered as an unconventional solid, whose chemical composition is defined by the ultimate and proximate analyses. HCOALGEN and DCOALIGT algorithms are used to calculate enthalpy and density of biomass and ash, using the data of proximate, ultimate and sulphur analyses. DCOALIGT refers to IGT (Institute of Gas and Technology) Coal Density Model:

$$\rho_i = \rho_i^{dm} / \left[\rho_i^{dm}(0.42\, w^d_{A,i} - 0.15 w^d_{Sp,i}) + 1.13 w^d_{A,i} - 0.5475\, w^d_{Sp,i}\right] \quad (11)$$

$$\rho_i^{dm} = 1/\left[a_{1i} + a_{2i} w^{dm}_{Hi} + a_{3i}(wW^{dm}_{Hi}) + a_{4i}(w^{dm}_{Hi})^3\right] \quad (12)$$

$$\rho_i^{dm} = 1/\left[a_{1i} + a_{2i} w^{dm}_{Hi} + a_{3i}(wW^{dm}_{Hi}) + a_{4i}(w^{dm}_{Hi})^3\right] \quad (13)$$

where $w^d_{x,i}$ is the fraction by weight, on a dry basis, of the j-th component, while the apex dm is used when the fraction is calculated on a dry basis and free of mineral substances. The capital letter W^{dm}_{Hi} refers to the mole fraction. HCOALGEN is the "General Coal Enthalpy Model" based on the general enthalpy model of coal and includes several correlations. The Peng-Robinson's equation of state with the Boston-Mathias modifications (PR-BM) is used to estimate all physical properties of the conventional components in the gasification process. The following assumptions are considered to simulate biomass gasification: i) steady-state and isothermal model; ii) zero-dimensional and kinetic-free model; iii) all the gases involved in the reactions are considered as ideal gases; iv) the char contains only Carbon; v) Sulphur and Nitrogen reactions are not considered; vi) tar

formation has been neglected; vii) instantaneous devolatilization of the biomass; vii) volatile products considered are H_2, CO, CO_2, CH_4 and H_2O; viii) the chemical reactions in the gasifier are in an equilibrium state; ix) the heat loss is neglected.
In the present work, the Restricted Chemical Equilibrium is chosen to specify the chemical reactions with different Temperature approaches (specifying a determined Temperature approach for each reaction). The RGibbs reactor evaluates the chemical equilibrium constant at T + ΔT approach, where T is the actual reactor temperature (specified in the settings as an input parameter obtained from experimental data and reported in the literature) and ΔT approach is the Approach Temperature, which represents the difference between the chemical equilibrium Temperature and the real reactor temperature. This method is used to modify the chemical equilibrium, to better simulate the non-equilibrium conditions of a real gasifier. As reactions, in addition to the methanation (5) and water gas shift (6), carbon combustion (19) and hydrogen combustion (20) are considered:

$$C + O_2 \rightarrow CO_2 \tag{19}$$

$$H_2 + 1/2\,O_2 \rightarrow H_2O \tag{20}$$

In Aspen Plus library there is not a gasification unit, but it is possible to simulate the fixed bed gasifier by combining different built-in units properly modified if needed. In the present work, the blocks reported in Table 1 are used. They represent biomass decomposition, gasification of biomass with air and the separation of the syngas from unconverted char and ashes. In the biomass decomposition phase, a RYield reactor (DECOMP) has been used to decompose the biomass (BIOMASS stream) into its constituent elements, including C, H_2, O_2, N_2, S and ash, by specifying the yield distribution according to the ultimate analysis of biomass, written by Fortran subroutine in a calculator block. The input parameters for this block are temperature, pressure, flow rate, as well as the ultimate and proximate analyses of the biomass used for the experimental campaign. The gasification phase is performed in a RGibbs reactor (GASIFIER) that calculates the composition of the produced gas by minimizing the Gibbs free energy and reaching a complete chemical equilibrium. The main parameters of the gasifier used are temperature and pressure obtained from the experimental campaign. In this block, the air is introduced as a gasification agent, through the AIR stream using temperature, pressure and flow rate of the air from experimental data.
Figure 1 shows the Aspen Plus flowsheet of the biomass gasification system and Table 1 gives summary descriptions of the unit operations of the blocks.

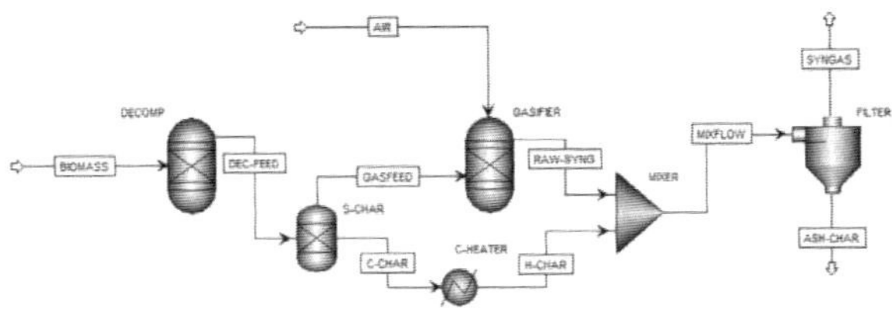

Figure 1. Scheme of the blocks used in the model.

Table 1. Scheme of the blocks used in the model.

BLOCK	Function
RYIELD (DECOMP)	Models a reactor by specifying the reaction yields of each component but the stoichiometry and the reaction kinetics are not known.
RGIBBS (GASIFIER)	Models a reactor in which equilibrium conditions are reached, and the Gibbs free energy minimization is applied. This block is useful when temperature and pressure are known.
SSPLIT (S-CHAR)	It is used to separate the flow into several specified fractions, each with the same composition and the same properties of the incoming flow.
HEATER (C-HEATER)	It is used to set the thermodynamic conditions of the outlet stream.
MIXER (MIXER)	Combines multiple streams into one stream.
SEP	Separates inlet stream components into any number of outlet streams.

Journal of Physics: Conference Series 2177 (2022) 012028 doi:10.1088/1742-6596/2177/1/012028

(FILTER)	

4. Results

The proposed thermodynamic equilibrium models have been validated using data from experimental campaigns reported in [16] and [17] that have been carried out on downdraft gasifiers. The chemical properties of biomass employed experimental campaigns and operating conditions of simulated case studies are summarized in Table 2 and Table 3.

Table 2. Ultimate and proximate analysis of biomass used for the validation of models.

		Ultimate analysis [wt.% Dry basis]				
Ref.	Biomass	C	H	O	N	S
[16]	Hazelnut shell	46.76	5.76	45.83	0.22	0.67
[17]	Rubber wood	50.60	6.50	42.00	0.20	-
		Proximate analysis [wt.% Dry basis]				
Ref.	Biomass	Volatile matter	Fixed carbon	Ash		
[16]	Hazelnut shell	71.62	27.50	0.88		
[17]	Rubber wood	80.10	19.20	0.70		

Table 3. Gasifier operating conditions of the experimental tests used for the validation of models.

Ref.	Biomass	No. Test	Biomass feeding rate [kg/h]	Air flow rate [Nm^3/h]	Equivalence ratio (ER)
[16]	Hazelnut shell	1	1.73	2.82	0.449
		2	2.15	3.53	0.451
		3	2.64	4.01	0.418
		4	3.19	4.40	0.380
[17]	Rubber wood	5	27.39	43.00	0.406
		6	25.27	43.00	0.425
		7	23.46	43.00	0.455
		8	28.37	43.00	0.380

The comparison between the numerical and experimental results has been carried out in terms of: i) molar composition of syngas (e.g., CH_4, H_2, CO, CO_2 and N_2) and ii) average gasification bed temperature. The analytical model is able to predict the average gasification bed temperature; therefore, the model has been validated also comparing the temperature values obtained from simulation with the experimental temperature values. The comparison has been carried out using the data reported in [16]. The results of such a comparison, as a function of ER, for the considered case studies, are shown from Figure 2 to Figure 4. In Table 4, the experimental results and the corresponding deviations between the experimental values and numerical results obtained using the developed models have been reported.

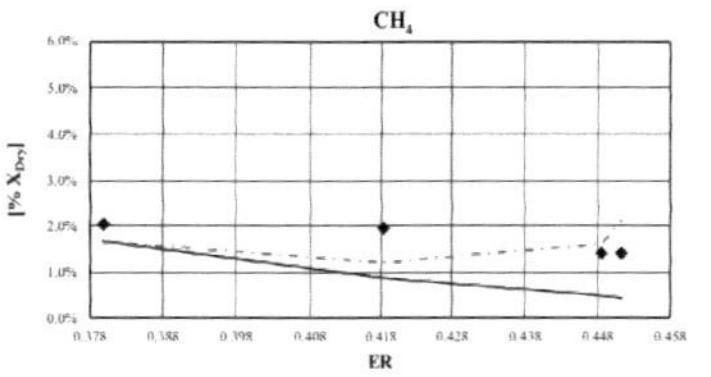

a)

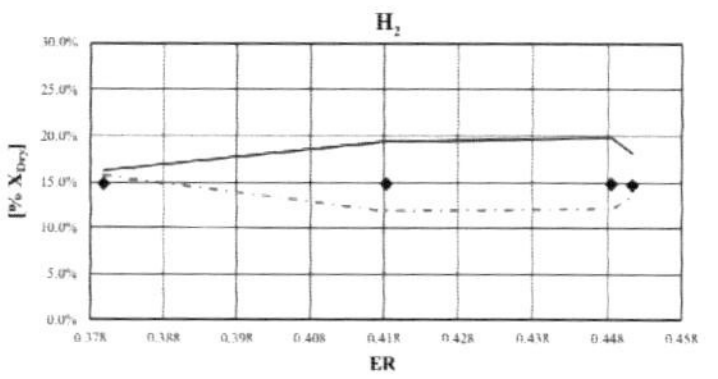

b)

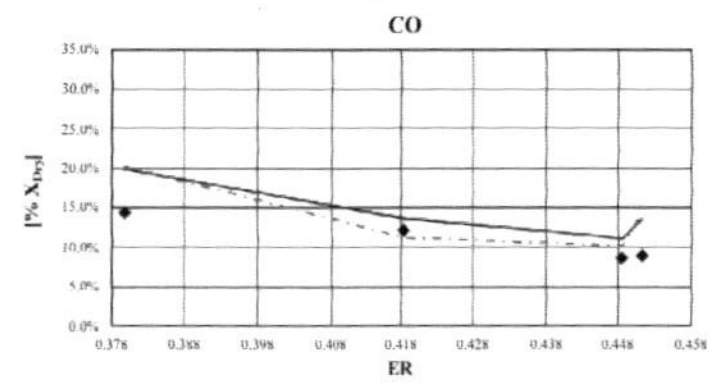

c)

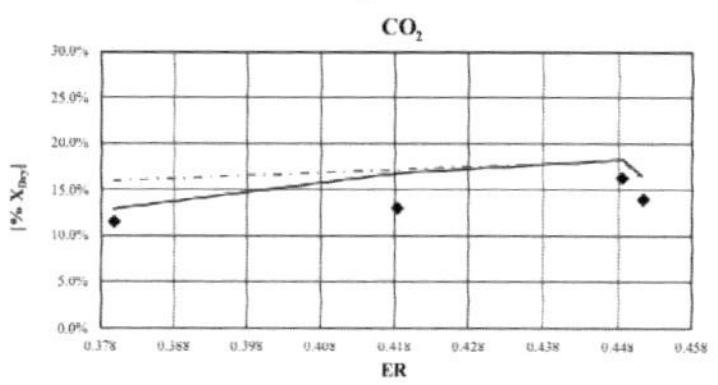

d)

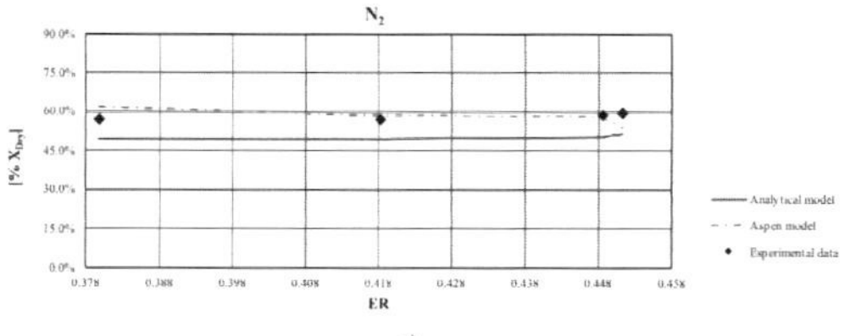

e)

Figure 2. Comparison of syngas mole fraction composition between results of analytical and Aspen models and experimental data reported in [16]: a) CH_4 mole fraction; b) H_2 mole fraction; c) CO mole fraction; d) CO_2 mole fraction; e) N_2 mole fraction.

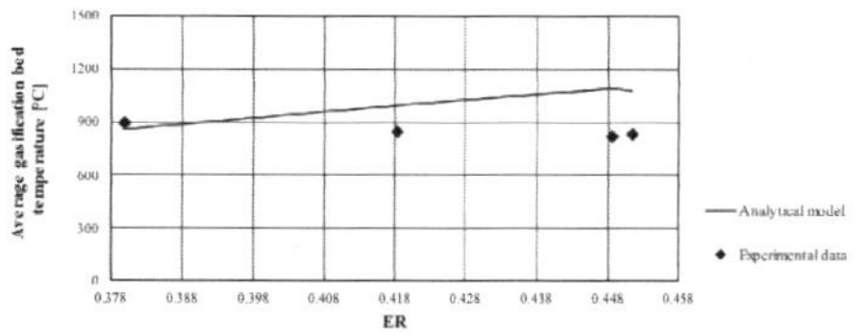

Figure 3. Comparison of average gasification bed temperature between the analytical model results and experimental data reported in [16].

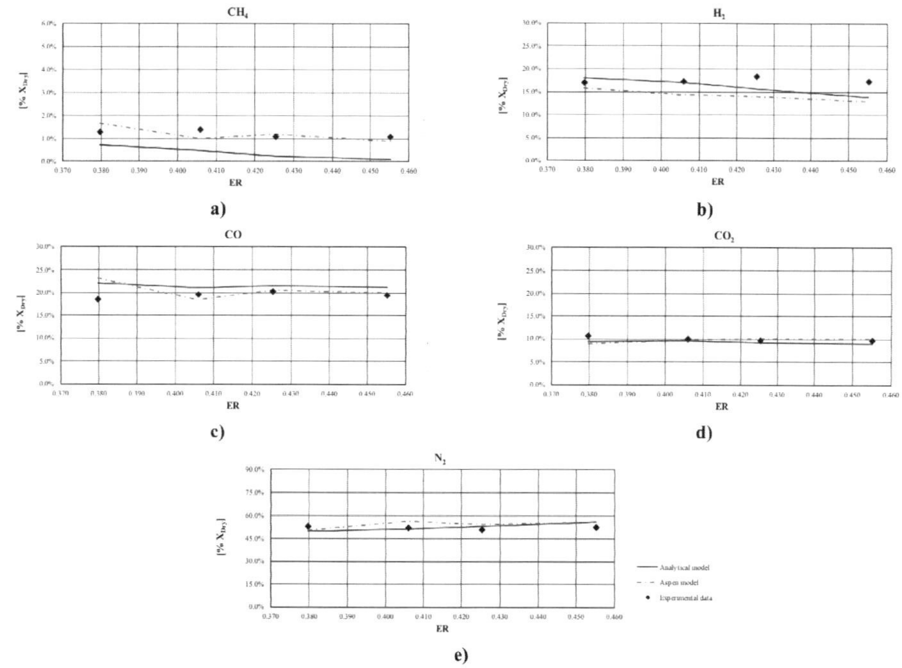

Figure 4. Comparison of syngas mole fraction composition between results of analytical and Aspen models and experimental data reported in [17]: a) CH_4 mole fraction; b) H_2 mole fraction; c) CO mole fraction; d) CO_2 mole fraction; e) N_2 mole fraction.

Table 4. Deviations between the experimental values and numerical results obtained using the developed models for the case studies analysed.

Biomass	No. Test		CH_4	H_2	CO	CO_2	N_2	Average gasification bed temperature [°C]
Hazelnut shell	1	Exp. data	0.0140	0.1477	0.0856	0.1633	0.5867	821.0
		Analytical model deviation	-64.3%	34.4%	29.0%	11.8%	-14.2%	33.0%
		Aspen model deviation	14.3%	-18.1%	19.2%	10.8%	-1.1%	-
	2	Exp. data	0.0140	0.1462	0.0892	0.1392	0.5967	833.0
		Analytical model deviation	-68.6%	24.1%	50.2%	18.4%	-13.6%	29.6%
		Aspen model deviation	50.0%	-7.7%	60.3%	15.7%	-9.5%	-
	3	Exp. data	0.0194	0.1477	0.1208	0.1302	0.5696	846.0
		Analytical model deviation	-55.7%	30.9%	13.4%	28.8%	-13.4%	17.8%
		Aspen model deviation	-38.1%	-19.4%	-7.3%	32.1%	2.5%	-
	4	Exp. data	0.0203	0.1477	0.1441	0.1142	0.5705	896.0
		Analytical model deviation	-17.7%	9.9%	37.8%	13.3%	-13.6%	-4.3%
		Aspen model deviation	-18.2%	6.4%	39.6%	40.0%	8.4%	-
Rubber wood	5	Exp. data	0.0140	0.1720	0.1960	0.0990	0.5190	-
		Analytical model deviation	-65.7%	-0.6%	7.4%	-3.2%	-0.8%	-
		Aspen model deviation	-28.6%	-16.9%	-6.1%	0.0%	8.9%	-
	6	Exp. data	0.0110	0.1830	0.2020	0.0970	0.5072	-
		Analytical model deviation	-78.2%	-14.0%	6.6%	-5.1%	5.0%	-
		Aspen model deviation	9.1%	-23.0%	1.5%	4.1%	6.7%	-
	7	Exp. data	0.0110	0.1720	0.1940	0.0970	0.5260	-
		Analytical model deviation	-90.0%	-19.4%	9.3%	-7.0%	6.1%	-
		Aspen model deviation	-18.2%	-24.4%	3.1%	4.1%	6.5%	-
	8	Exp. data	0.0130	0.1700	0.1840	0.1060	0.5270	-
		Analytical model deviation	-43.1%	6.1%	19.9%	-10.2%	-5.8%	-
		Aspen model deviation	27.7%	-7.2%	25.9%	-15.6%	-4.3%	-

From the analysis of the obtained results, it can be observed that CH_4 and H_2 molar fractions evaluated using the analytical model are generally characterized by a higher deviation from experimental values compared to the other synthesis gas components. In detail, CH_4 molar concentration is underestimated and decreases as ER increases for all case studies analysed. The analytical model overestimates the H_2 molar fraction in most of the cases. A good agreement between the results of the analytical model and experimental values can be observed as regards the CO and CO_2 molar concentrations for all case studies analysed. The N_2 content tends to be underestimated by the analytical model, probably because of the restrictive hypothesis of complete carbon conversion on which the model is based. Concerning the evaluation of average gasification bed temperature, as expected, as the ER increases the temperature increases. Since the gasifier is assumed to be adiabatic, analytical model generally overestimates the gasification temperature and the deviation between numerical results and experiments increases as the ER increases. As regards the Aspen model, a generally satisfactory agreement between the numerical results and the experimental data can be observed. More in detail, the concentration of CH_4, which slightly decreases with the ER, is well predicted by the model. Also the trend of H_2 concentration, which slightly varies with the ER, is fairly predicted by the model. The concentrations of CO and CO_2 usually show an inverse trend due to the water shift reaction in which they are involved. Their trend, as well as that of N_2, is in excellent agreement with the experimental results. As expected, obtained results evidence that the major limitation of both models relies in the thermodynamic equilibrium assumption and in neglecting tar and char.

5. Conclusions

In the present paper the authors present an analytical model based on the thermodynamic equilibrium approach and a numerical model using Aspen Plus in order to investigate the process of biomass downdraft gasification with air as gasification agent. The analytical and numerical models were compared with data from two experimental campaigns available in the scientific literature for downdraft gasifiers. Obtained results are shown in relation with ER. Globally, they are comparable with the test data for both models. Even though the proposed models are based on the simplifying assumption that the chemical reactions involving in the process reach the thermochemical equilibrium inside the gasifier reactor, the strengthens of the proposed methods consist of: i) not requiring details of system geometry or estimate the necessary time to reach that equilibrium; ii) employing a limited number of input data (e.g., biomass flow rate, oxidant agent flow rate, biomass chemical composition); iii) describing the biomass gasification process without the necessity of employing correlation for the calculation of the biomass gasification temperature. However, both developed models will have to improve by taking into account both char and tar production in the balance equations.

6. References

[1] Basu P 2010 *Biomass gasification and pyrolysis: practical design and theory* (Burlington, MA: Academic Press)

[2] Gagliano A, Nocera F, Patania F, Bruno M and Castaldo D G 2016 A robust numerical model for characterizing the syngas composition in a downdraft gasification process *Comptes Rendus Chim.* **19** 441–9

[3] Mansaray K G, Ghaly A E, Al-Taweel A M, Hamdullahpur F and Ugursal V I 1999 Air gasification of rice husk in a dual distributor type fluidized bed gasifier *Biomass Bioenergy* **17** 315–32

[4] Zaman S A and Ghosh S 2021 A generic input–output approach in developing and optimizing an Aspen plus steam-gasification model for biomass *Bioresour. Technol.* **337** 125412

[5] Sansaniwal S K, Pal K, Rosen M A and Tyagi S K 2017 Recent advances in the development of biomass gasification technology: A comprehensive review *Renew. Sustain. Energy Rev.* **72** 363–84

[6] Ramos A, Monteiro E and Rouboa A 2019 Numerical approaches and comprehensive models for gasification process: A review *Renew. Sustain. Energy Rev.* **110** 188–206

[7] Rodrigues R, Muniz A R and Marcilio N R 2016 EVALUATION OF BIOMASS AND COAL CO-GASIFICATION OF BRAZILIAN FEEDSTOCK USING A CHEMICAL EQUILIBRIUM MODEL *Braz. J. Chem. Eng.* **33** 401–14

[8] I. Petersen and J. Werther 2005 Experimental investigation and modeling of gasification of sewage sludge in the circulating fluidized bed *Chem. Eng. Process.* **44** 717–36

[9] J. S. Shuen, K. H. Chen, and Y. Choi 1993 A coupled implicit method for chemical non-equilibrium flows at all speeds *J. Comput. Phys.* **106** 306–18

[10] Vayenas C G, Debenedetti P G, Yentekakis I and Hegedus L L 1985 Cross-flow, solid-state electrochemical reactors: a steady state analysis *Ind. Eng. Chem. Fundam.* **24** 316–24

[11] Zoski C G 2007 *Handbook of electrochemistry* (Amsterdam ; Boston: Elsevier)

[12] Petersen I and Werther J 2005 Experimental investigation and modeling of gasification of sewage sludge in the circulating fluidized bed *Chem. Eng. Process. Process Intensif.* **44** 717–36

[13] Chapman S, Cowling T G and Park D 1962 The Mathematical Theory of Non-Uniform Gases *Am. J. Phys.* **30** 389–389

[14] Jarungthammachote S and Dutta A 2007 Thermodynamic equilibrium model and second law analysis of a downdraft waste gasifier *Energy* **32** 1660–9

[15] Channiwala S A and Parikh P P 2002 A unified correlation for estimating HHV of solid, liquid and gaseous fuels *Fuel* **81** 1051–63

[16] Dogru M, Howarth C R, Akay G, Keskinler B and Malik A A 2002 Gasification of hazelnut shells in a downdraft gasifier *Energy* **27** 415–27

[17] Jayah T H, Aye L, Fuller R J and Stewart D F 2003 Computer simulation of a downdraft wood gasifier for tea drying *Biomass Bioenergy* **25** 459–69

Computational Model For The Estimation Of Thermo-Energetic Properties In Dynamic Regime Of Existing Building Components

Costantino Carlo Mastino[1], Roberto Ricciu[1], Roberto Baccoli[1], Chiara Salaris[1], Roberto Innamorati[1], Andrea Frattolillo[1], Antonio Pacitto[2]

[1] D.I.C.A.AR University of Cagliari, Cagliari, Italy
mastino@unica.it, ricciu@unica.it, mastino@unica.it, rbaccoli@unica.it, chiara.salaris@unica.it, iroberto@unica.it andrea.frattolillo@unica.it,

[2] Department of Civil and Mechanical Engineering, University of Cassino and Southern Lazio, Cassino, Italy
antonio.pacitto@unicas.it

mastino@unica.it

Abstract. The guidelines of the European community towards a low-carbon economic society identify one of the most important scenarios in the energy efficiency of existing buildings. The discrepancy between the requirement and availability of free heat (endogenous heat, solar radiation) in certain hours of the day and operating conditions, makes the steady-state hypothesis generally inappropriate. In particular, the oscillating component of the transmitted flow, compared to the average temperature difference, is regressive in winter and dominant in summer. From this it follows the reliability of the stationary forecast models in winter and the need for dynamic forecast models in summer. The dominance of the continental climate in the EU, compared to the Mediterranean one, led to the actual delay in the development of dynamic models, especially at a regulatory level. In this paper, a methodology for assessing the dynamic properties of a building component is evaluated. The methodology, based on heat transmission equations implements a numerical model for existing building components whose input data can be obtained from experimental measurements. The developed model has been used to estimate the energetic and thermal behaviour of a building envelope subjected to energy efficiency measures.

1. Introduction

In the world the design of buildings is increasingly addressed to the use of digital procedures. The European Community has recently published directives 2014/24/EU [1] that provide for the digitization of processes. Italy accepted these directives has made Obligatory the use of the BIM (Building Information Modeling) for public works by providing for different time deadlines starting from 2019 up to 2025 where all public works concerning buildings must be managed digitally [2–7]. The flowchart of the procedures used in this work comprising: initial performance measurement, simulation of intervention solutions based on the use of the BIM model are described in Figure 1. The characterization from the thermal point of view of the components of the building envelope, of which there is often no information, and subsequently developing intervention solutions can be considered a reverse engineering problem as shown for numerous themes [8–11] . Energy design is among the digital procedures and the verification of dynamic energy parameters, it is fundamental to the realization of nZEB buildings. The Energy Performance of Buildings Is one of the Main Focuses of the European Union. In the year 2010, the Europe Council issued the Energy Performance of Buildings Directive [12] subsequently updated in 2018 [13], setting the minimum energy performance requirements for new buildings and major renovation projects [2]. In Italy the Directive was implemented with the national laws n. 90/13 and successive Technical Standard, such as the UNI/TS 11300 series (Italian National Unification/Technical Specification) [14] and ISO 52016 [15], were enacted. These technical specifications/standards are based on dynamic mathematical models that are closer to the measured thermal behavior of the building.

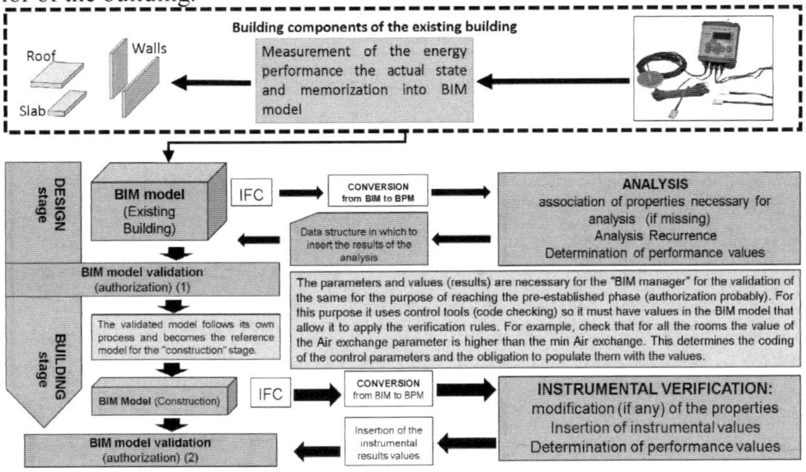

Figure 1 - Flowchart - Data measurement and BIM model

For the buildings components, these technical standards were developed the last ten years [15–17] and are more useful for temperate and sunny climates. The most important component of the building, which characterizes its dynamic behavior, is the building envelope. this It can be either be massive or lightweight with Solar shields or not, but should serve to guarantee the thermos-hygrometric comfort of its internal spaces [18]. This model utilizes the three mechanisms of heat transfer: radiative, conductive and convective. For most stratigraphy layers of the building components, for example the exterior walls of buildings, the heat transfer it happens through the conductive mechanism. The current use profiles of the private dwellings, public and services buildings by the people, highlight a dynamic variation of loads associated with the prevalence of the solar radiation in the Mediterranean climate, that also dependent on the variation of the different seasons. All of the abovementioned effects lead to a preference for dynamic physical models for energy analysis and simulation of energy behavior of the buildings in the Mediterranean area and to the realization of stratigraphies with different types of insulating materials [10]. One of the methods for dynamic analysis of the casing based on transfer matrices is explained in the ISO 13786 [17,19–21]. With the method described in ISO 13786 it is possible to use the transfer matrices method to evaluate the dynamic thermal heat transmittance, the phase shift that between the temperature peak and the incoming heat flow peak and the decrement factor (which indicates the ratio by the module of thermal transmittance and periodic thermal transmittance). In some works [22] is shown the wall placed in the climate chamber with the open shells and the sinusoidal. In Italy the SEN (National Energy Strategy) highlighted how the energy efficiency of existing buildings is one of the primary objectives as these buildings are the most energy-sized ones. Most of the existing buildings have no historical or technical documentation. UNI 10351 [23] and UNI 10355 [24] report thermal characteristics of the different component for old typical building walls or typical masonry stratigraphy.

Journal of Physics: Conference Series 2177 (2022) 012029 doi:10.1088/1742-6596/2177/1/012029

However, in general, it is impossible to establish exactly the correspondence between the existing walls and the recommended in UNI 10351 and 10355, unless it is through the trials of coring[25]. This test is highly invasive, and it is considered dangerous to the preservation of old buildings. In this work, the authors propose one reverse engineering methodology to model the existing state and find project solutions for energy efficiency of building components an inverse/indirect method to derive the thermal properties of building walls starting from thermal measurements. The inverse method is based on the direct method proposed in ISO 13786, which utilizes the harmonic analysis of integration of the Fourier heat equation probably published for the first time in the 1906 edition [18]. The results, carried out through a rebuilding of the reverse harmonic method of the Fourier heat equation, allow to rebuilt the thermal properties of the wall by simple measurement of the temperature and thermal flux both on the outside wall and the inside wall. Hence, in this work, two main sections are presented: the direct harmonic method and the in- verse harmonic method whit the use the BIM platform. In this case, to understand more clearly the second section, the first section is detailed with some flowcharts steps showed in figure 1 and figure 2. At the end of the work, a practical application of the proposed method can calculate the transfer matrix of the thermal property of a “dummy” wall because the homogeneous and isotropic material has the same thermal characteristics of a “real” wall with all of its layers when the magnitude of phase shift and the decrement factor are known. This “dummy” wall has the same properties of a wall with several layers. In the application of nondestructive methods, Brian H. and Mohamed T. K., or Malhotra V.M. and Carino N. J. and Pia G. et al. [26–28], agree with the statement that is possible to measure the average density ρ of the component “in situ”. In the procedure used by some autors [17,22], it is possible to have the specific heat c (J/kgK) of the wall in study-case. In this case, it will be possible to adopt the method for energy-retrofitting in a dynamic application. The model described by ISO 13786 [17] was implemented in the GEAR BIM Tools software, able to interact with the BIM models the interoperable data standard called IFC[29,30].

2. Energy benefit evaluation procedures through BIM Tools

The energy retrofit of existing buildings is a fundamental aspect to reason the objectives set by the European Board Has Recently Issued The EU Directive "Roadmap for Moving to Competitive -Low Carbon Economy in 2050" [31,32] Which Lays Down Recommendations to Achieve The Reduction of Greenhouse Gas Emissions Under 80 For Cent of the 1990 Level by Such a Date. Often for existing buildings the stratigraphies of building components and a way to ascertain the starting performance and to perform measures are often known. The procedure implemented, developed for the physical-mathematical part by Ricciu and others, allows through the use of the BIM model of the existing state and the data measured on the construction components the execution of intervention simulations to improve energy performance. This process is schematized in Figure 2

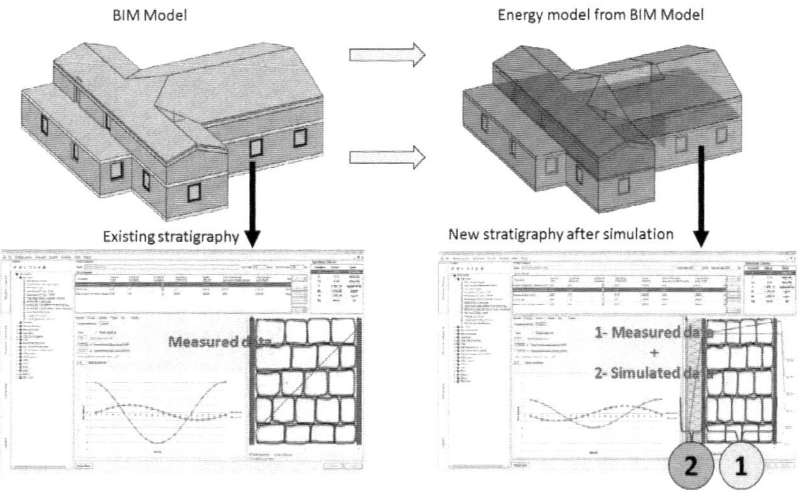

Figure 2- Generation of alternative solutions for stratigraphy based on the BIM energy model

The unified handling of a set of physical problems can be seen as an innovative aspect of the developed model presented here that integrates effectively with the three dimensional architectural CAD. The model does not need any ad hoc three dimensional geometric modelling, but can make use of the same model developed for the architectural project.
The physical, geometrical, environmental, characteristics, that are necessary for the simulation of the performance of a building, through the use of BIM are many. The tools that let use the geometrical data or more generally the data of the building from a three-dimensional model, modify and complete the information on the building, on climate, or on different environmental aspects with the need for energy simulations are numerous [33–35]. Several of them allow to import gbXML and IFC formats automatically or in a semi-automatic way to simplify the input of geometrical data and sometimes of those related to the physical properties of building components. A very important aspect when working with different data formats, is the congruence of geometric information data, which is not always guaranteed, taking into account the input mode required by the software or by the calculation model used for energy simulation [3]. This fundamental aspect, if not considered, can spearhead to results affected by errors [3]. The procedure tested in this work encountered a considerable advantage in using the data on the energetic performance of the measured construction components and entered within the BIM base data. The advantage of this approach consists in the fact that for the energy designer [4] it is faster to perform simulations and obtain scenarios of aid to decision-making. This aspect is reflected on the quality of the design result as it is possible to perform many more simulations and optimize the result compared, for example, to the beneficial cost ratio.

3. Case study

The methodology described was applied to a case study consisting of a building from the University of Cagliari shown in Figure 3 realized at the beginning of 1900.

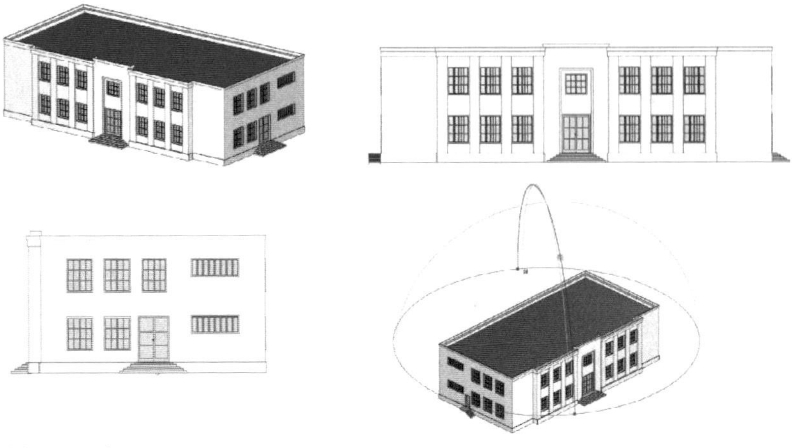

Figure 3- BIM model case study

On the study case various measures have been carried out to caress the walls of the building envelope from the thermo-energy point of view. The result of these measurements is shown in Figure 4. The graphs of the measurements show how the internal and external temperatures varies and how the relative thermal flow is varied to difference between internal and external temperature. The data have been reported raw as well as downloaded from the data logger and loaded into the GEAR model [33]. Performing the measures to characterize the external masonry of the building has been tested according to the method adopted by some autors to determine the oscocient thermal flow, surface temperatures completely referred to a period of 24 hours. After this it has been determined the transfer matrix to assign to the wall so as to be able to simulate the behavior of this with the new layers of insulating material. The new layers of insulating material are to be considered as an energy-efficient intervention on the existing building. In particular on existing walls.

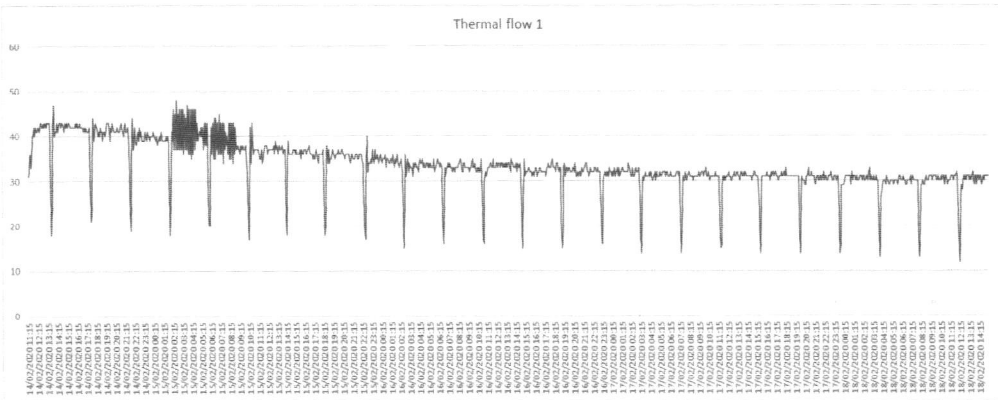

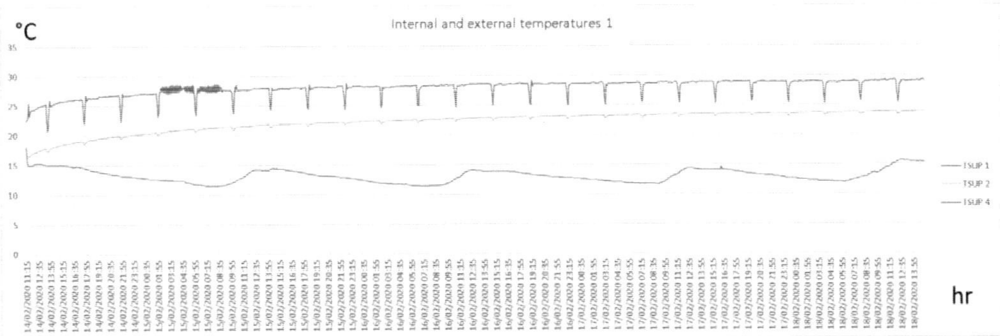
°C
Internal and external temperatures 1
TSUP 1
TSUP 2
TSUP 4
hr

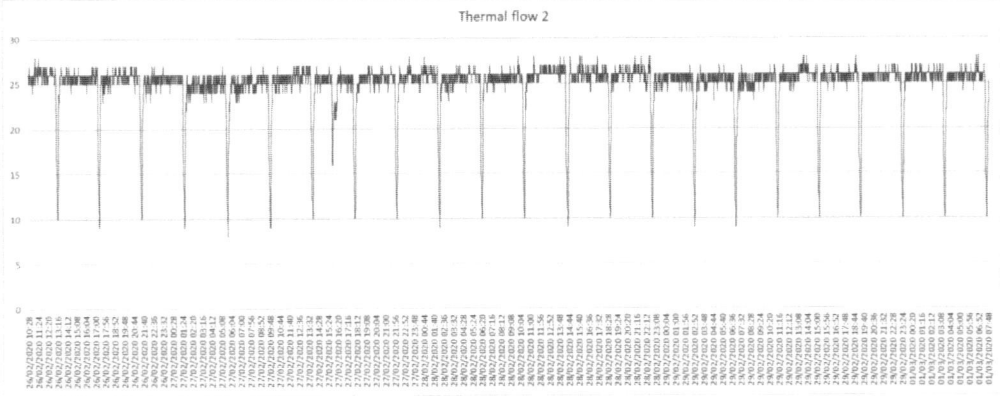
Thermal flow 2

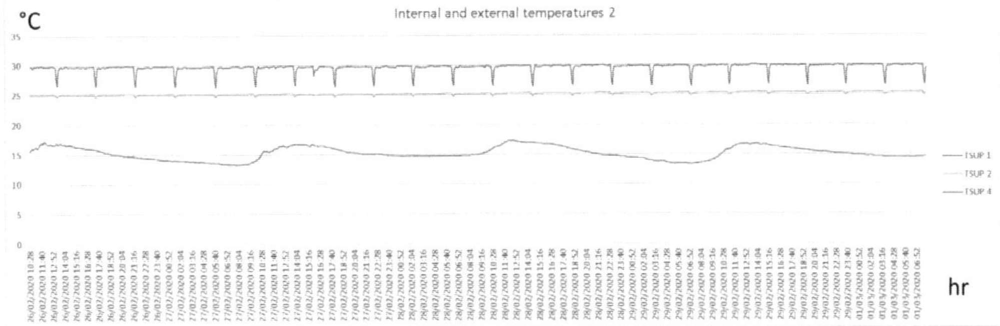
°C
Internal and external temperatures 2
TSUP 1
TSUP 2
TSUP 4
hr

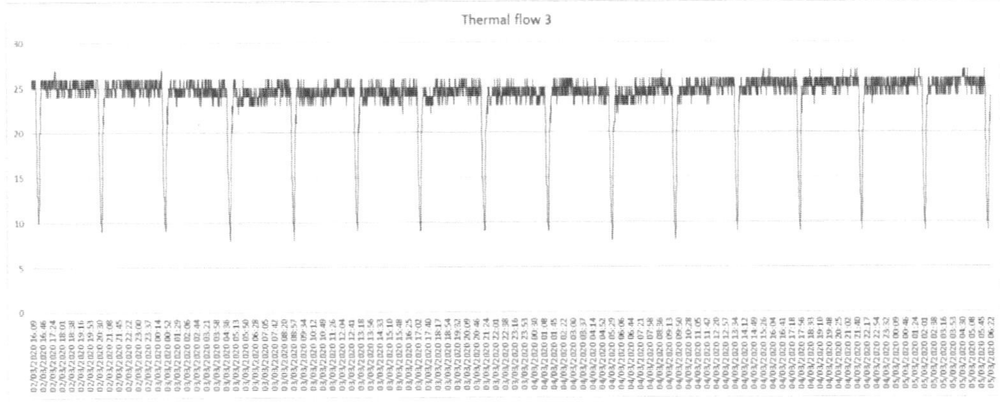
Thermal flow 3

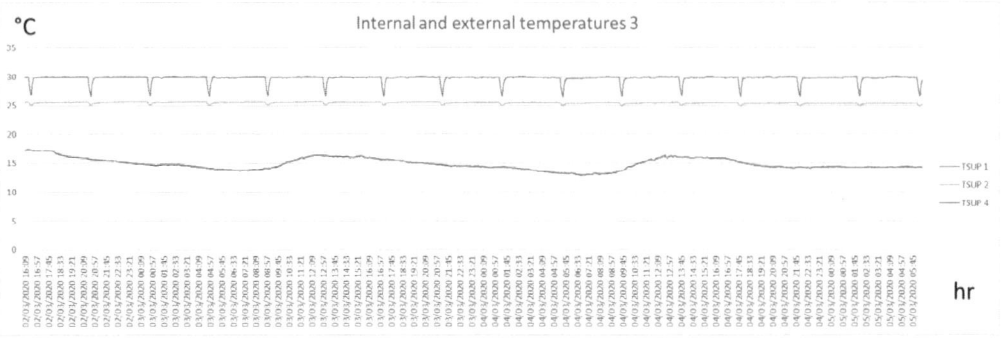

Figure 4- Thermal flow and temperatures measurements on the study case

In the BIM model created for the case the energy zoning has been defined. The zoning conscribed in creating groups of rooms with the themes of the energy point of view. In the Language BIM of the IFC standard, the Rooms are called spaces. The new layer of thermal insulator designed for the type of walls on which the measurement campaign was performed at all the walls that delimit the spaces. Figure 5 shows the graphic result of this zoning performed within the BIM platform.

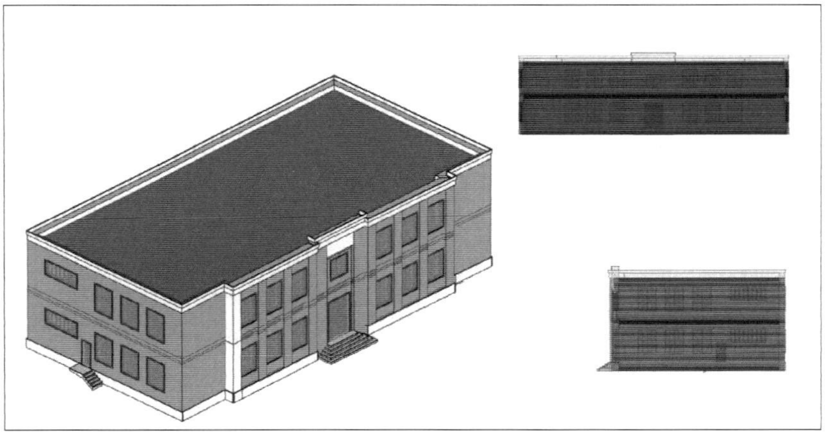

Figure 5- Energetic zones representation with BIM platform

According to ISO 13786 [17] and Carslaw H.S., and Jaeger J. Cresults, it is possible to use the direct matrix method to evaluate the periodic thermal heat transmittance; the phase shift be- tween the external temperature peak and the incoming heat flow peak; and the decrement factor. The method works with a single (or multi) layer wall, crossed by a sinusoidal heat flow generated by the sinusoidal variation of the temperature on the two sides of the wall, assuming known the thermal conductivity (W/mK), the specific heat (J/kgK) and the density (kg/m^3) of the entire wall. That is a typical thermal forcing on the building envelopes during the daily climatic conditions at middle latitudes. In this paper the authors propose the application of an inverse approach developed over ISO 13786 to obtain the thermal properties of the building components by measuring the sinusoidal periodic steady state (temperature and thermal flux) on the two faces of the wall. In this method the decrement factor and the phase shift between the input and output heat wave are measured and used to build the transfer matrix, from which the thermal conductivity (W/mK) and the specific heat (J/kgK) of the entire wall are carried out. Technically, the proposed method could be applied whenever there is a transfer of energy varying over time according to the sinusoidal law. If the variation was not sinusoidal, the Fourier harmonic analysis method could be applied. In this paper the works of some authors [17,22] method is mathematically solved and tested on a sample building component.

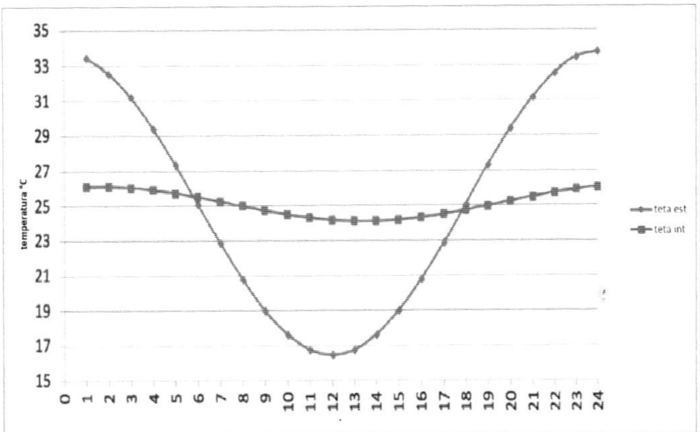

Figure 6- Energetic zones representation with BIM platform

Figure 6 shows the oscillations of external and internal temperatures in the period of 24hr considered. As can be observed the oscillation of the external temperature is much more accentuated than the internal one.

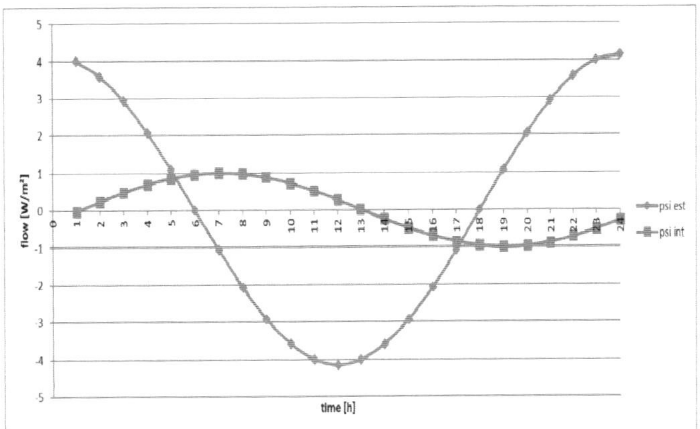

Figure 7- Energetic zones representation with BIM platform

In the same way in figure 7, the trend of the hypothesized thermal flow is shown on the internal and external surface of the building component. Finally, in figure 8 the workflow followed in the BIM Tool application in which it is one inverse methodology of method ISO 13786 explained by some authors [17,22], to apply it to the case study considered.

Journal of Physics: Conference Series 2177 (2022) 012029 doi:10.1088/1742-6596/2177/1/012029

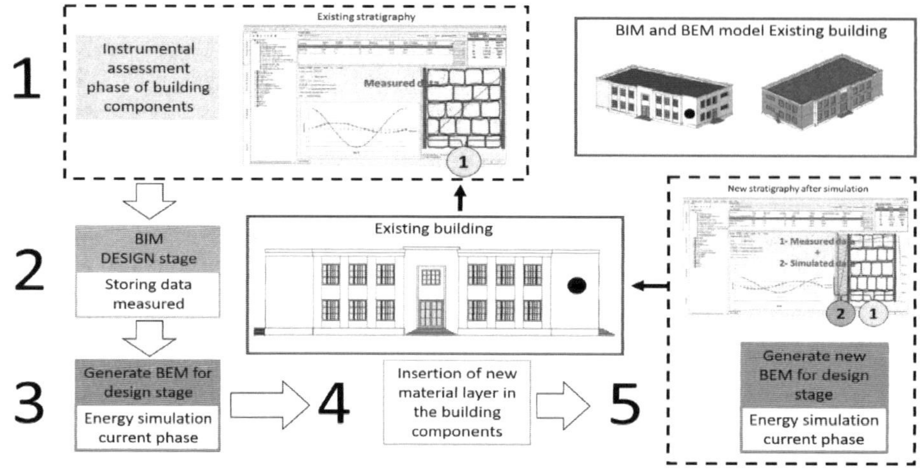

Figure 8- Energetic zones representation with BIM platform

The workflow described in Figure 8 In particular, steps 4 and 5 can be repeated several times to generate different intervention solutions. Everything can be used as an aid to the decision-making process to consider the different performance and economic aspects of the different proposed solutions.

4. Conclusions

In this work the calculation model provided by ISO 13786 was analyzed with the methodology proposed some autors. The inverse method based over ISO 13786 allows you to relate the data measured with data on the design phase.The model so described, useful in the planning phase of interventions on existing buildings makes it possible to operate project choices on the thermo-energy performance of the casing based on the measure of the provision of the individual building components. The model has been implemented within the Gear BIM Oriented software. The elaborations allowed to evaluate the utility of the model starting from the measurement of the ante-opera performance stored in the BIM model of the study case. On the BIM model the energy simulations that allowed the energy needs of the envelope according to the standards provided by the EPBD Two, following the designed thermal insulation intevents.

References

[1] Anon 2014 Directive 2014/24/EU of the European Parliament and of the Council of 26 February 2014 on public procurement and repealing Directive 2004/18/EC Text with EEA relevance

[2] Rezgui Y, Beach T and Rana O 2013 A GOVERNANCE APPROACH FOR BIM MANAGEMENT ACROSS LIFECYCLE AND SUPPLY CHAINS USING MIXED-MODES OF INFORMATION DELIVERY *J. Civ. Eng. Manag.* **19** 239–58

[3] Ivanova I, Kiesel K and Mahdavi A 2015 BIM-generated data models for EnergyPlus: A comparison of gbXML and IFC Formats BSA 2015, the Second IBPSA-Italy conference on "Building Simulation Applications (Bozen-Bolzano, Italy: bu,press) pp 407–14

[4] CIB W78 International Conference on Information technology R and Leonardo 2008 IFC BIM-Based Methodology for Semi- Automated Building Energy Performance Simulation Improving the management of construction projects through IT adoption (Santiago: Universidad de Talca, Departamento de Ingenier??a y Gesti??n de la Construcci??n)

[5] Borrmann A, Hochmuth M, König M, Liebich T and Singer D Germany's governmental BIM initiative – Assessing the performance of the BIM pilot projects 8

[6] Cumo F, Sferra A, Piras G, Mancini F, Tiberi M, Sforzini V, Vollaro B D L and Spiridigliozzi G LA METODOLOGIA BIM COME STRUMENTO PER UNA EFFICIENTE PROGETTAZIONE E GESTIONE DEGLI IMPIANTI DEGLI EDIFICI 78

[7] Czmoch I and Pękala A 2014 Traditional Design versus BIM Based Design *Procedia Eng.* **91** 210–5

[8] Khoukhi M, Abdelbaqi S, Hassan A and Darsaleh A 2021 Impact of dynamic thermal conductivity change of EPS insulation on temperature variation through a wall assembly *Case Stud. Therm. Eng.* **25** 100917

[9] Naganathan H, Chong W K and Ye N 2015 Learning Energy Consumption and Demand Models through Data Mining for Reverse Engineering *Procedia Eng.* **118** 1319–24

[10] Balaji N C, Mani M and Venkatarama Reddy B V 2019 Dynamic thermal performance of conventional and alternative building wall envelopes *J. Build. Eng.* **21** 373–95

[11] Lin C H and Ferng Y M 2016 Investigating thermal mixing and reverse flow characteristics in a T-junction using CFD methodology *Appl. Therm. Eng.* **102** 733–41

[12] Anon Directive 2010/31/EU of the European Parliament and of the Council of 19 May 2010 on the energy performance of buildings 23

[13] Anon Directive (EU) 2018/ of the European Parliament and of the Council of 30 May 2018 amending Directive 2010/31/EU on the energy performance of buildings and Directive 2012/27/EU on energy efficiency 17

[14] UNI 2014 UNI TS 11300-1Prestazioni energetiche degli edifici—Parte 1: Determinazione del fabbisogno di energia termica dell'edificio per la climatizzazione estiva ed invernale

[15] Anon ISO 52016 Energy performance of buildings—Energy needs for heating and cooling, internal temperatures and sensible and latent heat loads —Part 1: Calculation procedures

[16] Anon ISO 13790 "Energy performance of buildings - Calculation of energy use for space heating and cooling

[17] Anon ISO 13786 -Thermal performance of building components—Dynamic thermal characteristics—Calculation methods

[18] d'Ambrosio Alfano F R, Olesen B W and Palella B I 2017 Povl Ole Fanger's impact ten years later *Energy Build.* **152** 243–9

[19] Corrado V and Paduos S 2016 New equivalent parameters for thermal characterization of opaque building envelope components under dynamic conditions *Appl. Energy* **163** 313–22

[20] Aste N, Leonforte F, Manfren M and Mazzon M 2015 Thermal inertia and energy efficiency – Parametric simulation assessment on a calibrated case study *Appl. Energy* **145** 111–23

[21] Asan H and Sancaktar Y S 1998 Effects of Wall's thermophysical properties on time lag and decrement factor *Energy Build.* **28** 159–66

[22] Ricciu R, Ragnedda F, Galatioto A, Gana S, Besalduch L A and Frattolillo A 2019 Thermal properties of building walls: Indirect estimation using the inverse method with a harmonic approach *Energy Build.* **187** 257–68

[23] Anon UNI 10351:2021 Materiali e prodotti per edilizia - Proprietàtermoigrometriche - Procedura per la scelta dei valori di progetto

[24] Anon UNI 10355:1994 Murature e solai. Valori della resistenza termica e metodo di calcolo

[25] Di Bella A, Mastino C C, Barbaresi L, Granzotto N, Baccoli R and Morandi F Comparative study of prediction methods and field measurements of the acoustic performances of buildings made with CLT elements 10

[26] Hobbs B and Tchoketch Kebir M 2007 Non-destructive testing techniques for the forensic engineering investigation of reinforced concrete buildings *Forensic Sci. Int.* **167** 167–72

[27] V.M. Malhotra and N.J. Carino 2003 *Handbook On Non destructive Testing of Concrete, Second Edition* (CRC Press)

[28] Pia G, Casnedi L, Ricciu R, Besalduch L A, Cocco O, Murru A, Meloni P and Sanna U 2016 Thermal properties of porous stones in cultural heritage: Experimental findings and predictions using an intermingled fractal units model *Energy Build.* **118** 232–9

[29] Mastino C C, Baccoli R, Frattolillo A, Marini M and Bella A D 2017 The Building Information Model and the IFC Standard: Analysis of the Characteristics Necessary for the Acoustic and Energy Simulation of Buildings *3rd IBPSA-Italy conference Bozen-Bolzano* Building Simulation Applications BSA 2017 (Bozen-Bolzano: bu,press - Bozen-Bolzano University Press Free University of Bozen-Bolzano) pp 479–86

[30] Marini M, Mastino C C, Baccoli R and Frattolillo A 2018 BIM AND PLANT SYSTEMS: A SPECIFIC ASSESSMENT *Energy Procedia* **148** 623–30

[31] Anon COM/2011/112 Roadmap for moving to a competitive low-carbon economy in 2050 - COMMUNICATION FROM THE COMMISSION TO THE EUROPEAN PARLIAMENT, THE COUNCIL, THE EUROPEAN ECONOMIC AND SOCIAL COMMITTEE AND THE COMMITTEE OF THE REGIONS

[32] Anon COM/2011/885 Energy Roadmap 2050 the December

[33] Marini M, Baccoli R, Mastino C C, Da Pos V and Tóth Z 2015 A new computational model: G.E.A.R. Graphical Expert Analytical Relations *Proceedings of BSA 2015, the Second IBPSA-Italy conference on "Building Simulation Applications"* BSA 2015 (Bolzano: bu,press - Bozen-Bolzano University Press Free University of Bozen-Bolzano) pp 305–12

[34] Tobias Maile, Martin Fischer, and Vladimir Bazjanac 2007 Building Energy Performance Simulation Tools - a Life-Cycle and Interoperable Perspective

[35] Pinheiro S, O'Donnell J, Wimmer R, Bazjanac V, Muhic S, Maile T, Frisch J and van Treeck C MODEL VIEW DEFINITION FOR ADVANCED BUILDING ENERGY PERFORMANCE SIMULATION 9

Functionally-graded foams for volumetric solar receivers

M. Iasiello[1], W. K. S. Chiu[2], A. Andreozzi[1], N. Bianco[1] and V. Naso[1]
[1]Dipartimento di Ingegneria Industriale, Università degli Studi di Napoli Federico II, Piazzale Tecchio 80 - 80125, Napoli, Italy
[2]Department of Mechanical Engineering, University of Connecticut, 191 Auditorium Road, Storrs, CT, 06269, USA

Corresponding author: Marcello Iasiello – marcello.iasiello@unina.it

Abstract
The open volumetric receiver, one of the most important components of a Concentrated Solar Power (CSP) plant, is made up by a ceramic foam on which the concentrated solar radiation impinges. Ceramic foams are employed in volumetric solar receivers because of their high specific surfaces and their operating temperatures higher than those of metal foams. Thermo-fluid-dynamics in the graded ceramic foam of a volumetric solar air receiver for concentrated solar power is investigated numerically. Variable porosities and Pores Per Inch (PPI), according to different power laws, are accounted for. Governing equations are written with the Volume Averaging Technique (VAT) and are solved with the commercial software Comsol Multiphysics. The effects of different porosity and PPI laws, on the fluid velocity, pressure drop and temperatures, under different thermo-fluid-dynamic conditions, are highlighted.

Nomenclature

Symbol	Quantity	SI Unit	Symbol	Quantity	SI Unit
a	parameter in Eq.3	---	V	volume of the foam	m^3
C_f	inertial factor	---	z	axial coordinate	m
C_p	heat capacity	J/kg K	**Greek**		
d_c	cell size	m	β	extinction coefficient	---
e	emissivity	---	ε	porosity	---
h_c	convection heat transfer		μ	dynamic viscosity	kg/m s^2
	coefficient	W/m^2 K	ρ	density	kg/m^3
h_v	volumetric convection		σ	Stephan-Boltzmann constant	W/m^2 K^4
	heat transfer coefficient	W/m^3 K	ϕ	generic variable	
I	solar radiation intensity	W/m^2	**Superscript**		
k	thermal conductivity	W/m K	f	fluid	
K	permeability	m^2	s	solid	
L	length of the receiver	m	**Subscript**		
$\mathbf{n}$	normal vector	----	f	fluid	
p	pressure	Pa	in	inlet	
$\mathbf{q}_r$	radiative heat flux	W/m^2	max	maximum	
r	cylindrical coordinates	m	min	minimum	
R	radius of the receiver	m	out	outlet	
Re	Reynolds number	----	s	solid	
S	area of the foam struts	m^2	**Others**		
T	temperature	K	< >	volumetric average	
$\mathbf{u}$	velocity vector	m/s	—	average	
u'''	thermal power source	W/m^3			

1. Introduction

Concentrated solar power (CSP) plants are interesting alternatives to conventional fossil fuels in regions with high annual solar irradiation, because of their high efficiency in energy conversion. The open volumetric receiver, that is made up by a porous material onto which a concentrated solar radiation impinges, is one of the most important components of a CSP. Ceramic foams are promising materials in volumetric solar receivers thanks to their potential toward large specific surface area combined with beneficial pressure losses; besides, they can operate at temperatures higher than the melting temperature of the metal in metal foams.

The thermal performance of a silicon carbide ceramic foam as solar air receiver was investigated analytically by Bai [1]. The one-dimensional analysis of flow and heat transfer in unpressurized systems pointed out that an input solar energy flux limited the power capacity and the outlet air temperature enhancement. Wu et al. [2] predicted the temperature fields in the fluid and solid phases of the receiver using the Local Thermal Non-Equilibrium (LTNE) model. The effects of velocity, porosity, cell size and thermal conductivity of the solid on the temperature fields were also evaluated. The same authors [3] numerically investigated air convection heat transfer in the ceramic foam of a volumetric receiver. A sensitivity study on the heat transfer coefficient was carried out for different values of the porosity, air velocity and cell size. Sano et al. [4] investigated analytically the combined heat transfer in a silicon carbide ceramic receiver, using an LTNE model. The temperature and pressure fields were predicted. The pore diameter that optimized the receiver efficiency under the same pumping power was also found. The effect of geometry of the volumetric solar receiver on the temperature distribution and its efficiency were studied numerically by Fend et al. [5]. The comparison of predictions given by two numerical models with results of experimental tests exhibited a good agreement. The thermal performance of a ceramic receiver was investigated numerically by Wang et al. [6]. The effects of the heat flux boundary conditions, radiation heat loss, porosity, emissivity, flow mass and average particle diameter on the temperature distributions were pointed out. Roldan et al. [7], using a two-dimensional numerical model, predicted the thermal performance of the volumetric solar receiver of a solar furnace, for different configurations of the absorber. Simulations were compared to find the configuration that minimized flow instabilities and thermal stresses. The effect of thermal radiation in a SiC foam of a solar air receiver was investigated by Wang et al. [8]. The temperature distribution in solid and fluid phases was affected by the ratio of solid to fluid thermal conductivities and by the interphase convection. The impact of conduction and convection versus radiation was also analyzed. Forced convective heat transfer in a packed bed was investigated by Wang et al. [9], accounting for the collimated and diffusive thermal radiation transfer processes. The effects of porosity, pore diameter, ratio of the solid-to-fluid thermal conductivities, radiative characteristics on the temperature and heat flux distributions were analyzed. A significant effect of the thickness of the first porous layer on the temperature distribution and the pressure drop was found. The thermal performance of a cylindrical SiC ceramic foam in a volumetric solar receiver was investigated numerically by Andreozzi et al. [10], with a two-equation model for the energy equation. Temperature fields in the solid and fluid phases of the foam, fluid velocity and pressure drop, under different geometries and boundary conditions, were predicted. The parametric analysis of process parameters maximized the receiver efficiency.

In recent years, ceramic foam manufacturing can be more customized than in the past. Functionally-graded foams, having different morphological characteristics, such as porosity and Pores Per Inch (PPI), within the control volume, have been manufactured and investigated. Functionally-graded composite foams, with spatially variable characteristics, were overviewed by Udupa et al. [11]. Advances in the field were presented in a review by Xu and Zhang [12]. Xu and Zhao [13] manufactured functionally-graded metal foams by sintering different layers with Ag-Cu alloy sheets under a high-temperature muffle furnace environment. Chen et al. [14] numerically analyzed a double-porosity layer ceramic foam, finding higher air outlet temperatures and lower solid inlet temperatures for decreasing porosities. Heat transfer and pressure drop in a parallelepiped functionally-graded foam, with a uniform heat flux entering one of its sides, were investigated numerically by Iasiello et al. [15]. Porosity and cell size variable in the direction of the entering heat flux, according to different power-law functions, were considered. Nusselt numbers, friction factors and a Performance Evaluation Criteria for different power-laws, were predicted. Du et al. [16] optimized the radiative extinction coefficient in a volumetric solar receiver whose grade porosity varied in the flow direction. Du et al. [17] experimentally investigated heat transfer in a porous volumetric solar receiver with radially graded porosity and a non-uniform distribution of the solar irradiation onto the inlet section of the receiver.

The literature survey shows that graded foams enhance the thermal performance of solar receivers; however, the effects of different power laws functions of porosity and PPI on the thermal performance of graded foams deserve to be furtherly investigated.

Heat transfer and pressure drop in the functionally-graded ceramic foam of a cylindrical volumetric solar air receiver for concentrated solar power is numerically investigated in this paper. Both porosities and PPIs vary radially according to prescribed power laws are analyzed. Governing equations are written with the Volume Averaging Technique (VAT), under the assumption of different temperatures between solid and fluid phases. Radiation heat transfer is modeled by assuming the foam to be a semi-transparent medium with the radiative heat flux as a heat source in the solid phase equation; the heat source is modeled according to Beer-Lambert law. Equations are solved with the commercial code Comsol Multiphysics. The effects of graded porosity and PPI on the fluid velocity, pressure drop and temperatures, under different thermo-fluid-dynamic conditions, are highlighted.

2. Mathematical model

2.1. Geometry and variable characteristics of the receiver

The cylindrical volumetric receiver, filled with a silicon carbide (SiC) foam and a REFOS-like pressurized receiver are sketched in figure 1. The radius of the receiver is $R = 0.025$ m and its length is $L = 0.015$ m, which are values commonly used in ceramic foam volumetric solar air receivers [2]. Results reported in Subsection 3.3 will show that the chosen length of the receiver is larger than the thermal equilibrium length, that is the axial coordinate where the fluid temperature becomes nearly equal to the solid temperature [18]. The locally variable characteristics of the foam are: the porosity, $\varepsilon = V_f/V$, in the 0.80 – 0.90 range and Pores Per Inch (PPI) in the 10 – 40 range; both ranges are typical in commercial ceramic foams. PPI and ε are the independent parameters which uniquely define the foam morphology, all other characteristics depending on them. Air at uniform temperature, T_i, and velocity, $\mathbf{u}_i$, enters the cross section at $z = 0$, flows through the foam in the z direction, and exits the cross section at $z = L$.

Different values of the graded porosities and PPIs are analyzed in this work, in order to find the effects of the configuration on heat transfer and pressure drop. For the sake of comparison, when variable-porosity functions are taken into account, reference is made to the same average porosity, geometry and sizes of the volumetric receiver.

The dependence of the porosity, ε, on the radial coordinate, r, is taken into account with the correlations:

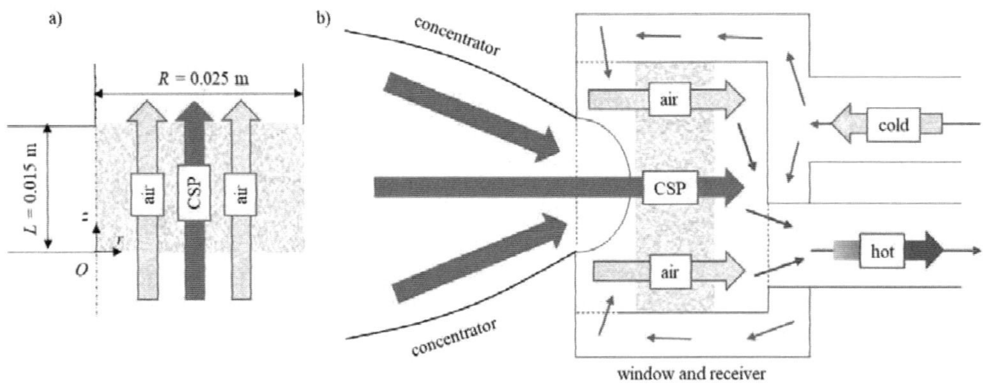

Figure 1. Sketch of a) the volumetric receiver; b) a REFOS-like pressurized receiver.

$$\varepsilon(r) = \varepsilon_{min} + (\varepsilon_{max} - \varepsilon_{min})\, r/R \tag{1}$$

for increasing porosity and

$$\varepsilon(r) = \varepsilon_{max} - (\varepsilon_{max} - \varepsilon_{min})\, r/R \tag{2}$$

for decreasing porosity, where ε_{min} and ε_{max} are the minimum and the maximum values of the porosity. In order to define the functions in equations (1, 2), $\varepsilon_{min} = 0.80$ and $\varepsilon_{max} = 0.90$ are set up, with the same value of the average porosity, $\bar{\varepsilon} = 0.85$, as it is shown in figure 2.

The dependence of PPI on the radial coordinate is taken into account with the correlation:

$$\mathrm{PPI}(r) = \mathrm{PPI}_{min} + (\mathrm{PPI}_{max} - \mathrm{PPI}_{min})\, (r/R)^{a} \tag{3}$$

where a is a positive dimensionless parameter, which prescribes different PPI(r) functions; $\mathrm{PPI}_{min} = 10$ and $\mathrm{PPI}_{max} = 40$ are set up. By constraining PPIs at the centerline, $r = 0$, and at the lateral surface, $r = 0.025$ m, of the receiver, different variable PPI functions are obtained. PPI as a function of the radius, according to equation (3), for different values of the a index, is presented in figure 3, which highlights a marked dependence of PPI on the a index.

2.2. Governing equations

The problem is solved averaging the governing equations for a continuous equivalent medium with volume-averaged characteristics. The local volume average of a generic variable ϕ and its intrinsic volumetric average, $\langle\phi\rangle$, averaged over V, are correlated as follows:

$$\langle\phi\rangle = \frac{1}{V}\int_V \phi \, \mathrm{d}V \tag{4}$$

Steady-state conditions and no effects of buoyancy are assumed.

The volume averaged mass conservation equation for the fluid phase is:

$$\nabla\cdot(\langle\rho u\rangle) = 0 \tag{5}$$

with ρ the density of the fluid and u the component of the velocity vector in the x direction.

A slip condition with no shear stresses on the symmetry axis, $r = 0$, is applied in the volume averaged momentum equation for the fluid phase:

$$\frac{\rho}{\varepsilon^2}\langle\mathbf{u}\cdot\nabla\mathbf{u}\rangle = -\nabla\langle p\rangle + \frac{\mu}{\varepsilon}\nabla^2\langle\mathbf{u}\rangle - \frac{\mu}{K}\langle\mathbf{u}\rangle - \frac{\rho\, C_f}{\sqrt{K}}|\langle\mathbf{u}\rangle|\,\langle\mathbf{u}\rangle \tag{6}$$

with p the pressure, μ the dynamic viscosity, K the permeability, C_f the inertial factor.

The volume averaged energy conservation equations are, for the fluid phase:

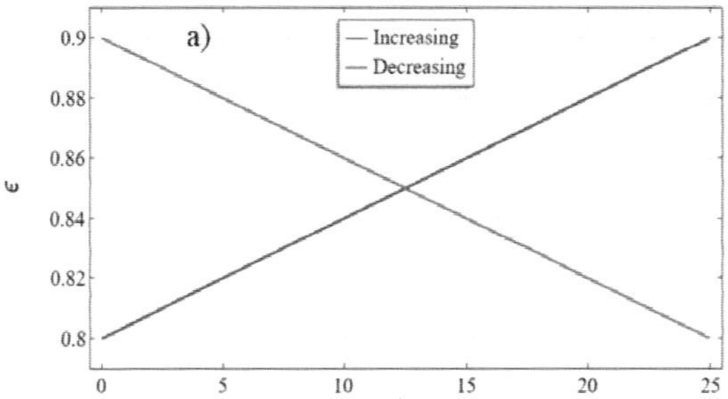

Figure 2. Increasing and decreasing porosities *vs.* the radial coordinate.

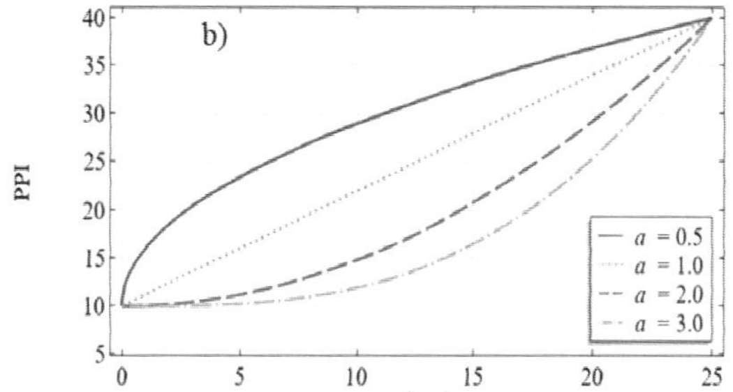

Figure 3. PPI *vs.* the radial coordinate, for different PPI power laws.

$$\left(\rho\, C_p\right)_f \nabla \cdot \langle \mathbf{u}\rangle^f \langle T_f\rangle^f = \nabla \cdot \left(\varepsilon\ k_f\, \nabla\langle T_f\rangle^f\right) + \overline{h_v}\ \left(\langle T_s\rangle^s - \langle T_f\rangle^f\right) \tag{7}$$

with T the temperature; k_f the thermal conductivity; $\overline{h_v}$ the average volumetric convection heat transfer coefficient, $\overline{h_v} = \overline{h_c}\ S/V$, where $\overline{h_c}$ is the convective heat transfer coefficient and S/V the specific surface area, that is the ratio of the total external surface area of the struts, S, to the volume of the foam, V, and for the solid phase:

$$\nabla \cdot \{[(1 - k_s)/3]\ \nabla\langle T_s\rangle^s\} + \nabla \cdot \mathbf{q}_\mathbf{r} + \overline{h_v}\left(\langle T_s\rangle^s - \langle T_f\rangle^f\right) = 0 \tag{8}$$

with $\mathbf{q}_\mathbf{r}$ the radiative heat flux. Andreozzi et al. [18] showed that, under the assumptions of no-emitted energy and no in-scattered energy, Bouguer–Beer–Lambert's law ($I = I_{in}e^{-\beta z}$) can be used:

$$\nabla \cdot \mathbf{q}_\mathbf{r} = \mathbf{n} \cdot \nabla I(r) = \mathrm{d}I(r)/\mathrm{dz}\ \ = -\langle\beta\rangle\, I_{in}(r)e^{-\langle\beta\rangle z} \tag{9}$$

where $\mathbf{n}$ is the unit vector in the direction of the radiation intensity field, $I_{in}(r)$ is the solar radiation intensity impinging onto the inlet section of the receiver, that depends on the radial coordinate, β is the extinction coefficient [19]:

$$\beta = [1 + 1.14\,(1 - \varepsilon)^{0.765}\, S/V]/4 \tag{10a}$$

with

$$S/V = [2.121\,\varepsilon^{-3.513}]/d_c \tag{10b}$$

as the specific surface area, taken from [20], and d_c the cell size. It is important to underline here that in the present paper both ε and d_c are a function of the radius (see equations (1, 2)).

2.3. Closing coefficients

The so-called porous media closing coefficients, permeability K, inertial factor C_f, and volumetric heat transfer coefficient h_v are required to solve the governing equations.

In the following, reference is made to the correlation proposed by Andreozzi et al. [20]:

$$d_c = -\,0.921\ \ln(\mathrm{PPI}) + 5.356 \tag{11}$$

and to the correlations proposed by Wu et al. [21]:

$$K = {d_c}^2/(1039 - 1002\,\varepsilon) \tag{12}$$

$$C_f = 0.5138\,\sqrt{K}\,\varepsilon^{-\,5.739}/d_c) \tag{13}$$

Wu et al. [21] proposed the correlation accounting for the dependence on porosity and cell size of the volumetric convection heat transfer coefficient, h_v, which, as it was reported in the previous subsection regroups h_c and S/V. Wu et al. [21] obtained the above said correlation rom computations performed on Kelvin-based open-cell foams:

$$h_v = (32.50\,\varepsilon^{0{,}38} - 109.94\,\varepsilon^{1{,}38} + 166.65\,\varepsilon^{2{,}38} - 86.98\,\varepsilon^{3{,}38})\ \mathrm{Re}^{0{,}438}\,\left(k_f/d_c^2\right) \tag{14}$$

with

$$\mathrm{Re} = \rho_f\,\langle u\rangle\, d_c/\varepsilon\ \mu_f \tag{15}$$

2.4. Boundary conditions

Boundary conditions for the volumetric receiver sketched in figure 1a are resumed in figure 4.

A Gaussian distribution of the solar irradiation onto the inlet cross section, $I_i\,(r)$ as in [10], which takes into account that the incident radiation is higher at the center of the receiver, is assumed:

$$I_{in}(r) = I_{in}\,(r = 0)\,\exp(-\,2560\,r^2) \tag{16}$$

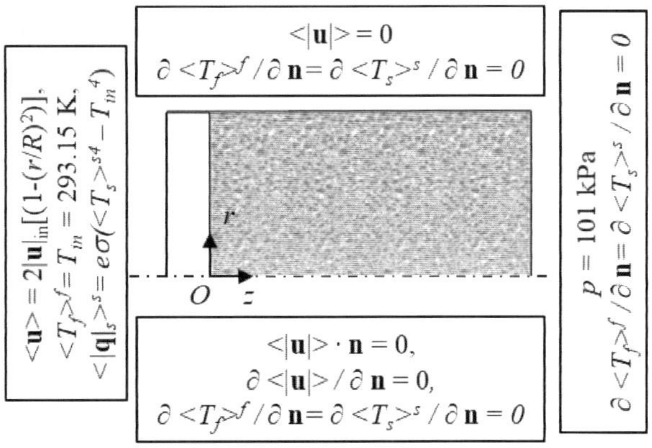

Figure 4. The boundary conditions for the cylindrical volumetric receiver.

where I_{in} $(r = 0)$ is the intensity of the solar radiation impinging onto the axis at the inlet section of the receiver, which is assumed to be equal to $1.5 \cdot 10^6$ W/m^2 [22]; it appears in the thermal power source term in equation (18).

A no slip and zero-gradient temperature boundary condition is imposed at the lateral surface of the receiver. An $e = 0.95$ emissivity of the solid phase at inlet and outlet cross sections is assumed to account for the radiative losses to the environment. The atmospheric pressure of the fluid, p_{in}, is assumed at the inlet section and an outflow condition for both fluid and solid temperatures is assumed at the outlet section. As to the solid phase, a radiative heat flux boundary condition is assumed both at the inlet ($z = 0$) and outlet ($z = L$) cross sections of the receiver:

$$- \mathbf{n} \cdot \mathbf{q_r} = e\,\sigma \left(\overline{T_{s,in}}^4 - \overline{T_{s,out}}^4 \right) \tag{17}$$

with $\overline{T_{s,in}}$ and $\overline{T_{s,out}}$ the average temperatures of the solid in the inlet and outlet sections, respectively.

Under the assumption of local thermal non-equilibrium, the convection heat transfer between the solid and fluid phases is accounted for as a thermal power source, u''':

$$u'''(z) = h_v(r) \left[\left| \left(\langle T_s \rangle - \langle T_f \rangle \right) \right| \right] \tag{18}$$

2.5. Numerical modeling

The numerical solution of the governing equations was carried out using the commercial code Comsol Multiphysics , employing a finite element scheme with a 10^{-4} convergence criterion. The grid contains 15,000 triangular and quadrilateral elements, with 150 elements in the z direction and 100 elements in the r direction. A large number of elements was not required, since the geometry of the computational domain is regular; besides computational times were reasonably small. Grid convergence was confirmed with a larger numbers of elements and showed negligible variation in the velocity profiles in the boundary layer. Equations are solved with streamline and crosswind stabilizations using the fully coupled PARDISO linear solver with a 10^{-4} relative tolerance in the equations. The model for foams with uniform characteristics was validated in [18].

3. Results and Discussion

For the sake of brevity, the volumetric average symbol < > will be dropped out in the following. The velocity and pressure drop in the fluid phase as well as the temperature in the solid phase are evaluated.

The cylindrical SiC receiver is $L = 0.015$ m long, with a $2\,R = 0.050$ m diameter. Thermophysical properties are assumed to be dependent on the temperature. The air density, ρ_f, is derived by the ideal gas law; the air viscosity, μ_f, the air heat capacity, $C_{p,f}$ at constant pressure, thermal conductivities of the air, k_f, and SiC, k_s, are calculated as follows [23]

$$\mu_f = 8.38\cdot10^{-7} + 8.36\cdot10^{-8}\,T - 7.69\cdot10^{-11}\,T^2 + 4.64\cdot10^{-14}\,T^3 - 1.07\cdot10^{-17}\,T^4 \qquad (19)$$

$$C_{p,f} = 1047.64 - 0.373\,T + 9.45\cdot10^{-4}\,T^2 - 6.02\cdot10^{-7}\,T^3 - 6.02\cdot10^{-7}\,T^3 - 1.29\cdot10^{-10}\,T^4 \qquad (20)$$

$$k_{f,} = 0.0023 + 1.16\cdot10^{-4}\,T - 7.90\cdot10^{-8}\,T^2 + 4.12\cdot10^{-11}\,T^3 - 7.44\cdot10^{-15}\,T^4 \qquad (21)$$

$$k_s = 191.92 - 0.33\,T + 2.74\cdot10^{-4}\,T^2 - 7.71\cdot10^{-8}\,T^3 \qquad (22)$$

Air at temperature T_{in} = 293.15 K enters the receiver with a uniform velocity $\mathbf{u}_i$ = 1.2 m/s.

A Gaussian distributed solar irradiation $I_{in}(r) = 1.5\cdot10^6 \exp(-2560\,r^2)$ impinges onto the inlet cross section. An e = 0.95 emissivity of the solid phase in the inlet and outlet cross sections of the receiver is assumed. Porosity and PPI either increase or decrease along the radial coordinate according to the following power laws, respectively: $\varepsilon(r) = \varepsilon_{min} + (\varepsilon_{max} - \varepsilon_{min})\,r/R = 0.80 + (0.90 - 0.80)\,r/R$; $\varepsilon(r) = \varepsilon_{max} - (\varepsilon_{max} - \varepsilon_{min})\,r/R = 0.90 - (0.90 - 0.80)\,r/R$; $\mathrm{PPI}(r) = \mathrm{PPI}_{min} + (\mathrm{PPI}_{max} - \mathrm{PPI}_{min})\,(r/R)^a = 10 + (40 - 10)\,(r/R)^a$ with a = 0.5, 1.0, 2.0, 3.0.

3.1. Fluid velocity

The fluid velocity as a function of the radial coordinate, at z = 0.0050 m, for square root (a = 0.5), linear (a = 1.0), quadratic (a = 2.0), cubic (a = 3.0) PPI functions and for radially increasing and decreasing porosities, are reported in figures 5a and 5b, respectively. Both figures exhibit large velocity gradients close to the side wall of the receiver, where the no slip condition holds. Figure 5a, for increasing porosities, points out slight decreases in the velocity outside of the boundary layer at increasing distances from the axis. This occurs because of the effects of porosity on the permeability and the inertial factor, which play a role in the momentum equation (equation (6)). On the contrary, figure 4b, for decreasing porosities, exhibits larger decreases in the velocity at increasing radial coordinates. In both cases the above noticed differences in the velocity profiles depend on the drag resistance, which, in turn, is affected by the permeability and inertial factor, which play a role in the momentum equation (see equation (6)) and depend on the porosity. It is also worth noticing that velocities in foams with decreasing porosity are far larger than in foams with increasing porosity.

As far as the PPI functions are concerned, both figures show that they poorly affect the velocity of the air, apart from the region around the axis of the receiver. One can also notice that the highest velocity of the air, 5.5 m/s, is attained for the square root PPI power law, at decreasing porosities.

3.2. Pressure drop

The pressure drop at the centerline as a function of the axial coordinate, for square root (a = 0.5), linear (a = 1.0), quadratic (a = 2.0), cubic (a = 3.0) PPI functions and for radially increasing and decreasing

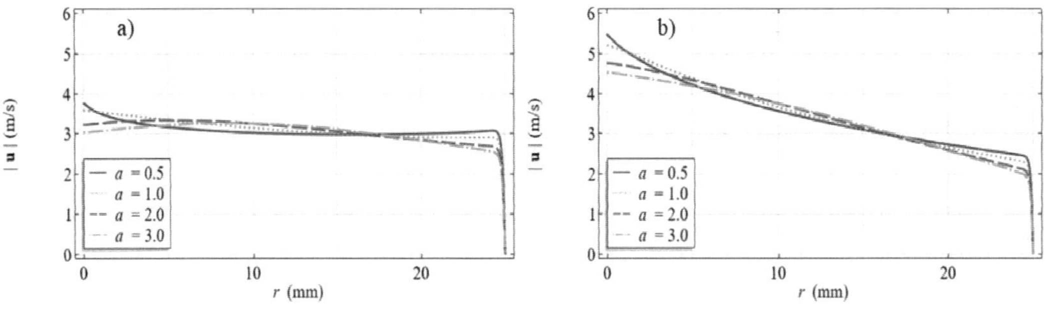

Figure 5. Fluid velocity *vs*. the radial coordinate, at z = 0.0050 m, for different PPI functions: a) increasing porosities; b) decreasing porosities.

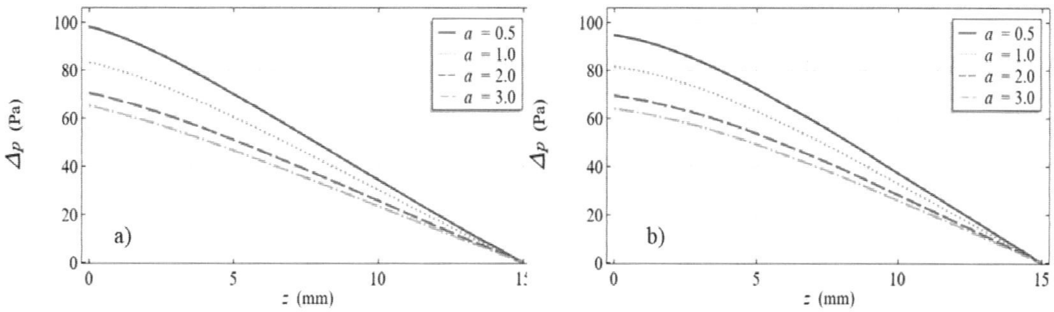

Figure 6. Centerline pressure drop *vs*. the axial coordinate, for different PPI functions: a) increasing porosities; b) decreasing porosities.

porosities, is reported in figures 6a and 6b, respectively. The figures show slightly smaller pressure drops for decreasing than for increasing porosities, since decreasing porosity decreases the permeability (see equation 11) and increases the inertial factor (see equation (13)), thus increasing pressure drop.

As to the effects of graded PPI, the cell size as a function of the radial coordinate, for square root (a = 0.5), linear (a = 1.0), quadratic (a = 2.0), cubic (a = 3.0) PPI functions, is reported in figure 7. The figure can be used to correlate the effect of the cell size on the drag resistance. The cell size is the smallest for the square root PPI function, for which the pressure drops were the largest, since the smaller the cell size the larger the drag resistance in the foam. Besides, smaller cell sizes make the permeability smaller and the inertial factor larger, leading to higher pressure drops.

3.3. Temperatures

Temperature fields in the solid phase of the receiver, for square root (a = 0.5), linear (a = 1.0), quadratic (a = 2.0), cubic (a = 3.0) PPI functions and for radially increasing and decreasing porosities, are reported in the upper and lower parts of figure 8, respectively. The figure shows that the lowest temperatures in the flow direction are attained at the lateral boundary of the receiver (r = 0.025 m), while the highest temperatures are reached in the region around its axis. The slight decrease in the solid temperature along the flow direction likely occurs because the convection heat flux released by the fluid overcomes the radiative energy absorbed by the solid phase. It is also worth noticing that temperature fields are two-dimensional because of the dependence on the radial coordinate of the irradiation intensity onto the inlet cross section of the receiver and the variability in porosity and PPI through the receiver.

The temperatures of the fluid and of the solid at the centerline as a function of the axial coordinate, for square root (a = 0.5), linear (a = 1.0), quadratic (a = 2.0), cubic (a = 3.0) PPI functions and for radially increasing and decreasing porosities, are reported in the left and right sides of figure 9, respectively. Figures show that the thermal equilibrium length, that is the axial coordinate where the fluid temperature becomes nearly equal to the solid temperature, is larger for decreasing porosities than

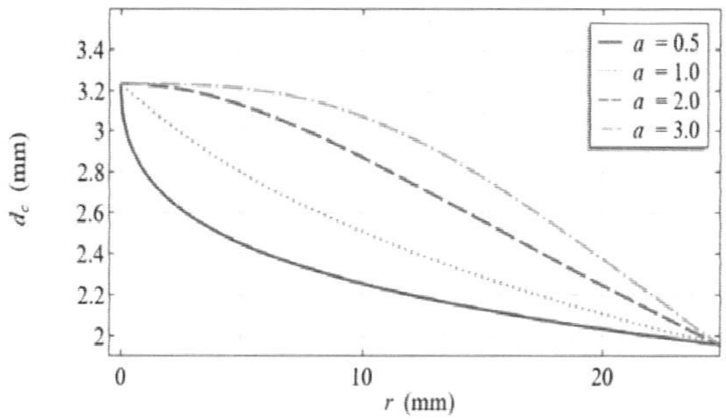

Figure 7. Cell size *vs*. the radial coordinate, for different PPI functions.

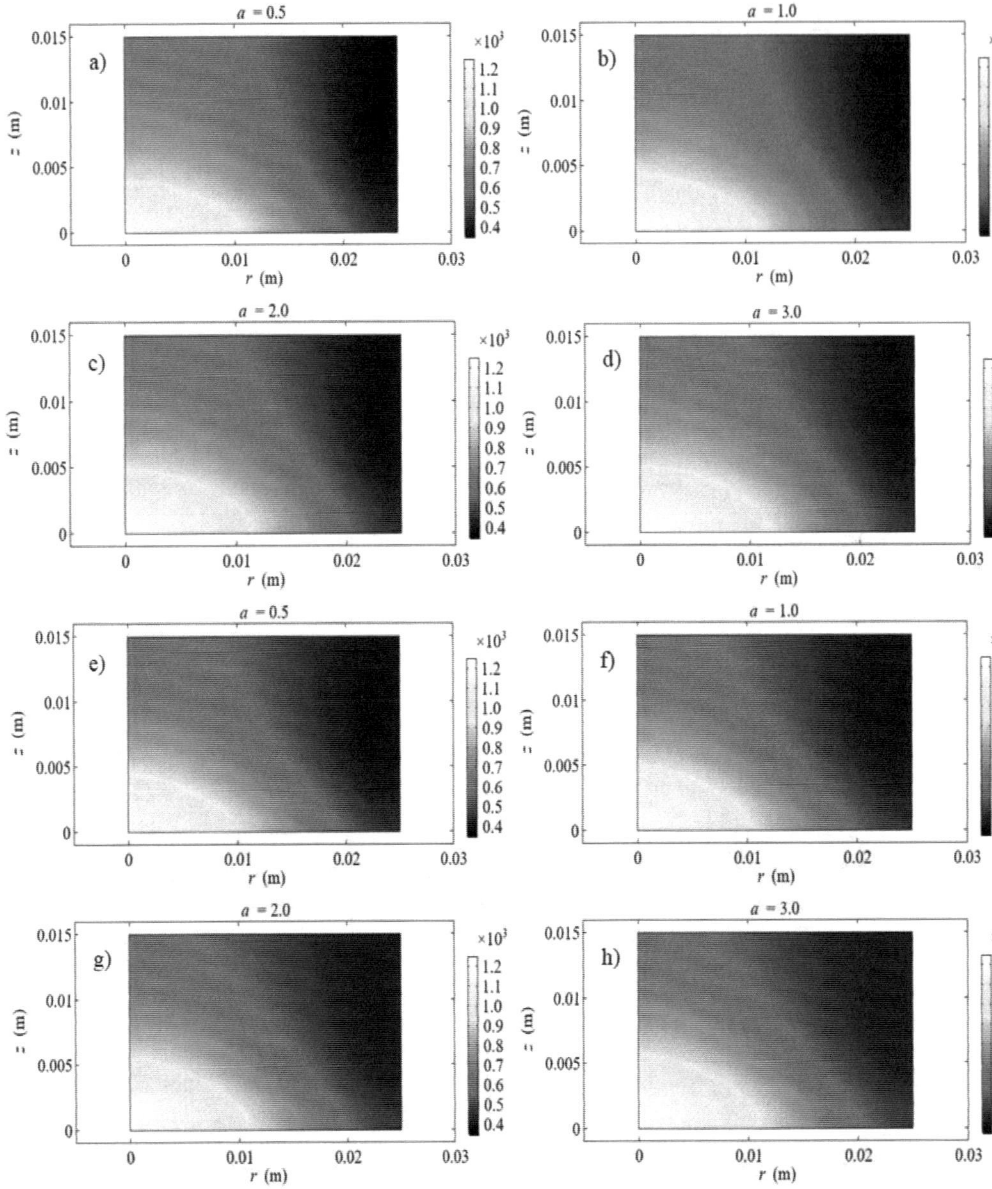

Figure 8. Temperature fields (°C) in the solid phase, for different PPI functions: a), b), c), d) increasing porosities; e), f), g), h) decreasing porosities.

for increasing porosities; this is consistent with the decrease in the volumetric convection heat transfer coefficient at decreasing porosity (see equation (14)). Figures also show that temperatures of both the solid and fluid phases are somewhat highest with the cubic PPI(r) function and increasing porosity. The regions of the receiver close to its inlet section absorb a large fraction of the impinging solar radiation, which is then released to the fluid by convection heat transfer. Thus, increasing the convection heat transfer coefficient by increasing the fluid velocity improves the thermal performance of the receiver.

The effects of the fluid velocity on the heat transfer in the foam are pointed out showing the dependence of the difference between the temperatures of the solid and fluid phases on the air velocity, for given sets of the foam characteristics. By way of example, for the foam configuration exhibiting the

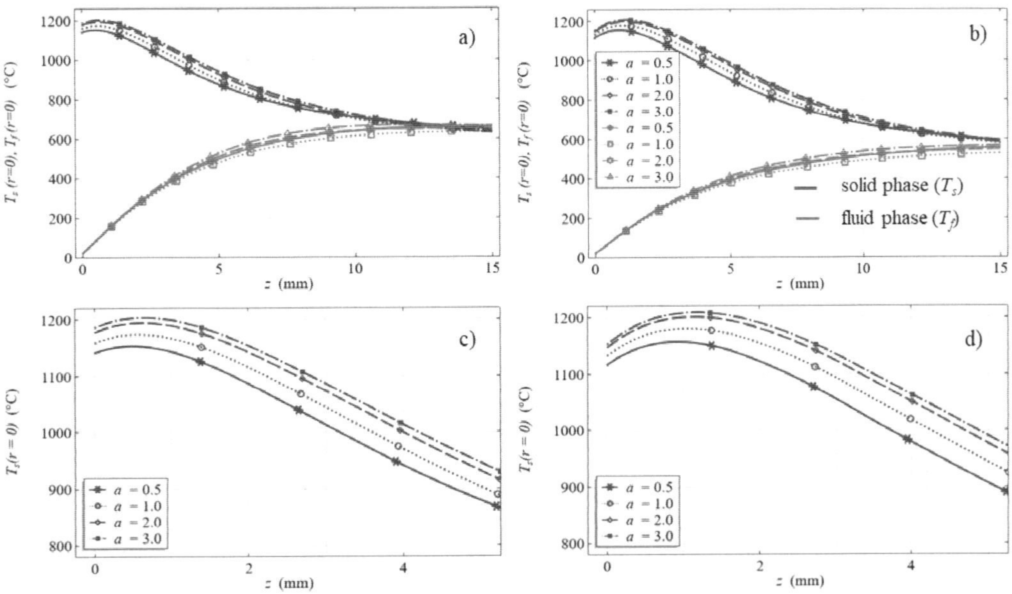

Figure 9. Centerline fluid and solid temperatures *vs.* the axial coordinate: a) and c) increasing porosities; b) and d) decreasing porosities.

lowest temperatures of the solid phase in the inlet section (square root PPI function and increasing porosity), the difference between the centerline temperatures of the solid and fluid phases as a function of the axial coordinate, for different inlet fluid velocities, is reported in figure 10. One can remark that the difference between the temperatures of the solid and fluid phases decreases at increasing inlet velocities because of the increase in the convection heat transfer and also because lower inlet velocity provides higher fluid temperatures.

4. Conclusions

Velocity and pressure drop in the air and temperatures of solid and fluid phases in an air volumetric solar receivers equipped with functionally-graded ceramic foams have been investigated numerically. Such foams are built up either by additive manufacturing or by placing different foam layers in series. Radially-variable porosity and PPI functions were considered: linearly increasing and decreasing porosities; square root, linear, quadratic, cubic PPI functions. The governing equations were solved with the Volume Averaging Technique by employing the local thermal non-equilibrium assumption. The effects of the radiative participating medium are taken into account by means of the Beer-Lambert law.

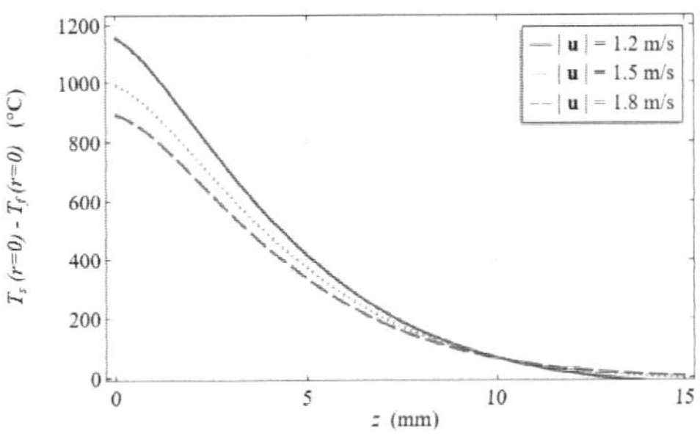

Figure 10. Difference between the temperatures of the solid and fluid phase *vs.* the axial coordinate, for the square root PPI function, increasing porosity and different inlet fluid velocities.

Results showed a poor effect of PPI on the fluid velocity, whereas varying porosity has a more marked effect. As to the pressure drop, it was slightly smaller with decreasing than with increasing porosity; the largest pressure drop was found for the square-root PPI function. Finally, the highest temperatures of both the solid and fluid phases were attained with the cubic PPI function and increasing porosity. This implies that receivers with a square-root function PPI look more promising since they reduce the temperature of the solid phase temperature in its entrance region and, as a consequence, radiative heat losses to the environment are reduced.

References

[1] Bai F 2010 *Int. J. Therm. Sci* **49(12)** 2400–4
[2] Wu Z, Caliot, C, Flamant G and Wang Z 2011 *Solar energy* **85(9)** 2374–85
[3] Wu Z, Caliot, C, Flamant G and Wang Z 2010 *J. Heat Mass Transfer* **54 (7–8**) 1527–37
[4] Sano Y, Iwase S and Nakayama A 2012 *J. Solar Energy* **134(2)** 021006
[5] Fend T, Schwarzbozl P, Smirnova O, Schollgen D and Jakob C 2013 *Renew. Energy,* **60** 655–61
[6] Wang F, Shuai F, Tan H, and Yu, C 2013 *Int. J. Heat Mass Transfer* **62(1)** 247–54
[7] Roldan M.I., Smirnova O, Fend T, Casas J.L. and Zarza E 2013 *Renew. Energy*, **62**, 116–28
[8] Wang P, Vafai K and Liu D Y 2014 *Num. Heat Transfer. A-Appl.* 65(10) 931–48
[9] Wang P, Vafai K, Liu D Y and Xu C 2015 *Int. J. Heat Mass Transfer* **80** 789–801
[10] Andreozzi A, Bianco N, Iasiello M and Naso V 2014 *Heat Transfer Eng.* **41(12)** 1085-99
[11] Udupa G, Rao S. S. and K.V. Gangadharan K.V. 2014 *Procedia Materials Science* **5** 1291-9
[12] Xu F and Zhang H 2018 *Eng. Struct.* **171** 309-25
[13] Xu Z.G. and Zhao C.Y. 2015 *Int. J. Heat Mass Transfer* **85**, 824-9
[14] Chen X, Xia X.L., Meng X.L. and Dong X.H. 2015 *Energy Conv. Manag.* **97** 282-9
[15] Iasiello M, Bianco N,. Chiu W.K. and Naso V 2020 *Int. J. Thermal Sciences* **160** 106696
[16] Du S, Ren Q and He Y.L. 2017 *Applied Energy* **207** 27-35
[17] Du S, Xia T, He Y.L, Li, Z.Y, Li D and Xi X.Q. 2020 *Appl. Energy*, **275**, 115343
[18] Andreozzi A, Bianco N, Iasiello M and Naso V 2016 *Int. J. Heat Technol.* **34** S489–9.
[19] Iasiello M, Cunsolo S, Bianco N, Chiu K.W. and Naso V., 2017 *Int. J. Thermal Sciences* **111** 129-37
[20] Andreozzi A, Bianco N, Iasiello M and Naso V. 2020 *Int. J. Physics: Conf. Series* **1599(1)** 012013
[21] Wu Y, Caliot C., Flamant G and Wang Z. 2011 *Int. J. Heat Mass Transfer* **54(7-8)** 1527-37
[22] Aichmayer L., Spelling J. and Laumert B. 2015 *Solar Energy* **114** 378-96
[23] COMSOL Multiphysics, Version 5.1. Burlington, MA: COMSOL Inc., Apr. 15, 2015.

Entropy Generation Analysis on Confined Impinging Slot Jets with Nanofluids in Metal Foam

B Buonomo[1], A di Pasqua[1], O Manca[1*], S Nappo[1]

[1] Dipartimento di Ingegneria, Università degli Studi della Campania "*Luigi Vanvitelli*", Via Roma 29, Aversa (CE) 81031, Italy
Corresponding author e-mail: oronzio.manca@unicampania.it

Abstract. In this paper a numerical entropy generation analysis on confined slot impinging jets made up of a totally filled configuration of metal foam in mixed convection is presented. A two-dimensional domain is assumed and several Peclet numbers (equal to 350, 750, 1500) are considered. Rayleigh numbers is imposed equal to 30000. The employed working fluids are pure water or Al_2O_3/water based nanofluids. The nanoparticle volume concentrations are in a range from 0% to 4% and the nanoparticle diameter is varied in a range from 20 nm to 80nm. The target surface is characterized by a constant temperature, estimated starting to the value of Rayleigh number. The distance of the target surface is five times greater than the slot jet width. The thermal and fluid dynamic properties of nanofluids are evaluated by a single-phase model method. The model of local thermal non equilibrium (LTNE) is considere in order to simulate the behavior of the metal foam characterized by a number of pores per inch (PPI) equal to 5, 10, 20, 40 and a porosity around 0.90. The data in terms of total global entropy generation related to the thermal and viscous effects are given Results show increasing values of the total entropy generation for increasing values of Peclet number and nanoparticle concentration. This behaviour is more evident at higher Peclet number values. In addition, the total entropy generation increases with the increase of PPI in corresponding to the same porosity value.

1. Introduction

The heat transfer sector has the aim to promote and to produce applications with a higher efficiency in many systems in several fields, such as of automotive, aerospace, electronic and process industry. The application of confined impinging jets represents a good solution that maintains huge cooling efficiency for thermal industrial applications. For their high cooling effectiveness, uniformity and controllability, the slot jets are widely used. The application of a cover of porous medium on the impinged surface is becoming more common because the metal foam is characterized by high conductivity and porosity that represent a further enhancement of heat transfer behavior. Furthermore, the application of working fluids with the addition of nanoparticles represents a promising solution in order to improvement the thermal behaviour of the base fluids; in fact, nanofluid is preferred due to the better thermal conductivity and therefore superior thermal performance in comparison to base fluid. The combination of metal foams and nanofluids with high thermal conductivity is an interesting and promising technique. Entropy generation minimization analysis is important for system optimizations, where pumping power and rate of heat transfer should be optimized for the best performance of the industrial applications.

A numerical analysis on the entropy generation in impinging jets with nanofluids was accomplished by Mahmoudabadbozchelou et al. [1]. The authors executed an optimization with the application of artificial neural network and genetic algorithm to find the improved heat transfer in nanofluids respect to pure water. The results show that increasing the size and concentration of nanoparticles improves the heat transfer. Furthermore, it is possible to note that the entropy generation rates due to the temperature difference and friction are equal for air, while the rate due to the heat transfer is major for water and Al_2O_3/water nanofluids. Salimi et al. [2] conducted a numerical analysis on a three-dimensional impinging jet porous heat sink under local thermal non-equilibrium hypothesis. The authors employed the lattice Boltzmann method (LBM) into the porous section, while finite volume method (FVM) is used around it. The results show that the most important parameter is the porous layer value. The average viscous entropy generation number increases with the increase of the Reynolds number or porous layer thickness. Furthermore, in corresponding to a porosity equal to 0.67, the thermal entropy generation number presents a maximum value in the center of the porous media and it is possible to assume the local thermal equilibrium condition. However, for a porosity equal to 0.86, the LTNE is around the hot wall. In addition, the total entropy generation number is greater than the entropy presented in the case without porous section.

Ekiciler at al. [3] executed an investigation on nanofluid jet impingement cooling to investigate the thermal behavior and entropy generation numerically. The nanofluids employed were Al_2O_3/water nanofluid with different nanoparticle shapes and volume fractions. The results indicate that, for a nanoparticles concentration equal to 3% and triangular-corrugated impinging surface at Re = 500, the heat transfer increases about 29.5%. Platelet shape of nanoparticle presents an increment equal to 43% respect to the pure water. Furthermore, the lowest entropy for flat impinging surface is presented for the nanofluid with nanoparticle volume concentration of 3.0% and blade shape of nanoparticles. A numerical investigation on pulsed imping jet on metal foam employing the local thermal non-equilibrium model is carried out by Hosseinalipour et al. [4]. The authors found that higher amplitudes of pulsation increases the thermal performance of the impingement jet. In fact, a pulsative impingement jet with step function velocity and the frequency of 150 Hz and amplitude of 80% enhances the thermal performance of 15.4% respect to a steady jet. Furthermore, the results show that the rate of entropy generation is usually lower in corresponding to the critical frequency.

A numerical study on microjet made of photopolymer resin was carried out by Paniagua-Guerra and Ramos-Alvarado [5]. The authors found that the mini-fins presents a little effect on the hydraulic performance, but the thermal behavior is improved. In addition, the entropy generation is higher in the copper plate. Alabdaly and Ahmed [6] analyzed the thermal behavior of a confined slot impinging jet with SiO_2-water nanofluid numerically. The results highlight that the average Nusselt number and total entropy generation increases with increasing Reynolds number, nanoparticles volume fraction and ribs height. In addition, the best configuration is characterized by circular ribs and SiO_2-water with a nanoparticle concentration equal to 4%. A numerical study on heat transfer and entropy generation on nanofluid confined impinging slot jet is executed by Yousefi-Lafouraki et al. [7]. The authors used Al_2O_3-water nanofluids with two-phase mixture model approach. The entropy generation, mainly due to the temperature difference, is higher in the stagnation and the wall jet region. Furthermore, the exergy loss value decreases with the increase of the height and with the decrease of particle concentration.

In this paper, a numerical entropy generation study is accomplished on a mixed convective heat transfer problem in a confined slot impinging jet totally filled with aluminium foam. Al_2O_3/water based nanofluids characterized by particle volume concentrations from 0% to 4% and the particle diameters equal to 20, 30 and 80 nm are employed. A two-dimensional model is investigated with different Peclet number values. Rayleigh numbers is considered equal to 30000. The metal foam is characterized by a number of pores per inch (PPI) equal to 5, 10, 20, 40 and a porosity around equal to 0.90. Results are presented in terms of total global entropy generation related to the thermal and viscous effects and in terms of Bejan number.

2. Governing equations and physical domain

The physical domain is constituted by a confined slot impinging jet in a metal foam with nanofluids. The 2D sketch of the system under analysis is presented in the Figure1.

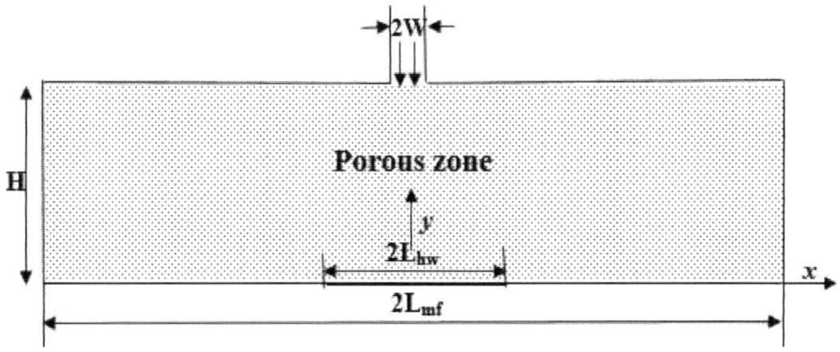

Figure 1. Physical Domain

The impinging slot jet is characterized by H equal to 30 mm and a length $2L_{mf}$ of 420 mm. The length of the impinged heated wall $2L_{hw}$ is 60 mm and the width of jet, indicted as $2W$, is equal to 6 mm. The microscopic flow equations are written using the local volume averaging process [8-9] to replicate the thermal and fluid dynamic behavior into the domain. The metal foam is considered homogeneous and isotropic. The properties of the fluid phase and solid matrix are assumed temperature independent. The Darcy–Forchheimer-Brinkman model and LTNE (Local Thermal Non-Equilibrium) assumptions are considered to simulate the presence of the aluminum foam in the impinging slot jet. Using the hypothesis above indicated, the governing equations are:

-Conservation of mass

$$\frac{\partial u}{\partial x}+\frac{\partial v}{\partial y}=0 \quad (1)$$

-x- momentum equation

$$\rho_{nf}\left(\frac{u}{\varepsilon^2}\frac{\partial u}{\partial x}+\frac{v}{\varepsilon^2}\frac{\partial u}{\partial y}\right)=-\frac{\partial p}{\partial x}+\frac{\mu_{nf}}{\varepsilon}\left(\frac{\partial^2 u}{\partial x^2}+\frac{\partial^2 u}{\partial y^2}\right)-\left(\frac{\mu_{nf}}{K}u+\frac{C_F}{K^{1/2}}\rho_{nf}\sqrt{u^2+v^2}u\right)+\rho_{nf}g\beta_{nf}(T_{nf}-T_0) \quad (2)$$

-y- momentum equation

$$\rho_{nf}\left(\frac{u}{\varepsilon^2}\frac{\partial v}{\partial x}+\frac{v}{\varepsilon^2}\frac{\partial v}{\partial y}\right)=-\frac{\partial p}{\partial y}+\frac{\mu_{nf}}{\varepsilon}\left(\frac{\partial^2 v}{\partial x^2}+\frac{\partial^2 v}{\partial y^2}\right)-\left(\frac{\mu_{nf}}{K}v+\frac{C_F}{K^{1/2}}\rho_{nf}\sqrt{u^2+v^2}v\right) \quad (3)$$

where ε is the porosity, ρ_{nf} and μ_{nf} are nanofluid density and viscosity, u and v are the velocity components in Cartesian coordinates, K and C_F are the permeability and inertial coefficient of the metal foam, respectively.

-Fluid phase energy equation

$$\left(\rho c_p\right)_{nf}\left(u\frac{\partial T_{nf}}{\partial x}+v\frac{\partial T_{nf}}{\partial y}\right)=\varepsilon k_{nf}\left(\frac{\partial^2 T_{nf}}{\partial x^2}+\frac{\partial^2 T_{nf}}{\partial y^2}\right)+\eta h_{sf}\alpha_{sf}\left(T_s-T_{nf}\right) \quad (4)$$

-Solid phase energy equation

$$(1-\varepsilon)k_s\left(\frac{\partial^2 T_S}{\partial x^2}+\frac{\partial^2 T_S}{\partial y^2}\right)-\eta h_{sf}\alpha_{sf}\left(T_s-T_{nf}\right)=0 \quad (5)$$

in which c_p is the nanofluid specific heat, k_{nf} and k_s are the fluid phase and solid matrix thermal conductivity, T_{nf} and T_s are the temperature of fluid phase and solid phases of metal foam,

respectively. The terms α_{sf} and h_{sf} are the specific surface area density and the interfacial heat transfer coefficient between the fluid phase and solid matrix, due to the LTNE model.
The equations for the evaluation of the entropy generation for local thermal non-equilibrium porous model are the following [10]:

-Fluid phase entropy generation

$$\dot{S}'''_{gen,f} = \underbrace{\frac{\varepsilon k_f}{T_f^2}\left[\left(\frac{\partial T_f}{\partial x}\right)^2+\left(\frac{\partial T_f}{\partial y}\right)^2\right]+\frac{\alpha_{sf}h_{sf}(T_s-T_f)^2}{T_sT_f}}_{\dot{S}'''_{gen,\Delta T}}+\underbrace{\frac{\mu_f}{T_f}\Phi+\frac{\mu_f\varepsilon}{KT_f}\left(u^2+v^2\right)+\frac{C_F\rho\varepsilon^2}{T_f\sqrt{K}}\left(u^2+v^2\right)^{3/2}}_{\dot{S}'''_{gen,\Delta p}} \quad (6)$$

where Φ in Eq. (6) is the well-known dissipation function; its form for incompressible fluid at a two-dimensional rectangular coordinate is:

$$\Phi = 2\left[\left(\frac{\partial u}{\partial x}\right)^2+\left(\frac{\partial v}{\partial y}\right)^2\right]+\left(\frac{\partial u}{\partial y}+\frac{\partial v}{\partial x}\right)^2 \quad (7)$$

The fourth term on the right-hand side of Eq. (6) considers the irreversibility rate of internal energy per unit volume by viscous drag in the porous section, and the last term in Eq. (6) represents the irreversibility rate of internal energy per unit volume by form drag in the porous section.

-Solid phase entropy generation

$$\dot{S}'''_{gen,s} = \underbrace{\frac{(1-\varepsilon)k_s}{T_s^2}\left[\left(\frac{\partial T_s}{\partial x}\right)^2+\left(\frac{\partial T_s}{\partial y}\right)^2\right]+\frac{\alpha_{sf}h_{sf}(T_f-T_s)^2}{T_sT_f}}_{\dot{S}'''_{gen,\Delta T}} \quad (8)$$

The evaluation of the parameters K and C_F was carried out with the correlations of Bhattacharya et al. [11]. The relationships of the work [12] are used to estimate the values of α_{sf} and h_{sf}. The key parameters of the metal foams, employed in this study, are listed in the Table 1 [11, 12]. Consequently, pore density and porosity values are chosen from the data given in [11,12].

Table 1. Parameters of the employed aluminum foams [11,12]

PPI	ε	K(m^2)	C_F
5	0.905	1.7x10^{-7}	0.078
10	0.8991	9.4x10^{-8}	0.068
20	0.9005	9.0x10^{-8}	0.088
40	0.9132	5.3x10^{-8}	0.084

In the present paper the laminar flow and its convective heat transfers of pure water/ Al_2O_3 nanofluids under steady-state conditions are studied. The nanoparticle volume concentrations ϕ range from 0% to 4%. The diameter of nanoparticles (d_{np}) is equal to 20, 30 and 80 nm. A single-phase model has been assumed in order to illustrate the nanofluid presence. The density and the specific heat are estimated by the expressions of the work [13]. For the dynamic viscosity of the nanofluid, the used correlation is [14].

The correlation of [15] is employed to estimate the thermal conductivity of the nanofluid. The nanofluid thermal expansion coefficient β is evaluated as [13]. The properties of the alumina nanoparticles are to: ρ_{np} = 3880 kg m^{-3}, c_{pnp} = 729 J kg^{-1} K^{-1}, k_{np} = 42 W m^{-1} K^{-1}.
The thermophysical properties of the nanofluids employed in the analyses are listed in Table 2. As the diameter of the nanoparticles changes, the only thermophysical quantity that changes is the nanofluid thermal conductivity.

Table 2. Parameters of the of the working fluids [13-15]

ϕ	ρ (kg/m^3)	μ (kg/ms)	c_p (J/kgK)		k (W/mK)		β (1/K)
				d_{np}=20 nm	d_{np}=30 nm	d_{np}=80 nm	
0%	998.2	1.0x10^{-3}	4182		0.600		2.1x10^{-3}
1%	1027	1.4x10^{-3}	4052	0.605	0.607	0.613	2.0x10^{-3}
4%	1113	3.4x10^{-3}	3701	0.620	0.630	0.651	1.8x10^{-3}

3. Numerical model

The finite volume method is employed to resolve the governing equations. The code Ansys-Fluent 18.2 is used to execute the numerical investigations. The SIMPLE algorithm is accomplished for the pressure-velocity coupling; the least square cell is used to evaluate the gradient evaluation for the spatial discretization. The PRESTO algorithm is employed to obtain the pressure calculation; the second order upwind scheme is applied for energy and momentum equations. Convergence criteria are equal to 10^{-5} for the continuity and the velocity components while for the energy equal to 10^{-8}.
The numerical domain is showed in the Figure 2. The details of grid analysis and validation are given in [16] and not repeated in the present paper.

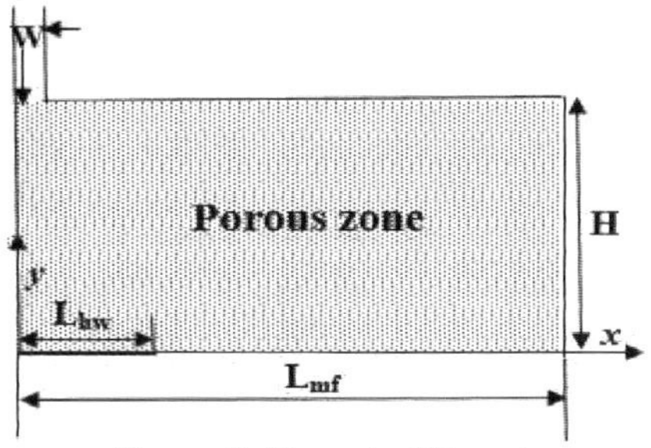

Figure 2. Numerical Domain

The boundary conditions of the problem are the following: the left edge is symmetric, the upper and bottom edges are adiabatic wall, and the heated wall is an isothermal edge. The impinging jet is characterized by a temperature T_0 equal to 293 K and a velocity equal to u_0. The impinging jets velocities u_0 are estimated from the definition of Pe. The outflow condition is imposed to the exit. The reference length is the width of the jet $2W$ equal to 6 mm for the dimensionless number.

4. Results and Discussions

The entropy generation method analyses are accomplished for different Peclet values equal to 350, 750, 1500, and for Rayleigh number equal to 30000. The total global entropy generation $\dot{S}_{gen,tot}$ is evaluated as:

$$\dot{S}_{gen,tot} = \dot{S}_{gen,f} + \dot{S}_{gen,s} \quad (9)$$

where $\dot{S}_{gen,f}$ and $\dot{S}_{gen,s}$ are the fluid phase and solid matrix global entropy generation, respectively. The global total entropy generation results, evaluated in the porous section, for different nanoparticles volume concentrations and nanoparticles diameter values, are plotted in the Figure 3 only for 5 PPI and 40 PPI because the behavior of the $\dot{S}_{gen,tot}$ is the same for 10 PPI and 20 PPI.

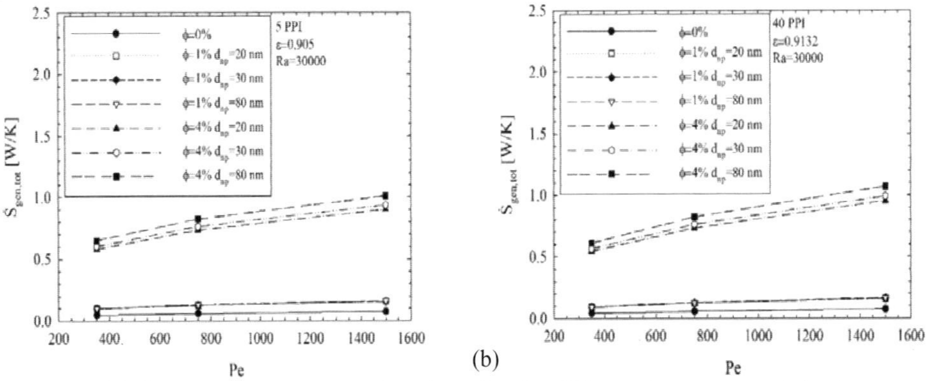

(a) (b)

Figure 3. Global entropy generation as a function of *Pe* for different ϕ and nanoparticles diameters and: (a) 5 PPI, (b) 40 PPI

From the Figures 3(a) and 3(b), one can observe as the total global entropy generation increases slightly with increasing of Peclet number for all values of nanoparticles concentration. Furthermore, it is possible to note that as ϕ increases there is an increment in terms of $\dot{S}_{gen,tot}$ both for 5 PPI and 40 PPI. In addition, the variation of nanoparticles diameters does not change the value of total global entropy generation in corresponding to ϕ=1%, while, for ϕ=4%, the effect of the difference of d_{np} is more visible; in fact; both for 5 PPI and for 40 PPI, the difference in terms of $\dot{S}_{gen,tot}$ is around 12% between d_{np}=20 nm and d_{np}=80 nm.

In the Figure 4, the total global entropy generation is presented for several PPI values approximately to the same porosity around 0.90.

It is possible to observe that $\dot{S}_{gen,tot}$ increases with the increase of number of pores per inch in corresponding to 5, 10 and 20 PPI because this metal foam samples are characterized by the same value of porosity equal to 0.90; while, for 40 PPI the trend is different because this foam has a porosity equal to 0.9132. One can notice that, for *Pe*=350, the global entropy generation presents the lowest value for 40 PPI; while, for *Pe*=1500, the minor value is in corresponding to 5 PPI.

Bejan number *Be*, a typical parameter employed in the entropy generation method, is defined as:

$$Be = \frac{\dot{S}_{gen,\Delta T}}{\dot{S}_{gen,\Delta T} + \dot{S}_{gen,\Delta p}} \tag{10}$$

where in the denominator the first term indicates the entropy generation due to the heat transfer between finite temperature difference $\dot{S}_{gen,\Delta T}$ and the second term indicates the entropy generation by the pressure drop due to friction $\dot{S}_{gen,\Delta p}$. The sum of two terms of denominator of Eq. (10) represents the total global entropy generation. The Bejan number is showed for different ϕ and PPI values in Figures 5(a) and 5(b), respectively.

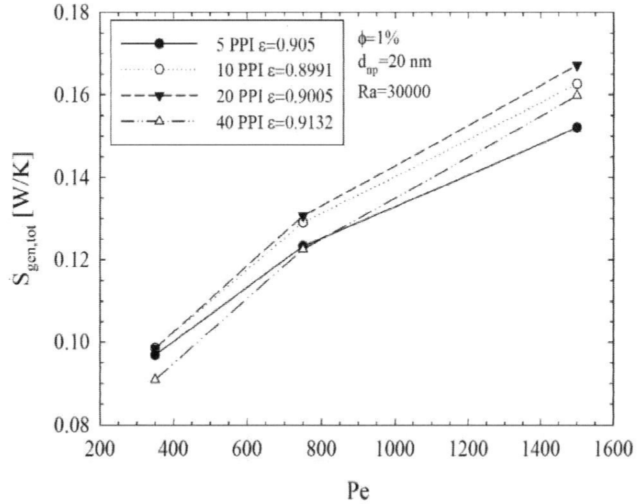

Figure 4. . Global entropy generation as a function of *Pe* for different PPI values

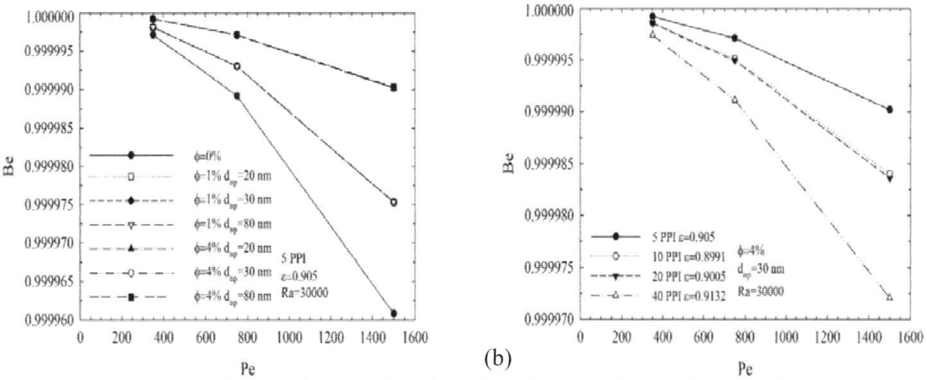

(a) (b)

Figure 5. . Bejan number as a function of *Pe* for several (a) ϕ, (b) PPI values

From the Figure 5(a), the Bejan number *Be* increases with increasing nanoparticles volume concentrations and there is not variation when nanoparticles diameter values change. Furthermore, *Be* decreases with increasing number of pores per inch (Figure 5(b)).

5. Conclusions

The total global entropy generation increases with increasing of Peclet number and nanoparticles volume concentrations. Furthermore, in corresponding to the same value of porosity, $\dot{S}_{gen,tot}$ increases with the increase of PPI. The data in terms of Bejan number show that *Be* increases with increasing ϕ and with decreasing PPI.

Acknowledgements

This research was partially funded by MIUR (Ministero dell'Istruzione, dell'Università e della Ricerca), grant number PRIN- 2017F7KZWS and by Università degli Studi della Campania "Luigi Vanvitelli" with the grant number D.R. n. 138 under NanoTES project - V:ALERE program 2020.

References

[1] Mahmoudabadbozchelou M, Eghtesad A, Jamalia S, Afshin H 2020 Entropy analysis and thermal optimization of nanofluid impinging jet using artificial neural network and genetic algorithm *Int. Comm. Heat Mass Transf.* Vol. **119** 104978.

[2] Reza Salimi M, Taeibi-Rahni M, Rostamzadeh H 2020 Heat transfer and entropy generation analysis in a three-dimensional impinging jet porous heat sink under local thermal non-equilibrium condition *Int. J. Therm. Sc.* Vol. **153** 106348.

[3] Ekiciler R, Ali Çetinkaya M S, Arslan K 2020 Effect of shape of nanoparticle on heat transfer and entropy generation of nanofluid-jet impingement cooling *Int. J. Green En.* Vol. **17 (10)** pp. 555-567.

[4] Hosseinalipour S M, Rashidzadeh S, Moghimi M, Esmailpour K 2020 Numerical study of laminar pulsed impinging jet on the metallic foam blocks using the local thermal non-equilibrium model *J. Therm. Analysis Calorimetry* Vol. **141** pp. 1859–1874.

[5] Paniagua-Guerra L E, Ramos-Alvarado B 2020 Efficient hybrid microjet liquid cooled heat sinks made of photopolymer resin: thermo-fluid characteristics and entropy generation analysis *Int. J. Heat Mass Transf.* Vol. **146** 118844.

[6] Alabdaly I K, Ahmed M A 2019 Numerical investigation on the heat transfer enhancement using a confined slot impinging jet with nanofluid *Propulsion Power Research* Vol. **8 (4)** pp. 351-361.

[7] Yousefi-Lafouraki B, Ramiar A, Ranjbar A A 2019 Numerical Investigation of Laminar Forced Convection and Entropy Generation of Nanofluid in a Confined Impinging Slot Jet Using Two-Phase Mixture Model *Iran J Sc. Technol. Trans. Mech. Eng.* Vol. **43 (1)** pp. 165–179

[8] Nield D A, Bejan A, "Convection in Porous Media", 4th Edition, Springer, New York, 2013.

[9] Whitaker S, "The Method of Volume Averaging", Springer, Netherlands, 1998.

[10] Baytas A C 2007 Entropy Generation for Thermal Nonequilibrium Natural Convection with a Non-Darcy Flow Model in a Porous Enclosure Filled with a Heat-Generating Solid Phase *J. Porous Media* Vol. **10(3)** pp. 261-275.

[11] Bhattacharya A, Calmidi V V, Mahajan R L 2002 Thermophysical properties of high porosity metal foams *Int. J. Heat Mass Transf.* Vol. **45** pp. 1017-1031.

[12] Calmidi V V, Mahajan R L 2000 Forced Convection in High Porosity Metal Foams *ASME J. Heat Transf.* Vol. **122** pp. 557-565.

[13] Khanafer K, Vafai K, Lightstone M 2003 Buoyancy-driven heat transfer enhancement in a two dimensional enclosure utilizing nanofluids *Int. J. Heat Mass Transf.* Vol. **46 (19)** pp. 3639–3653.

[14] Pak B C, Cho Y I 1998 Hydrodynamic and heat transfer study of dispersed fluids with submicron metallic oxide particles *Exp. Heat Transf. Int. J.* Vol. **11 (2)** pp. 151–170.

[15] Nan C W, Birringer R, Clarke D R, Gleiter H 1997 Effective thermal conductivity of particulate composites with interfacial thermal resistance *J. Appl. Physics* Vol. **81 (10)** pp. 6692–6699.

[16] Bianco V, Buonomo B, di Pasqua A, Manca O 2021 Heat Transfer Enhancement of Laminar Impinging Slot Jets by Nanofluids and Metal Foams *Therm. Science Eng. Progr.* Vol. **22** art. 100860.

Comparison of heat transfer processes during selective and whole body newborn's therapeutic hypothermia – case report

Bandoła D.[1], **Ostrowski Z.** [1], **Rojczyk M.** [1], **Wagstyl D.** [1], **Tumidajski J.** [1], **Bojdoł M.** [1], **Walas W.** [2], **Halaba Z.** [3], **Nowak A.J.** [1]

[1] Silesian University of Technology, Department of Thermal Technology, Poland
[2] Paediatric and Neonatal Intensive Care Unit, Institute of Medical Sciences, University of Opole, Poland
[3] Department of Paediatrics, Institute of Medical Science, University of Opole, Poland

andrzej.j.nowak@polsl.pl

Abstract. In this paper selected heat transfer issues related to the therapeutic hypothermia of newborns are presented and discussed. The therapeutic hypothermia is currently the only well-established treatment of newborns suffering for perinatal asphyxia (PA). Considered heat transfer issues are closely associated with the non-invasive thermal measurements carried out at the Department of Anaesthesiology and Intensive Therapy of Children and Newborns of the University Clinic Hospital in Opole, Poland. Obtained results allow for preliminary comparison of the heat transfer processes occurring during both types of therapeutic hypothermia.

1. Introduction

Despite recent advances in perinatal care, neonatal hypoxic-ischaemic encephalopathy (HIE) is one of the most common causes of severe neurological deficits in children, affecting approximately 15 per 10,000 live births. According to the World Health Organization, perinatal asphyxia (PA) represents the third most common cause of neonatal death [1]. Therapeutic hypothermia (TH) is a method of treatment that has been proven effective in protecting the brain against the effects of ischaemia/hypoxia in neonates after PA [2].

Generally, there are two main types of TH: selective TH and whole body TH. The essence of TH is to maintain the newborn's core temperature (or deep body temperature) at a reduced level, which in the case of selective head hypothermia is 34-35°C, and 33-34°C in the case of whole body hypothermia, for certain time. This can be achieved in different ways, depending on the type of cooling device. In both cases, the newborn is placed in an open incubator, and in both cases, the treatment lasts 72 hours.

The selective TH is carried out with a help of special device, for instance the Olympic Cool-Cap [3], in which medical staff manually controls the patient's temperature. The cooling element is a water-cooled cap that is put on the newborn's head. The patient's temperature is measured in three places: the rectal sensor measures the core temperature, and two sensors measuring the surface temperature are located on the abdomen and scalp, respectively. Patient's data should be defined and entered into the device memory before cooling begins. Based on the weight of the newborn and parameters of indoor environment, the device pre-sets the temperature of the water flowing through the cap.

The whole body TH is again used with the help a special device, for instance the Arctic Sun® 5000 (Medivance, Inc., USA) [4], in which the patient's temperature is controlled automatically. The cooling element is a water-cooled pad (actually two connected pads with separate water lines) which the patient is wrapped in. The patient's core temperature is measured with a rectal sensor and displayed on the device screen. Other measurement results are usually not utilised by neonatologists for treatment. Both cooling techniques of neonates cooling are schematically demonstrated in Fig.1.

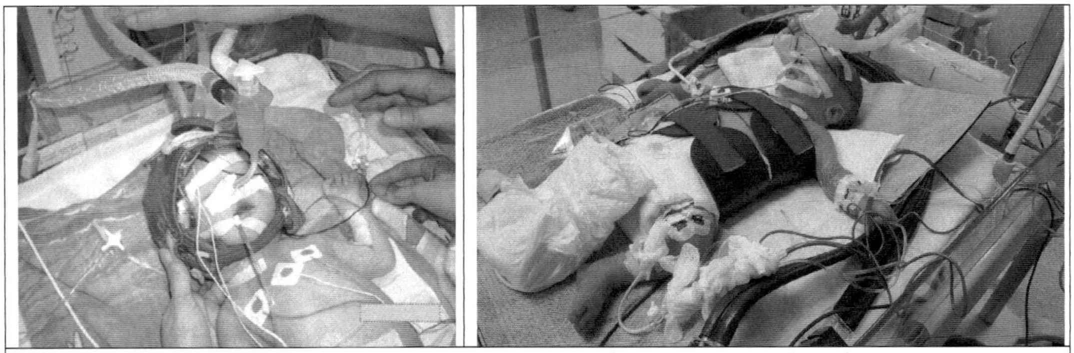

Fig.1. The selective and whole body Therapeutic Hypothermia

The main aim of this study is a preliminary comparison of the heat transfer processes occurring during both, the selective as well as the whole body TH. The presented results are based on the non-invasive thermal measurements carried out at the Department of Anaesthesiology and Intensive Therapy of Children and Newborns of the University Clinic Hospital in Opole, Poland [5, 6]. Although part of these measurements are directly recorded by cooling device (i.e. Cooling Cap or Arctic Sun) as the technical files of csv type, access to them is unfortunately not straightforward. Moreover, due to the authors' best knowledge comparison of selected temperature histories during TH carried out by two analysed techniques is not available yet. Therefore in this work emphasises are put on essential differences in temperature histories in those two TH types.

Because in this project a new method of assessing the thermogenesis of infants with HIE undergoing TH is also investigated [7], additional measurements (still non-invasive) are collected using specially designed measuring device shown in Fig. 2.

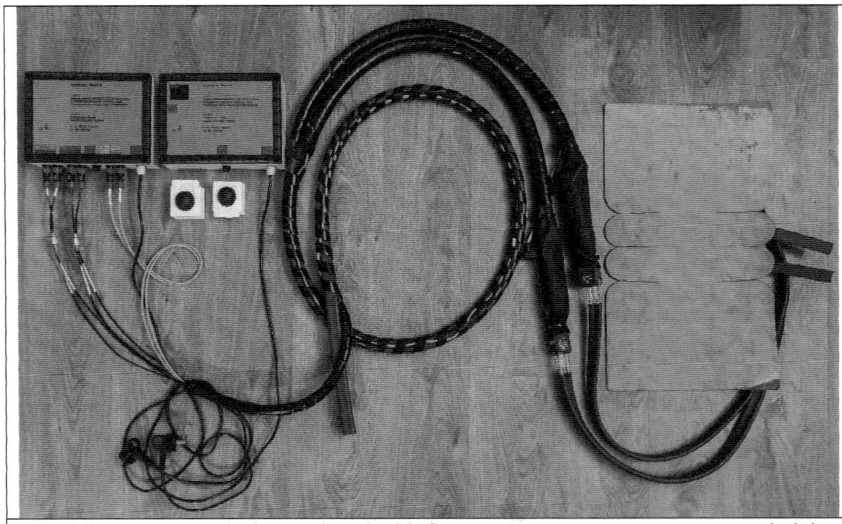

Fig.2. Measurement device equipped with flow meters, temperature sensors and globe thermometers

This device consisting two the water flow meters, temperature sensors, two globe thermometers and registration and control units, is connected to one of IN/OUT ports of the cooling element. The device is also equipped with a GSM system and allows one to monitor fluctuations of measured quantities during therapy even remotely using an internet browser. In this way all measurements carried out during therapy can be made without any disturbance to medical personnel. It should be stressed that all recorded data are transferred via an encrypted connection and are secured by user account and password. Hence, from the therapeutic point of view, there are no differences and medical staff can operate in a standard way.

Temperature difference in each water line is measured using paired thermocouples within the Wheatstone's bridge which guarantee of temperature uncertainty below 0.05K. More details regarding measurement device can be found in [6] and the practical use of non-invasive thermal measurements to assess the degree of brain injury in newborns treated with therapeutic hypothermia [7].

2. Results of selected measurements

To demonstrate the main features of the cooling processes occurring during selective and whole-body hypothermia, results of non-invasive measurements described briefly in section 1 are discussed now for two neonates undergoing TH.

The first patient (a female infant, internal ID – patient no 7) was born at 40 weeks of gestation. The body mass was 3.130 kg while the length was 54 cm. After relevant medical examination patient has been qualified for selective head TH (using the Olympic Cool-Cap device). Treatment started on the birth day and lasted for 72 hours. The following standard temperatures characterising treatment have been recorded: the temperature in the rectum, the temperature of the skin over the abdomen and temperature of the skin on the head. Radiant temperature representing of the mean temperature of all objects irradiating neonate as well as convective temperature representing a mean temperature of local indoor environment have also been recorded using a custom-made globe thermometer unit [6]. All these temperatures, together with the mean temperature of the water flowing through the cooling cap (setting manually by the medical personnel) have been recorded and are shown in Fig. 3.

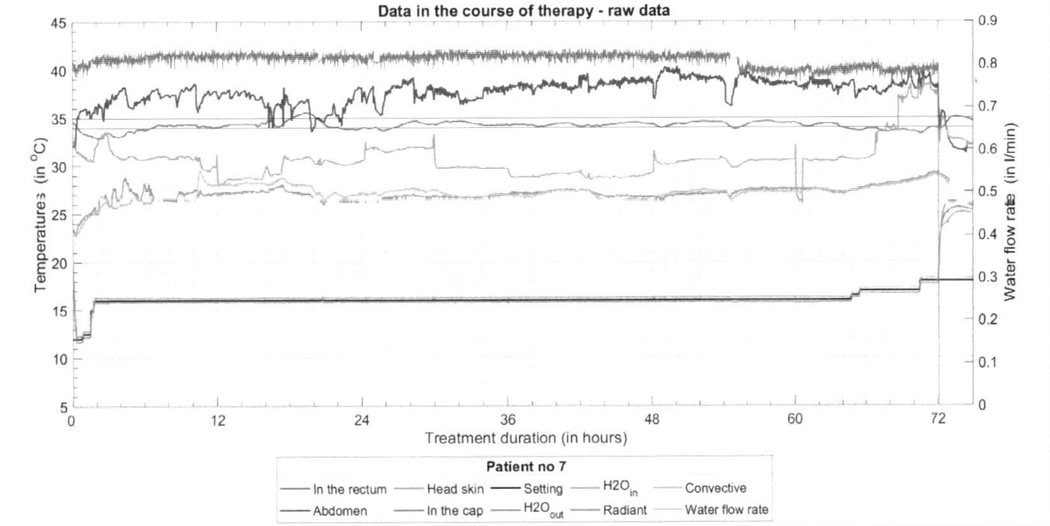

Fig.3. Temperatures recorded during therapeutic hypothermia using Cool Cap

It should be noted that the core temperature, represented by the red curve, generally falls within the required range, i.e. 34°C-35°C, although some small escapes from this range are also visible. Required corrections and/or modifications of the core temperature are made by medical personnel as changes of the radiant warmer power and should be linked with changes of the radiative and convective temperatures. In the end of the treatment, (hours 64+) the alteration of the water mean temperature was needed. It should also be noted that the temperature of the skin on the head (indicated by greenish lines) generally oscillates around 30°C. It has regular (approx. every 12 hours) temporary drops of this temperature. This is the result of a routine procedure of removing the cooling cap and examining the state of head skin every 12 hours. Other sudden decreases of temperature, visible also on blue lines, might be results of the application of some fluid boluses at room temperature.

Analogous hypothermic treatment, but using the Arctic Sun® 5000 (Medivance, Inc., USA), was also offered to the second patient (a male infant, internal ID – patient no 21). This patient was born at 38 weeks of gestation. The body mass was 3.680 kg while the length was 55 cm. After relevant medical examination patient has been qualified for the whole-body TH. Treatment started on the birth day and lasted analogously for 72 hours.

As already mentioned, the patient's temperature in case of cooling using the Arctic Sun® 5000 is controlled automatically. This means that the cooling process is organised in significantly different way. Patient's core temperature, measured by sensor in the rectum, depends on temperature of cooling fluid flowing through coils of the pad. The heat transfer process between patient's body and cooling fluid is additionally supported by the heater located inside control unit. Its thermal power, which is graded into 32 levels, is adjusted automatically. The overview of the whole therapy is shown in Fig. 4, while details of selected 4 hours (between 36th and 40th hour) of therapy are presented in Fig. 5.

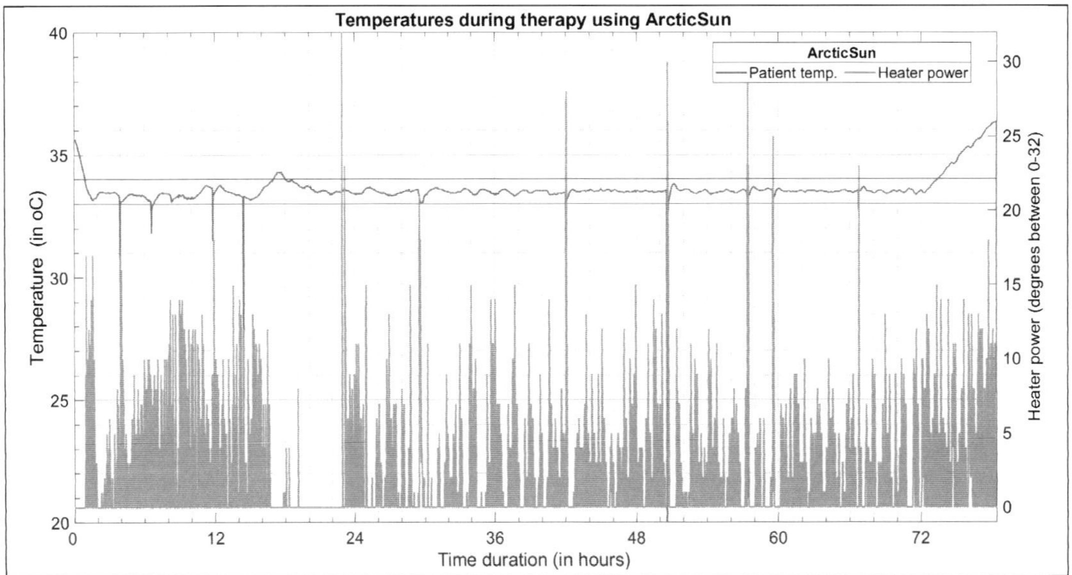

Fig.4. Core temperature and heater power during therapeutic hypothermia using Arctic Sun® 5000

It should be noted that patient's core temperature (red lines in Fig. 4 and 5) is quite stable, more stable than in case of using Cool Cap. As already mentioned, the cooling process, no matter it is selective hypothermia or the whole body one, is not the only treatment applied to the neonate. Some of them like medical examinations, application of some boluses, routine inspection of the skin condition under the cooling element, injections, etc. may even require unwrapping of cooling element or disconnection of measurement sensor for certain period. The latter one directly results in a sudden drop of red line for a quite short or for a bit longer time and this is what happened in the vicinity of 15 h of therapy. The only core temperature escape above target range (between 17th and 18th hours in Fig. 4) was a result of some additional medical treatment.

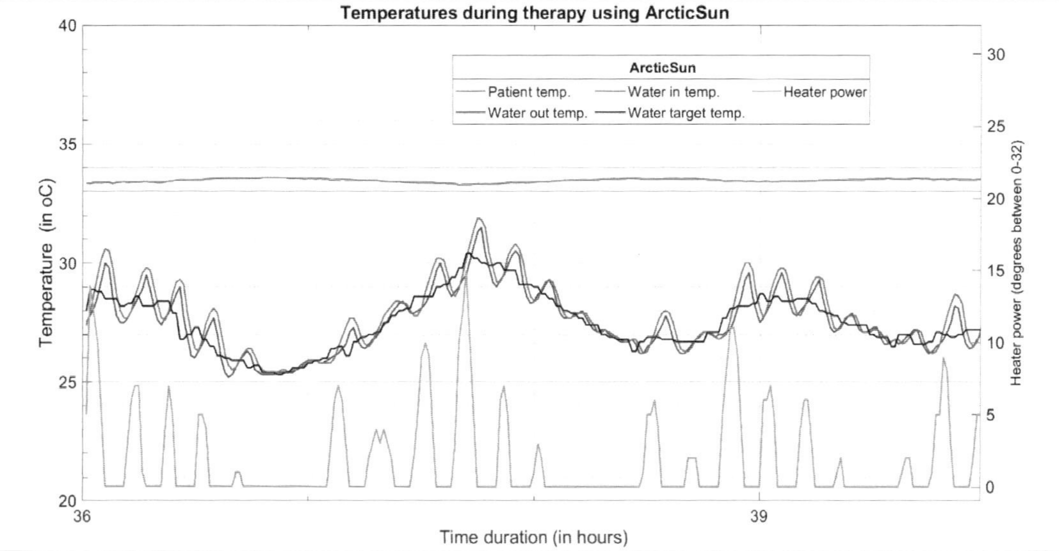

Fig.5. Details of therapy between 36th and 40th hour (using Arctic Sun® 5000)

In Fig. 5 it is clearly visible that temperature of the inlet and outlet cooling water (magenta and blue lines) is subject of low frequency as well as high frequency oscillations. Nevertheless these temperatures follow target temperature (yellow line). It is also worth noting that those oscillations are closely related to portions of heat delivered by the internal heater (green line). The green peaks themselves are responsible for the high frequency oscillations of the inlet and outlet cooling water but varying heights of green peaks are responsible for their low frequency oscillations.

Volume flow rates of the cooling water flowing through both parts of the pad and temperature differences which water experiences within each part of the pad are shown in Fig. 6. Those data allows one to make appropriate energy balance and determine amount of heat $Q_{cooling}$ retrieved from neonate's body within certain time period defined by the number of time steps $\Delta\tau_i$, i.e.

$$Q_{cooling} = c_w \, \rho_w \sum_i \dot{V}_{w,i} \, \Delta T_{w,i} \, \Delta\tau_i \tag{1}$$

where $\dot{V}_{w,i}$ is the volume flow rate of the cooling water measured at every i^{th} time step and $\Delta T_{w,i}$ stands for a temperature increase of the cooling water at this time step. The water specific heat and water density are marked by c_w (4186.8 J kg^{-1} K^{-1}) and ρ_w (1000 kg m^{-3}), respectively. Comparing both discussed TH methods, the amount of heat $Q_{cooling}$ retrieved from neonate's body within the first hour of the therapy have been calculated. In case of selective TH 43.6 kJ per 1kg of body mass while for the whole body TH analogous quantity was equal to 42.1 kJ per 1kg of body mass. These results need to be verified on larger sample of patients.

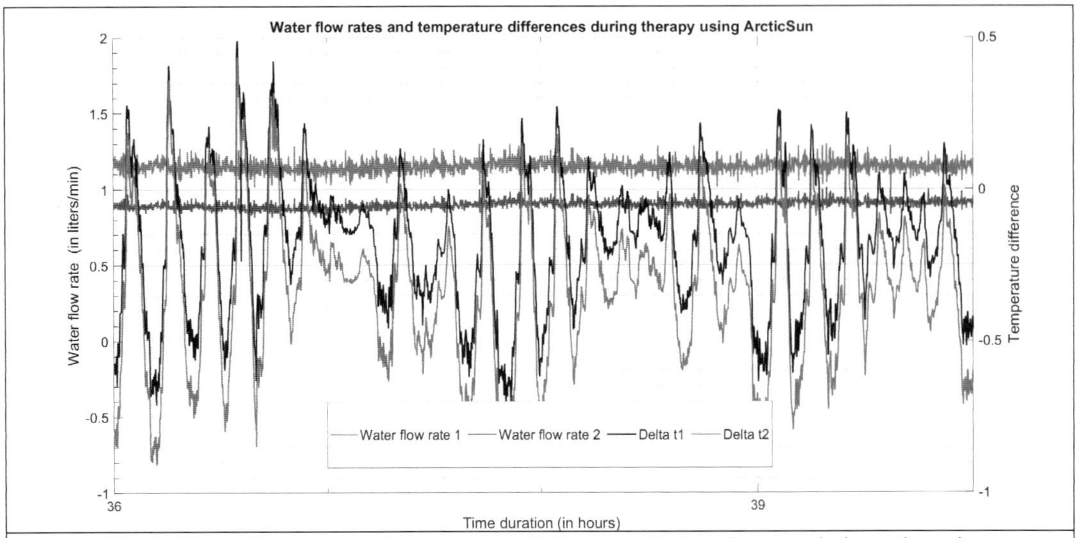

Fig.6. Water flow rates and temperature differences during therapeutic hypothermia

3. Concluding remarks

In this paper some thermal non-invasive measurements carried out at the Department of Anaesthesiology and Intensive Therapy of Children and Newborns of the University Clinic Hospital in Opole, Poland are described and discussed. Data recorded for two selected newborns are used to preliminary compare characterisation of the heat transfer processes occurring during performed therapies. Considering this comparison the following remarks can be formulated:

- The patients' core temperatures have been more stable in case of the whole-body hypothermia. This conclusion is drawn comparing courses of the red lines in presented graphs.
- In case of the selective hypothermia (i.e. Cool Cap) temperature of the cooling water is set manually by the medical personal and changes rather rarely. In case of the whole-body hypothermia (i.e. Arctic Sun) temperature of the cooling water strongly fluctuates, following the target temperature - the yellow line in Fig. 5. The target temperature, as a result of internal algorithm of the control system, also strongly oscillates.
- Portions of the heat delivered by the heater (green line in Fig. 5) cause the low- and also high-frequency oscillations of cooling water temperature.
- In case of the selective hypothermia (i.e. Cool Cap) core temperature is adjusted to the required range by changing power of the radiant warmer. Results of these corrections are visible on course of radiative temperature – greyish line in Fig. 3.

Described measurements are being continued. At the same time medical partners in the project carry out follow-up assessment of treated patients.

Acknowledgment

The study is supported by the National Science Centre (NCN), Poland, within the project no UMO-2018/29/B/ST8/01490.

References

[1] World Health Organization. http://www.childmortality.org/. Global Health Observatory (GHO); WHO 2016. Accessed April 10, 2020.

[2] Hayashi N., Dietrich D.W.. (2004), Brain hypothermia treatment, 1 ed., Springer-Verlag, Tokyo, Japan.

[3] Olympic Cool-Cap System Trainer, (2007), The operating instruction Olympic Medical, a division of Natus, USA.

[4] Thc Arctic Sun® 5000 Temperature Management System, (2020), Medivance, Inc., USA.

[5] Bandoła D., Rojczyk M., Ostrowski Z. et al., Expermental setup and measurements of the heat transfer rate during newborn brain cooling process. Arch Thermodyn. 2018;39(3):85-96.

[6] Walas W., Bandoła D., Ostrowski Z. et al., Theoretical basis for the use of non-invasive thermal measurements to assess the brain injury in newborns undergoing therapeutic hypothermia. Sci Rep, **10,** 22167 (2020). https://doi.org/10.1038/s41598-020-79009-3.

[7] Walas W, Mączko A, Halaba Z, Bekiesińska-Figatowska M, Miechowicz I, Bandoła D, Ostrowski Z, Rojczyk M, Nowak AJ. Thermal Index for early non-invasive assessment of brain injury in newborns treated with therapeutic hypothermia: preliminary report. *Sci Rep.* 2021 Jun 15;**11**(1):**12578**. doi: 10.1038/s41598-021-92139-6.

Glass beads retro-reflective coating for building application: albedo assessment in urban canyon configurations

Marta Cardinali[1*], Alessia Di Giuseppe[1], Alberto Maria Gambelli[2], Mirko Filipponi[1,2], Beatrice Castellani[1,2], Andrea Nicolini[1,2], Federico Rossi[1,2]

[1] CIRIAF, Interuniversity Research Centre on Pollution and Environment "M. Felli", Via G. Duranti 63, 06125, Perugia, Italy

[2] University of Perugia, Department of Engineering, Via G. Duranti 93, 06125, Perugia, Italy

[*] Corresponding author: cardinali@crbnet.it

Abstract. Materials for urban pavements and building coatings have a direct impact on urban thermal environment: low-reflective and highly absorptive materials, in fact, could contribute to the exacerbation of the Urban Heat Island (UHI) phenomenon, which negatively affects building energy consumptions and indoor and outdoor thermal comfort. These effects could be worsened especially in Urban Heat Canyon configurations, where the inter-building effect becomes more relevant. As regards UHI mitigation strategies, high-reflective (HR) and retro-reflective (RR) materials have been studied for their application as urban coatings, in order to provide a cooling effect inside urban canyons. In this framework, this study aims at comparing the albedo performance of HR and RR materials applied on a small scale urban canyon facility. To this aim, an in-field experimental campaign was conducted in two days during summer 2020. Different canyon coating combinations and height/width (H/W) ratios were investigated and compared in terms of albedo. Results show that an albedo worsening occurs when the RR coating was applied to the canyon's surfaces, because of the lower global reflectance of the RR with respect to the HR. Nevertheless, the application of RR coating on walls, coupled with HR paint on pavement, lead to higher albedo values of up to 4.7%, with respect to the opposite mixed coating combination. Future developments may concern the investigation of the optimum design of a glass bead RR material for building application, by studying the correlation between its optical capabilities and the dimensions and density of the glass beads.

1. Introduction

The ongoing world population increase represents one of the major issues of the last decades, although it is occurring with a lower growing rate [1]. As a consequence, the continuous urbanization leads to several environmental problems, as e.g. the higher building density, the lack of green areas and open spaces, the rise of anthropogenic heat and pollution levels. These factors contribute to the occurrence of the Urban Heat Island (UHI) phenomenon, a well-known environmental issue consisting in higher air temperatures in cities than in the neighboring rural areas [2].

UHI negatively affects building energy consumptions, indoor and outdoor thermal comfort [3], especially in Urban Heat Canyon configurations, where the inter-building effect becomes more relevant. One of the main aspects affecting UHI is represented by low-reflective and highly absorptive construction materials used as building coatings, streets, sidewalks, etc., which results in a high amount of solar radiation being absorbed by urban surfaces [4]. In this context, cool materials (i.e. cool roofs and cool pavements) and urban vegetation (i.e. green and wet roofs) have been widely investigated as effective strategies for UHI mitigation, by increasing surface solar reflectance and evapotranspiration, respectively [5,6].

As regards cool surfaces, in addition to high-reflective (HR) materials, also retro-reflective (RR) ones have emerged for use on innovative urban coatings, as they are able to reflect the incident solar radiation mainly toward the same direction. Several studies in Literature focused on the benefits provided by RR materials on urban environment, in terms of cooling potential inside urban canyons [7-9]. Rossi et al. [7] revealed that, when the vertical surfaces inside the urban canyon configurations were treated by RR materials with respect to the diffusive one, the air temperature was up to 1% lower as well as the temperature reduction for the pavements was between 5% and 22% for canyon's height/width (H/W) ratio equal to 1, from 1% to 2% for canyon's H/W ratio equal to 0.5, and between 5% and 10% for canyon's H/W ratio equal to 2. Yuan et al. [8] analyzed the UHI mitigation potential of two different types of glass beads RR coatings with different refractive index (i.e. 1.5 and 1.9), through a 2D urban canyon model, showing that only one type of glass beads RR sample (the one with refractive index of 1.9) is able to increase the urban albedo. Morini et al. [9] exhibited that façades covered by RR materials lead to an increase of equivalent albedo equal to 7% in comparison with façades made by traditional concrete, characterized by a comparable global reflectance. In this framework, this study aims at evaluating the albedo changes on an urban canyon facility, with different coatings combinations and H/W ratios. A HR coating, composed by a high-reflective white paint and a RR coating, made of glass beads embedded on the same HR paint, were applied in different ways, in order to create four different canyon's coating combinations: HR (RR) on both walls and pavements, HR (RR) on walls and RR (HR) on pavement. Thus, an in-field experimental campaign was conducted in two days in summer 2020, by comparing the albedo measured inside the urban heat canyon facility. The experimental campaign involved the albedo investigation by changing also the canyon H/W ratio (i.e. equal to 0.5 and 1, respectively) and the combination of HR and RR coatings on the canyon surfaces (i.e. on the horizontal and vertical ones).

2. Motivation

Given the contribution of RR materials in terms of directional reflectivity, this paper deals with an in situ experimental analysis of the albedo performance of HR and RR coatings applied in an urban canyon facility. The aim of this paper is to investigate the effect of different coatings combination on the canyon's albedo, with different H/W ratios. The HR and RR coating were combined in different ways, in order to create four different coating combinations inside the canyon, i.e. HR (RR) on both walls and pavement, HR (RR) on walls and RR (HR) on pavement. Therefore, this study could provide useful information about the interaction between walls and pavement on an urban canyon, in order to assess the optimal coating combination of HR and RR.

3. Materials and Methods

In this section, the urban canyon facility with the HR and RR tested coatings is firstly described and then the applied methodology is presented.

3.1. Urban Canyon Facility

An inner courtyard located in the CIRIAF building, University of Perugia (43° 7' 9.449" N, 12° 21' 27.451" E) is used to simulate a small-scale urban canyon. It consists of four facing walls with a rectangular layout (1,82 x 2,30 m) oriented in NO-SE direction, useful to represent a dense urban context, and an horizontal surface pushed down inside the inner courtyard at different levels according to the simulated H/W ratio (equal to 0.5 and 1, respectively). Two different plywood panels covered by the HR and RR coatings, were used to simulate the pavement in the urban canyon, resulting in

Journal of Physics: Conference Series 2177 (2022) 012033 doi:10.1088/1742-6596/2177/1/012033

different combinations of coatings on the canyon surfaces. The small-scale urban canyon facility and its location are shown in Figure 1a and 1b.

Figure 1. An aerial view of the case-study location, at CIRIAF (University of Perugia), with the shape of the small-scale urban canyon facility evidenced in red (Figure 1a) and a picture of the urban canyon facility (Figure 1b).

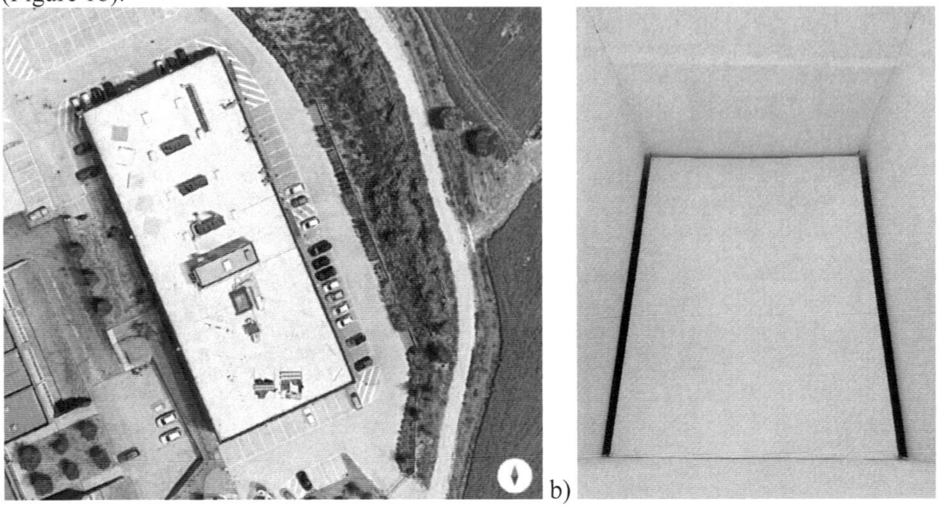

a) b)

3.2. High-reflective and Retro-reflective coatings

A HR and a RR coating have been applied on the vertical and horizontal canyon surfaces in order to study their influence on the albedo assessment inside the urban canyon. The HR coating consisted of a high-reflective white paint, whereas the RR coating was made of glass beads with an average diameter 0.1-0.2 mm, embedded on the same high-reflective paint (i.e. the ratio between glass beads and white paint was 1:1). In a previous work [10], the optical performance of the abovementioned HR and RR coatings was analyzed in lab with both spectrophotometric and angular reflectivity characterization. Results from this previous contribution [10] showed that the glass beads modified the global reflectance of the high-reflective paint, but they provided a directional reflectance capability to the coating. It was demonstrated that the HR coating exhibited the highest global reflectance (80.1%), whereas for the RR coating it decreases to 72.8% [10]. Despite this lower global reflectance, the RR coating provides a strong retro-reflective component for incident directions near the normal to the surface (i.e. from 90° to -40°), with the maximum retro-reflective capability equal to 16.2% at 90° angle of incident radiation [10].

In the present work, the HR and RR coatings, whose optical features have already been assessed [10], are further investigated as coatings of an urban canyon facility. The canyon's vertical surfaces were firstly coated by the HR paint and then by the RR coating, applying them directly to the walls of the inner courtyard. In order to simulate the horizontal urban canyon surface, two plywood panels were realized and covered with the HR and RR coatings, respectively. In this way, it was possible to investigate different combinations of coatings, along with two canyon's H/W ratios. Figure 2 shows a picture of the surface's appearance of the HR (Figure 2a) and RR (Figure 2b) coatings, when the light hits them. For detailed information about the methodology, please see section 3.4. of the present work.

Figure 2. The appearance of the HR (Figure 2a) and RR (Figure 2b) coatings when an incident light beam strikes their surface.

a)  b) 

3.3. Equipment for albedo monitoring

The urban canyon facility was equipped with an albedometer, supplied by Delta Ohm Srl (model LP PYRA 05) and constituted by two pyranometers (Class I). The albedo is calculated as the ratio between the reflected solar radiation and the incident one. The technical datasheet of the albedometer is provided in [11]. The albedometer was positioned at the center of the canyon's facility opening, as shown in Figure 3, in order to calculate the contribution of all canyon surfaces. A data logger by Campbell Scientific was connected to the albedometer for recording data. The data logger was set to record data every 1 second during the tests. Then, the system was connected to a processor in which the collected data were recorded.

A fixed weather station settled above the rooftops level at the CIRIAF building was used to perform the environmental monitoring. Data concerning global radiation (W/m^2) and air temperature (°C) were collected during the monitoring days, in order to compare the microclimate boundary conditions.

Figure 3. Delta Ohm Srl Albedometer, model LP PYRA 05 [11], positioned at the center of the canyon's opening.

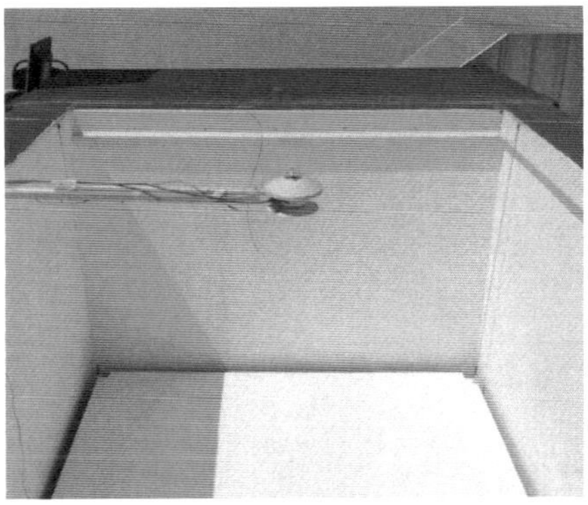

3.4. Methodology

The albedo monitoring campaign on the urban canyon facility was carried out in two different days between the end of June and the beginning of July 2020, by varying the canyon coatings and geometry, i.e. the canyon's height/width ratio (H/W). In particular, two days near the summer solstice were selected for the monitoring campaign: 24th of June 2020 and the 1st of July 2020. For each selected day, albedo measurements have been performed indicatively from 11:00 a.m. to 2:00 p.m., when the solar radiation reaches its peak values. Each albedo measurement was 5 minutes long, with data acquisition every 1 second. For the specific geographical location, the maximum solar elevation is about 70° at 1:15 p.m., with no significant changes between the two monitoring days.

For each day of albedo measurements, several canyon configurations have been investigated, both changing the coating of canyon's vertical and horizontal surfaces, and also varying the depth of the base panel, in order to study two different H/W ratios, equal to 0.5 and 1. On the 24th of June, the walls of the urban canyon facility were coated with the HR paint, whereas alternatively the HR and RR coating was applied to the base panel, in order to simulate the canyon's pavement. On the 1st of July 2020, the RR coating was applied on walls and both HR and RR materials were considered for pavement. In this way, it was possible to compare the matching of the two types of coatings in terms of albedo changes. In Figure 4, global radiation and air temperature variations during the monitored days are presented, as they affect the albedo measurement in a predominant way. Global radiation is similar between this two days, with the maximum difference of 109.7 W/m^2 at 2:00 p.m., whereas the maximum and average air temperature variations are 2.2 °C and 1 °C, respectively. Table 1 shows the details of the selected albedo monitoring days, the number of measurements, the typology of canyon's coatings (i.e. HR or RR) for walls and pavement during each observation and the investigated H/W ratios.

Figure 4. Global Radiation (GlobR) and Air Temperature (AirT) on 24th (solid lines) of June and 1st (dotted lines) of July 2020.

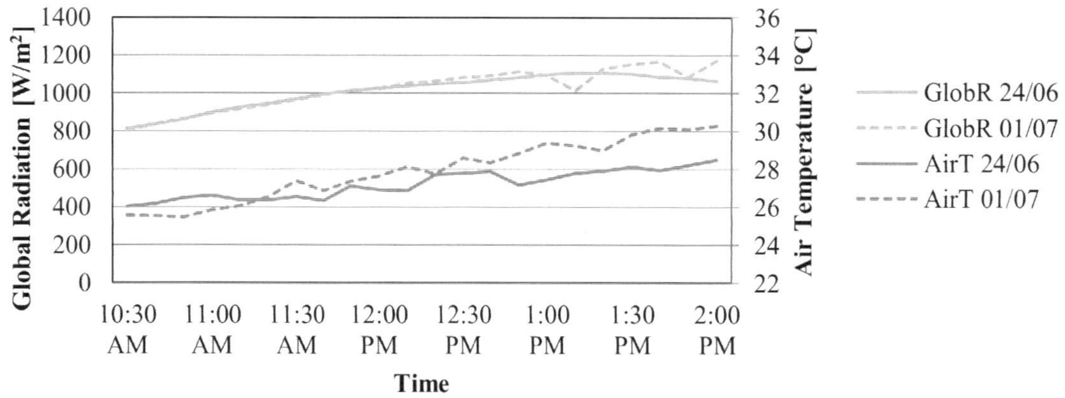

Table 1. Albedo monitoring campaign: selected days, number of performed measurements, types of walls and pavement coatings and the investigated canyon's configurations.

Day	Number of measurements	Canyon's walls coating	Canyon's pavement coating	Canyon's H/W ratio
24/06/2020	4	HR	HR	1
	4	HR	HR	0.5
	4	HR	RR	1
	4	HR	RR	0.5
01/07/2020	4	RR	RR	1
	4	RR	RR	0.5
	4	RR	HR	1
	4	RR	HR	0.5

4. Results and Discussion

Albedo results are presented as an average over the 5 minutes monitoring range, for each configuration. Since it was not possible to monitor simultaneously different canyon configurations, the comparison has been made between configurations whose albedo measures have been performed nearly at the same time, with a maximum time difference of about 19 minutes. For that reason, times showed in the graphs below are only representative of the slight time intervals within which two performed albedo measures were compared.

As already mentioned in section 3.2 and resulting from a previous study [10], the RR coating, made of glass beads embedded on the high-reflective paint, exhibits a lower global reflectance with respect to the HR coating, made of the same paint without glass beads. This difference in terms of global reflectance results in an albedo worsening when the RR coating was applied both to walls and

Journal of Physics: Conference Series 2177 (2022) 012033 doi:10.1088/1742-6596/2177/1/012033

pavement, even if the addition of glass beads provides an angular reflectivity to the coating. In Figure 6, on the 24th of June, the total HR canyon's configuration shows higher albedo values with respect to the mixed configuration (HR on walls and RR on pavement), for both H/W ratios and in all the monitoring intervals. This means that the RR pavement affects albedo in a negative way, with respect to the total HR configuration. In Figure 7, the opposite mixed configuration (where RR coating was applied to walls and the pavement was covered with HR paint) exhibits higher albedo values, with respect to the total RR configuration. Despite the strong directional reflectivity exhibited by the RR coating, the higher global reflectance of the HR paint provides better albedo results in all the investigated configurations. Although this, the comparison between the mixed coating combinations could represent a key issue to be investigated, in order to analyze the optimal coating combination in an urban canyon. To this aim, the mixed coating configurations have been compared between the 24th of June and the 1st of July, and presented in Figure 8 and Figure 9. In all graphs is clear that the combination between RR and HR materials is more efficient when vertical surfaces are treated with RR coating and horizontal ones are made of HR paint. It is also worth noting that the highest albedo differences occurred when the solar elevation was around its maximum, i.e. about 70° at 1:15 p.m. The maximum albedo increase, due to the application of RR on walls, is equal to 4.7% referred to the canyon's H/W=0.5, at 1:30 p.m., with respect to the opposite mixed configuration.

Figure 6. Averaged albedo values on 24th of June 2020. Comparison between the total HR configuration and mixed configuration (HR on walls and RR on pavement), for both H/W ratios.

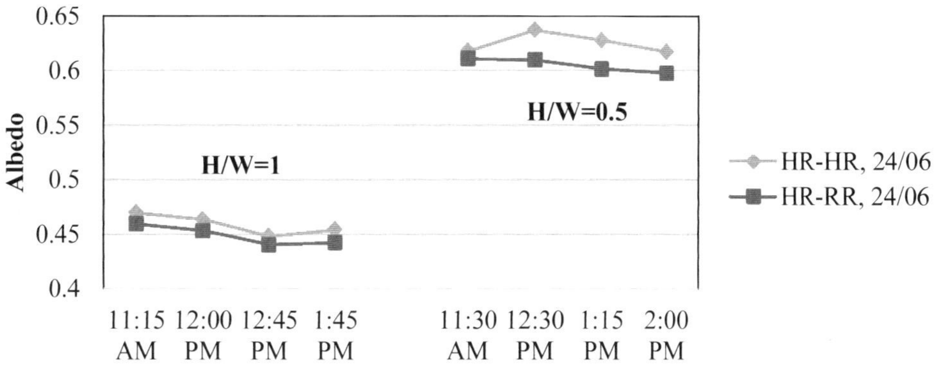

Figure 7. Averaged albedo values on 1st of July 2020. Comparison between the total RR configuration and mixed configuration (RR on walls and HR on pavement), for both H/W ratios.

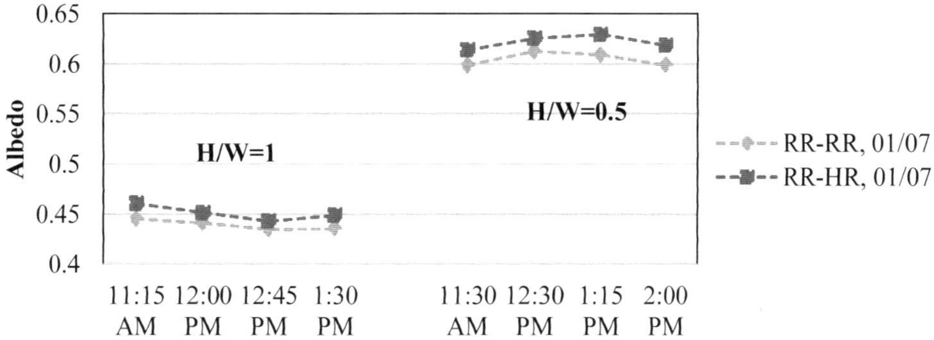

Figure 8. Averaged albedo values for mixed coating combinations, on 24th of June and 1st of July 2020, for canyon's H/W=1.

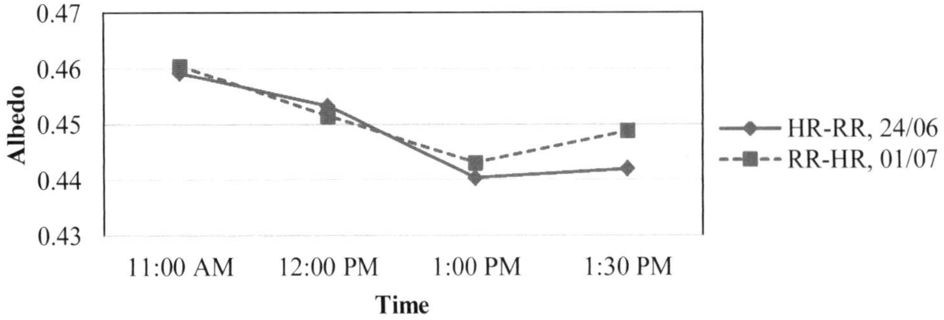

Figure 9. Averaged albedo values for mixed coating combinations, on 24th of June and 1st of July 2020, for canyon's H/W=0.5.

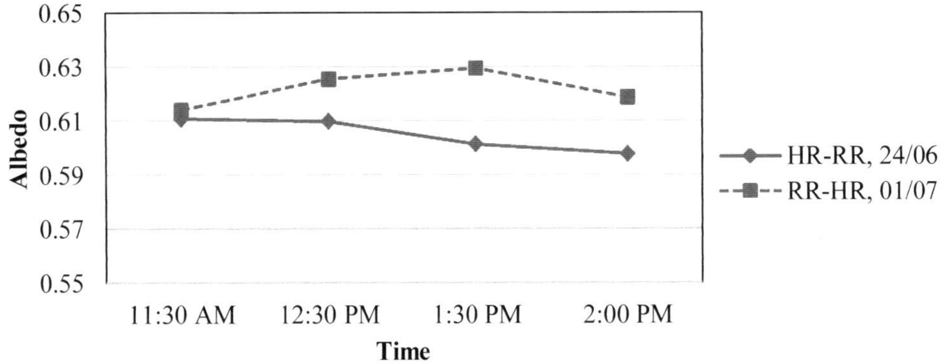

5. Conclusions and Future Developments

In this study, albedo changes on an urban canyon facility, with different coatings combinations and H/W ratios, were assessed. A HR and a RR coating, made of glass beads embedded on the same high-reflective paint, were combined in order to create four different coating combinations, i.e. HR (RR) on both canyon's walls and pavement, HR (RR) on walls and RR (HR) on pavement. Given that the HR coating's global reflectance is higher that the RR one, the mere directional reflectivity provided by the RR coating is not enough to enhance the canyon's albedo. This results in a canyon's albedo worsening when the RR coating was applied both to walls and pavement, with respect to the corresponding HR configuration. In order to evaluate the optimum combination of HR and RR coatings on the canyon's surfaces, a comparison between the mixed configurations has been carried out. Results show that the application of RR coating on walls, coupled with HR paint on pavement, lead to higher albedo values, with respect to the opposite combination. The highest albedo increase, due to the combination of RR on walls and of HR on pavement, is equal to 4.7% for canyon's H/W=0.5 at 1:30 p.m., when the solar elevation reaches its maximum. Further research efforts should be made in the assessment of RR materials effectiveness in urban canyons, as this study presents some limitations. Firstly, the addition of the glass beads to an extremely high-reflective coating may not be as advantageous as if the diffusive coating had a medium global reflectance. Secondly, the urban canyon facility is not a real-scale model of an urban canyon, as there is also the reflection contribution of the two additional walls. Furthermore, the application of the glass beads on the canyon's walls should be further investigated in order to prevent the detachment of the glass beads from the vertical surfaces, along with ageing effects.

Acknowledgements

The authors would like to thank both Prochima S.r.l. for supplying the glass beads, and INDEX Construction Systems and Products S.p.A., which has provided the high-reflective white paint for the realization of RR and HR coatings. Authors would also like to thank EAP LAB (CIRIAF, Interuniversity Research Centre on Pollution and Environment "M. Felli", University of Perugia) for providing data from the weather station.

References

[1] United Nations, Department of Economic and Social Affairs, Population Division (2019). World Population Prospects 2019: Highlights (ST/ESA/SER.A/423)

[2] Taha H., (1997) "Urban climates and heat islands: albedo, evapotranspiration, and anthropogenic heat", Energy and Buildings, 25, pp. 99-103. https://doi.org/10.1016/S0378-7788(96)00999-1

[3] Santamouris M., (2020) "Recent progress on urban overheating and heat island research. Integrated assessment of the energy, environmental, vulnerability and health impact. Synergies with the global climate change", Energy and Buildings, 207, 109482. https://doi.org/10.1016/j.enbuild.2019.109482

[4] Santamouris M., (2015) "Analyzing the heat island magnitude and characteristics in one hundred Asian and Australian cities and regions", Science of the Total Environment, 512, pp. 582–598. doi: 10.1016/j.scitotenv.2015.01.060

[5] Akbari H. and Kolokotsa D., (2016) "Three decades of urban heat islands and mitigation technologies research", Energy and Buildings, 133, pp. 834–842. doi: 10.1016/J.ENBUILD.2016.09.067

[6] Hendel M., Parison S., Grados A., Royon L., (2018) "Which pavement structures are best suited to limiting the UHI effect? A laboratory-scale study of Parisian pavement structures", Building and Environment, 144, pp. 216-229. https://doi.org/10.1016/j.buildenv.2018.08.027

[7] Rossi F., Morini E., Castellani B., Nicolini A., Bonamente E., Anderini E., Cotana F., (2015) "Beneficial effects of retroreflective materials in urban canyons: results from seasonal monitoring campaign", 33 rd UIT Heat Transfer Conference, IOP Publishing, Journal of Physics: Conference Series 655 (2015) 012012. doi: 10.1088/1742-6596/655/1/012012

[8] Yuan J., Emura K., Farnham C., Sakai H., (2016) "Application of glass beads as retro-reflective facades for urban heat island mitigation: Experimental investigation and simulation analysis", Building and Environment, 105, pp. 140-152. https://doi.org/10.1016/j.buildenv.2016.05.039

[9] Morini E., Castellani B., Presciutti A., Anderini E., Filipponi M., Nicolini A., Rossi F., (2017) "Experimental analysis of the effect of geometry and façade materials on urban district's equivalent albedo", Sustainability, 9, 1245. doi: 10.3390/su9071245

[10] Di Giuseppe A., Cardinali M., Castellani B., Filipponi M., Gambelli A.M., Postrioti L., Nicolini A., Rossi F., (2021) "The Effect of the Substrate on the Optic Performance of Retro-Reflective Coatings: An In-Lab Investigation", Energies, 14, 2921. https://doi.org/10.3390/en14102921

[11] Delta Ohm. Available online: https://www.deltastrumenti.it/images/pdf/deltaohm/meteorologia/LP_PYRA05_06_D_it.pdf (accessed on 09 May 2021)

Liquid holdup measurement for gas-liquid stratified flows by means of resistive probes and image processing

I M Carraretto[1], L P M Colombo[1] and M Guilizzoni[1]

[1]Politecnico di Milano, Dept. of Energy, via Lambruschini 4, 20156 Milan, Italy

igormatteo.carraretto@polimi.it

Abstract. Flow patterns exert a fundamental influence on the behaviour of multiphase flows, and they must be brought into play when dealing with their modelling. This is usually done by means of summarizing quantities as the phase holdups and the interfacial area concentration. Many techniques have been designed during the years to measure them, among which the use of probes relying on electrical resistance is one of the simplest and less expensive. While having these points of strength, resistive probes are intrusive devices. This work is therefore devoted to a comparison between liquid height (and derived quantities) measurements – for stratified and stratified-wavy air-water flows – performed using a conventional resistive probe and by means of an image-based technique. Validation of the latter was performed using computer-generated flow images. Then, an experimental campaign was carried out for flows with liquid superficial velocities in the range 0.03 ÷ 0.06 m/s and gas superficial velocities in the range 0.77 ÷ 2.31 m/s. Results showed that the two methods give answers within very few percent of difference, which is more than satisfactory in this field. The results are also in good agreement with some of the most credited literature models and correlations.

1. Introduction

In the study of multiphase flows, the distribution of the phases within the duct plays a fundamental role and it is usually taken into account in the models by means of quantities as the phase density function, the phase holdups and the interfacial area concentration [1]. Sampling of such quantities is far from being an easy task and many techniques have been designed during the years, ranging from very complex and expensive ones (e.g., gamma-ray attenuation, high speed X-ray tomography, magnetic resonance imaging and wire mesh sensors; reference to some fundamental papers about these techniques can be found in [2]) to much simpler devices as the different types of optical and electrical impedance probes which have been revised during the years ([3]–[8]).

In fact, in many multiphase flows, the phases have significantly different refraction indices and electrical impedances, and local, section- or volume- averaged measurements based on these quantities can reveal information about their spatial and temporal distribution within the flow. With a suitable selection of the probe geometry and appropriate calibration, such probes have been used to measure a variety of multiphase flow quantities, including film thickness in annular flows and liquid height in stratified flows (e.g., measuring the liquid layer depth by detection of the interface, thanks to a resistive probe in flush-wire configuration, as described in [8]). In fact, they can generate a practically "Boolean" voltage signal when the probe tip switches from a phase to the other, crossing the interface.

However, they are an intrusive method, i.e., it is needed to have holes on the ducts under analysis and the flow may be affected – even if minimally –by the presence of the probe. Hence image-based methods were developed, based on the processing of external views of the flow. (typically, from the side, e.g., [9]). In this work, a technique suitable for stratified flows is presented and its results are compared with those from resistive probes.

2. Experimental set-up and operating conditions

The experimental activity was carried out in the laboratory of Multiphase Thermal-Fluid Dynamics at the Department of Energy, Politecnico di Milano.

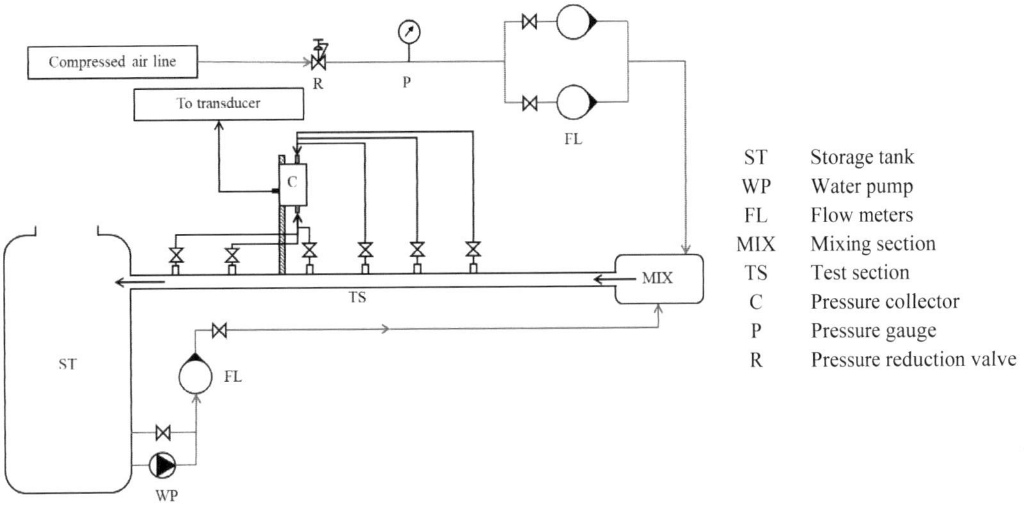

Figure 1. Layout of the experimental setup

The liquid is supplied from the bottom of a 4.0 m^3 storage tank by means of a CALPEDA centrifugal pump (volume flow rate = 0.12 ÷ 0.75 m^3/h; head = 6.5 ÷ 20 m). The liquid flow rate is measured by a float-type flow meter (whose characteristics are reported in Table 1) and set through a bypass valve upstream of the flow meter. Air flow rate is provided by the Department compressed air line at 0.8 MPa and measured by a float-type flow meter, the reading of which is suitably corrected to account for pressure and temperature deviation from standard conditions; the working point is set both through a pressure reducing valve and the flow meter valve. The liquid and the gas are injected into a mixing section, subsequently the two-phase flow enters the test section, that is made of transparent PMMA ducts having 60 mm inner diameter and 5 mm thickness. More details about the setup, reported in Figure 1, can be found in [10] and in [11].

Table 1. Flow meters characteristics

Name	Fluid	Full Scale (FS)	Error	T_c [° C]	P_c [Pa]
ASAMETRO P13-2800	Water	0.1 ÷ 1 m^3/h	± 3 % FS	20	-
ASAMETRO N5-2008	Air	2.5 ÷ 23.5 m^3/h	± 2.5 % FS	20	101 325

The superficial velocities of the phases for the flow conditions investigated in this work were J_L = 0.03 ÷ 0.06 m/s for the liquid and J_G = 0.77 ÷ 2.31 m/s for the gas. Within these ranges, the flow was perfectly stratified or stratified with small-amplitude waves. Therefore, the liquid holdup, ε_L was evaluated by measuring the liquid layer depth h_L and by making use of equations (1) and (2), linking the geometrical quantities shown in Figure 2. Similarly, the interfacial area concentration in its 2D cross-sectional definition can be very easily calculated as the ratio between the length of the chord representing the interface and the cross-section area of the duct.

Uncertainties of the main quantities are reported in Table 2. Further details about repeatability and uncertainty analysis for the same setup and similar flow conditions can be found in [10].

$$\varepsilon_L = \frac{\Omega_L}{\Omega} = \frac{\gamma - \sin\gamma}{2\pi} \text{ (1) and } \gamma = 2\cos^{-1}\left(1 - \frac{2h_L}{D}\right) \text{ (2)}$$

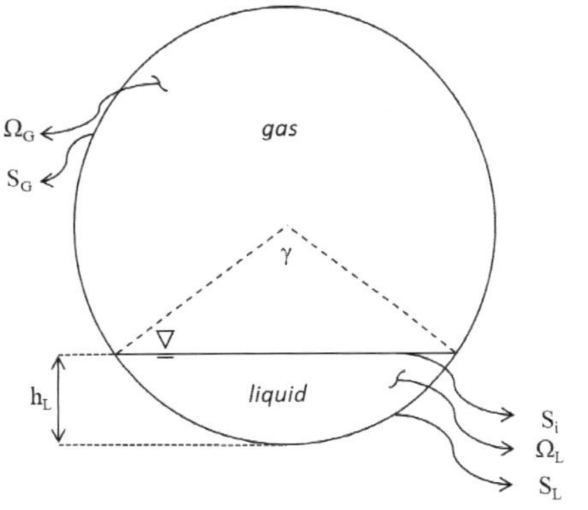

Figure 2. Sketch of the phase distribution in the pipe cross section

Table 2. Relative uncertainties of the main quantities

Quantity	J_G [m/s]	J_L [m/s]	ε_L [–]
Uncertainty [%]	2 ÷ 7	2 ÷ 3	1 ÷ 6

As anticipated in the Introduction, to measure the liquid height the use of a resistive probe and an image-based, non-intrusive, methods were compared.

Concerning the resistive probe, it is a single-tip, external-electrode resistive probe that consists of the sensing needle, fully insulated apart from the tip, and introduced radially from above into the flow, and of a second wire, submerged into the liquid phase, that acts as a reference electrode. The latter is connected to a +5 V power supply, while the needle is connected to an analogue input of the acquisition board and grounded through a resistor. Since air is not electrically conductive, such a probe allows to sample the state density function (at a frequency of 10 kHz): the voltage signal is zero until the needle touches the liquid phase, then the circuit closes and the signal jumps to a positive value. As the probe is inserted into the duct by means of a probe support endowed with a ruler to provide the immersion depth, when the probe tip touches the liquid the height of the latter can be easily determined. Further details can be found in [10].

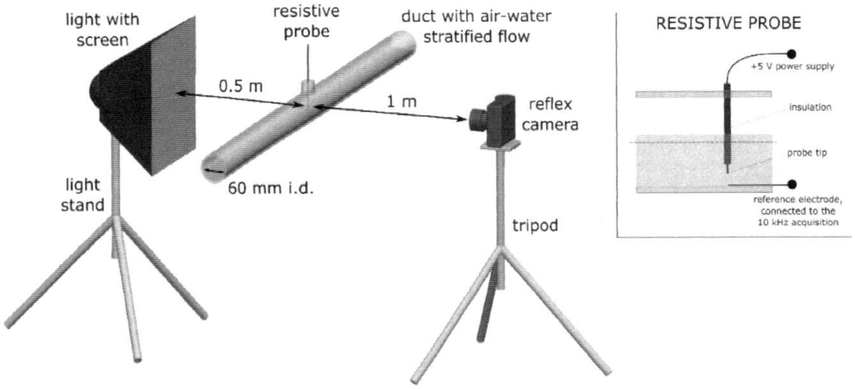

Figure 3. Scheme of the experimental setup for the image acquisition, including the duct with an air-water stratified flow, the reflex camera on its tripod and the light with the screen, on its light stand. The resistive probe is also shown in red and sketched in further details in the inset.

As for the image-based measurement, the experimental photographs were acquired using a conventional back-illumination technique, using a photography lamp (800 W) with a diffusing screen and a Nikon D90 single-lens reflex camera with an AF-S Micro NIKKOR 60mm lens. The camera was placed at 1 m distance from the portion of the pipeline under analysis, whereas the light with the screen was placed at 0.50 m distance from the latter. Both were aligned at the same height of the duct axis. Figure 3 shows a scheme of the experimental setup for the image acquisitions and of the resistive probe. The acquired images were analysed to determine the liquid height by direct measurement of the pixel distance between the duct bottom generatrix and the air-water interface. The interface position is manually identified, as the image characteristics and in some cases the interface waviness make the application of automated interface detection techniques impossible. The duct external diameter is known and provides the pixel-to-meter conversion factor. The liquid height, the liquid holdup and the interface area can be then calculated. As the camera was vertically aligned with the duct axis, while the liquid height was always lower than the latter, due to perspective a small difference between the interface heights in the front and back parts of the duct is present in all the acquired images. Therefore, liquid height was measured at both the front and back triple lines, and the two values were averaged. For each of the images, measurement was also performed in three axial positions, again averaging the results, and by different operators. The latter proved to have a negligible influence on the results.

3. Validation of the image-based technique

In the use of the proposed image-based technique, many aspects may in theory originate trueness issues. First of all, the capillary effects, leading to water rise with formation of a meniscus at the duct wall, and the optical distortions due to the changes in refraction indices between water, PMMA and air. Then, in decreasing order of likely importance, also the potential effects of lighting (adequate and in proper position/orientation), field of view and depth-of-field, lens distortion, and sensor resolution should be considered.

Concerning the optical distortion, a preliminary analysis was performed in Matlab® [12] considering 1D optical paths according to the Snell law, as shown in Figure 4, where a ray is traced from an ideal pinhole camera (on the left) to the duct. The real liquid level is drawn in cyan. For better visibility in scheme, the camera in this example is placed at only 0.1 m from the duct axis and the liquid height is set at 10 mm, so the discrepancy between real and apparent levels is much larger than in real experimental conditions. The real results evidenced that given the relatively small thickness of the duct, the distance of the camera from the duct itself and the liquid heights in the investigated flow conditions, the correction is by far lower than the overall experimental uncertainty. As for the capillary effects, Figure 5 shows the shape of the interface and the meniscus for different liquid levels. The interface shape was obtained by CFD simulations of the system, using the Volume-of-Fluid approach in a finite volume framework, by means of the *interFoam* solver of OpenFOAM® [13].

The settings for such simulations are the same as those described in [14]. A value of the water-air surface tension equal to 72 mN/m and a contact angle between water, air and PMMA equal to 70° were considered. These are likely to be not fully accurate values for the investigated flows, so the results should be considered mainly as qualitative ones. Nonetheless, it can be seen how the combined effect of the contact angle and of the inclination of the duct wall with respect to the viewing direction reduces very much the meniscus apparent height for the liquid levels of interest. With the additional consideration that the liquid height is manually identified by the operator, it is possible to conclude that the meniscus is not an issue for the measurement.

As no-truly "reference" technique is available for measurement in this field, validation of the proposed image processing approach was made by using computer-generated images.

Virtual ducts with stratified and stratified-wavy air-water flows were created by the already cited numerical simulations and by the 3D modelling software Art of Illusion [15], exported in Wavefront OBJ format (a format widely used for computer graphics) and rendered with the physically-based rendering engine Appleseed [16]. Duct dimensions (diameter, thickness, etc.) and the camera and light positions with respect to the duct were set to replicate the experimental ones. PMMA and water refractive indices were set to 1.51 and 1.33, respectively. Photorealism in a strict sense was not of interest, the aim was to obtain images including the major optical features of the experimental ones. Figure 6 and Figure 7 show two examples of rendered images compared with the corresponding experimental photographs of flows with the same liquid height, for a case with a practically flat interface and for a stratified-wavy flow pattern. A small drop, floating on the air-water surface, was also included. Similar drops were occasionally observed in the experimental photographs, and they help in identifying the correct level of the air-water interface.

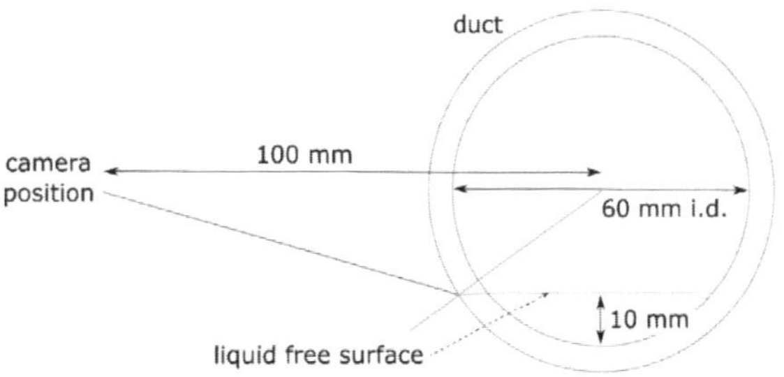

Figure 4. An example of the optical ray deviation analysis.

Figure 5. Menisci formed by the air-water interface.

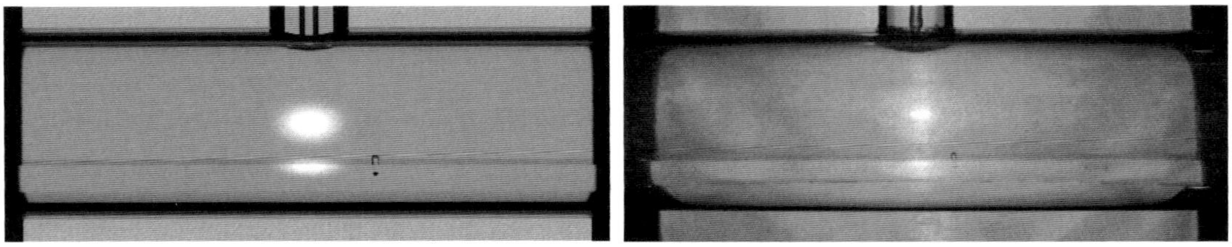

Figure 6. An example of rendered flow image (on the left) compared with an experimental photograph of flow with the same liquid height (on the right), for a flow with a practically flat interface

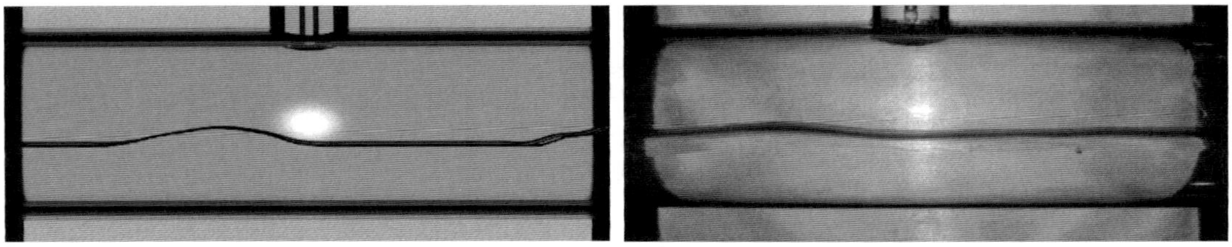

Figure 7. An example of rendered flow image (on the left) compared with an experimental photograph of flow with the same liquid height (on the right), for flows with a wavy interface

Measurement of the liquid height on the rendered images was carried out with the same procedure used on the experimental pictures. In each picture, liquid height was measured at both the front and back triple lines, and the two values were averaged. As the liquid was modelled/simulated, its exact height was known, and the measurement error could be calculated. Table 3 reports the results for the investigated liquid heights, both when considering the camera perfectly aligned with the duct and when intentionally introducing a misalignment, to assess the sensitivity of the measure to this parameter. The percentage error shown in the table is calculated as (measured value – true value) / true value · 100. As it can be seen, there are regularities in the errors, as it was expected being the optical paths influenced by the changes in the refraction indices in a deterministic way (they are less affected when the interface is near to the duct mid-plane). A couple of odd values are present, since for those condition the interface was more difficult to identify, and a single pixel of variation significantly changes the error for the rendered images. In any case, the error is always very small, independently from the eventual camera misalignment, and particularly in the range of liquid heights (> 16 mm) of interest for this work.

The vertical shift of the camera with respect to the duct axis that may occur in the experimental campaign is very small, as confirmed by the very thin stripe representing the interface in the experimental photographs. In any case, vertical shifts up to 30 mm – which is an exceedingly large value equal to one half of the duct internal diameter – were investigated, but as they are always very small with respect to the camera distance from the duct (1 m), they have practically no influence on the results. As already said, the measurement is obviously sensitive to errors in the manual selection of the interface, but the high resolution of the experimental pictures (the duct diameter occupies around 830 px, with respect to 260 px in the rendered images) effectively mitigates this issue.

Table 3. Results of the measurements on the rendered images

h_L [mm]	**Camera vertical position**	**Percentage Error** [-]
11	aligned with the duct axis	-0.83
11	10 mm higher than the duct axis	-0.83
11	30 mm higher than the duct axis	0.38
16	aligned with the duct axis	-0.33
16	20 mm higher than the duct axis	-0.33
18.5	aligned with the duct axis	-0.18
21	aligned with the duct axis	-0.07
21	5 mm higher than the duct axis	-0.07
21	30 mm higher than the duct axis	-0.07
23.5	aligned with the duct axis	0.58
27	aligned with the duct axis	0.31
30	aligned with the duct axis	0.00

Based on the rendered images, a correction function could also be obtained to account for the effects of the duct curvature and change in the refractive index along the optical path between the interface and the camera. Given the very small values of the errors that can be obtained without any correction, this step would change the final values of the phase fractions by a negligible amount (fractions of percent), so it can be safely skipped in the measurement. In summary, the validation on rendered images proved that the image-based measurement technique can return accurate values of the liquid height.

4. Results

Figure 8 shows some representative examples of the acquired images, at the different values of the liquid superficial velocity J_L and gas superficial velocity J_G. It can be seen how the liquid height (and therefore the liquid holdup and the interfacial area concentration) increase when both the liquid and the gas superficial velocities are increased. The flow pattern is always stratified, even if small waves appear at the highest superficial velocities. Figure 9 reports the liquid holdup measured with image processing technique, with respect to the liquid holdup obtained using the classical resistive probe method. The values obtained are included in the ± 15 % band, showing a good match for the lower values of the liquid holdup, whereas a worse accuracy is envisaged at the higher liquid holdup values. This is likely due to the presence of small waves in the latter cases. In fact, the liquid height with the image-based method is measured averaging the samples at three different axial coordinates, so it returns a value related to the mean liquid height. On the contrary, with the resistive probe it is assumed that the liquid region has been stably entered when the signal is dominantly zero, so the measured liquid height is nearer to the wave bottom. This results in a lower value of the estimated liquid holdup with respect to the image processing method.

Journal of Physics: Conference Series 2177 (2022) 012034 doi:10.1088/1742-6596/2177/1/012034

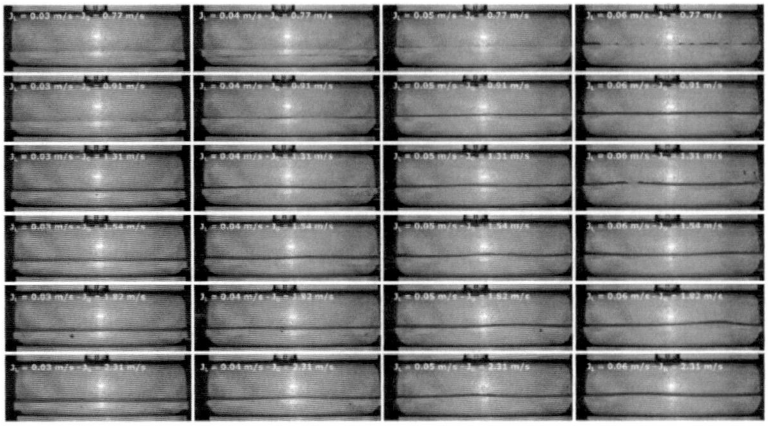

Figure 8. Examples of the acquired flow images, at the different values of J_L and J_G.

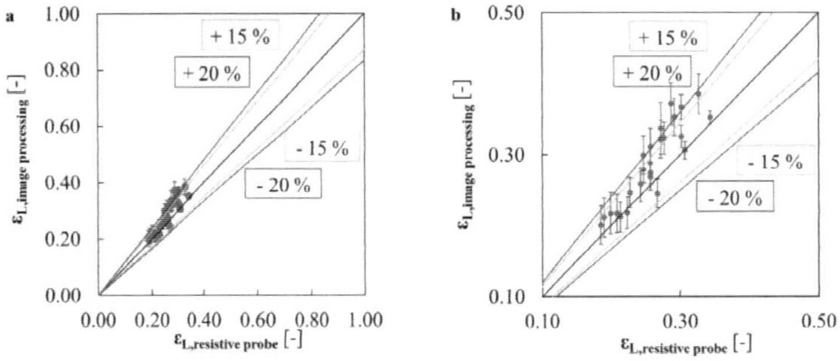

Figure 9. (a) Parity plot of liquid holdup obtained through image processing vs. liquid holdup obtained with resistive probe and (b) focus on the region of interest

The obtained results were also compared to three well-known models available in the open literature, the Armand [17], Toshiba [18] and Lockhart-Martinelli [19] models (Figure 10 (a)). Also, in this case the results well agree with the correlations, showing a variation included in ± 15 %, which is an acceptable range for multiphase flows. Whereas in Figure 10 (b) is reported the interfacial area concentration for the various superficial velocities.

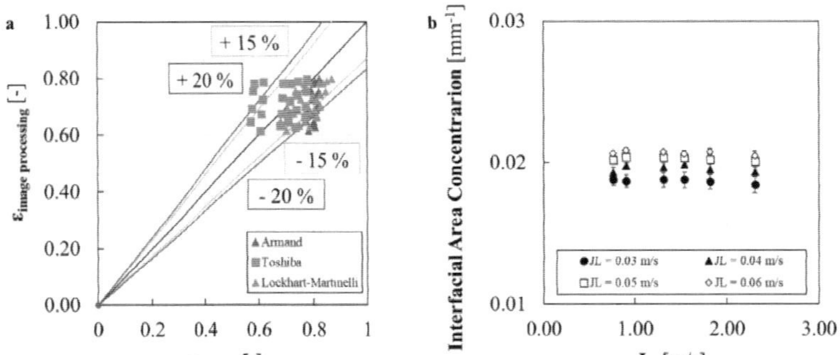

Figure 10. (a) Parity plot of gas holdup obtained through image processing vs. gas holdup obtained through models and (b) Interface area concentration vs. Superficial gas velocity

5. Conclusions

An image-based technique was tested for the measurement of the liquid height, liquid holdup and interfacial area concentration in stratified flow conditions. The technique was validated on images created by computer graphics and then used on 24 flow conditions at different values of the liquid and gas superficial velocities. The results were compared with those obtained with resistive probes. A very good agreement was found both between the two measurement methods and with the predictions of some widely used literature correlations.

References

[1] M. Ishii and T. Hibiki, *Thermo-fluid dynamics of two-phase flow*. Springer Science & Business Media, 2010.

[2] M. Guilizzoni, "Flow pattern identification in gas-liquid flows by means of phase density imaging," *Int. J. Multiph. Flow*, vol. 51, pp. 1–10, 2013.

[3] O. C. Jones Jr and J. M. Delhaye, "Transient and statistical measurement techniques for two-phase flows: a critical review," *Int. J. Multiph. Flow*, vol. 3, no. 2, pp. 89–116, 1976.

[4] G. F. Hewitt, "Measurement of two phase flow parameters," *STIA*, vol. 79, p. 47262, 1978.

[5] S. Banerjee and R. T. Lahey, "Advances in two-phase flow instrumentation," in *Advances in nuclear science and technology*, Springer, 1981, pp. 227–414.

[6] N. P. Cheremisinoff, "Instrumentation for complex fluid flows," *CRC Press*, 1986.

[7] O. C. Jones, J. T. Lin, and L. Ovacik, "Investigation of electrical impedance imaging relative to two-phase, gas-liquid flows," *Chem. Eng. Commun.*, vol. 118, no. 1, pp. 299–325, 1992.

[8] S. L. Ceccio and D. L. George, "A review of electrical impedance techniques for the measurement of multiphase flows," *J. Fluids Eng.*, vol. 118, no. 2, pp. 391–399, 1996.

[9] M. Guilizzoni, B. Baccini, G. Sotgia, and L. P. M. Colombo, "Image-based analysis of intermittent three-phase flow," *Int. J. Multiph. Flow*, vol. 107, pp. 256–262, 2018.

[10] I. M. Carraretto, L. P. M. Colombo, D. Fasani, M. Guilizzoni, and A. Lucchini, "Pressure drop and void fraction in horizontal air-water stratified flows with smooth interface at atmospheric pressure," *Fluids*, vol. 5, no. 3, 2020.

[11] L. P. M. Colombo, I. M. Carraretto, A. G. Di Lullo, C. Passucci, and A. Allegrucci, "Experimental study of aqueous foam generation and transport in a horizontal pipe for deliquification purposes," *Exp. Therm. Fluid Sci.*, vol. 98, pp. 369–380, 2018.

[12] "MATLAB." [Online]. Available: www.mathworks.com.

[13] "OpenFOAM." [Online]. Available: https://openfoam.org.

[14] M. Guilizzoni, G. Salvi, G. Sotgia, and L. P. M. Colombo, "Numerical simulation of oil-water two-phase flow in a horizontal duct with a Venturi flow meter," *J. Phys. Conf. Ser.*, vol. 1224, no. 1, 2019.

[15] "Art of Illusion." [Online]. Available: http://www.artofillusion.org.

[16] "Appleseed." [Online]. Available: https://appleseedhq.net.

[17] A. A. Armand, "The resistance during the movement of a two-phase system in horizontal pipes," *Izv Vse Tepl Inst. 1*, pp. 16–23, 1946.

[18] P. Coddington and R. Macian, "A study of the performance of void fraction correlations used in the context of drift-flux two-phase flow models," *Nucl. Eng. Des.*, vol. 215, pp. 199–216, 2002.

[19] R. W. Lockhart and R. C. Martinelli, "Proposed correlation of data for isothermal two-phase, two component flow in pipes," *Chem. Eng. Progr.*, vol. 45, pp. 39–48, 1949.

Journal of Physics: Conference Series 2177 (2022) 012035 doi:10.1088/1742-6596/2177/1/012035

Effect of Different Cooling Methods on Thermal Performance of Single Loop Pulsating Heat Pipe

N Iwata[1], F Bozzoli[1,2], L Cattani[3], L Pagliarini[1], P Vocale[3]

[1] Department of Industrial Engineering, University of Parma, Parco Area delle Scienze 181/A I-43124 Parma, Italy
[2] SITEIA.PARMA Interdepartmental Centre, University of Parma, Parco Area delle Scienze 181/A, I-43124 Parma, Italy
[3] CIDEA Interdepartmental Centre, University of Parma, Parco Area delle Scienze 181/A, I-43124 Parma, Italy

Corresponding author e-mail: naoko.iwata@unipr.it

Abstract. A pulsating heat pipe (PHP) is a promising two-phase passive thermal device that consists of a capillary tube meandering between an evaporator and a condenser. Heat is transferred from the evaporator to the condenser by the self-excited oscillation of vapor and liquid. The complete knowledge of the heat transport mechanisms, particularly at the operating limit of PHPs, is far from satisfactory. In PHP, the phase-change acts as a driving force of the self-excited oscillation of the fluid. In this context, the condensation highly affects the performance of PHPs, while few studies have investigated the effect of the sink temperature or the cooling methods. In this study, a single metallic loop is tested to understand the local phenomena focusing on the impact of cooling methods of the condenser section: natural convection and forced air convection. Regardless of the cooling methods, the PHP showed the intermittent and unstable oscillation at low power input just after the start-up, and then continued to operate stably until the heater turned off after the evaporator temperature exceeds 140 °C. The resistance of natural convection cooling is more than 1 K/W lower than that of forced convection cooling, for the same heat input conditions. However, the resistance of both cooling methods at the same evaporator temperature shows a good agreement. The resistance is inversely proportional to the evaporator temperature as well as the heat input.

1. Introduction

In the last two decades, thermal management has become a critical issue for industry due to the widespread introduction of microelectronics in almost every processing line. The heat fluxes of many electronic devices are now much higher than those predicted in the 2004 International Electronics Manufacturing Initiative technology roadmap [1]. Murshed and De Castro report linearly continuing increase of the flux with the consequent problem of high temperatures [2]. A high operating temperature reduces the reliability and performance of electronics, and eventually causes fatal failure. Thus, the efficient thermal management is essential to remove heat from heat sources and prevent heat-related failures. To satisfy the growing needs of different industrial and technological sectors, the development of innovative heat transport devices with high thermal performance, compactness, and low-cost has become a top priority.

Two-phase thermal devices are more efficient than single-phase devices because some or all of the heat transport occurs via latent heat. A pulsating heat pipe (PHP) is a promising two-phase passive thermal device that consists of a capillary tube meandering between an evaporator and a condenser. The PHP is partially filled with a working fluid that exists as a mixture of liquid slugs and vapor plugs. Heat is transferred from the evaporator to the condenser by the self-excited oscillation of vapors and liquids.

PHPs were invented in the 1990s [3-5] and various experimental studies have been conducted to investigate the effect on thermal performance of varying the PHP configuration (shape, number of turns, pipe diameter, and length) and working fluid [6]. The results showed that gravity and number of turns strongly affect the performance [7]. In addition, the heat transfer performance is maximized at working fluid filling ratios of 40–60%, and a working fluid with a small latent heat and a large sensible heat has a higher heat transport capability [8]. Theoretical studies have modeled the heat transfer phenomena and models based on a wave equation of pressure oscillation [9] or thin-film evaporation and condensation [10] have been proposed. However, the unified modeling for hydrodynamic and heat transport mechanisms is far from being satisfactory.

In PHPs, the phase-change acts as a driving force of the self-excited oscillation of the fluid. In this context, the condensation highly affects the performance of PHPs, while few studies have investigated the effect of the sink temperature or the cooling methods. Kim and Kim [11] experimentally investigated the effect of the condenser temperature on the thermal performance of a micro-PHP. They report that the thermal resistance of the MPHP reduces as the condenser temperature rises due to the enhanced latent heat transfer resulting from the increase of the average volumetric fraction in the condenser. Ahmad and Jung [12] compared the thermal performance and flow regimes of a closed-loop PHP considering different cooling methods, i.e. forced air convection and free convection. Their results indicate that the free convective PHP exhibits delayed start-up pulsation and early evaporator dry-out.

In this study, a single metallic loop is tested focusing on the effect of the adopted cooling methods in the condenser section. Single loops are the simplest configuration of PHP, therefore they have been used in research to understand their thermal and fluid behaviour [13-15]. Ethanol is used as the working fluid due to suitable operating temperature and availability. The experiments are carried out considering two different cooling methods, i.e. natural convection and forced air convection.

2. Experimental setup

The single loop is made of stainless-steel (SUS304) tube with the inner and the outer diameters of 1.59 mm (1/16 inches) and 3.15 mm (1/8 inches), respectively. In the temperature range from 0 °C to 210 °C the inner diameter is less than the so-called critical diameter d_{crit} defined as the following equation [16]:

$$d_{crit} = 2\sqrt{\frac{\sigma}{g(\rho_l - \rho_v)}} \tag{1}$$

where σ, g, ρ_l and ρ_v are surface tension (N/m), gravitational acceleration (m/s^2), liquid density (kg/m^3), and vapor density (kg/m^3), respectively. The tube was bent with a bending radius of 35 mm as shown in Fig. 1(a). The loop is constituted by an evaporator and a condenser and was placed in a vertical position with the evaporator at the bottom. A copper wire heater (Pirelli, electrical resistance 1 Ω) is wrapped around the evaporator and fixed by aluminium tape so that the electrical heat load was uniform in the section. The loop is heated up non-symmetrically with respect to the gravitational force to induce a circulation flow and improve the overall thermal performance as the past studies have suggested [14, 15]. The electrical power was applied and increased in stepwise by the DC power supply (JO POWER ALP6010) until the operating limit are reached . After the maximum power input was applied, the input was decreased in stepwise again and eventually was turned off. Each heat input step was kept for several minutes to reach the pseudo-steady state.

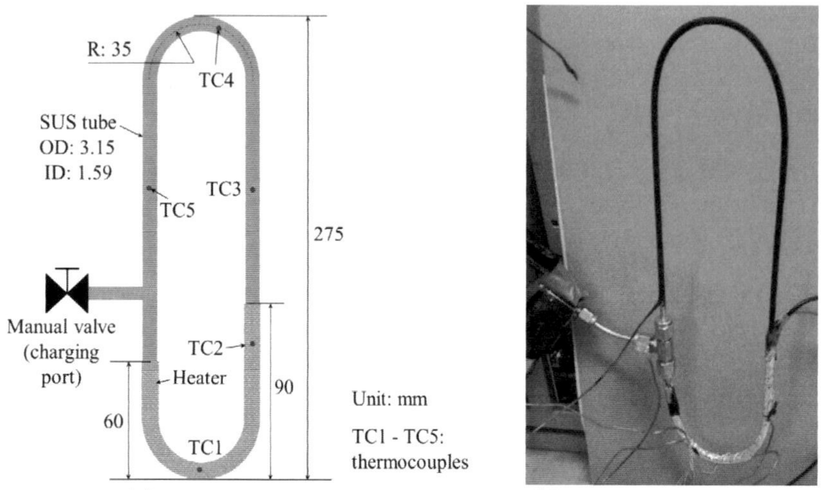

(a) (b)

Fig. 1 (a) Schematic diagram and (b) photo of the single loop PHP.

The tube of the condenser section was cooled in two ways, i.e natural convection and forced convection. Atmospheric air was moved by a 120 mm square DC fan (DELTA ELECTRONICS, DSB1212LThe fan was placed facing the condenser section with the distance of 120 mm to have an air flow at about 2.2 m/s near the condenser. The space below the fan was partitioned to prevent the air flowing to the evaporator as shown in Fig.2.

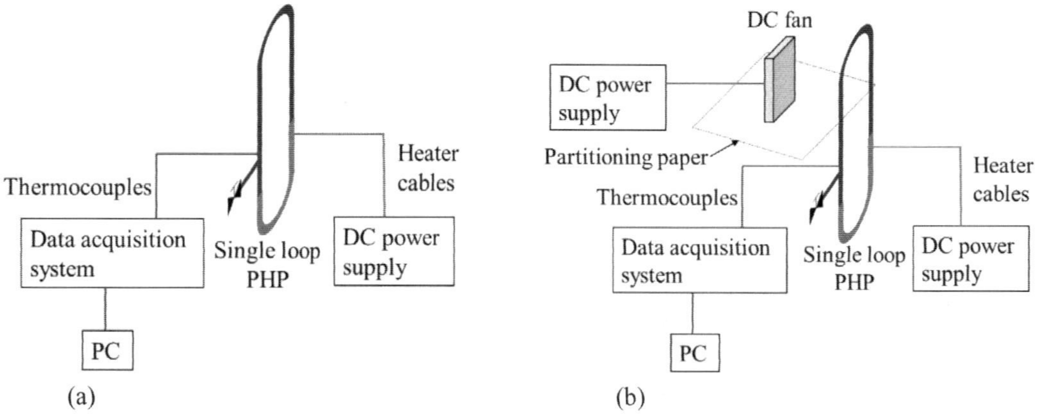

(a) (b)

Fig. 2 Experimental setup of (a) natural convection cooling and (b) forced convection cooling cases.

Ethanol (CH3CH2OH, Purity ≥ 99.8 %, ACS Reagent grade) was used as working fluid. The following steps were taken to charge the device with the fluid, using the setup shown in Fig. 3. First, the loop and the tubes were vacuumed to below than 30 Pa by vacuum pump (AGILENT DS40M). The pressure was measured by a vacuum gauge (AGILENT 0531TC) and a controller (AGILENT RGC-150). Then the valves V2 and V4 were closed and the valve V3 was opened to completely fill the loop with the fluid. After that, the charging tubes (in green in figure 3) were vacuumed again and the valve V1 was slightly opened to discharge the unnecessary fluid from the loop and achieve the target filling ratio. The fluid is degassed beforehand in the tank by means of continuous boiling.

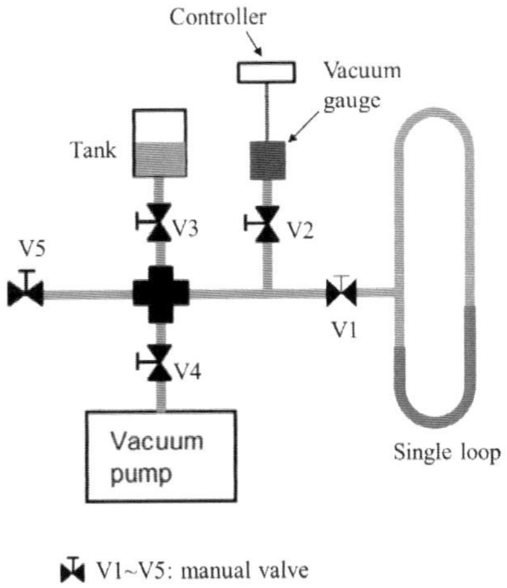

Fig. 3 Setup of fluid charging

The filling ratio is defined as the ratio of the liquid working fluid volume to the total PHP volume at room temperature. The measured filling ratio was 68%. The filling ratio was checked by measuring the weight of the loop with and without the fluid on a precision electronic balance (KERN EG620-3NM) which has a readability of 0.001 g and a verification value of 0.01 g. A fluid mass of 0.01 g corresponds to 0.7 % of the fluid mass in the loop at 100% filling ratio at room temperature. To provide a comparison with the purely conductive mode, the empty loop was also tested.

Two T-type thermocouples were attached to the evaporator. One of them (TC 1) was attached directly on the tube and was in a gap between the heater wires while the other one was attached on the heater. Three thermocouples were attached on the condenser. Moreover, the air temperature was measured. The temperature signals were acquired by the acquisition system (AGILENT 34970A) with a sampling frequency of 1 Hz.

3. Experimental results

3.1. Start-up

The PHP started up with a power input of about 11-12 W. Figure 4 shows the temperature histories of the PHP under natural convection cooling and forced convection cooling, during the start-up and during the subsequent pseudo-steady state. The electrical power input is 12 W for the former and 11 W for the latter. The difference of 1 W between the two is due to the experimental procedure and therefore does not exactly indicate the difference in heat input required for the PHP start-up. The fluid oscillation is activated throughout the PHP regardless of the cooling methods since the condenser temperatures show intermittent and unstable oscillations at low heat input cases immediately after the start-up. It is considered to be a stop-over phenomenon in which the oscillation of the fluids temporarily stops for several seconds to several tens of seconds due to the insufficient heat load to maintain the stable oscillation.

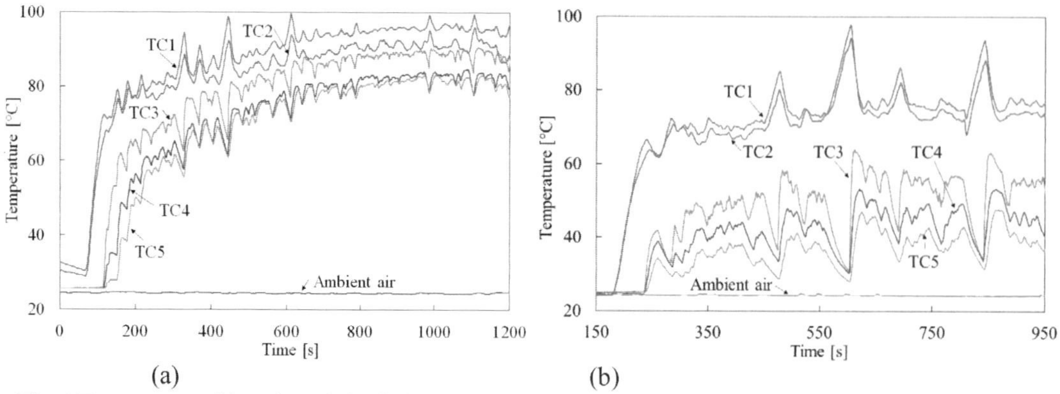

Fig.4 Temperature histories of single loop PHP during start-up under (a) natural convection cooling condition and (b) forced convection cooling condition.

3.2. Stable operation

After the power input was increased to around 14-15 W, the temperatures show the stable oscillation as shown in Fig. 5. The electrical power input in Fig. 5 is 15 W for natural cooling condition and 14 W for forced convection cooling. The temperature difference between the evaporator and the condenser under forced convection cooling is more than 10 °C larger than that under natural convection. Moreover, the temperature gradient in the condenser under forced convection is also higher than that under natural convection. It is assumed to be caused by the supercooled liquid generated by forced convection.

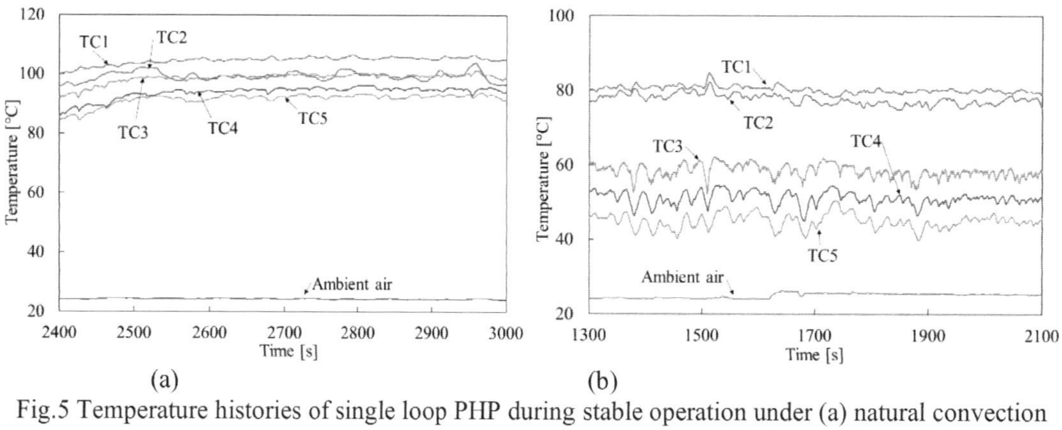

Fig.5 Temperature histories of single loop PHP during stable operation under (a) natural convection cooling condition and (b) forced convection cooling condition.

3.3. Operating limit

The PHP still operated even when the evaporator temperature exceeded 140 °C regardless of the cooling methods. Figure 6 shows the temperature histories of all power input conditions. In both cooling method cases, not only the evaporator temperatures but also the condenser temperatures increase with the power input enhancement. Since the larger heat amount is dissipated in forced convection than in natural convection, the power input is larger at the same evaporator temperature in the former condition than in the latter condition. In any case, the PHP does not reach the operating limit. It is assumed that the loop will reach the limit at the higher evaporator temperature or at the higher power input than those measured in this study. Iwata et al. [18] proposed that the operating limit of PHP is determined by the operating temperature and, at maximum, PHP could operate until the operating temperature reaches the critical temperature of the working fluid in case of the optimum filling ratio. The critical temperature of the ethanol is 243.9 °C. In order to observe the operating limit of PHP in this study, further investigation at higher temperature, which is impractical for practical use, is required.

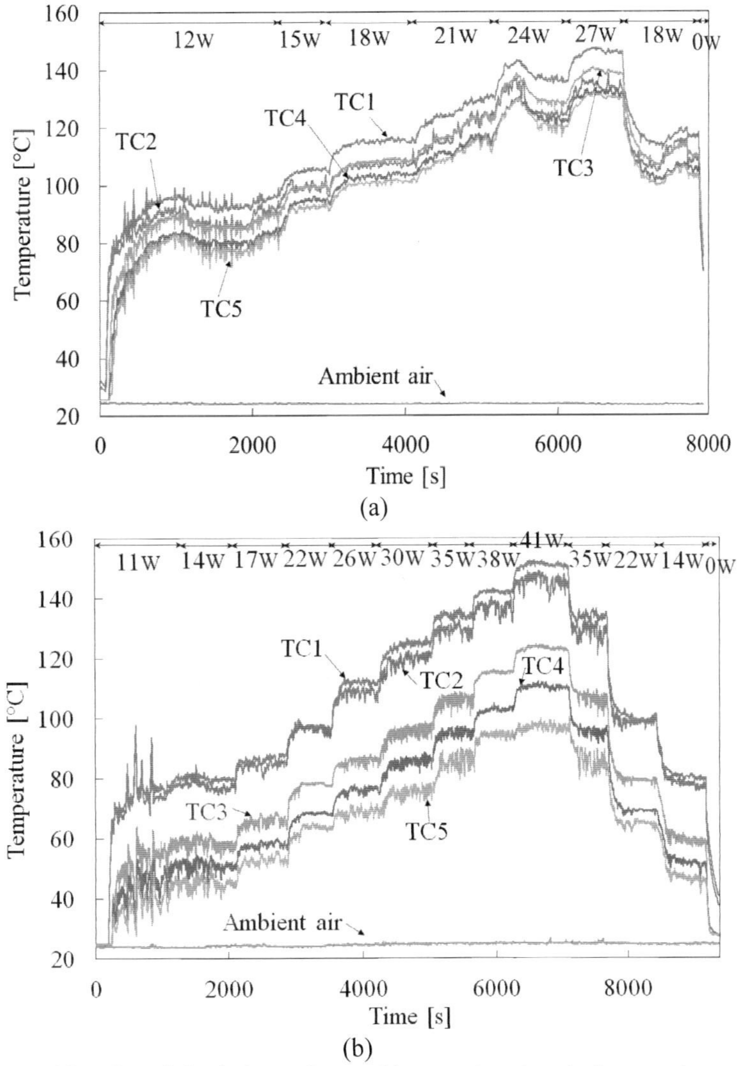

Fig.6 Temperature histories of single loop along with stepwise electrical power increments under (a) natural convection cooling and (b) forced convection cooling. re.

3.4. Thermal resistance

In general, the heat transfer performance of the PHP is evaluated by means of thermal resistance [13, 14, 17]. The thermal resistance R is evaluated adopting the following equation:

$$R = \frac{T_e - T_c}{Q_{net}} \tag{2}$$

where T_e and T_c are the evaporator and condenser temperatures, respectively. They are evaluated by averaging the temperatures measured with thermocouples for 120 seconds in pseudo-steady state, as expressed by the following equations:

$$T_e = T_{TC1} \quad (3)$$

$$T_c = \frac{\sum(T_{TC3}, T_{TC4}, T_{TC5})}{3} \quad (4)$$

where TC1 and TC 3-5 corresponds to the thermocouples shown in Fig.1(a). Q_{net} in Eq (2) indicates the net heat input to the evaporator, which is calculated by the following equations:

$$Q_{net} = Q_{ele} - Q_{loss} \quad (5)$$

$$Q_{ele} = V \cdot I \quad (6)$$

where Q_{ele}, V , and I are the electrical power input, the voltage, and the current, respectively. Q_{loss} is the heat loss from the parts other than the condenser (i.e, the evaporator and the charging system including the valves, T-junctions, and the pipe) to ambient air. The heat loss is calculated by the following equation:

$$Q_{loss} = hA(T_w - T_a) \quad (7)$$

where h , A, T_w, and T_a are the convective heat transfer coefficient, the surface area, the surface temperature, and the ambient air temperature, respectively. The heat transfer coefficient was estimated experimentally to be 95 W/m^2/K for forced air convection and to be 14.5 W/m^2/K for natural convection. The ratio of heat loss to power input is less than 31 % for forced convection cooling and around 46 % for natural convection cooling.

The measurement uncertainties of the T-type thermocouples and of the electrical power supply are ± 0.2 °C and ± 1 % FS (full scale), respectively. The full scale of the current and the voltage of the power supply is 60 V and 10 A, respectively. The measurement uncertainty of the data acquisition system for the temperature measurement is 1.0 °C. The uncertainties of direct measurement parameters, i.e., T_e, T_c, V, and I, are synthesized by the system uncertainty δ_S due to the accuracy of the instruments and the random uncertainty δ_R due to the repeatability of the data as follows:

$$\delta = \sqrt{{\delta_S}^2 + {\delta_R}^2} \quad (8)$$

The overall uncertainty of the thermal resistance was estimated based on the error propagation methods described by Moffat [19] and was found to be less than 17 % for all the input power levels. The uncertainty ranges are shown by vertical error bars in the thermal conductance plots of Fig. 7 and 8.

In Fig. 7, the thermal resistance of the PHP is reported versus the net heat input (Q_{net}). The thermal resistance of the empty loop is determined experimentally around 200 K/W, which corresponds to heat conduction of the empty stainless-steel pipe. The resistance of the PHP decreases along with the heat input increases as shown in Fig. 7. It is the typical tendency of the PHP as reported in the previous studies [13, 14, 17]. The resistance of the PHP with natural convection cooling is, at least, 1 K/W lower than that in forced convection cooling, under the same heat input conditions. This is because the PHP of natural convection cooling has a higher operating temperature and smaller temperature difference between the evaporator and the condenser than that of forced convection cooling. Furthermore, this implies that forced convection cooling has a higher maximum heat transfer rate, i.e. the heat dissipation amount just before the operating limit, than that of natural convection.

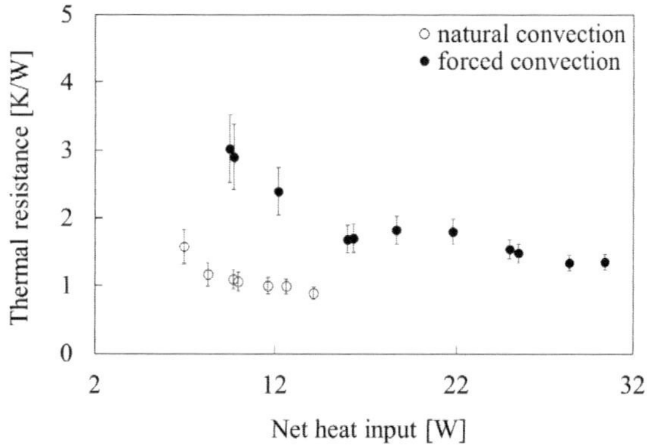

Fig.7 Correlation between net heat input and thermal resistance of the PHP.

In Fig. 8, the thermal resistance of the PHP with both FR is reported versus the evaporator temperature ($TC1$ in Fig. 1(a)). The resistances of the PHP of the both cooling methods at the same evaporator temperature show a good agreement. The resistance is inversely proportional to the evaporator temperature as well as the heat input. The resistance could be determined by the evaporator temperature, same as the operating limit. To validate the assumption and deeper understand the behaviour, the further investigation is planed: the visualization of the flow pattern by the high-speed and high-resolution infrared camera and the estimation of local heat transfer by the inverse heat conduction problem, as presented in [20,21] for different types of PHP.

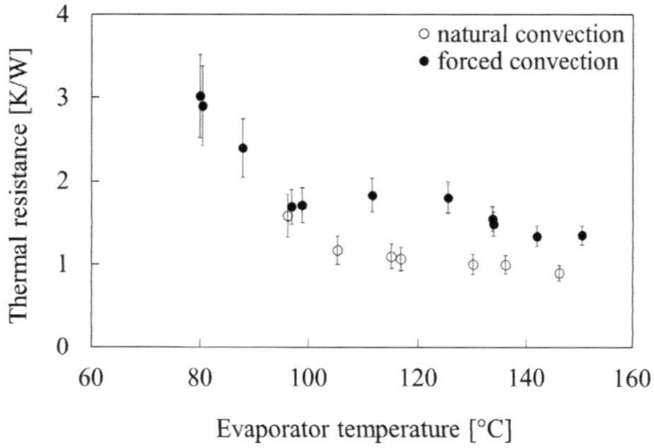

Fig.8 Correlation between evaporator temperature and thermal resistance.

4. Conclusions

A single metallic loop PHP was tested to understand the local phenomena, focusing on the effect of the cooling methods of the condenser section. Ethanol was adopted as working fluid with a filling ratio of 68 %. The condenser of the bottom-heated loop was cooled by two ways: natural convection and forced air convection. The PHP started up at the input of around 11-12 W. Regardless of the cooling methods, the PHP showed the intermittent and unstable oscillation at low power input just after the start-up, and then continued to operate stably until the heater was turned off after the evaporator temperature exceeded 140 °C. Therefore, the operating limit was not observed in this study. The resistance of natural convection cooling is more than 1 K/W lower than that of forced convection cooling for the same heat input conditions because the temperature difference between the evaporator and the condenser of the former is smaller than that of the latter. However, the resistance of both cooling methods at the same evaporator temperature shows a good agreement. The resistance is inversely proportional to the evaporator temperature as well as the heat input.

Acknowledgement
This work has received funding from the European Union's Horizon 2020 research and innovation programme under the Marie Skłodowska-Curie grant agreement No 894750.

References
[1] Electronics manufacturing initiative technology roadmap, International Electronics Manufacturing Initiative (2004).
[2] Murshed, S.S. and De Castro, C. N., A critical review of traditional and emerging techniques and fluids for electronics cooling, *Renew. Sust. Energ. Rev.*, 78, pp. 821–833 (2017)
[3] Akachi, H., U.S. Patent, Patent Number 4921041, May 1 (1990).
[4] Akachi, H., U.S. Patent, Patent Number 5219020, June 15 (1993).
[5] Akachi, H., Polasek H., and Stluc P., Pulsating heat pipes, *Proc. 5th Intl. Heat Pipe Symp.*, pp. 208–217 (1996).
[6] Han, X., Wang, X., Zheng, H., Xu, X., and Chen, G., Review of the development of pulsating heat pipe for heat dissipation, *Renew. Sust. Energ. Rev.*, 59, 692-709 (2016).
[7] Charoensawan, P., Khandekar, S., Groll, M., and Terdtoon, P., Closed loop pulsating heat pipes: Part A: parametric experimental investigations, *Appl. Ther. Eng*, 23(16), pp. 2009–2020 (2003).
[8] Zhang, Y. and Faghri, A., Advances and unsolved issues in pulsating heat pipes, *Heat Transf. Eng.*, 29 (1), 20-44 (2008).
[9] Miyazaki, Y. and Arikawa, M., Oscillatory flow in the oscillating heat pipe, *Proc. 11th Intl. Heat Pipe Conf.*, pp. 143–148 (1999).
[10] Shafii, M. B., Faghri, A., and Zhang, Y., Analysis of heat transfer in unlooped and looped pulsating heat pipe, *Intl. J. Numer. Meth. Heat Fluid Flow*, 12(5), pp. 585–609 (2002).
[11] Kim, J. and Kim, S., J., Experimental investigation on working fluid selection in a micro pulsating heat pipe, *Energ. Conv. Manag.*, 205, pp. 112462.1-112462.13 (2020).
[12] Ahmad, H. and Jung, S. Y., Effect of active and passive cooling on the thermo-hydrodynamic behaviors of the closed-loop pulsating heat pipes, *Intl. J. Heat Mass Transf.*, 156, pp. 119814.1-119814.19 (2020).
[13] Khandekar, S. and Groll, M., An insight thermo-hydrodynamic coupling in closed loop pulsating heat pipes, *Intl. J. Ther. Sci.*, 43, pp. 13–20 (2004).
[14] Mangini, D., Marengo, M., Araneo, L., Mameli, M., Fioriti, D., and Filippeschi, S., Infrared analysis of the two phase flow in a single closed loop pulsating heat pipe, *Exp. Ther. Fluid Sci.*, 97, pp. 304–312 (2018).
[15] Ilincal, A., Mangini, D., Mameli, M., Fioriti, D., Filippeschi, S., Araneo L., Roth N., Marengo, M., Fluid-flow pressure measurements and thermofluid characterization of a single loop two-phase passive heat transfer device, *J. Phys.: Conf. Ser.* 923, pp.012022.1-012022.13 (2017).
[16] Ma, H., *Oscillating Heat Pipe*, Springer, (2015).
[17] Yoon, A. and Kim, S. J., Understanding of the thermo-hydrodynamic coupling in a micro pulsating heat pipe, *Intl. J. Heat Mass Transf.*, 127, pp. 1004–1013 (2018).
[18] Iwata, N., Ogawa, H., and Miyazaki, Y., Maximum heat transfer and operating temperature of oscillating heat pipe, *J. Heat Transf.,* 138(12), pp. 122002.1-122002.5 (2016).
[19] Moffat, R. J., Describing the uncertainties in experimental results, *Exp. Ther. Fluid Sci.*, 1, pp.3-17 (1988).
[20] Cattani, L., Mangini, D., Bozzoli, F., Pietrasanta, L., Mameli, M., Filippeschi, S., Rainieri, S. and Marengo, M., 2019. An original look into pulsating heat pipes: Inverse heat conduction approach for assessing the thermal behaviour, *Ther. Sci. Eng. Prog.*, 10, pp.317-326 (2019).
[21] Pagliarini, L., Cattani, L., Bozzoli, F., Mameli, M., Filippeschi, S., Rainieri, S. and Marengo, M.,

Thermal characterization of a multi-turn pulsating heat pipe in microgravity conditions: Statistical approach to the local wall-to-fluid heat flux, *Intl. J. Heat Mass Transf.*, 169, p.120930.1-120930.16 (2021).

Development of an innovative temperature control system in a fermenter: application to the case of the wine industry

Matteo MALAVASI [(a)], Luca CATTANI [(b)], Fabio BOZZOLI [(a, c)], Sara RAINIERI[(a, c)]

(a) Department of Engineering and Architecture, University of Parma, Parco Area delle Scienze 181/A I-43124 Parma, Italy
(b) CIDEA Interdepartmental Centre, University of Parma, Parco Area delle Scienze 181/A, I-43124 Parma, Italy
(c) SITEIA.PARMA Interdepartmental Centre, University of Parma, Parco Area delle Scienze 181/A, I-43124 Parma, Italy

* Corresponding author e-mail: Matteo Malavasi (matteo.malavasi@unipr.it)

Abstract. Temperature is one of the most important parameters influencing the fermentation in wine making process. Indeed, its control during this phase is fundamental to obtain a high-quality product. Fermentation is an exothermic process and for this reason a good cooling system in fermenter tanks is required. The temperature control systems usually adopted are represented by an external cooling jacket in which cold water flows. This system is based on the water circulation which is activated only when a critical temperature in the fermenter is exceeded making the control less precise. In this paper an alternative solution based on heat pipe technology is designed and its feasibility is investigated through the application of its ideal variant. This innovative temperature control system has the advantage to work on continuous mode, thus keeping the temperature in the fermenter more uniform during the whole process. Moreover, the proposed system is fully passive reducing the costs of this process phase. The results of this work demonstrated that the device application could be feasible and that it represents a promising solution.

1. Introduction

The food industry consumes a high amount of energy to transform raw materials into safe, attractive, and economically valuable food. Due to the rising of energy costs and the need to reduce the increasing environmental pollution, it is necessary to develop techniques that improve energy efficiency in this sector. The optimization of the thermal treatments of the food industry is one of the aspects that contributes most to increase energy efficiency in this sector since they represent almost 60% of the energy consumption of the entire food industry [1]. At the same time, it is necessary to do not compromise the effectiveness of these processes: since thermal treatments are the most adopted solutions to make food safe and stable over time their efficacy is fundamental for the health of the consumer. During a thermal treatment, the uniformity of the temperature distribution within the food product is a critical factor both for the efficacy of the treatment itself and for the optimization of the energy efficiency of the process. However, the need to obtain large production volumes often forces the adoption of poorly controlled process conditions. A food application that presents these problems is represented by wine fermenters. Alcoholic fermentation, carried out by yeasts, is one of the metabolic process used in the food industry to produce wines. Chemical reactions that take place during this process are exothermic, which means that a certain amount of heat is released into the grape must, causing its temperature to rise also of 10°C from the beginning to the end of the process.[2] Temperature is one of the main factors that can influence the fermentation phase during wine-making process [3]. If not properly controlled, it can lead to problematic fermentations [4]. Temperature indeed has an impact both on yeast development and on fermentation kinetics and consequently on the final product aromatic compounds [3]. For this reason, temperature control accuracy and uniformity are extremely important [3] and thus, efficient cooling and automatic control systems must be designed [3]. Currently, the more common refrigeration methods for fermenters include circulating cold fluid through the double lining of metallic tank [6].

In this work an original temperature control system, based on heat pipe technology, is proposed to keep the product temperature more uniform and reduce energy consumption. Heat pipes are two-phase passive heat transfer devices that are particularly attractive for their high heat transfer capability, efficient thermal control, adaptability, simplicity, and very low cost [7]. The objective of this study is to evaluate the feasibility of applying a thermosyphon heat pipe for the implementation of a passive temperature control system in a fermentation tank for winemaking process, and to compare it with the standard temperature control methodology. For this reason, a fermenter was designed and realized to simulate grape fermentation; in this device it was possible to use both the new fermentation temperature control system with heat pipe and the standard one based on water flow in the jacket.

2. Design and implementation of the experimental setup

2.1. *The fermenter tank*

A fermenter tank in AISI 316L with an internal volume of 130 l was realized. It was equipped with a classical cooling system composed by an external cooling jacket where water circulates from the bottom to the upper side for temperature control of the product.

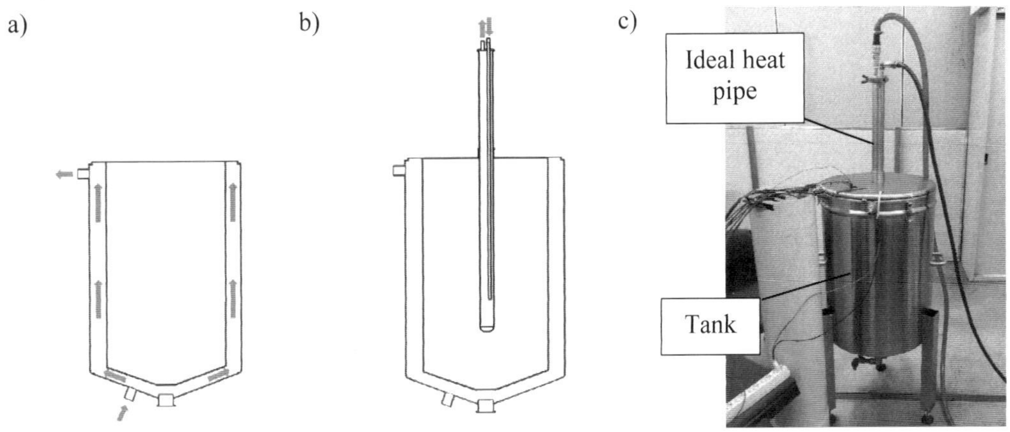

Figure 1: (a) Schematic section of the fermenter and the flow of the cooling fluid inside the jacket; (b) schematic section of the ideal heat pipe installed in the tank and (c) a picture of the experimental setup

Inside the jacket a baffle imposes to the cooling fluid an obliged path with a spiral shape, wrapping the entire tank and creating turbulence to enhance the heat transfer. Figure 1a shows the schematic section of the fermenter and the representation of the principal direction of the cooling water circulating in the jacket. To permit the application of the innovative cooling system to the fermenter, in the tank's top a supporting device was realized to allow the insertion of the thermosyphon heat pipe in the tank. It consents to keep the heat pipe in a perfect vertical position with its bottom part (≈ 50 cm) immersed in the product contained in the tank. In this way it was possible to test the two different cooling systems (i.e., cooling jacket and heat pipe cooling) on the same fermenter allowing a fair and robust comparison between the two technologies.

2.2. *The Heat Pipe*

To investigate whether the thermosyphon heat pipe would permit a satisfactory control of the product temperature inside the fermentation tank, it was decided to experimentally simulate an ideal heat pipe with zero overall thermal resistance. This device consisted of an outer AISI 316L pipe, in which water circulated as the cooling fluid. The water entered in correspondence of the heat pipe top through a small diameter tube, which almost entirely crossed the heat pipe length, conveying the water to the lower part of the device (figure 2). Refrigerating water then exit from another section on the top of the heat pipe. Providing a sufficient volumetric flow rate of cooling water, the ideal heat pipe presented a uniform temperature on its entire surface: it means that the thermal resistance between the bottom (evaporator) and the top (condenser) is null. To provide the device with the required flow rate and the constant required temperature of the water, a chiller (Haake Kryo thermal 140) was used in recirculation mode.

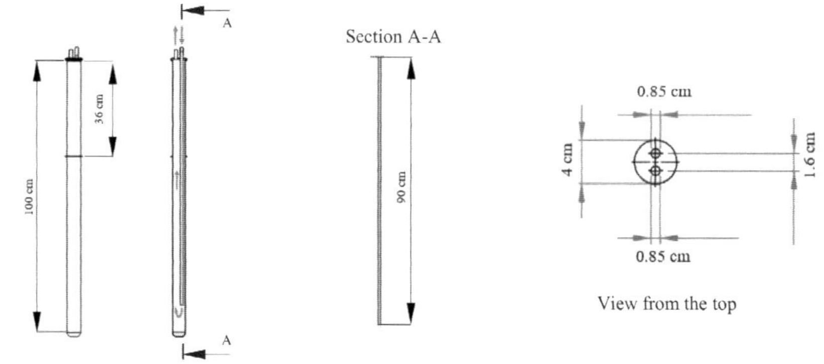

Figure 2: Schematic representation of the ideal heat pipe and its internal flow

2.3. The experimental program and the acquisition system

The aim of the work was to investigate the behaviour of an innovative cooling system based on the heat pipes technology and understand if it could control the product temperature during the fermentation process. In particular, the system was used to maintain the temperature of the product inside the tank in the range 22-25°C, considering it as the optimum for the wine must under fermentation process [6].

For this reason, different simulations of the fermentation process were performed and, to simplify the system, water was used as surrogate of the fermentation substrate. Since grape must is mainly composed by water, the thermodynamic behaviour of the surrogated can be quite representative of the real one. From a stoichiometric point of view, it is possible to summarize the alcoholic fermentation process in this way:

$$C_6H_{12}O_6 \rightarrow 2\ CH_3CH_2OH + 2\ CO_2 + 106.3\ kJ \quad (1)$$

Therefore, in good approximation, during a typical alcoholic fermentation process, each mole of fermented glucose ($C_6H_{12}O_6$) releases about 106.3 kJ [6]. For the experimental tests, it was simulated the fermentation of 100 l of grape. This amount contains about 300 g/l of glucose, corresponding to a total of 166.6 moles, which means that about $18\cdot10^6$ J are generated from their fermentation.

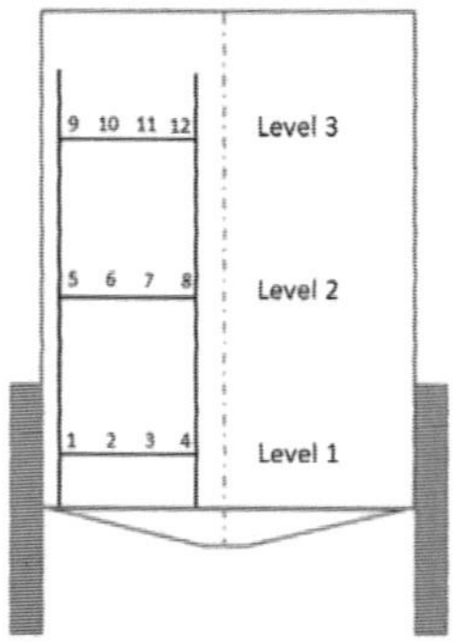

Figure 3: Thermocouples position inside the fermenter tank

Fermentation process normally lasts about 1 week but the fermentation kinetics is principally concentrated in the first 2 days and consequently great part of the heat is released in 48 hours (~80-90%) [6]. Considering that, the associated thermal power is approximately 100 W for every 100 l of fermenting wort. To provide the heat power that simulates the heat generated by fermentation, an immersion wire thermoelectric resistance was uniformly disposed inside the tank. During the tests, the temperature of the product was acquired in 12 different positions (Figure 3) inside the tank by means of T-type thermocouples, at 4 radial distances for 3 axial coordinates. The three height levels were spaced 15 cm one from the other with the first level about 15 cm from the bottom of the tank. For each level, the thermocouples were located at the radial distances of 6 cm, 11.5 cm, 17 cm, and 22.5 cm from the central axis, respectively. Before starting the tests, the uncertainty of the thermocouples was evaluated to be around 0.1°C.

3. Results

Preliminary tests were carried out to understand the behavior of the innovative system compared to the traditional cooling jacket. As said before during the first 2 days a heat power of about 100W is released during the fermentation process: this is an average value since the fermentation kinetics varies also during the first two days reaching peaks that could be significantly higher than the mean value. To test the ability of the proposed system to control the temperature during the most adverse conditions, a security coefficient of 3 has been considered for the heat generated: tests were conducted also adopting a resistance of 300 W inside the product. In these tests water at 13°C was used as refrigerant fluid for both the cooling systems. This value has been chosen since it is in the typical range of temperatures at which city water is provided and therefore adopted in the traditional system with the jacket. In every test the product starting temperature was about 22°C and the duration of the experiment was about 3 hours. During fermentation, depending on the type of grape vinified, the pumping over of the must can be applied or not: to cover the whole case scenario and to compare the two systems under any conditions, tests were executed under two different conditions, i.e., with and without the mixing of the product inside the tank. In the classical refrigeration systems, the cooling water does not flow continuously in the jacket, but the system is activated every time that the product temperature overcomes the optimal fermentation upper limit of 25°C and is switched off when it drops below the optimal fermentation lower limit of 22°C. The activation logic is simply governed by usually one single temperature sensor, generally a thermistor that is placed as much as possible in a central position inside the tank. To simulate this condition, in the present investigation, the refrigeration is activated once the average temperature of the second level of thermocouples exceeds 25°C, considering it the one that best approximates the average temperature of the 100 l of water in the tank. In tests with the application of the ideal heat pipe as cooling system the cold water at 13 °C flows inside the ideal thermosyphon heat pipe with a flow rate high enough to avoid differences of temperature between the inlet and outlet sections. The maximum temperature difference was found to be 0.2 °C: in this way it was possible to consider the temperature at the evaporator and at the condenser equal and therefore the thermal resistance as almost null, as it was supposed for an ideal heat pipe operating between the hot product and the cold water. The afore mentioned conditions were applied both for the tests in which the mixing was applied and in those where it was not. During tests with the simulation of the pumping over of the grape, mixing action was applied every half an hour.

3.1. Comparison between the two systems without product mixing

The distribution of the measured temperatures in the case of absence of product mixing is reported as a function of time in figure 4, for the cooling jacket system and the heat pipe cooling system. It is possible to notice for both the cooling systems a thermal stratification of the product inside the tank that makes distinguishable the three levels of thermocouples. It appears slightly amplified in the case of the cooling jacket: despite the double activation of the cooling system (at 80 mins and at 180 mins) allowed the temperatures of the second level to be kept within the set limits (22-25 °C), this cannot be said for the other levels. In particular, the first level is cooled too much reaching about 19.5 °C, while the third level reached excessive temperatures of about 27 °C. In a real process with such a temperature distribution, the yeast, responsible of fermentation, could be negatively influenced, worsening the fermentation results. In the real case probably, this pronounced thermal stratification will be smoothen by some factors as higher density of the product, more uniform generation of heat, etc. Nevertheless, it could seriously represent a source of deterioration of the fermentation process, and it must be considered.

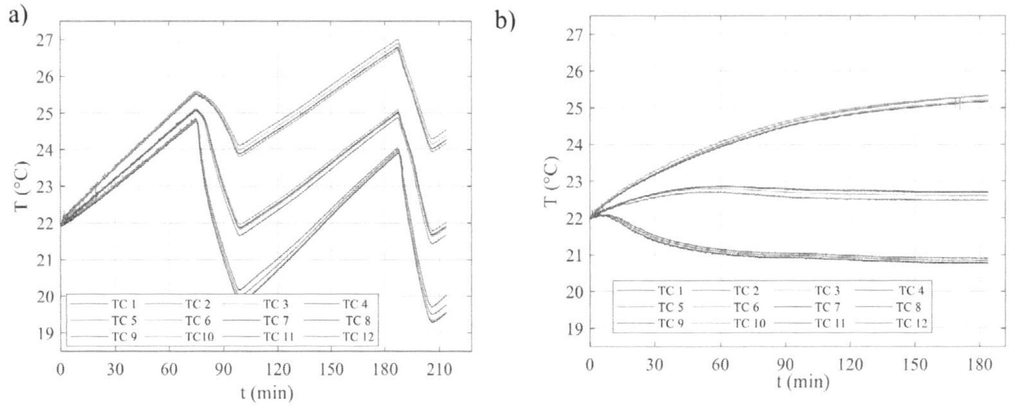

Figure 4: Temperature distribution for tests with 300 W input power without product mixing a) with cooling jacket; b) with ideal heat pipe

To better describe the thermal uniformity inside the product during the tests, the deviation σ of the measured temperatures respect to the target one was calculated in this way:

$$\sigma = \sqrt{\frac{\sum_{i=1}^{N}(T_i - T_{target})^2}{N}} \quad (2)$$

where, N is the total number of measurements during the whole test, T_i (°C) is the *i-th* measured temperature, T_{target} (°C) is the optimal temperature for the product: in the range 22-25 °C any value can be considered optimal but for the computation of the parameter of Eq. (2) one single temperature must be chosen and in this case 22°C was selected. In Table 1 the average values of σ for any test condition are reported together with the maximum value reached by σ. It is possible to observe that for the test with the cooling jacket, the average σ is about ± 2.04 °C, while for the ideal heat pipe it reaches ± 1.35 °C confirming what was observed in figure 4 where the jacket presented a slightly more intense stratification of the temperature.

This is also underlined by the fact that σ_{max} for the jacket is more than 1.5 °C higher than that for the ideal heat pipe. Therefore, despite there is a temperature stratification for both the cooling systems, the ideal heat pipe seems to perform better even if from figure 4 it could be seen that at the end of the test the top level exceeds 25 °C and the bottom one is cooled below 22 °C.

Cooling system	σ	σ_{max}
Jacket	± 2.04 °C	5.02 °C
Jacket with product mixing	± 1.86 °C	4.17 °C
Ideal heat pipe	± 1.35 °C	3.36 °C
Ideal heat pipe with product mixing	± 0.78 °C	1.87 °C
Heat pipe in typical fermentation conditions	± 0.25 °C	1.50 °C

Table 1: σ for any test conditions.

In figure 5 it is reported the spatial distribution of the temperature in correspondence of the instant at which the maximum thermal stratification occurred: it is possible to observe that the ideal heat pipe produced a smaller temperature difference between the top and the bottom levels. It is also possible to observe that the stratification occurred in both cases only along the axial coordinate, while the radial stratification is negligible.

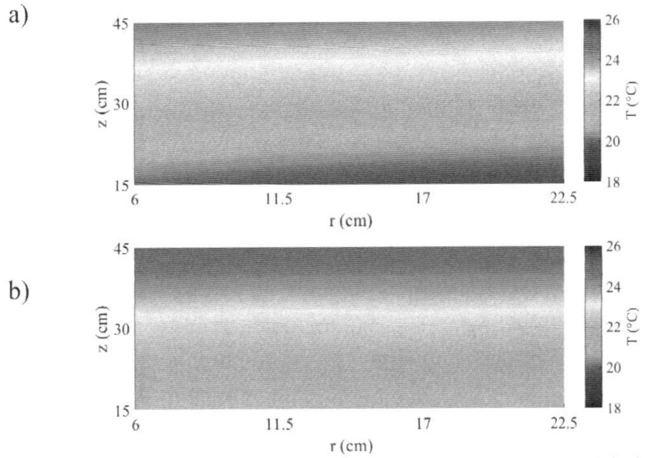

Figure 5: Spatial temperature distribution in correspondence of the instant at which the maximum thermal stratification occurred for the a) case of the jacket (t=210 min) and the b) case of the ideal heat pipe (t=180 min)

3.2. Comparison between the two systems with product mixing

The distribution of the measured temperatures in the case of product mixing is shown as a function of time in figure 6, for both the cooling systems. It is possible to observe that the mixing helped to prevent an excessive thermal stratification for both the cases, allowing to obtain a more homogeneous product temperature distribution. For the jacket case, it is possible to observe greater temperature oscillations respect to the case of the ideal heat pipe: this behavior is ascribable to the intermittent activation of the cooling system. Moreover, it is possible to observe, also in the presence of product mixing, the exceeding of the upper optimal fermentation temperature limit.

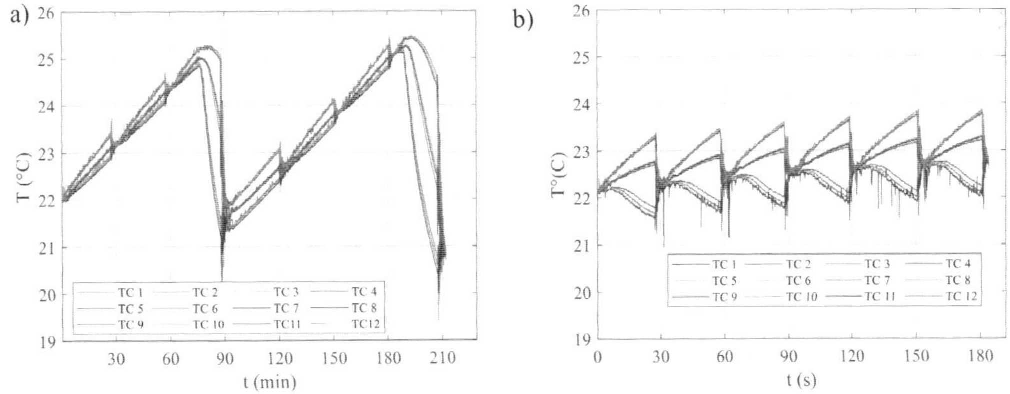

Figure 6: Temperature distribution for tests with 300W input power applying the mixing of the product a) with the jacket cooling system; b) with the ideal heat pipe as cooling system

The value of σ, calculated as reported in Eq. (2), with the cooling jacket is about ± 1.86 °C, while for the ideal heat pipe it reaches ± 0.78 °C: the better performance in terms of temperature uniformity for the ideal heat pipe are also confirmed by the fact that σ_{max} for the jacket is more than 2.0 °C higher than for the ideal heat pipe. It is possible to state that the best combination in terms of uniformity of temperature inside the product is the application of the ideal heat pipe with the mixing of the product. At the conclusion of the preliminary tests, it is possible to say that the application of a heat pipe is a possible solution for controlling the temperature inside the tank during fermentation, especially in the case of product mixing.

3.3. Ideal heat pipe tested in environmental cooling configuration

Since the ideal heat pipe demonstrated that could better operate respect to the cooling jacket in terms of uniformity of product temperature, it was decided to test the innovative heat pipe cooling system while operating between the product and the air of the wine cellar: it could enable to get rid of the cold water and make the system simpler and cheaper. In the conceived configuration of the real heat pipe the cooling at the condenser section is performed by natural convection with air at the temperature of the wine cellar. To simulate this working condition the temperature of the water flowing in the ideal heat pipe was set at 18°C: it could be considered a representative temperature of wine cellars. The power provided by the wire resistance was 100W to simulate the average fermenting condition as mentioned in section 2. The duration of the test was about 3 hours, and the product was mixed every hour.

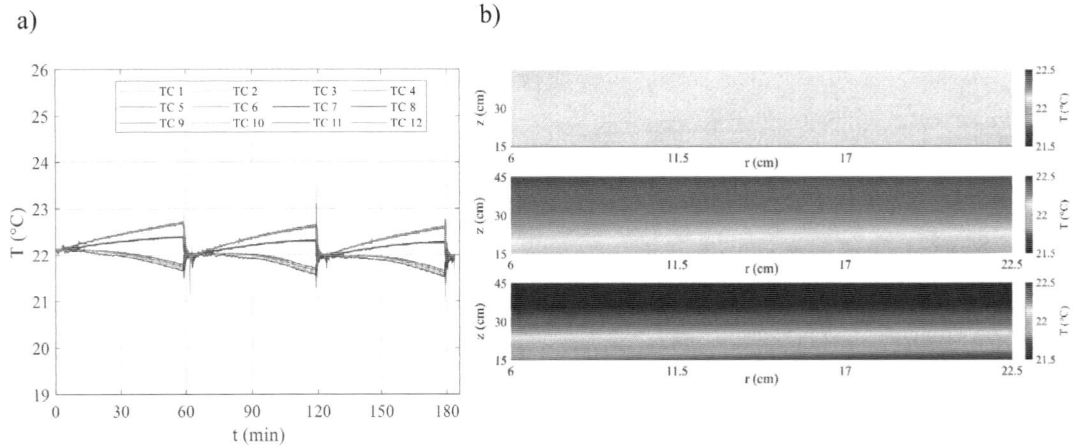

Figure 7: Temperature distribution of the ideal heat pipe tested in environmental cooling configuration a) over the time and b) over the space at 0 s, 1800 s and 3600 s (from top to bottom).

In figure 7 it is shown the temperature distribution as a function of time: it is possible to see that the temperature limits have been respected. The mixing temperature is equal to the initial one, which means that in this condition the power dissipated by the ideal Heat Pipe is equal to the one provided by the fermentation. It is a very important outcome since it demonstrates that this device could perform well in the typical fermenting conditions during wine making, permitting to dissipate the heat released by the exothermic reaction of fermentation. Maintaining the product inside the optimal temperature range for the whole fermenting process, could bring significant advantages to the final product. In figure 7b it is reported the spatial distribution of the temperature for three different instants, i.e., 0 s, 1800 s and 3600 s. They correspond to the beginning of the test, to the middle time between the starting and the first activation of the product mixing and few instants before the first activation of the mixing, respectively. It is possible to observe that, also in this case the stratification occurred only in the axial direction, and it was emphasised at the instant immediately before the mixing action. It suggests again that the mixing of the fluid could improve the effectiveness of the new cooling system. The value of σ was evaluated also in this test and it was found to be ± 0.25°C: it confirms the good result in terms of temperature uniformity of the new cooling system combined with a mixing action on the product.

4. Conclusions

This work constitutes the feasibility study for the realization of an innovative control system of the grape must temperature during fermentation process in wine making. The realized system is based on the heat pipe technology.
To verify the applicability of the proposed innovative control system, the process that occurs during the fermentation of the wine was simulated by thermal resistances and it was verified that the optimal fermentation temperatures were respected. An ad hoc fermenter tank was therefore designed and built for allowing the performance comparison of the traditional cooling setup (cooling jacket) with the innovative temperature control system here proposed. An ideal thermosyphon heat pipe with thermal resistance equal to zero was experimentally simulated to reproduce the best possible heat transfer conditions and to verify the feasibility of the project.
The experimental campaign demonstrated that:

- in terms of uniformity of product temperature, the ideal heat pipe could better perform respect to the cooling jacket. Indeed, for the cases without product mixing, the deviation σ of the measured temperatures respect to the target one for the ideal heat pipe was 30% lower than for the jacket, while for the cases with product mixing it was almost 60% lower ;
- under environmental cooling configuration simulating wine cellar conditions, the ideal heat pipe was able to maintain the product temperatures within the optimal temperature limits of the fermentation, keeping a good uniformity (σ =± 0.25°C).

For this reason, it is possible to affirm that this kind of device could constitute a significant improvement in the temperature control systems for the winery industries: its adoption in place of the cooling jacket, could reduce costs for the fermentation phase and for all the wine making process and could improve the uniformity of product temperature allowing to obtain higher quality products.
The positive results achieved, suggest that a future step for this work could be the design and the realization of a real heat pipe that operates at the same conditions verified for the ideal one and its testing during a real fermentation phase.

Acknowledgments

The Authors would like to acknowledge the Emilia-Romagna Region support through the grant "Piano triennale alte competenze per la ricerca, il trasferimento Tecnologico" as well as the European Space Agency (ESA) support through the grant 4000128640/19/NL/PG/pt, ESA MAP project TOPDESS.
MBS Srl (Italy) is gratefully acknowledged for setting up the experimental apparatus.

References

[1] L. Wang, "Energy efficiency technologies for sustainable food processing," *Energy Effic.*, vol. 7, no. 5, pp. 791–810, 2014, doi: 10.1007/s12053-014-9256-8.
[2] A. Goelzer, B. Charnomordic, S. Colombié, V. Fromion, and J. M. Sablayrolles, "Simulation and optimization software for alcoholic fermentation in winemaking conditions," *Food Control*, vol. 20, no. 7, pp. 635–642, 2009, doi: 10.1016/j.foodcont.2008.09.016.
[3] S. Colombié, S. Malherbe, and J. M. Sablayrolles, "Modeling of heat transfer in tanks during wine-making fermentation," *Food Control*, vol. 18, no. 8, pp. 953–960, 2007, doi: 10.1016/j.foodcont.2006.05.016.
[4] M. I. Zenteno, J. R. Pérez-Correa, C. A. Gelmi, and E. Agosin, "Modeling temperature gradients in wine fermentation tanks," *J. Food Eng.*, vol. 99, no. 1, pp. 40–48, 2010, doi: 10.1016/j.jfoodeng.2010.01.033.
[5] G. H. Fleet, "Yeast interactions and wine flavour," *Int. J. Food Microbiol.*, vol. 86, no. 1–2, pp. 11–22, 2003, doi: 10.1016/S0168-1605(03)00245-9.
[6] P. Ribéreau-Gayon, D. Dubourdieu, B. Donèche, and A. Lonvaud, *Handbook of Enology: Volume 1, The Microbiology of Wine and Vinifications*, vol. 1. 2006.
[7] Faghri A. 2014. Heat pipes: review, opportunities, and challenges. Frontiers in Heat Pipes 5,1

Heat fluxes distribution and uncertainty in low-Reynolds flow regimes inside a ribbed channel

L Vitali, P Gramazio, D Fustinoni and A Niro
Politecnico di Milano, Department of Energy,
Campus Bovisa, Via Lambruschini 4, 20156 Milano, Italy

Corresponding author email: alfonso.niro@polimi.it

Abstract. Experimental studies on the local convective heat transfer enhancement induced by ribs in channels are generally carried out in steady state, by imposing a known heat flux on the ribbed surface and subsequently retrieving the convective heat flux by means of a local heat balance, i.e. by subtracting all the non-convective heat fluxes from the source term. Such heat fluxes are retrieved from temperature measurements by means of models, whose complexity can range from simple algebraic expressions to complex simulations. In this context, an experimental technique based on FEM simulations, a radiative model for semi-transparent enclosures, and a filtering technique to retrieve the heat fluxes across the plane of the heated surface has been developed and tested in a 1:10 AR channel, for Reynolds number ranging from 650 to 7500 and a flow of air. This paper presents a discussion on the heat flux distribution as a function of the Reynolds number for a streamwise-pointing chevron (V) rib configuration with 60° angle of attack, with P/e = 20, showing the increasing weight of non-convective heat dissipations as the Reynolds number decreases, and their effect on the uncertainty of the Nusselt number measurement.

1 Introduction

Heat transfer enhancement in closed channels by means of passive turbulators is still a topic actively researched, not only as this apparently simple technology has a wide range of applications, e.g. the cooling of turbine blades, of the first wall in fusion reactors and of photovoltaic modules, or the efficiency enhancement of solar air heaters, all characterized by different geometrical and fluid-dynamics parameters, but also because the overall heat transfer and pressure drop increment are underlain by complex fluid-dynamics phenomena, and the path towards a complete agreement between experimental and numerical analyses has yet to be fully traveled.

In scientific literature, the problem is usually analyzed from two main standpoints. The first is focused on measuring the overall heat transfer augmentation and pressure drop increment induced by specific configurations. Studies on single families of configurations are then compared in review articles, as done for example by Kumar and Kim [1], typically in terms of Nusselt number enhancement with respect to the smooth channel, i.e. Nu/Nu_0, and a thermohydraulic performance parameter η, often defined as a function of Nu and the friction factor f as follows:

$$\eta = (\mathrm{Nu}/\mathrm{Nu}_0)/(f/f_0)^{0.33} \quad (1)$$

The second approach consists of analyzing the local heat transfer augmentation given by simpler configurations, which are usually paired with CFD simulations for cross-validation purposes, see as example the work by Tanda and Satta on angled and intersecting ribs [2]. An analysis of the techniques employed for such experiments is beyond the scope of this work, but, with few notable exceptions, as the experiments carried out by Chung et al. [3] which exploit the heat-mass transfer analogy applied to the measurement of naphthalene sublimation, they are generally based on the retrieval of the temperature field on the surface under investigation by means of optical techniques, which is fed to appropriate thermal models to retrieve the desired convective heat flux.

The two approaches are complementary, and have been both used at the ThermALab Laboratory at Politecnico di Milano in a long-term study on heat transfer enhancement in closed channels, see for example [4] on the global approach. In particular, a channel with one heated wall has been built for improving the detail of the local convective heat transfer measurement, and a new analysis procedure has been developed by Gramazio, Vitali, Fustinoni and Niro, see [5] [6], in the effort to validate new CFD simulations by means of DNS. To this end, it is also important to study not only high-Reynolds regimes, i.e. Re > 10000, which are typical of applications like turbines and solar air heaters, but also low-Reynolds regimes, namely from 600 to 10000, as they can comprehend operational flow regimes of a wide array of compact heat exchangers. In this Reynolds range, there is a general lack of results in scientific literature, and the models required to retrieve the convective heat flux maps from temperature measurements should adequately take into account heat fluxes other than convective, namely conductive and radiative, as their relative contribution to the heat balance can be non-negligible.

Moreover, in many papers, see for example [7], [8], [9] and [10], uncertainties on the local Nusselt number are declared, usually based on analyses according to Moffat [12], which proposes an uncertainty estimation procedure for single-sample measurements, but the resulting values are hardly circumstantiated.

All that considered, the purpose of this paper is to analyze the local convective heat transfer results of a V-down rib configuration, in particular concerning the minimum uncertainty of the Nusselt number and the heat flux relative contribution to the total energy balance, highlighting the difference between high- and low- Reynolds regimes.

2 Materials and methods

2.1 Experimental apparatus and procedures

The measurements have been carried out on an open-loop channel operated in suction, shown in figure 1. Summarizing its main features, the channel lower surface is heated by means of two Printed Circuit Boards (PCBs) independently powered and controlled. The PCBs have a substrate made of FR-4 fiberglass-reinforced epoxy laminate material, whereas copper tracks are 0.017 mm thick, 7.8 mm wide, spaced 0.2 mm apart each other, and covered by a black lacquer which also fills up the gaps between them, making the heating surface flat, smooth and with a high emissivity. A double-glazing 130 mm x 130 mm germanium window grants the optical access to a FLIR T650sc IR-camera, with a 640 x 480-pixel resolution and a 20 mK NETD at 30 °C. The main experimental features are listed in table 1. Further details on the calibration procedures for the used sensors are reported in [5].

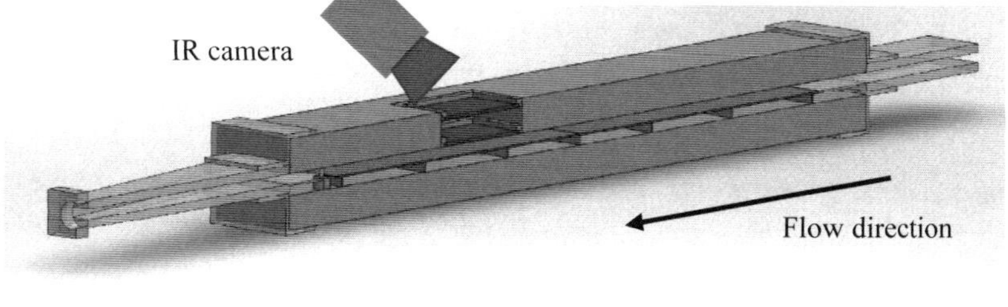

Figure 1. Lengthwise section of the test channel for local heat transfer measurements

Table 1: main geometrical parameters

H	Channel height	12	mm	L_h	Heated length	840	mm
W	Channel width	120	mm	*e*	Rib side dimension	4	mm
AR	Aspect ratio	10		*P*	rib-to-rib distance (pitch)	80	mm
D_h	Hydraulic diameter	21.8	mm	*e/H*	Blockage	0.33	
L	Channel length	960	mm	*P/e*	Adimensional pitch	20	

While the details of the experimental procedures and models are explained in details in [5] and [6], the objective of the test is to retrieve the local Nusselt number at the bottom surface, which has an imposed heat flux as boundary condition, henceforth called $Nu_{loc,q}(x,y)$ and defined as in equation 2:

$$Nu_{loc,q}(x,y) = \frac{h_{loc}(x,y)D_h}{k_{air}} \tag{2}$$

Being the hydraulic diameter Dh a constant and since the air's thermal conductivity k_{air} is considered to be so within the experimental temperature range, the uncertainties regarding $Nu_{loc,q}(x,y)$ are contained in the local convective heat transfer coefficient $h_{loc}(x,y)$, defined in equation 3 as:

$$h_{loc,q}(x,y) = \frac{q_c''(x,y)}{T_w(x,y) - T_b(x)} \tag{3}$$

Where $T_w(x,y)$ is the wall temperature, measured by IR thermography, and $T_b(x)$ the bulk fluid temperature, which is assumed to have a linear trend and therefore is determined by measuring the inlet and outlet bulk temperature, respectively named T_{in} and T_{out}.
The local convective heat flux $q_c''(x,y)$ is obtained by applying the energy balance to an element of the copper tracks of the bottom heater, i.e. where the heat generation by means of Joule's effect takes place. Considering a spanwise slice of the heater, as in figure 2, the energy balance is written as in equation 4:

$$q_c''(\bar{x},y) = \sigma_{el}s_{Cu} - q_{rad}''(\bar{x},y) - q_{k,back}''(\bar{x},y) + q_{k,y}''(\bar{x},y) - q_{k,y+dy}''(\bar{x},y) + q_{k,x}''(\bar{x},y) - q_{k,x+dx}''(\bar{x},y) \tag{4}$$

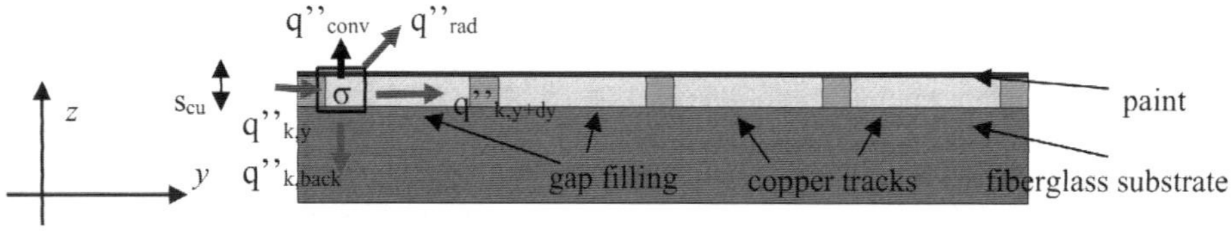

Figure 2. PCB scheme of a slice in x-y direction, with the energy balance on the copper layer

where σ_{el} is the volumetric heat dissipation by Joule effect, which is considered uniform in the copper tracks, s_{cu} the copper layer thickness, $q''_r(x,y)$ the heat flux radiated by the coating, $q''_{k,back}(x,y)$ the heat flux diffusing in the negative *z* direction from the rear of the copper track layer, and finally $q''_{k,y}(x,y)$ and $q''_{k,y+dy}(x,y)$ the heat fluxes transmitted by conduction in *x-y* direction through the layer itself. The conductive fluxes in *x* direction (out of plane) are also taken into account, but are not shown in Figure for clarity. The detailed description of the methods used to retrieve all the heat fluxes appearing in equation 4 can be found in [6] and is beyond the scope of this paper. Briefly, the conductive heat flux towards the backside of the breadboard is obtained by means of a 2D FEM model developed in COMSOL®, which reproduces the geometry of the channel spanwise section and calculates the heat flux using as boundary condition the measured temperature on the channel's bottom surface and the convective and radiative heat losses towards the environment, the conductive term along the plane of the heater by applying the laplacian to the filtered temperature field, and the radiative heat flux by means of a radiative model which takes into account the presence of a double semi-transparent glazing in the IR wavelength band.

The main experimental parameters and measured temperatures of the trials presented in this paper are listed in table 2, where Q_{in} is the total power input of the heaters, which is varied as a function of Reynolds so that steady-state temperatures reached by the experimental set-up are comparable between trials and safe for the equipment.

Table 2: experimental parameters

Re	Q_{in} / W	T_{in} / °C	T_{out} / °C	$\Delta T_{in\text{-}out}$ / °C
670	11.7	25.46	35.86	10.40
800	14.2	25.50	37.19	11.69
1080	19.6	25.46	38.75	13.29
1350	24.2	25.06	38.85	13.79
2150	37.3	23.63	40.46	16.83
3200	52.7	25.93	40.89	14.96
5400	81.8	25.74	40.83	15.08
7500	115.8	26.68	42.69	16.01

2.2 Uncertainty analysis

Since the dependence of the measurand, i.e. the local heat flux, from the measurements is given by non-linear models, it is not possible to perform an uncertainty propagation as described by the first part of the GUM [11]. A Monte Carlo analysis, on the other hand, while theoretically more feasible, requires the definition of the Probability Distribution Functions for all the input quantities, which appears a feat beyond the purpose of this work. In addition to these points, also in the light of the considerations by Moffat on single-sample experiments [12], in studies like the one presented here, it can be totally misleading to provide a "certain" uncertainty value, based on uncertainties of the single measurement devices, as many uncertainty sources in complex experiments, like focus position, environmental conditions, and the positioning of the IR camera, can be easily overlooked, and the best way to validate the results is by cross-comparison with different numerical and experimental techniques applied to the same problem. Also, literature data in simple test cases can be a valuable benchmark, but is more useful to verify trends rather than exact values, which may depend on the specific experimental setup, as shown for example in figure 3 from Sun et al [13], where various trends for the heat transfer characteristics of rectangular channels are shown in a single graph.

Measurement uncertainties are due to temperature sensors, IR-camera, and mass-flow-rate meter. The standard deviation of thermocouples and thermo-resistances measurements has been evaluated by a calibration procedure in a temperature-controlled water bath, resulting in 0.02 K and 0.01 K respectively. The declared NETD of the IR camera, i.e. 0.02 K at 300 K, has been also verified in the entire measurement range, however the filtering strategy adopted can produce local temperature differences with respect to the measurement up to 1 K, with an average of 0.1 K. The effect of possible reflections is not considered in this context. The mass-flow-rate meter has been calibrated by the Italian "Istituto Nazionale di Ricerca Metrologica (INRiM)", and a 2 % accuracy is proven for the range of Reynolds numbers exploited in this research. Thermal conductivities of all test-section materials have been measured in the ThermALab laboratory with an uncertainty at most of ±4 % and radiative properties have been measured with an uncertainty of ±1 % up to 20 μm, and a constant behavior is assumed after that wavelength.

Journal of Physics: Conference Series 2177 (2022) 012037 doi:10.1088/1742-6596/2177/1/012037

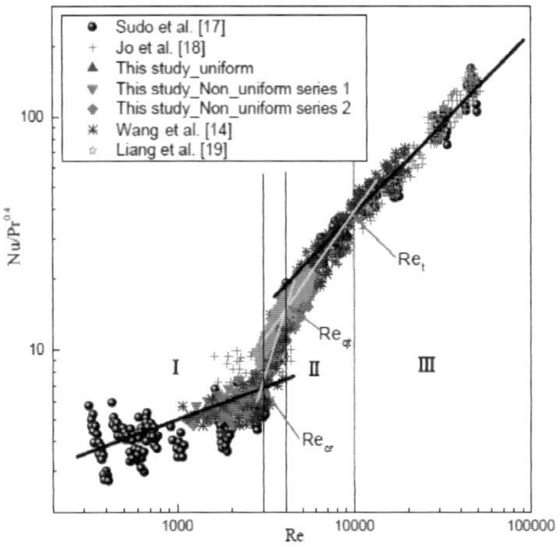

Figure 3. Experimental values of $Nu/Pr^{0.4}$ as function of Re for narrow smooth rectangular channels from Sun et al. [13], database collected from various studies carried out with different boundary conditions. They propose 2 correlations to fit data with ± 25% accuracy for laminar and transition regime and show that the Dittus-Boelter correlation has ± 20% accuracy for fully-turbulent regime (Re > 10000)

Considering this data, a minimum uncertainty on the local Nusselt number, based from the definition of equation 3, and considering only the uncertainties induced by the temperature measurements explicitly appearing in the formula, i.e. the bulk temperature $T_b(x)$ and the wall temperature $T_w(x,y)$, can be given as follows:

$$\frac{\delta Nu}{Nu} = \frac{T_w(x,y) - T_b(x)}{q_c''(x,y)} + \sqrt{\frac{q_c''(x,y)^2}{(T_w(x,y) - T_b(x))^2}\delta T_w + \frac{q_c''(x,y)^2}{(T_w(x,y) - T_b(x))^2}\delta T_b} \tag{5}$$

The minimum average uncertainty value for the Nusselt number calculated by means of equation 5 for all configurations is equal to 12%. This value does not take into account the contribution given by the term $q_c''(x,y)$, which is affected by all the models mentioned in paragraph 2.1. As will be shown in paragraph 3, while for high-Reynolds flows $q_c''(x,y)$ can be, with good approximation, considered equal to the input heat flux, for low Reynolds other contributions have to be taken into account.

Finally, deviations from the assumed linear behaviour of $T_b(x)$, which may be caused by local radiative losses in the region of the germanium window, could introduce an estimated average bias of 5% on *h*, i.e. a slight underestimation, for low-Re measurements.

3 Results

Before showing the results of the tested configurations in terms of Nu/Nu_0, common trends and features of the results are analyzed for one configuration. In particular, let us consider the V-down configuration, as it is, among those tested, the one with the highest local Nu increment as the Reynolds number increases.

The first consideration regards the absolute value of the heat fluxes. Since an experimental constraint is to maintain the heated surface at an average temperature high enough to appreciate local differences induced by the convective pattern, but also to avoid overheating, the power input is set to be comparable to previous experiments in the constant-temperature channel, for each Reynolds, as in table 2. The implication of this choice is that the temperatures of the channel show very little significant variability between experiments at different Reynolds number and, consequently, all the non-convective heat fluxes have comparable absolute values. The only one that changes, along with the increasing power input at higher Reynolds, is the convective one, as shown in Figure . Figure shows the same information, but directly compares the heat flux breakdown relative to the input power for the lowest and maximum tested Re.

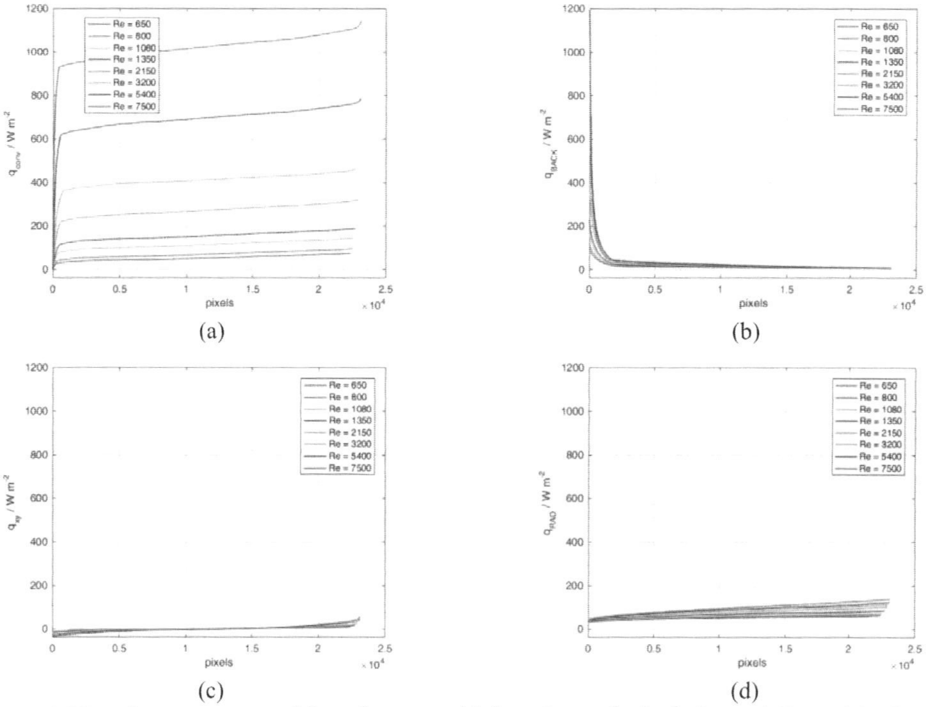

Figure 4. Heat flux maps, sorted from lowest to highest for each pixel, for each Reynolds. Convective heat flux (a), backside heat flux (b), in-plane heat flux (c) and radiative heat flux (d)

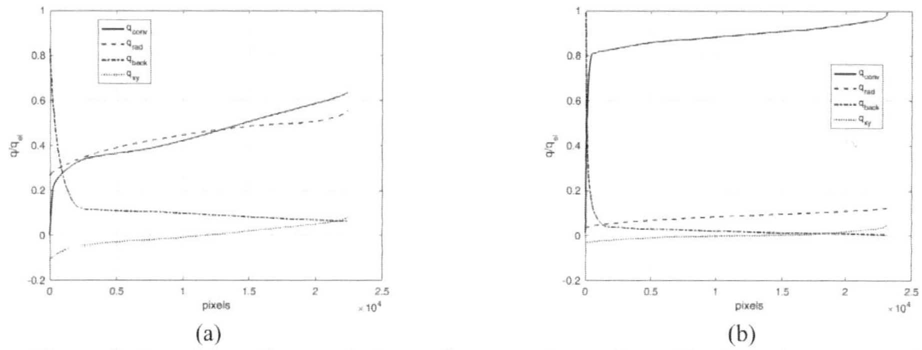

Figure 5. Sorted heat fluxes relative to the power input, Re = 670 (a) and Re = 7200 (b)

The consideration that follows is that, as the Reynolds number increases, the convective heat flux increases relatively to the other heat fluxes, with the only exception of the backside component towards the very edge, which is limited to impose a null convective flux at the corner. That implies that, as Re tends to infinity, the local convective heat flux tends to be uniformly distributed and equal to the input power. Consequently, the uncertainty on the local convective heat flux converges to that of the input power, as Reynolds number increases. Otherwise, the uncertainty linked to all the other models, in this case the radiative one, can affect the absolute measurement at lower Reynolds.

On the other hand, the local convective heat transfer coefficient $h(x,y)$ and, consequently, the local Nusselt number $Nu(x,y)$, defined by equations 2 and 3, are dependent not only on the local convective heat flux, but also on the difference between the local wall temperature $T_w(x,y)$ and the fluid bulk temperature $T_b(x)$, which is considered linearly increasing from the inlet to outlet. In the zones of efficient heat transfer enhancement, the local wall temperature has the tendency to reach the fluid bulk temperature, thus causing a hyperbolic increment of the local heat transfer coefficient, as visible in Figure . This behavior is better seen on 2D maps, showing the percentage difference from the average heat flux and convective coefficient, as in figure 7.

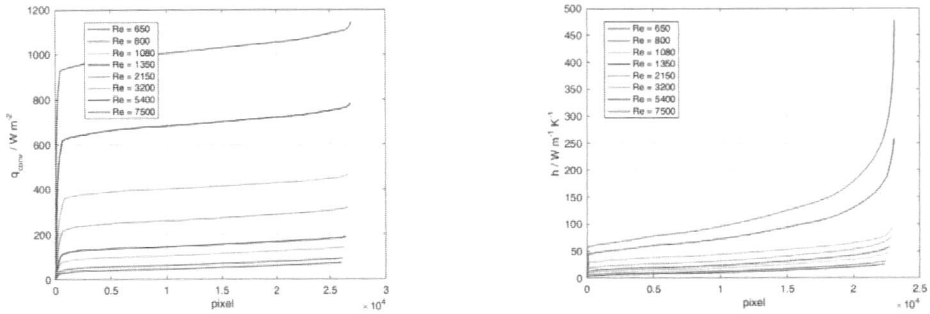

(a) (b)

Figure 6. Sorted heat flux maps (a) and sorted local convective coefficient maps (b)

Re	$\Delta Q_{conv}(x,y)$ %	$\Delta h(x,y)$ %
670	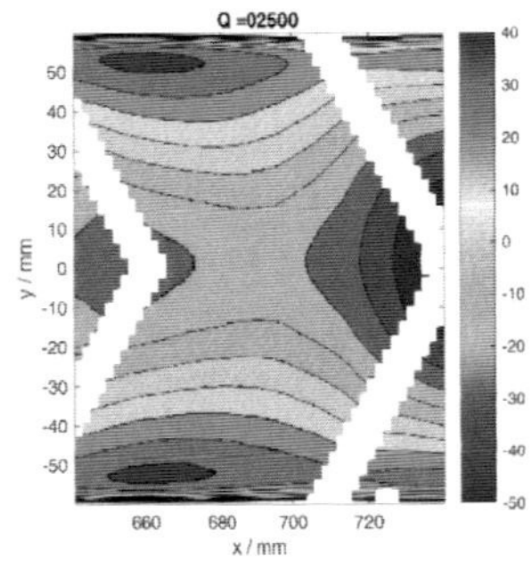	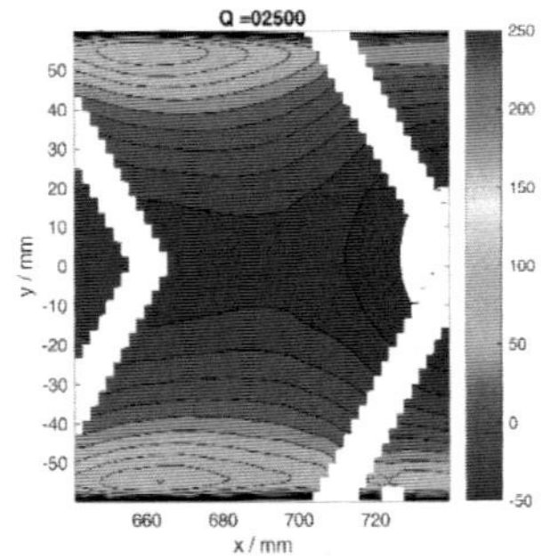
7500	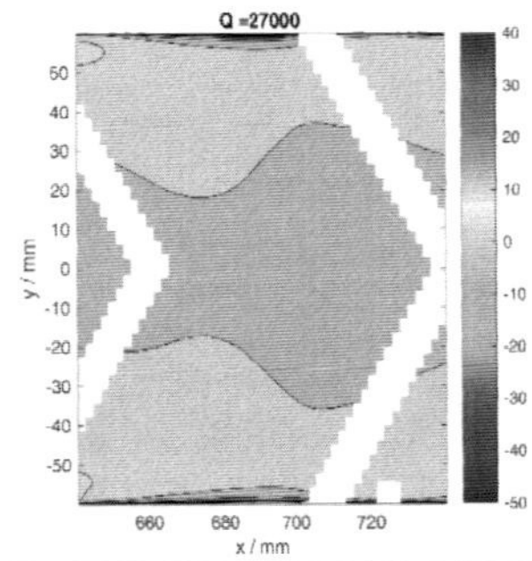	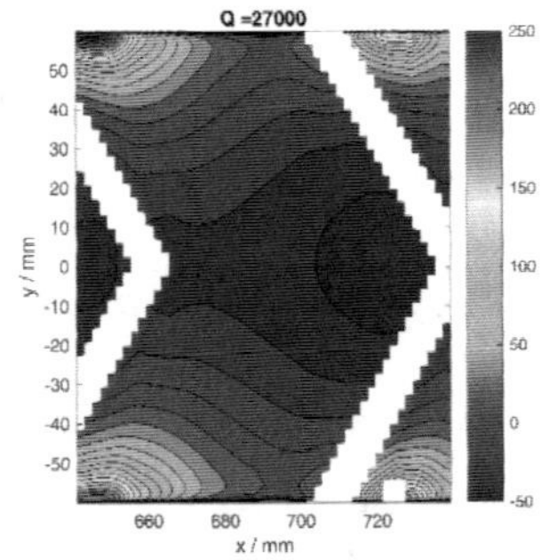

Figure 7. Maps of percentage difference from the average, convective heat fluxes (left) and convective coefficient (right), V down configuration. Q refers to the volumetric flow rate. Flow from left to right

While being partially beyond the scope of the paper, the resulting Nu/Nu_0 maps with respect to the smooth configuration, given by the correlations reported in [13], is shown in figure 8 for Re equal to 670, 2150 and 7500.

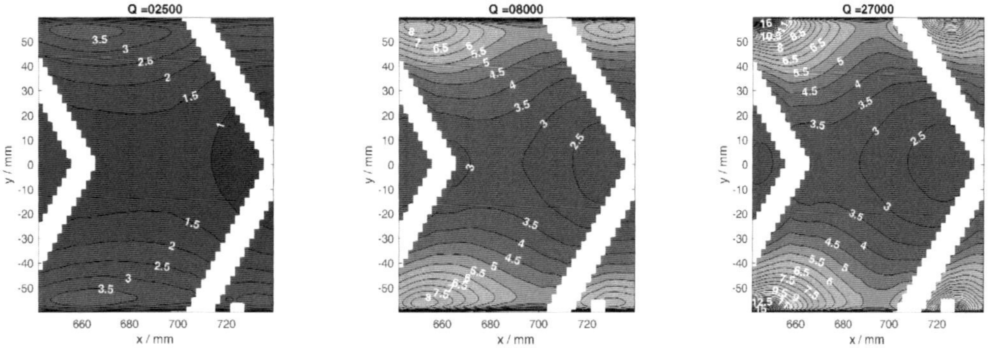

Figure 8. Nu/Nu_0 maps, for Reynolds 670, 2150 and 7500 (left to right). Flow from left to right.

4 Conclusions

This paper illustrates some considerations on the uncertainty associated with the measurement of the local heat transfer coefficient in a ribbed closed channel. In particular, the analysis shows that for high Reynolds number, i.e. Re > 10000, most of the input heat flux, for this particular experimental set-up, is transferred to the air flow by convection. As a consequence, locally, the uncertainty on the Nusselt number is mainly linked to the uncertainty of the temperature measurements, in particular to the temperature of the surface under observation and the inlet and outlet bulk temperature.
For the minimum Reynolds under investigation, i.e. Re = 650, the radiative term weighs from 40% to 60% of the heat balance, and conductive terms can account for up to 20%. The uncertainty associated with the models for their calculation can therefore be taken into account for a sounder total uncertainty definition. Finally, heat transfer increment maps for the analyzed configuration are reported, to illustrate the context of this analysis.

References

[1] Kumar A, Kim M-H. 2015 *App. Therm. Eng.* **89** 239–61
[2] Tanda G, Satta F 2021 *Int. J. of Heat and Mass Trans.* **169** 120906
[3] Chung H, Park JS, Park S, Choi SM, Rhee D-H, Cho HH 2015 *Int. J. of Heat and Mass Trans.* **88** 357–67
[4] D Fustinoni, P Gramazio, L P M Colombo and A Niro 2012 *J. Phys.: Conf. Ser.* **395** 012042
[5] D Fustinoni, P Gramazio, L Vitali and A Niro 2017 *J. Phys.: Conf. Series* **796** 012015
[6] P Gramazio, L Vitali, D Fustinoni and A Niro 2017 *J. Phys.: Conf. Series* **923** 012052
[7] Gao X, Sundén B 2001 *Heat and Mass Trans.* **37** 315–20
[8] Cavallero D, Tanda G. 2002 *Exp. Therm. and Fluid Sci.* **26** 115–21
[9] Gupta A, SriHarsha V, Prabhu SV, Vedula RP. *Exp. Therm. and Fluid Sci.* **32** 997–1010
[10] Peng W, Jiang P-X, Wang Y-P, Wei B-Y. 2011 *App. Therm. Eng.* **31** 2702–8.
[11] BIPM 2010 JCGM 100:2008, Evaluation of measurement data
[12] Moffat R J 1988 *Exp. Therm. And Fluid Sci.* **1** 3-17
[13] Sun R, Song G, Zhang D, Deng J, Su GH, and Kulacki FA 2020 *Exp. Therm. and Fluid Sci.* **114** 110055.

Journal of Physics: Conference Series 2177 (2022) 012038 doi:10.1088/1742-6596/2177/1/012038

Study on the local heat transfer behaviour in a flat plate pulsating heat pipe

Luca Pagliarini[a], Luca Cattani[c], Fabio Bozzoli[a,b], Maksym Slobodeniuk[d], Vincent Ayel[d], Cyril Romestant[d], Yves Bertin[d], Sara Rainieri[a,b]

[a] Department of Engineering and Architecture, University of Parma, Parco Area delle Scienze 181/A I-43124 Parma, Italy
[b] SITEIA.PARMA Interdepartmental Centre, University of Parma, Parco Area delle Scienze 181/A, I-43124 Parma, Italy
[c] CIDEA Interdepartmental Centre, University of Parma, Parco Area delle Scienze 181/A, I-43124 Parma, Italy
[d] Pprime Institute, CNRS – ENSMA – Université de Poitiers, UPR 3346, Futuroscope-Chasseneuil, France

Corresponding author e-mail: Luca Pagliarini (luca.pagliarini@unipr.it)

Abstract. A closed-loop Flat Plate Pulsating Heat Pipe, filled with a water-ethanol mixture (filling ratio = 50% vol.), is tested in horizontal orientation at different heat loads to assess local wall-to-fluid heat fluxes exchanged within its adiabatic section. In addition, the thermal interactions between adjacent channels are investigated since they could be responsible for poorer thermal performances in such kinds of cooling systems. The studied device is obtained from a copper plate (width: 80 mm, length: 200 mm, thickness: 3.5 mm) machined with a single square shaped groove (3 x 3 mm^2) forming a series of 8 turns in the evaporator zone and covered with a second copper plate having same length and width, and thickness equal to 0.5 mm. During the Flat Plate Pulsating Heat Pipe operation, the temperature of the external wall, coated with a high-emissivity paint, is monitored within the whole adiabatic section by an InfraRed camera during the pseudo-steady state. The thermographic measurements are post-processed by adopting the Inverse Heat Conduction Problem resolution approach, adequately validated by numerical simulations. The resulting wall-to-fluid heat fluxes are studied in terms of amplitude and oscillation, thus providing novel pieces of information regarding the working behaviour of Flat Plate Pulsating Heat Pipes in terms of oscillatory flow and thermal interactions by conduction between adjacent channels.

1. Introduction

Pulsating Heat Pipes (PHPs) are two-phase passive heat transfer devices which have been achieving resound interest within the scientific community due to their low manufacturing cost and high heat transfer capabilities. Depending on their geometrical configuration, PHPs can be generally classified as tubular/capillary PHPs and Flat Plate PHPs (FPPHPs) [1]. In particular, the latter, may be effectively adopted in such applications where an optimal thermal contact between the cooled section and FPPHPs is required or when compactness coupled with structural reliability are necessary. The most promising fields of application are represented by standard and micro-electronics cooling, thermal management of batteries in the automotive field and space applications [2]. Although FPPHPs have been already intensively studied in terms of global performance, i.e. their capability of transferring heat from a hot source to a dissipating section, their working principles are still not fully clarified, especially due to the complex interactions occurring by transient conduction between adjacent channels. In fact, the geometrical continuity and high thermal inertia of such devices result in lower temperature gradients over the whole solid domain. This may lead to lower pressure differences between adjacent channels and, consequently, to weaker driving forces for self-sustained oscillations of the slug-plug flow in saturation conditions between the evaporator and the condenser [3].

Khandekar et al. [4] were among the firsts to speculate about the adverse effects of conduction between adjacent channels in FPPHPs. Also, from the investigation of a 10-turn FPPHP in horizontal orientation, Ayel et al. [5] highlighted the occurrence of dry-outs, i.e. lack of liquid in the evaporator section, due to the coupling effect of orientation and transverse conduction, which results in low pressure gradients for the fluid motion. Other interesting considerations were presented by Takawale et al. [6], who compared in same test conditions the thermofluidic behaviour of a FPPHP with that of a tubular PHP, in which no conductive phenomena occur between adjacent channels. In particular, space amplitudes of fluid oscillations in the FPPHP were much lower than those observed in the tubular PHP, thus proving the hampering effect of transverse conduction in FPPHPs in terms of both fluid motion and global performance of the device. Yet, no results regarding the heat transfer due to transverse conduction were provided.

The present work proposes to describe the thermal interactions between adjacent channels in metallic FPPHPs by means of a novel technique, based on the application of the Inverse Heat Conduction Problem (IHCP) resolution approach. To this aim, an 8-turn, closed loop FPPHP, previously investigated at varying tilt angles with water, aqueous mixtures, and surfactants [7], is tested in horizontal orientation at different heat loads provided at the evaporator. Firstly, the adopted method is presented and validated by means of numerical simulations. The local heat flux exchanged within the adiabatic section between the working fluid and the device wall is finally evaluated for the considered study case to provide useful pieces of information regarding the oscillatory flow and spatial thermal interactions occurring in the solid domain and between the solid and the fluid. Such wall-to-fluid heat flux was only evaluated in the adiabatic section, and not in the other zones of the studied device.

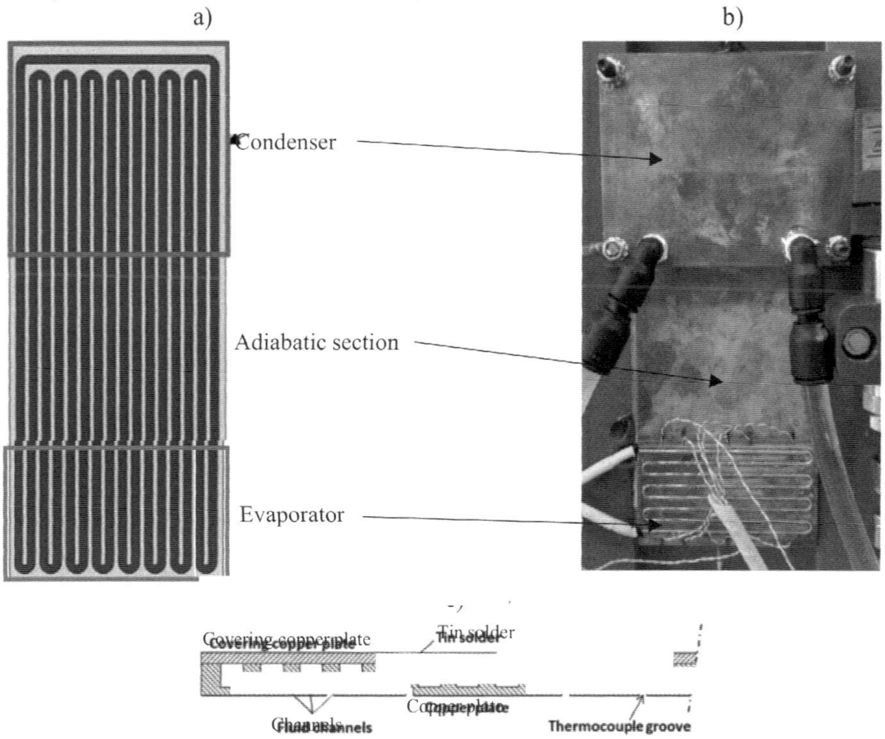

Figure 1: Grooved path, with reference to the evaporator, condenser and adiabatic section positions (a) and back-side of the FPPHP (b), resulting in a series of channels for the working fluid (c).

2. Experimental apparatus

The studied FPPHP (Fig. 1a) [7] is constituted by a copper plate (width: 80 mm, length: 200 mm, thickness: 3.5 mm) machined with a single square shaped groove (3 x 3 mm²), forming a series of 8 U-turns in the evaporator zone (Fig. 1b-c). The copper plate is brazed with tin to a second copper plate having same width and length, and thickness equal to 0.5 mm, to guarantee a perfect sealing from the outside and between channels. Two connection pipes are brazed on top and bottom of the device for evacuation and filling procedures. The evaporator, covered with insulating material, is heated by a wire electrical heater (Thermocoax® Type NcAc15, 1 mm external diameter), embedded in a copper plate (40 x 80 x 2 mm) by means of a serpentine groove machined on one side of the plate,. The power input is provided to the heater by an electrical power supply (ELC® ALR3220, ±10 mV). On the other hand, the condenser section is cooled by a grooved copper plate heat exchanger (80 x 80 mm), directly brazed on the rear face of the FPPHP and attached to a thermal bath (Huber® CC240 wl) controlling the temperature of the cooling medium, i.e. an ethylene-glycol/water mixture. The FPPHP outer front surface was uniformly coated with a high-emissivity paint ($\varepsilon = 0.92$) and its outer wall temperature was acquired within the adiabatic section by means of a high-speed Medium Wave Infrared (MWIR) camera (FLIR® SC5500, sampling frequency: 50 Hz, pixel resolution: 320 x 256, uncertainty: ± 0.2°C).

The FPPHP was first evacuated and sealing tested by means of an ultra-high vacuum system (ASM 142, Adixen by Pfeiffer Vacuum®) coupled with a helium detector. The device was therefore partially filled with a water-ethanol mixture (20 wt. % of ethanol, filling ratio: 50% at 20°C), previously degassed by heating-cooling cycles. The FPPHP was positioned in horizontal orientation, and the condenser temperature was set to 20°C. The power input given to the evaporator ranged from 50 W up to 200 W. Two IR acquisitions of 1 minute each were carried out within the device pseudo-steady state, i.e. after about 30 minutes from each heat load step applied to the evaporator.

3. Local heat flux evaluation

The local convective heat flux exchanged at the fluid-wall interface, q, was evaluated starting from the temperature distribution on the outer surface of the adiabatic section, i.e. the wall-environment interface, used as input parameter for the inverse heat conduction problem resolution approach. With reference to a single FPPHP branch, the three-dimensional solid domain was outlined as a plane-symmetrical geometry, as shown in Fig. 2a in cross sectional view, where x_{int} corresponds to half of the channel height. In the present analysis, only the area delimited by the red dotted lines, i.e. having thickness $s = x_{ext} - x_{int} = 0.5$ mm, was considered.

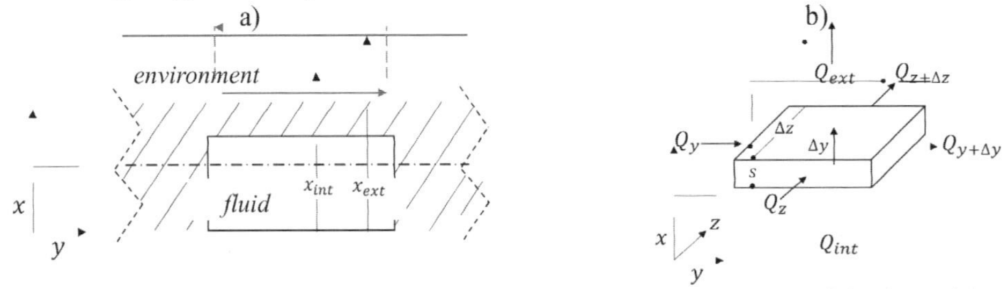

Figure 2: Cross sectional view of the three-dimensional domain within the adiabatic section (a) and generic wall element (b). Red dotted lines delimit the section of interest for the present analysis.

For such a portion, by assuming no volumetric heat generation in the wall, the transient energy balance equation reads as:

$$k\nabla^2 T = \rho c_p \frac{\partial T}{\partial t} \tag{1}$$

where k, ρ and c_p are the solid wall thermal conductivity, density, and specific heat, respectively. Eq. (2) is completed by the following boundary conditions, applied on the inner and outer surface of the channel, respectively:

$$-k\frac{\partial \overline{T}}{\partial x}\bigg|_{x=x_{int}} = q \tag{2a}$$

$$-k\frac{\partial T}{\partial x}\bigg|_{x=x_{ext}} = h_{env}(T - T_{env}) \tag{2b}$$

where q is the exchanged heat flux, T_{env} is the environmental temperature and h_{env} is the overall heat-transfer coefficient between the channel wall and the surrounding environment by both natural convection and radiation. The thermal conductivity k of copper was certified by the manufacturer to be equal to 401 W/mK at 300 K and the overall heat-transfer coefficient h_{env} was assumed equal to 10 W/m^{2}K according to the average temperature field obtained from measurements. By considering the thin-wall approximation, the temperature referred to the inner surface, i.e. for $x = x_{int}$, was assumed to be equal to that on the outer surface, i.e. for $x = x_{ext}$. Hence, the 3D solid domain was approximated as a 2D domain, where the temperature varies only as a function of the two spatial coordinates (y and z) and time. For the infinitesimal wall section having dimensions Δy and Δz, the local energy balance equation can be expressed in terms of internal energy variation over time as follows:

$$\rho c_p \frac{\partial T}{\partial t} s\Delta y\Delta z = \left(Q_y - Q_{y+\Delta y}\right) + (Q_z - Q_{z+\Delta z}) + Q_{int} - Q_{ext} \tag{3}$$

The terms appearing in eq. (3) are defined as:

$$\left(Q_y - Q_{y+\Delta y}\right) = \frac{k\partial^2 T}{\partial y^2} s\Delta y\Delta z \tag{4a}$$

$$(Q_z - Q_{z+\Delta z}) = \frac{k\partial^2 T}{\partial z^2} s\Delta y\Delta z \tag{4b}$$

$$Q_{int} = q\Delta y\Delta z \tag{4c}$$

$$Q_{ext} = h_{env}(T - T_{env})\Delta y\Delta z \tag{4d}$$

By substituting eqs. (4a-d) in eq. (3) and rearranging, the heat flux q results:

$$q = s\left(\rho c_p \frac{\partial T}{\partial t} - k\frac{\partial^2 T}{\partial y^2} - k\frac{\partial^2 T}{\partial z^2}\right) + h_{env}(T - T_{env}) \tag{5}$$

During the tests, T_{env} varied from 18°C to 24°C. Eq. (5) can be solved by means of the finite difference method. However, being experimental temperature data unavoidably noisy, eq. (5) gives unstable results because the derivative operators are very sensitive to small perturbations in the input data when coupled with the negative effect of noise [8]. A convenient way to overcome these difficulties is filtering out the unwanted noise from the raw temperature data. The effectiveness of the Gaussian kernel in this kind of approach was experimented by Murio et al. [9], Delpueyo et al. [10] and Bozzoli et al. [11]. The application of a Gaussian filter reduces the data high-frequency components, thus behaving alike the regularization term in the Tikhonov regularization method [12]. According to the filtering approach, the heat flux on the inner wall surface can be obtained by using in Eq. (5) the filtered temperature T_f instead of the measured temperature T. In particular, the spatial discretization for T_f was equal to the spatial resolution of the MWIR camera (≈ 4e-4 m).

4. Method validation and application to experimental data

The adopted method was validated by using synthetic temperature data, generated by solving the direct problem with a known distribution of the wall-to-fluid heat flux at the internal wall surface. The numerical model for the direct problem was implemented within the Comsol Multiphysics® environment. A portion of the plane-symmetrical adiabatic section, constituted by 4 channels numbered in Fig. 3a, was here considered since the thermal interactions between adjacent channels were assumed to similarly replicate in the whole adiabatic section.

The resulting temperature distributions were then spoiled by a Gaussian noise level, characterized by a standard deviation σ equal to 0.04 K, and used as input data for the IHCP resolution approach, i.e. adopted in eq. (5). According to the expected physical behaviour in real cases, the convective heat flux was assumed to vary between the two surfaces A and B of the generic channel, outlined in Fig. 3b, as follows:

$$q_A = a \cdot \sin(2\pi t \cdot f + \varphi) \quad (6a)$$

$$q_B = b \cdot \sin(2\pi t \cdot f + \varphi) \quad (6b)$$

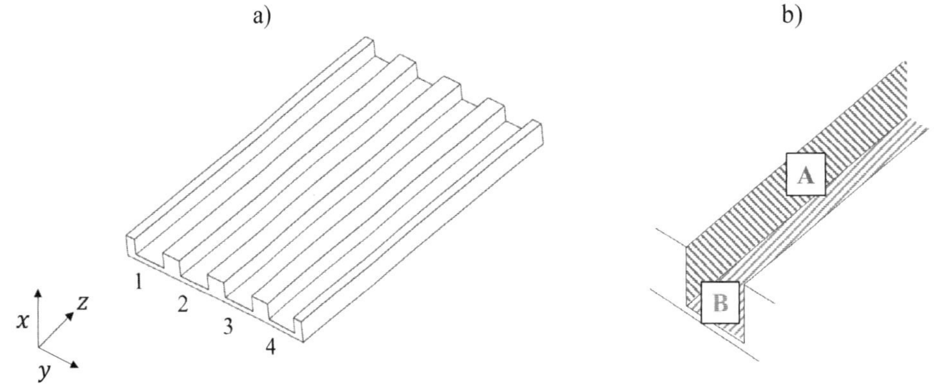

Figure 3: Solid plane-symmetrical domain for the direct problem (a). The wall-to-fluid heat flux is considered to vary between the surfaces A and B (b).

The amplitudes and frequency a, b and f were set to 10000 W/m^2, 2500 W/m^2 and 0.8 Hz, respectively, while the phase φ was set to 0 for channels 1 and 3, and to $-\pi$ for channels 2 and 4. From the direct problem, solved by using the heat fluxes of eq. (6a-b), the maximum residual between the inner and outer surfaces was equal to about 0.01 K. The thin-wall approximation of eq. (3) was therefore validated, since the maximum residual was much lower than the noise level of the experimental apparatus (σ = 0.04 K).

The exact heat flux q, defined in eq. (6b), was finally compared with the restored heat flux, i.e. the one obtained by eq. (5), to prove the reliability of the here adopted inverse method. In Fig. 4, both q_{exact} and $q_{restored}$ are shown for a single spatial point referred to the third channel (Fig. 3a) over 20 seconds. As noticeable, besides some fluctuations around the exact value due to residual noise, the restored heat flux follows with good accuracy the instantaneous trend of the exact heat flux in terms of both amplitude and frequency. Such comparison was successfully replicated for every channel and every spatial point, thus validating the employed procedure. In particular, the restoration error for the presented procedure, was quantified by means of the following equation:

$$E_q = \frac{\|q_{estimated} - q_{exact}\|_2}{\|q_{exact}\|_2} \quad (7)$$

It has been found to be equal to about 12% of the wall-to-fluid heat flux actually exchanged in the real device.

The method described in Section 2 was therefore applied to the experimental cases by post-processing the thermographic acquisitions. In Fig. 5, all the terms constituting the overall heat flux q defined in eq. (5) are separately shown for a single central channel at varying heat loads to the evaporator to better understand the differences from spatial to time heat transfer behaviour within the adiabatic section. In particular, the terms related to the two space directions y and z, i.e. $q_z = -k\frac{\partial^2 T}{\partial y^2}s$

and $q_y = -k \overline{\frac{\partial^2 T}{\partial z^2}} s$, are individually considered as well as the storage term $q_{storage} = \rho c_p \overline{\frac{\partial T}{\partial t}} s$, while the heat dissipated towards the environment $q_{env} = h_{env}(T - T_{env})$ was found to be negligible with respect to the other terms for every test case. It must be pointed out that, as suggested by Cattani et al. [13] and Pagliarini et al. [14] for tubular PHPs, the storage term is capable of providing some information about the fluid motion inside the device, even without a direct fluid visualization.

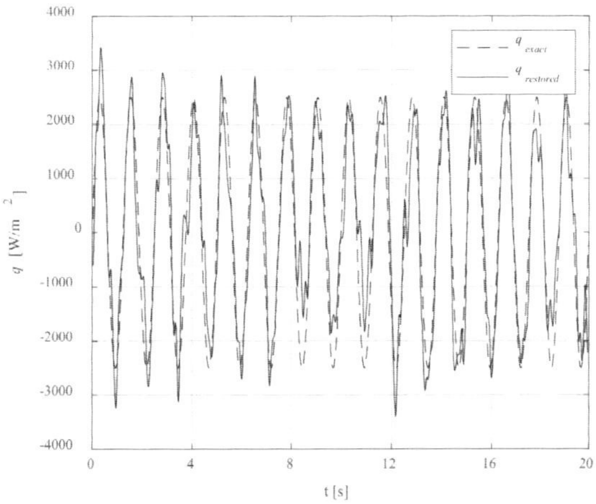

Figure 4: Comparison between the exact and the restored heat flux, obtained from the IHCP resolution approach, for a single spatial point referred to channel 3.

In fact, positive values of the storage term correspond to the passage of hot vapour plugs or liquid slugs from the evaporator to the condenser. Negative values denote instead flow reversal phenomena, during which cold fluid flows from the condenser to the evaporator at a lower temperature than that of the channel wall, previously heated by hot fluid coming from the evaporator. Hence, the oscillating motion of the working fluid between the two heat transfer areas in the considered channel is represented by the alternance of positive and negative values of the storage term over time. On the contrary, the space term along the y direction in the present FPPHP is referred to the thermal interactions by transient conduction between adjacent channels.

For power input equal to 50 W (Fig. 5a), the storage term presents slight variations during the first seconds of acquisitions. Such evidence denotes the occurrence of fluid oscillations between the evaporator and the condenser through the analysed channel, which stop after about 5 seconds from the beginning of the observation window. Hence, the provided power input cannot guarantee a continuous fluid motion. On the other hand, the space term along the y direction is almost null over the entire 60 seconds of acquisition due to very weak thermal interactions between adjacent channels, while the term along the z direction always exhibits negative values.

According to eq. (5), this is due to the unbalancing of the incoming and outgoing heat fluxes along z through the wall element, probably linked to the fact that a portion of the overall local heat flux is yielded by the wall to the working fluid in the considered channel. By increasing the power input up to 100 W (Fig. 5b), the storage term presents more regular variations over time, despite the absence of fluid motion, i.e. storage term oscillations, in the range 30 – 45 seconds.

The space term along the y direction similarly fluctuates, denoting that adjacent channels start operating and thermally interacting with the analysed branch, while the space term along the z direction still assumes negative, almost constant values. At high heat loads, i.e. 150 W (Fig. 5c) and 200 W (Fig. 5d), the device exhibits strong and fast local thermal interactions, in accordance with previous works [15]. For the highest considered power input, the storage term presents absolute peaks of about 5000 W/m^2 and fluid oscillation frequencies around 0.6 Hz, the latter qualitatively evaluated by counting the number of peaks occurring over the whole acquisition. In addition, for such test conditions, the space term along y exhibits peaks of about 7000 W/m^2, highlighting the presence of significant thermal interactions between adjacent channels. However, it has to be clarified that, since the present analysis only focuses on the thin-wall surfaces of the device, the thermal interactions represented by the q_y term could be either interactions between adjacent channels or between the thin-wall sections and the side, thick walls of each channel.

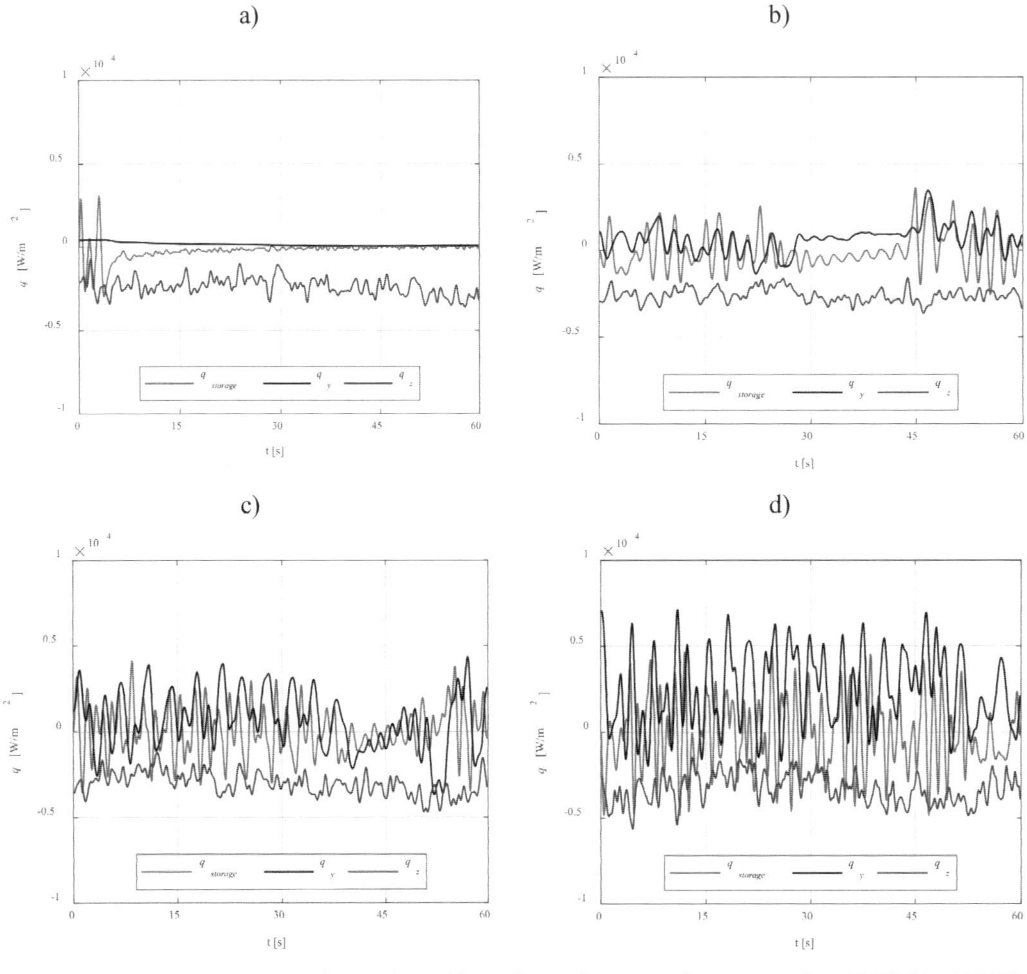

Figure 5: Terms constituting the evaluated heat flux q for power inputs equal to 50 W (a), 100 W (b), 150 W (c) and 200 W (d).

Conclusions

An 8-turns Flat Plate Pulsating Heat Pipe made of copper and partially filled with a water-ethanol mixture was tested in horizontal orientation with the aim of better understanding its local heat transfer behaviour in terms of oscillatory flow and thermal interactions between adjacent channels. The temperature referred to the outer wall surface within the adiabatic section was monitored, during the pseudo-steady state of the device, by means of an infrared camera. The raw thermographic measurements were therefore filtered and used as input parameters for the inverse heat conduction problem resolution approach, previously validated by numerical simulations. The resulting wall-to-fluid heat fluxes were investigated term by term at different power inputs to the evaporator, providing a better insight into the oscillatory flow typical of such heat transfer devices. Specifically, the storage term reached maximum peaks and oscillation frequencies, at high heat loads, of about 5000 W/m^2 and 0.6 Hz, respectively, while the space term along y strongly varied with peaks of about 7000 W/m^2, denoting the occurrence of significant interactions by transverse conduction. On the other hand, the space term along the z direction exhibited almost constant, negative values, suggesting that the device wall constantly provides heat to the working fluid in the considered channel. Further investigations on the presented layout are required to fully verify the provided results, especially focusing on both global and local heat transfer performances at different test conditions.

Acknowledgments
The Authors would like to acknowledge the European Space Agency (ESA) support through the grant 4000128640/19/NL/PG/pt, ESA MAP project TOPDESS.

Nomenclature

Symbol	Quantity	SI unit
a,b	amplitudes	W/m^2
c_p	Specific heat at constant pressure	J/kgK
f	Frequency	Hz
h	Heat transfer coefficient	W/m^2K
k	Thermal conductivity	W/mK
Q	Power	W
q	Heat flux	W/m^2
s	Wall thickness	m
T	Temperature	K, °C
t	Time	s
U	Internal energy	W
x, y, z	Spatial coordinates	m
φ	Phase	rad
ρ	Density	kg/m^3
Subscripts/superscripts		
env	Environment	
int/ext	Internal / external	
out	Outer	
f	Filtered	

References

[1] D. Bastakoti, H. Zhanga, D. Lia, W. Caia, F. Li, An overview on the developing trend of pulsating heat pipe and its performance, App. Therm. Eng. 141 (2018) 305–332.

[2] V. Ayel, M. Slobodeniuk, R. Bertossi, C. Romestant, Y. Bertin, Flat plate pulsating heat pipes: a review on the thermohydraulic principles, thermal performances and open issues, Appl. Therm. Eng. 197 (2021), 117-200.

[3] V. Ayel, L. Araneo, A. Scalambra, M. Mameli, C. Romestant, A. Piteau, M. Marengo, S. Filippeschi, Y. Bertin, Experimental study of a closed loop flat plate pulsating heat pipe under a varying gravity force, Int. J. Therm. Sc. 96 (2015) 23-34.

[4] S. Khandekar, M. Schneider, P. Schafer, R. Kulenovic, M. Groll, Thermofluid dynamic study of flat-plate closed-loop pulsating heat pipes, Microsc. Thermophys. Eng. 6 (2002) 303-317.

[5] V. Ayel, C. Romestant, Y. Bertin, V. Manno, S. Filippeschi, Visualisation of flow patterns in flat plate pulsating heat pipe: influence of hydraulic behaviour on thermal performances, Int. J. Heat Pipe Sci. Tech.. 5 (2014) 1-4

[6] A. Takawale, S. Abraham A. Sielaff P. S. Mahapatra, A. Pattamatta, P. Stephan, A comparative study of flow regimes and thermal performance between flat plate pulsating heat pipe and capillary tube pulsating heat pipe, Appl. Therm. Eng. 149 (2019) 613-624

[7] V. Ayel, M. Slobodeniuk1, R. Bertossi, A. Karmakar, F. Martineau, C. Romestant, Y. Bertin, S. Khandekar, Thermal performances of a Flat-Plate Pulsating Heat Pipe tested with water, aqueous mixtures and surfactants, Int. J. Th. Sc. (under review).

[8] J. V. Beck, B. Balckwell, C. R. St. Clair Jr, Inverse Heat Conduction – Ill-posed problems, New York: John Wiley & Sons Inc. 1985.

[9] A. D. Murio, The Mollification Method and the Numerical Solution of Ill-Posed Problems, New York: John Wiley & Sons Inc. 1993.

[10] D. Delpueyo, X. Balandraud, M. Grédiac, Heat source reconstruction from noisy temperature fields using an optimised derivative Gaussian filter, Infr. Phys. Tech. 60 (2013) 312–322.

[11] F. Bozzoli, G. Pagliarini, S. Rainieri, Experimental validation of the filtering technique approach applied to the restoration of the heat source field, Exp. Th. Fl. Sc. 44 (2013) 858-867.

[12] V. Morozov, Methods for Solving Incorrectly Posed Problems, Springer-Verlag, New York, 1984.

[13] L. Cattani, D. Mangini, F. Bozzoli, L. Pietrasanta, N. Miché, M. Mameli, S. Filippeschi, S. Rainieri, M. Marengo, An original look into pulsating heat pipes: Inverse heat conduction approach for assessing the thermal behaviour, Th. Sc. Eng. Prog. 10 (2019) 317-326.

[14] L. Pagliarini, L. Cattani, F. Bozzoli, M. Mameli, S. Filippeschi, S. Rainieri, M. Marengo, Thermal characterization of a multi-turn pulsating heat pipe in microgravity conditions: Statistical approach to the local wall-to-fluid heat flux, Int. J. Heat Mass Tr. 169 (2021), 120930.

[15] V. Nikolayev, M. Marengo, Pulsating heat pipes: basics of functioning and numerical modeling, ISBN 978-981-3234-36-9, in: J.R. Thome (Ed.), Encyclopedia of Two-Phase Heat Transfer and Flow IV, vol. 1 Modeling of Two-Phase Flows and Heat Transfer, World Scientific, 2018.

Performance Evaluation Criteria for Triangular Microchannels with Smoothed Corners

Nicola Suzzi, Marco Lorenzini

Università di Udine, DPIA – Dipartimento Politecnico di Ingegneria e Architettura, Via delle Scienze 206, 33100 Udine (UD), Italy, Università di Bologna, DIN – Dipartimento di Ingegneria Industriale, Via Fontanelle 40, I-47121, Forlì (FC) Italy

marco.lorenzini@unibo.it

Abstract. The manufacturing capabilities available today for microchannels make it possible to produce in a comparatively easy, quick and cheap way several cross-sections, possibly allowing modifications of the macro-geometry. Another way is smoothing the corners of polygona cross-sections: this eliminates low-gradient areas and increases transport phenomena, i.e. frictional losses and heat transfer. Several ways of assessing the relative performance of the new shape in comparison to the original have been suggested over the years, among which are Performance Evaluation Criteria (PEC), as proposed by Bergles and Webb. PEC are based on a first-law analysis and aim at extremising the thermal power, heat exchange area, inlet temperature difference or pumping power under varying constraints. In this work equilateral triangular microchannels with uniform wall temperature are considered, through which a liquid flows in fully-developed, steady laminar regime. The cross-sectional area has its corners progressively rounded, and the velocity and temperature profiles are determined, in order to compute the Poiseuille, Nusselt and Stanton numbers, which are then employed in computing the objective functions for some PEC.

1. Introduction

Micro heat exchangers (MHXs) consist of several microchannels of various shape and surface texture which exhibit high area-to-volume ratios and high heat transfer coefficients, which makes them candidates for the removal of high specific heat fluxes, as is the case e.g. in microchips [1-3]. The capability of such microchannels to be manufactured in several shapes and their modularity provide a wide portfolio of options for enhancement of the heat transfer process, albeit at the expenses of high pressure drops, even in the higher range of the laminar regime, so much more so if the fluid is a liquid [4-6] Several strategies may be adopted for heat transfer enhancement in heat exchangers, as thoroughly discussed in [7], but promoting turbulence through inserts of various shape inside microchannels makes the modification of channel morphological characteristics a much more viable option [8-9]. This work aims at investigating the effect of changing the base cross-sectional geometry (triangular) of a microchannel by smoothing its corners. Thus, optimization of channel cross-sectional geometry is employed in this work. No viscous dissipation is considered, and Performance evaluation criteria (PEC) are used in order to determine heat transfer enhancement under defined constraints [7,10], and are sometimes coupled to second-law analyses [11]. This work complements other investigations by the same authors which dealt with other micro-geometries or boundary conditions [12-14], considered viscous dissipation [15,16] or electro-osmotic effects [17-19].

The effect of smoothing the corners of a square duct under H1 and H2 thermal boundary conditions on the Poiseuille and Nusselt numbers was investigated by [8, 9] and the numerical results used to evaluate the optimal geometrical configuration according to PECs criteria and entropy generation. In [12] the Poiseuille and Nusselt numbers was numerically computed as a function of the corners rounding radius of various cross-sectional geometry subjected to H1 boundary conditions, assuming negligible viscous heating. The effect of viscous heating on heat transfer performance was investigated, in terms of Nusselt number, in [4, 6, 15]. Here, a microchannel with a triangular cross-section and sides of equal length with rounded corners is considered. The flow is assumed Newtonian, laminar, fully developed and with negligible viscous heating. Uniform temperature is imposed at the walls.

The velocity and temperature fields are computed to obtain the Poiseuille and Nusselt numbers, which are in turn employed in the objective functions of the PEC, following the same approach as [9-13], in order to estimate the influence of the geometry of the cross-section. Results for a selection of PEC are shown and commented.

2. Statement of the problem

The geometry investigated is that of a triangular microchannels with sides of equal length, a, whose sharp corners are progressively rounded, with a radius of curvature r_c, to a limiting non-dimensional value $R_{c,max}={}^{r_c}/_a=1/(2\sqrt{3})$, Fig.1. Ten different values of R_c are considered. The walls of the duct have a uniform temperature T_w. The fluid flowing within the channel is incompressible, Newtonian, with thermodynamic and transport properties independent of temperature and pressure. The flow regime is pressure-driven, laminar and fully developed and viscous heating is of no significance. Forced convection is the predominant heat transfer mode and no axial conduction occurs in the fluid.

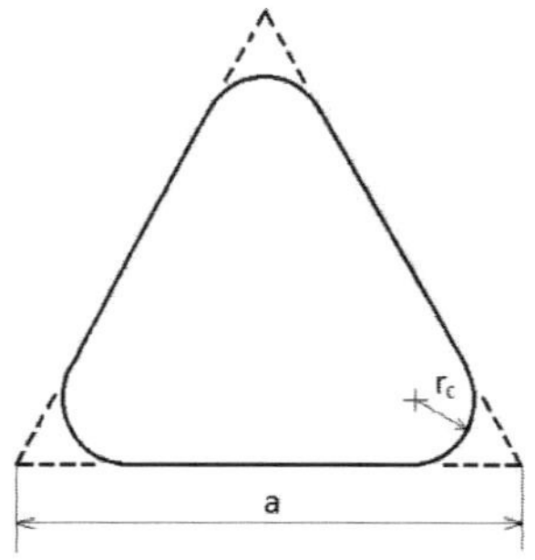

Figure 1: Triangular duct with rounded corners.

In order to compute the Poiseuille and Nusselt numbers, Po and Nu respectively, which are needed by the objective functions to be employed in the first-law optimization, the velocity and temperature fields for the fully developed flow must be determined. This is accomplished by casting the governing equations for momentum and energy conservation in non-dimensional form, Eqs. (1) and (2)

$$\frac{\partial^2 u^*}{\partial y^{*2}}+\frac{\partial^2 u^*}{\partial z^{*2}}=p^* \tag{1}$$

$$\frac{\partial^2 \vartheta^*}{\partial y^{*2}}+\frac{\partial^2 \vartheta^*}{\partial z^{*2}}=0 \tag{2}$$

Using the non-dimensional quantities defined below, Eq. (3):

$$y^*=\frac{y}{D_h},\ z^*=\frac{z}{D_h},\ u^*=\frac{u}{u_b},\ p^*=\frac{D_h^2}{\mu u_b}\frac{dp}{dx},\ \vartheta^*=\frac{T-T_w}{T_b-T_w} \tag{3}$$

Where x is the axial coordinate, y and z the coordinates related to the channel's cross-section, D_h the hydraulic diameter. The bulk velocity is u_b, whereas u is the velocity component along the channel axis, p is the pressure and μ the fluid viscosity in Pa·s. The local fluid temperature is T, whilst T_b and T_w are the bulk and wall temperatures respectively. The boundary conditions are no slip of the fluid at the wall, i.e. $u_w^*=0$ and uniform wall temperature, i.e. $\vartheta_w=0$.

The equations are then solved numerically with an approach detailed in [19]. Knowledge of u* and ϑ^* allows computation of the Nu and Po; details can be found in [13,16,19]; these quantities are then employed to compute the objective functions needed by the PEC chosen for the optimization. In general, the use of PEC allows the achievement of one of the goals listed below:

- maximise heat transfer, $\dot{Q}$;
- minimise heat transfer surface, A;
- minimise the inlet temperature difference between the bulk fluid and walls, $\Delta T_i=T_w-T_{b,i}$;
- minimise the pumping power P, which is the same as minimising the pressure drop, Δp.

The objective functions are the functional expressions for $\dot{Q}$, A, ΔT_i and P, which are dependent on six quantities, namely the shape of the channel cross-section, the channel length, L, the mass flowrate $\dot{m}$, the pumping power, P, the heat flux, $\dot{Q}$, and inlet temperature, ΔT_i.The number of channels, N, would also be an independent variable, but is not relevant for the case of a single microchannel.

Objective functions compare the value of a reference configuration to the one obtained by changing the optimisation parameter (the non-dimensional radius of curvature, in the present case) and the ratios of the two quantities are denoted by a superscript. For the heat flux the expression is

$$\dot{Q}'=\frac{\dot{Q}}{(\dot{Q})_{ref}}=\Delta T_i'\cdot\dot{m}'\frac{1-e^{-4\frac{Nu}{Re\,Pr}\frac{L}{D_h}}}{1-e^{-4\left(\frac{Nu}{Re\,Pr}\frac{L}{D_h}\right)_{ref}}}=\Delta T_i'\cdot\dot{m}'\frac{1-e^{-4\left(St\frac{L}{D_h}\right)_{ref}Nu'\frac{P_h'}{\dot{m}'}\frac{L'}{D_h'}}}{1-e^{-4\left(St\frac{L}{D_h}\right)_{ref}}} \tag{4}$$

Where P_h is the heated perimeter, Re the Reynolds number, St is the Stanton number, $St=\frac{Nu}{RePr}$. For the pumping power, the objective function, where A_c is the cross-sectional area, is

$$P'=\frac{P}{(P)_{ref}}=\frac{(f\,Re)'\ \dot{m}'^2\ L'}{A_c'\ D_h'^2} \tag{5}$$

2.1. Geometrical constraints

The reference configuration corresponds to a cross-section with sharp corners (R_c=0). It is easily seen that the expression for D_h, A_c,and P_h, all depend on both the non-dimensional radius of curvature and the length of the cross-section side, a'. This behaviour is not shared by the Poiseuille and Nusselt numbers, which are independent of the length a. If the objective functions are to be optimised with respect to R_c, four different cases must be treated, each with a different geometrical constraint which eliminates any dependence on the length a, see Table 1.

Geometric constraint	a/a_{ref} ratio
$\frac{a}{a_{ref}}=1$	$\frac{a}{a_{ref}}=1$

$D_h'=1$	$\frac{a}{a_{ref}}=\frac{1}{2\sqrt{3}}\frac{\left(3+2R_c\left(\pi-3\sqrt{3}\right)\right)}{\left(\frac{\sqrt{3}}{2}+R_c^2\left(\pi-3\sqrt{3}\right)\right)}$
$P_h'=1$	$\frac{a}{a_{ref}}=\frac{1}{\left(1+R_c^2\frac{2}{3}\left(\pi-3\sqrt{3}\right)\right)}$
$A_c'=1$	$\frac{a}{a_{ref}}=\sqrt{\frac{1}{\left(1+R_c^2\frac{2}{\sqrt{3}}\left(\pi-3\sqrt{3}\right)\right)}}$

Table 1 – Constrained geometrical parameters (left) and corresponding expression of a' (right).

For each PEC the choice of the geometrical constraint determines one expression of the objective function.

In the following section, some examples of PECs are given, after showing examples of velocity and temperature fields and the Poiseuille and Nusselt numbers as a function of the radius of curvature.

3. Results and discussion

Figure 2a shows the velocity distribution over the cross-section in the case of sharp corners, whilst Fig. 2b shows the same quantity when the corners are smoothed at half the maximum allowable radius of curvature.

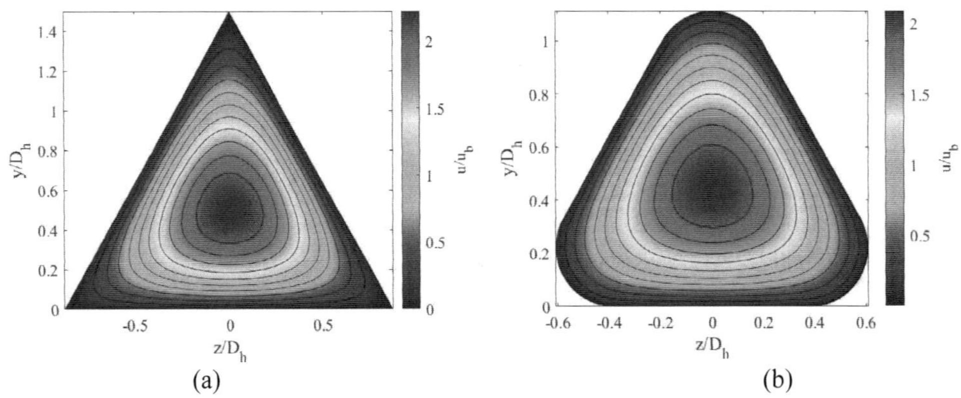

Figure 2: Velocity profile for R_c=0 (a) and R_c=$R_{c,max}$/2 (b).

The maximum velocity is in the barycentric position in both cases, but when R_c=0 the value is higher. This is due to fluid stagnation at the sharp corners, which does not occur when they are smoothed; this implies that local gradients are larger in the latter case, as confirmed by the temperature distribution, Fig. 3. The larger gradients at the smoothed corners of the cross-section signal an increase in transport phenomena, which is captured by the trends of the Poiseuille and Nusselt numbers, plotted in Fig. 4. It should be noted that the maximum increase in Po is below 20%, and that the Poiseuille number remains almost constant for $R_c>0.15$, whereas the Nusselt number increases monotonically with R_c up to about 45% more than the value it attains for the base cross-section. It should also be pointed out that the local value of the Nusselt number along the heated perimeter varies sharply, with maxima at the midpoint of each side and minima at the corners, but for sharp corners the minima are close to zero and increase steadily with R_c; the maxima, conversely, increase at first, reaching a maximum at around half $R_{c,max}$,.

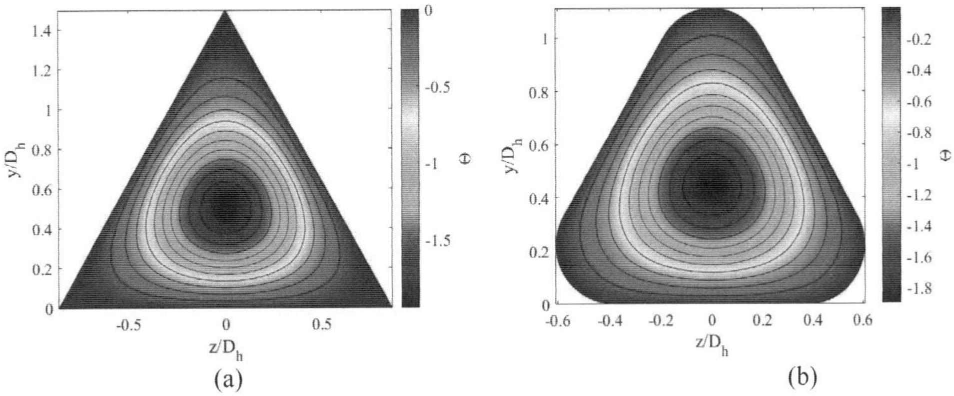

Figure 3: Temperature profile for $R_c=0$ (a) and $R_c=R_{c,max}/2$ (b).

For larger radii, the straight portion of the sides shortens and the maximum Nusselt drops below the value for $R_c=0$, but the average value increases, hence the trend of Fig. 4b.

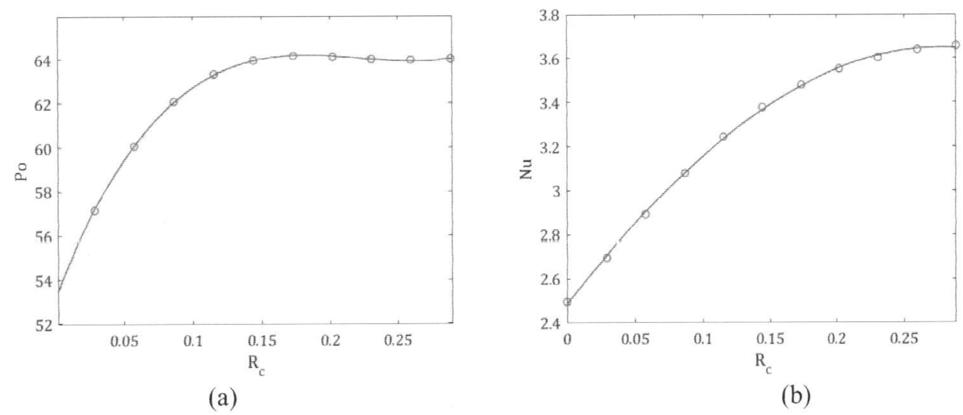

Figure 4: Poiseuille number (a) and Nusselt number (b) as a function of R_c.

3.1. *PEC optimisation*

Once the Poiseuille and Nusselt numbers are determined, PEC optimization can be carried out using algebraic expressions such as, Eqs.(4) and (5) [13], some cases are presented below. If most quantities are normalized using the corresponding values for the base geometry ($R_c=0$), some reference values had to be assumed for the Reynolds number ($Re_{ref}=5$), the Prandtl number (Pr=7 throughout), the channel length (L=5 mm) and its hydraulic diameter ($D_{h,ref}=100$ μm). Since the two objective functions depend on five different quantities, three of them must be kept constant in turn. The triplet of variables which do not change as R_c is varied determines the kind of PEC which is applied.

3.1.1. FG2a. Maximisation of the heat transfer (so-called FG1a PEC) under fixed inlet temperature difference, length and mass flowrate and minimisation of ΔT_i if $\dot{Q}$, L and $\dot{m}$ are kept constant (called FG1b PEC) show that the radius of curvature has no practical influence and is therefore neglected here. The PEC FG2a involves maximisation of the heat transfer with ΔT_i, L and P constant. After some calculations it can be demonstrated that Eq.(4) becomes

$$\dot{Q}'=C_1\left(\frac{a}{a_{ref}}\right)^2\frac{1-e^{-\frac{C_2}{\left(\frac{a}{a_{ref}}\right)^2}}}{C_3} \qquad (6)$$

Where C_1, C_2 C_3 are functions of the radius of curvature only, and therefore independent of a/a_{ref}. Because of this, the value of $\dot{Q}'$ is dependent on the geometrical constraint chosen from those in Table 1. This is clearly shown in the four plots of Fig. 5a, which represent the different trends of the heat flux as the radius of curvature is varied: when the side length is kept constant, the increase is moderate at best, with a maximum around 9% at $R_c \cong 0.25$. For all other cases the trend is monotonous and has its extreme value (a minimum for D_h= constant, a maximum in the two remaining instances) for the highest radius of curvature; it should be remarked that the cross section becomes circular then.

3.1.2. FG2b. The mirror case to FG2a is FG2b, where the minimum inlet temperature difference is sought for fixed heat flux, channel length and pumping power. Again, after some algebraic manipulations, Eq.(4) can be cast into the form of Eq. (7):

$$\Delta T_i'=\frac{1}{C_1\left(\frac{a}{a_{ref}}\right)^2}\frac{C_3}{1-e^{-\frac{C_2}{\left(\frac{a}{a_{ref}}\right)^2}}}=\dot{Q}'^{-1} \qquad (7)$$

As is to be expected, the trends shown in Fig. 5b are the reverse of the previous criterion, and the configuration yielding the best results (for maximum radius of curvature) is the one with constant heated perimeter. In this case the reduction is dramatic, with ΔT_i dropping to 50% of the reference value for $R_c \cong 0.2$ and decreasing further to 35% of the original for the maximum radius of curvature.

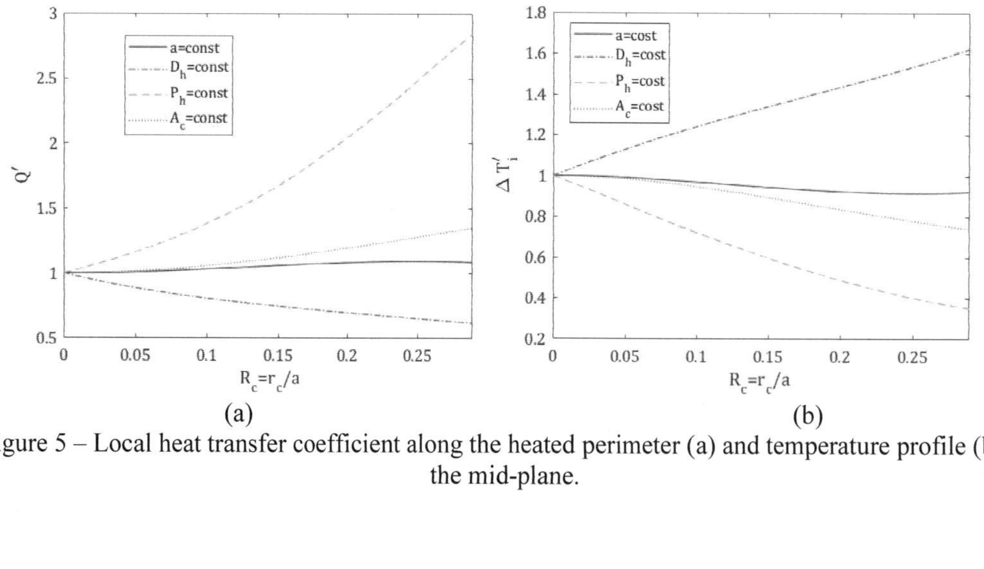

(a) (b)

Figure 5 – Local heat transfer coefficient along the heated perimeter (a) and temperature profile (b) on the mid-plane.

Journal of Physics: Conference Series 2177 (2022) 012039 doi:10.1088/1742-6596/2177/1/012039

3.1.3. FN1. The FN-type criteria have minimisation of either the channel's length or the pumping power as their goals. The FN1 criterion has the heat transfer, inlet temperature difference and pumping power as constraints, i.e. quantities that are held fixed throughout the optimisation and L as its objective function. Contrary to the previous cases, this leads to an expression which is not explicit in L', as shown in Eq. (8):

$$L'=\left[C_1\left(\frac{a}{a_{ref}}\right)^2\frac{1-e^{-C\frac{C_2}{\left(\frac{a}{a_{ref}}\right)^2}(L')^{\frac{3}{2}}}}{C_3}\right]^2 \qquad (8)$$

In this case, care must be exerted to limit the non-dimensional quantities to physically significant values, lest the results become negative, therefore meaningless. This was considered when choosing the reference values mentioned in the opening paragraph of this section.

Equations (6) and (8) only differ for $(L')^{\frac{3}{2}}$ in the exponential term; unsurprisingly, that the trend of L' is similar to that of $\dot{Q}$, as, results from the comparison of Figs. 5a and 6a. It is therefore concluded that only when the hydraulic diameter is fixed does smoothing the corner bring a benefit, and that this shortening of the channels is larger, the more the cross-section approaches the circular channel. Among the remaining constraints, geometries with fixed side exhibit little sensitivity to the radius of curvature, whilst a fixed heated perimeter causes a stronger increase in the channel's length the higher the value of R_c.

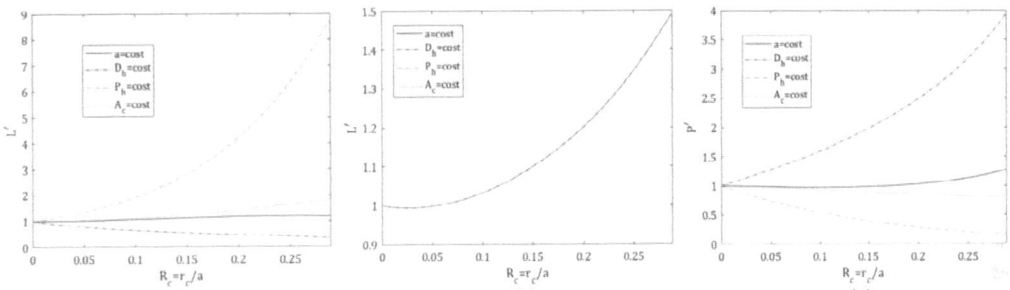

(a) (b) (c)

Figure 6 Change in channel length for the FN criteria: FN1 (a), FN2 (b) and FN3 (c).

3.1.4. FN2. Another criterion to minimize the channel's length is known as FN2; in this case the constraints are fixed heat flux, inlet temperature difference and mass flowrate, $\dot{Q}', \Delta T_{in}', \dot{m}' = 1$. When theys are applied, the objective function becomes Eq. (9):

$$L'=\frac{D_h'}{P_h'}\frac{1}{Nu'} \qquad (9)$$

Since both D_h' and P_h' are proportional to the a/a_{ref} ratio, L' is a function of Nu', which is dependent on R_c only: as a consequence, the four curves for the geometrical constraint all collapse onto one, as shown in Fig. 6b, which demonstrates how L' grows with the radius of curvature for R_c>0.1 and is insensitive to it for values below this threshold.

3.1.5. FN3. The criterion FN3 strives to minimize the pumping power, P', under the same constraints as FN2, i.e. fixed heat flux, inlet temperature and mass flowrate. Again, the objective function can be recast accounting for the constraints, into the form provided by Eq. (10)

$$P' = C_1 \left(\frac{a}{a_{ref}}\right)^{-4} \tag{10}$$

Where C_1 is a constant which depends on R_c only. The corresponding trend for the four geometrical constraints is shown in Fig. 6c and are essentially monotonous with the radius of curvature, with a sharp increase in the case of a fixed hydraulic diameter and a strong decrease in the case of fixed heated perimeter, the remaining cases having much smaller changes, with an increase in pumping power for fixed side and a decrease for fixed cross-sectional area.

4. Conclusions

In this work, the Poiseuille and Nusselt numbers for the fully-developed, laminar flow of a Newtonian fluid through a triangular microchannel with smoothed corners and uniform wall temperature have been computed, and the results used in the PECs for microchannel optimisation, using the heat flux and the pumping power as the objective functions. It has been demonstrated that smoothing the corners enhances the transport phenomena, which result in augmented momentum transport (hence Po goes from around 53 to 64 when $R_c \cong 0.10$ and remains almost constant thereafter) and heat transfer (with Nu increasing from around 2.5 to 3.64 at $R_c \cong 0.26$). The values of Po and Nu have then been used to analyse different PECs. It has been demonstrated that one further geometrical constraint must be applied in most cases if the objective functions (enhanced heat flux and pumping power) are to be optimised as a function of the radius of curvature only. For several criteria the results have been presented quantitatively and commented. It is hoped that the paper can contribute some general indications at design level and help comprehension of the transport phenomena in triangular microchannels.

References

[1] Ohadi M, Choo K, Dessiatoun S and Cetegen E E, *2013 Next Generation Microchannel Heat Exchangers*; (Springer New York, NY USA; Berlin/Heidelberg)
[2] Lorenzini M, Fabbri G and Salvigni S. *2007 Appl. Therm. Eng.* **27**(5-6) pp 969-975
[3] Kew P and Reay D, *2011 Appl. Therm. Eng.* **31**(5) pp 594-601
[4] Yang Y, Chalabi H, Lorenzini M, and Morini G L, *2014 Heat Tr. Eng.* **35**(2), pp 159–170
[5] Morini G, Lorenzini M, Colin S, and Geoffroy S, *2006 Proc. of ICNMM2006*, **A,** (ASME, New York, NY) pp. 411–418.
[6] Morini G L, Yang Y and Lorenzini M, *2012 Experimental Heat Transfer* **25**(3), pp 151–171
[7] Webb R L, *1994 Principles of Enhanced Heat Transfer*, (Wiley, New York)
[8] Ray S and Misra D, *2010 Int. J. of Th. Sci.* **49**, pp 1763–1775
[9] Chakraborty S. and Ray S, *2011 Int. J. of Th. Sci.* **50**, pp 2522–2535
[10] Saha S K, Ranjan H, Emani S M and Barthi A K, *2020 Performance Evaluation Criteria in Heat Transfer Enhancement* (Springer New York, NY USA; Berlin/Heidelberg)
[11] Petkov V M and Zimparov V D, *2013 Int.l Review of Chem. Engineering* **5**, pp 74–87
[12] Lorenzini M and Morini, *2011 Heat Transfer Engineering* **32(13-14)**, pp 1108–1116
[13] Lorenzini M and Suzzi N, *2016 Heat Transfer Engineering* **37** (13-14), pp 1096–1104
[14] Lorenzini M, Daprà I, and Scarpi G, *2017 Applied Thermal Engineering* **122**, pp 118–125
[15] Lorenzini M, *2013 Houille Blanche* **4**, pp 64–71
[16] Suzzi N and Lorenzini M, *2019 Int. J. of Th. Sci.,* **145** 106032
[17] Morini G L, Lorenzini M, Salvigni S, and Spiga M, *2006 Int. J. of Th. Sci.* **45**(10), pp 955–961
[18] Lorenzini M, *2020 Thermal Science Engineering Progress* **19** 100617
[19] Suzzi N and Lorenzini M *2021 Fluids* **1,** 1010000

Thermo-fluid dynamic resonance in cancer cells

Giulia Grisolia and Umberto Lucia

Dipartimento Energia "Galileo Ferraris", Politecnico di Torino, Corso Duca degli Abruzzi 24, 10129 Torino, Italy

giulia.grisolia@polito.it (Giulia Grisolia)

Abstract. In the third decade of XX century, Warburg pointed out that cancer cells follow a fermentative respiration process, as a consequence of a metabolic injury. In this paper, we consider this statement in the following way: any cell process requires energy, so, in the cell, a control of the energy conversion can represent a possible control of the cell processes. Engineering thermodynamics is the science that studies the conversion of energy into work. So, thermodynamics could represent a powerful approach to analyse of the energy conversion in the biosystems, for their control. Cells regulate their metabolisms by energy and mass (ions included) flows, and the heat flux occurs by the convective interaction with their environment. Here, we consider fluxes through the biosystems border, their shapes and the characteristic time of thermal interaction with the blood and water, in the cell environment. Moreover, just in relation to time, it is possible to consider the resonance phenomena. Resonance forces natural behaviours of systems, when a wave of a frequency, related to the characteristic time, income to a system. Here, we introduce the biothermodynamic characteristic frequency, which is the characteristic frequency of a biosystem, evaluated by a thermo-fluid dynamic approach, in order to control the fluxes through the cancer membrane, and to force it towards an optimal behaviour, by changing the concentrations of ions, inside and outside of the membrane itself. The result consists in a control of the cellular metabolic processes, and also of the energy available to cancer, for its growth. In this way, the cancer growth rate can be reduced.

Nomenclature

Latin letters

A	Area [m^2]
c	Specific heat [J kg^{-1}K^{-1}]
C	Capacitance [F]
G	Gibbs free energy [J]
H	Enthalpy [J]
Q	Heat [J]
Q	Thermal power [W]
r'	Volume/Area ratio [m]
R	Resistance [Ω]

S	Entropy [J K^{-1}]
t	Time [s]
T	Temperature [K]
V	Volume [m^3]

Greek letter

α	Convection coefficient [W m^{-2} K^{-1}]
χ	Ion concentration [mol m^3]
ϕ	Electric potential [V]
ρ	Mass density [kg m^{-3}]
τ	Characteristic time [s]

Symbols

pH	Potential of Hydrogen

Constants

R_u	8314 J $kg^{-1}K^{-1}$
F	96485 A s mol^{-1}

1. Introduction

In 2018, eighteen million people have contracted cancer worldwide, and more than four hundred million people are expected to contract cancer by 2025 [1].

In medicine and biology, cancer is defined as a disease caused by an uncontrolled division of abnormal cells, in a part of the body. So, it is an irregular behaviour of cells, and it is considered a disorder, caused by genetic or epigenetic alterations in the somatic cells. The unregulated growth of cells may also be spread in other parts of the body [2],[3].

A great number of behavioural risks allows the cancer to begin, even if its first cause is considered genetic [2]; indeed, human body contains about 10^{14} cells, and all kinds of cancer begin its growth, by changing some cells [4]. Cells in human body are specialized for different tasks, but they all have a similar structure, and use biochemical reactions for life and growth, proliferation, etc. At any cell division, mutation occurs. Mutation is the mechanism by which genes are copied twice, damaged or lost: mutation is a possible pathway for the cancer to start [2],[4]. But, up today, the trigger mechanism isn't still completely understood. What is clear is that interdisciplinary and holistic approach must be considered.

In any thermodynamic approach, a cell can be modelled as an open system which converts heat (its metabolic energy) into work. In this context, we must consider the Warburg results, as a starting point of our research. Indeed, in 1931, the Nobel laureate Otto Warburg showed that cancer cells, if compared with the normal ones, follow a different respiration pathway, which is characterized by a glucose fermentation, even when there is no lack of oxygen. He highlighted that the variation on the cancer cells metabolism was caused by a metabolic injury [5]. In healthy cells, mitosis and cell growth are phenomena which occur in synch, in order to preserve the cell size, during its replication. The Warburg result embodies the bioenergetic base to relate the lactic acid production, and the extracellular-intratumoral acidification, to the cancer growth and proliferation (including metastasis formation) [6]. Furthermore, the pH of the cytoplasmatic cells, and the extracellular environment, are directly linked to the cells membrane potential [7]. Comparing the polarization of quiescent cells with the one of the differentiated cells, the latter results hyperpolarized [8], and this hyperpolarization increases the outflow of some ions, such as Ca^{2+}, K^+, Zn^{2+}, etc. Consequently, recent researches have been addressed towards the adoption of specific channel inhibitors, as anticancer therapies [7],[9]. They have highlighted both the role of inhibition of proliferation in healthy and in neoplastic cells, and also the role of ion channels in the signalling network [10]. One of the key issues of carcinogenesis is the study of the derangement of mitosis and of the mitosis/apoptosis ratio [11],[12]. However, all the cellular processes are driven by fluxes of mass and energy; indeed, in Nature, all the spatial and the temporal structures are the result of the fluxes optimization, related to the cell living conditions [13].

But, energy conversion always causes irreversible production of entropy [14], as well known from the second law of thermodynamics, also in biological systems [15]. In this context, in the history of thermodynamics, two scientists introduced a new viewpoint for the evaluation of irreversibility, introducing a global analysis of a general thermodynamic system: the French physicist Louis Georges Gouy (1854-1926) [16] and the Slovak engineer and physicist Aurel Boleslav Stodola (1859-1942) [17], who, independently, proved that the loss of exergy can be evaluated by the product of the environmental temperature and the entropy generation [14], the entropy due to irreversibility. Today, this theorem represents out powerful way to evaluate irreversibility in real processes and systems; indeed, it is the basis of thermodynamic optimization in engineering design.

Resonance is defined as the phenomenon in which an oscillating system drives another system to oscillate with a greater amplitude at specific frequencies, the resonant ones. At these frequencies, also small harmonic forces can produce large effects, due to the storage of vibration energy. Recently, at a frequency of about 1 MHz, the neurons microtubules have been shown to oscillate spontaneously, generating the characteristic shape of the electrical oscillations of the brain, at the electroencephalographic signal of 4-40 Hz nested gestalts, called beat frequencies [18],[19]. This result highlights the fundamental role of resonance in cell behaviour.

In relation to the previous considerations on the energy conversion in a cell, we highlight that the fundamental thermodynamic aspect of life is the reorganization. It can be achieved by dissipating heat towards environment. In this context, cells can exchange heat with the fluid flows around them, and this can be studied by the thermo-kinetic lumped model. This model considers the cell as an isothermal system, which exchanges heat only at its surface [20],[21]. In this way, we try to control the behaviour of the cell by a thermo-magnetic resonance. This is a thermodynamic resonance, induced by an electromagnetic wave, at the frequency of the thermal characteristic time. In this paper, we review our previous theoretical and experimental results [22],[23],[24],[25].

2. The thermo-kinetic lumped biophysical model

Life is characterized by organizational and thermodynamic biophysical processes, based on the "natural" principle of the maximum conversion of available energy [21]. Indeed, the cellular biochemical reactions involve external metabolites to perform work (replication, transcription and translation), and related heat outflow. Thus, a cell exchanges energy and matter through its membrane [26], driven by endogenous electric fields [27] and source of other reactions within cells and tissues.

The Warburg result, related to the metabolic injury, highlights the fundamental role of the energy conversion in cells, and it allows us to introduce the thermo-kinetic lumped model in biophysical analysis of cells. Indeed, we can analyze the cell system as a black box, which is the usual thermal physical approach for the open systems, and consider the genetic regulation as a control system of the cell. Moreover, the genetic regulation cannot affect the thermodynamic balances. Cells exchange heat power with their environment and, the heat flux is related to their metabolism. This heat outflux occurs by convection with the fluids around any cell, and it results [13]:

$$\dot{Q} = \rho_{cell} V c_{cell} \frac{dT_{cell}}{dt} = \alpha A\left(T_{cell} - T_0\right) = \alpha \frac{V}{\langle r \rangle}\left(T_{cell} - T_0\right) \tag{1}$$

where ρ_{cell} is the mass density of the cell, V is its volume, c_{cell} is its specific heat, T_{cell} is the cell temperature, α is the convection coefficient, $A = V <r>^{-1}$ is the surface area of the cell, which varies during the phases of the cellular development, $<r> = V/A$ is the volume-area ratio, that is a characteristic parameter of the heat exchange through the cell membrane, and ($T_{cell} - T_0$) is the temperature difference between the cell temperature and the environment temperature (T_0). Now, introducing the first principle of thermodynamics:

$$\dot{Q} = \rho_{cell} V c_{cell} \frac{dT_{cell}}{dt} \tag{2}$$

into the Equation (1), it follows the differential equation:

$$\frac{1}{T_{cell} - T_0} \frac{d\left(T_{cell} - T_0\right)}{dt} = \frac{\alpha}{\rho_{cell} c_{cell}} \frac{1}{\langle r \rangle} \tag{3}$$

with

$$\frac{\alpha}{\rho_{cell} c_{cell}} \frac{1}{\langle r \rangle} = \frac{1}{\tau} \tag{4}$$

where τ is the characteristic time of the convective heat exchange [13].

The cell membrane is a double lipid layer that separates the cytoplasm from the cell environment. In real cell membranes, some proteins perform a function of channels, in order to allow the inflows and outflows of mass and ions. But, this function reduces the value of the electric resistance of the membrane. So, it is possible to develop an electric model of cell membranes. It is the *RC* circuit analogy, as represented in Figure 1.

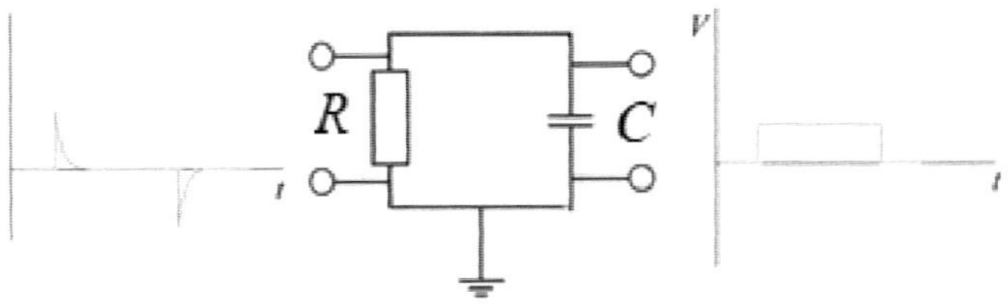

Figure 1. Electric analogy of a cell membrane. The cell membrane can be considered as a parallel *RC* circuit [28].

If the *RC* circuit analogy is considered [28], the current flows across the resistor of resistance *R* can be evaluated during the charge and the discharge of the capacitor of capacity *C* [29]:

$$I(t) = \frac{\Delta\phi}{R} e^{-t/\tau_{circ}} \tag{5}$$

where $I(t)$ is the current, $\Delta\phi$ is the value of electric potential applied to the capacitor, R is the value of the electric resistance, and $\tau_{circ} = RC$ is the characteristic time of the system (the circuit *RC*). If the circuit interacts with a harmonic wave, it presents a resonant frequency $\nu \sim \tau_{circ}^{-1} = (RC)^{-1}$ [29], to which is related the heat power $Q^{\cdot}$, dissipated by the Joule effect in the resistor, during the resonant phenomenon, as follows:

$$\dot{Q}(t)=\frac{R\,\phi_M^2}{\sqrt{R^2+\tau_{circ}^2\left(2\pi C\right)^{-2}}}\sin^2\left(\frac{2\pi}{\tau_{circ}}t\right) \tag{6}$$

where ϕ_M is the maximum value of the electric potential of the electromagnetic wave.

In accordance with the *RC* analogy for the cell membrane, we consider that, in the thermal interaction with its environment (Eq. (4)), the cell presents also a characteristic time. Consequently, we can argue that an electromagnetic wave could generate a resonant effect, at a frequency proportional to this time. So, we could obtain the maximum heat exchange of the cell towards the environment. In accordance with the Warburg approach, the result is a more efficient management of the energy conversion. Indeed, in this way, the cell doesn't need to storage energy in lipidic molecules, with the result of decreasing its proliferation.

So, we evaluate the characteristic time τ in Equation (4) and generate an electromagnetic wave at a frequency τ^{-1}, named characteristic frequency. The related heat flow results:

$$\dot{Q}=\frac{Q}{\tau}=T_0\frac{\Delta S_{cell}}{\tau} \tag{7}$$

where Q is the heat wasted by the cell towards its environment, T_0 is the environmental temperature and ΔS_{cell} is the entropy variation of the cell. As a consequence of the definition of Gibbs free energy G [30]:

$$\frac{\Delta G}{\tau}=\frac{\Delta H}{\tau}-T_0\frac{\Delta S_{cell}}{\tau}\Rightarrow \Delta G=\Delta H-\dot{Q}\,\tau \tag{8}$$

where H is the enthalpy, and T_0 is the environmental temperature. But, the cell membrane potential can be related to the Gibbs free energy by the following equation [31],[32],[33]:

$$\Delta G=\Delta\phi-2.3\frac{R_uT_0}{F}\Delta\text{pH} \tag{9}$$

where ϕ is the cell membrane electric potential, R_u is the universal gas constant, F is the Faraday constant, and pH is the potential of hydrogen. Consequently, the application of an electromagnetic wave, at the characteristic frequency, determines a change in the membrane electric potential:

$$\Delta\phi=\Delta H-\dot{Q}\,\tau+2.3\frac{R_uT_0}{F}\Delta\text{pH} \tag{10}$$

with a related variation in the ion concentration at the membrane forced in accordance to the relation [34]:

$$\chi_{out}=\chi_{in}\exp\left(\frac{\Delta\phi}{R_uT_0}\right) \tag{11}$$

where χ_{out} and χ_{in} are the concentrations of any ion species outside and inside of the cell membrane. Consequently, a regulation of the cell functions is related to the regulation of the membrane electric potential.

This effect, here theoretically derived, has been experimentally confirmed in Refs. [22],[23],[24],[25]. In Table 1, we summarize only some results, in order to show the effect of the exposure, of each cancer cell line, to its characteristic resonant frequency. The setup, used to develop the experimental verification of the model, consists of two independent couples of coaxial coils wound into a frame of cylindrical shape, with an outer radius of 8 cm and a distance between the two coaxial coil couples of 8 cm. The experimental setup was placed inside an incubator, in order to maintain the cells living conditions. The cells were placed in the center of this apparatus. The volume/area ratio of any cell line, involved in the experimental analysis, was measured by microscope imaging in order to evaluate the characteristic frequency. The cells were exposed to an electromagnetic wave, generated by the device itself, at their characteristic frequency. In Table 1, three independent cancer cell lines are presented, originated from different tumors, different from each other in morphology, growth rate and oncogene expression. In each experiment, three different cell lines were exposed all together to the electromagnetic field, but at the characteristic frequency of only one of them, in order to prove the resonance phenomenon. Moreover, control samples of each cell line were incubated, without being exposed to the electro-magnetic field. After four days, the proliferation of untreated and exposed cells was analyzed by a colorimetric assay. In each experimental condition, a reduced cell number (growth variation, compared with the control samples) was found only for the cell line for which the applied frequency was the characteristic one, for the line itself. So, all the experiments have confirmed the resonance phenomenon.

Table 1. ELF-EMF frequencies and the cellular biophysical quantities.
For each cell line, the evaluation of 30 cells in different fields was carried out. Cell resonant frequency was used to calculate the frequency of the forcing Extremely Low Frequency Electric and Magnetic Fields (ELF-EMF). The result is the decrease in the growth of the cancer cells [22].

Cell line	**Evaluated frequency** [Hz]	**Applied frequency** [Hz]	**Growth variation** [%]
MCF7 (breast cancer)	5.0 ± 0.7	5 ± 1	-22 ± 2
SKBR3 (breast cancer)	8.0 ± 2.0	8 ± 1	-18 ± 3
GTL16 (gastric carcinoma)	14.0 ± 3.0	14 ± 1	-24 ± 1

3. Discussion

The cell cycle has been pointed out to be monitored by a check system, which controls DNA integrity, before the cell has any transition to its next living phase. In eukaryotic cells, the main control processes occur at the G1/S transition, in late S (DNA synthesis) phase, at mitosis (M) entry and at the metaphase to anaphase transition. All the processes are controlled by the cyclin-dependent kinases, which is regulated by the oscillatory expression of G1 and G1/S-cyclins, S-cyclins, and M-cyclins. The anaphase-promoting complex/cyclosome (APC/C) triggers the transition between metaphase to anaphase. The entry into the cell cycle from a quiescent (G0) phase is stimulated by mitogens. The exit from mitosis can bring to differentiation, apoptosis, or return to quiescence [8]. These mechanisms result altered in neoplastic cells.

All the biochemical processes require energy, and any energy conversion process generates outflows of energy, by virtue to the second law of thermodynamics. Thus, it is possible to analyse the cells system behaviour, by following a thermodynamic approach, considering the inflows and outflows of energy and masses, ions included.

Consequently, it follows that the surface phenomena are transduced by ion channels and energy transfer, to the protein machineries in the cytosol, by combining the sensitivity to electrical and chemical signals - proper of cooperative allosteric proteins - and the ability to exploit the energy release, that occurs during the exchanges of mass and energy across the membrane, which amplify the biochemical signalling. In cancer cells, an alteration on some processes related to energy and ion channelling occurs, reducing their proliferation control.

We can control the variation of the biochemical quantities, and consequently the behaviour of the cancer forcing heat exchange, and the related concentrations of ions inside and outside the membrane.

In summary, a characteristic frequency wave interacts with the cancer system with the effect of forcing ions fluxes, which change the membrane electric potential and the pH, conditioning the biochemical reactions, and, consequently, forcing them towards a normal behaviour.

4. Conclusions

In this paper we have introduced a thermodynamic approach to cancer, based on the energy balance related to the Warburg results on metabolism. The basis of this approach is the electric analogy of the living cell membrane and the resonant effect in relation to thermal outflow from the cell towards its environment. The model proposed has also been experimentally confirmed by *in vitro* experiments.

The result could be useful as support to present anticancer therapies because it allows the control of the cancer growth, improving the effect of the anticancer therapies.

The limit of this approach is the knowledge of the volume/area ratio of the cell: this is a limit, because it is difficult to obtain this information in clinical applications and it is not unique for all the cells, but its value is in a range for any cancer cell line. However, this quantity is fundamental to obtain the characteristic frequency of the therapeutic electromagnetic field.

References

[1] Bray F, Ferlay J, Soerjomataram I, Siegel R L, Torre L A and Jemal A 2018 Global cancer statistics 2018: GLOBOCAN estimates of incidence and mortality worldwide for 36 cancers in 185 countries. *CA: A Cancer Journal for Clinicians* **68** 394
[2] Saini A, Kumar M, Bhatt S, Saini V and Malik A 2020 Cancer causes and treatments *International Journal of Pharmaceutical Sciences And Research* **11** 3109
[3] Sitki-Copur M 2019 State of cancer Research around the globe *Oncology Journal* **33** 181
[4] Stewart B W and Wild C P 2014 *World cancer report 2014* (Lyon: International Agency for Research on Cancer)
[5] Warburg O, Wind F and Negelein E 1927 The metabolism of tumors in the body Journal of General Physiology **8** 519
[6] Alfarouk K O, *et al.* 2014 Glycolysis, tumor metabolism, cancer growth and dissemination. A new pH-based etiopathogenic perspective and therapeutic approach to an old cancer question *Oncoscience* **1** 777
[7] Binggeli R and Cameron I L 1980 Cellular potentials of normal and cancerous fibroblasts and hepatocytes *Cancer Research* **40** 1830
[8] Becchetti A 2011 Ion channels and transporters in cancer. 1. Ion channels and cell proliferation in cancer *American Journal of Physiology-Cell Physiology* **301** C255
[9] Tameire F, Verginadis I I, Koumenis C 2015 Cell intrinsic and extrinsic activators of the unfolded protein response in cancer: Mechanisms and targets for therapy Seminars in Cancer Biology **33** 3
[10] Arcangeli A, Crociani O, Lastroioli E, Masi A, Pillozzi S and Becchetti A 2009 Targeting ion channels in cancer: a novel frontiers in antineoplastic therapy *Current Medical Chemistry* **16** 66
[11] Comen E A, Norton L and Massague J 2011 Breast Cancer Tumor Size, Nodal Status, and Prognosis: Biology Trumps Anatomy *Journal of Clinical Oncology* **29** 1
[12] Lucia U, Ponzetto A, Deisboeck T S 2015 A thermodynamic approach to the 'mitosis/apoptosis' ratio in cancer *Physica A* **436** 246
[13] Bejan A 2000 *Shape and Structure, from Engineering to Nature* (Cambridge: Cambridge

University Press)
[14] Bejan A 2006 *Advanced Engineering Thermodynamics* (New York: John Wiley)
[15] Katchalsky A and Curran P F 1967 *Nonequilibrium Thermodynamics in Biophysics* (Cambridge: Harvard University Press)
[16] Picard E 1926 Annonce la mort de M. Georges Gouy *Comptes Rendus des Sèances de l'Acadèmie des Science* (*Paris*) **182** 293
[17] Martin J, Klein A J and Kox R S (Eds) 1993 *The Collected Papers of Albert Einstein, Volume 5: The Swiss Years: Correspondence, 1902-1914* (Princeton: Princeton University Press)
[18] Ghosh S, Sahu S and Bandyopadhyay A 2014 Evidence of massive global synchronization and the consciousness *Physics of Life Review* **11** 83
[19] Sahu S, Ghosh S, Fujita D and Bandyopadhyay A 2014 Live visualizations of single isolated tubulin protein self-assembly via tunneling current: effect of electromagnetic pumping during spontaneous growth of microtubule *Scientific Reports* **4** 7303
[20] Lucia U and Grisolia G 2020 Thermal resonance and cell behaviour *Entropy* **22** 774
[21] Lucia U and Grisolia G 2020 Resonance in thermal fluxes through cancer membrane *Atti dell'Accademia Peloritana dei Pericolanti* **98** SC1
[22] Lucia U, Grisolia G, Ponzetto A, Bergandi L and Silvagno F 2020 Thermomagnetic resonance affects cancer growth and motility *Royal Society Open Science* **7** 200299
[23] Bergandi L, Lucia U, Grisolia G, Granata R, Gesmundo I, Ponzetto A, Paolucci E, Borchiellini R, Ghigo E and Silvagno F 2019 The extremely low frequency electromagnetic stimulation selective for cancer cells elicits growth arrest through a metabolic shift *BBA - Molecular Cell Research* **1866** 1389
[24] Lucia U, Grisolia G, Ponzetto A and Silvagno F 2017 An engineering thermodynamic approach to select the electromagnetic wave effective on cell growth *Journal of Theoretical Biology* **429** 181
[25] Lucia U, Grazzini G, Montrucchio B, Grisolia G, Borchiellini R, Gervino G, Castagnoli C, Ponzetto A and Silvagno F 2015 Constructal thermodynamics combined with infrared experiments to evaluate temperature differences in cells *Scientific Reports* **5** 11587
[26] Caplan S R and Essig A 1983 *Bioenergetics and Linear Nonequilibrium Thermodynamics. The Steady State* (Cambridge: Harvard University Press)
[27] Bustamante C, Chemla Y R, Forde N R and Izhaky D 2004 Mechanical processes in biochemistry *Annual Review of Biochemistry* **73** 705
[28] Johnston D and Miao-Sin Wu S 1994 *Foundations of Cellular Neurophysiology* (Cambridge: MIT Press)
[29] Purcell E M and Morin D J 2013 *Electricity and Magnetism* (Cambridge: Cambridge University Press)
[30] Ansermet j-P and Brechet S D 2019 *Principles of Thermodynamics* (Cambridge: Cambridge University Press)
[31] Morowitz H J 1979 *Energy flow in Biology* (Woodbridge: Ox Bow Press)
[32] Santillàn M 2014 *Chemical Kinetics, Stochastic Processes, and Irreversible Thermodynamics* (Berlin: Springer)
[33] Sears F W and Salinger G L 1975 *Thermodynamics, Kinetic Theory, and Statistical Thermodynamics* (New York: Norosa Addison-Weslay)
[34] Denbigh K G 1956 The second-law efficiency of chemical processes *Chemical Engineering Science* **6** 1

Experimental Analysis of Steam Condensation Heat Transfer Coefficient in Water Pool at Sub-Atmospheric Pressure

A. Pesetti, C. Merello, G. Giambartolomei, M. Raucci, D. Aquaro

Department of Civil and Industrial Engineering, DICI, University of Pisa - Largo Lucio Lazzarino 2, 56126 Pisa (Italy)

alessio.pesetti@for.unipi.it

Abstract. The experimental analysis of steam direct contact condensation in a water pool at sub-atmospheric condition was carried out in relevant configurations for the International Thermonuclear Experimental Reactor (ITER) Vacuum Vessel Pressure Suppression System, during a postulated Ingress of Coolant Event category IV accidental scenario. This transient accident was experimentally simulated in a reduced scale (1:22) facility at University of Pisa, adopting a defined scaling law. The steam jet plumes were video recorded and image analysis was performed at stable and interfacial oscillation condensation regimes, providing steam plume average length and interfacial area. The average heat transfer coefficient was therefore computed and correlated to the condensation driving potential. Also, the steam jet length divided by orifice diameter (*L/D*) was characterised as function of driving potential. Empirical correlations available in literature at atmospheric conditions were modified for evaluating vacuum conditions peculiarities. The remarkable higher dependence of heat transfer coefficient (*HTC*) to driving potential parameter, at vacuum conditions, was highlighted by the driving potential exponent equal to 1.2 instead of 0.04 at ambient pressure. The obtained *L/D* and *HTC* correlations are able to predict with ± 15% and ± 30% of error the corresponding experimental data, respectively.

1. Introduction

The pressurization of the International Thermonuclear Experimental Reactor (ITER) Vacuum Vessel (VV) [1] is limited to the maximum value of 0.15 MPa, due to Ingress of Coolant Event (ICE) (i.e., Loss of Coolant Accident, LOCA) from in-vessel cooling systems and Loss of Vacuum Accident (LOVA), and represents a crucial safety issue for the ITER plant. The pure (ICE) or mixed with air (ICE and LOVA) steam released into the VV need to be discharged into the Vacuum Vessel Pressure Suppression System (VVPSS), working at sub-atmospheric pressure conditions.

The steam Direct Contact Condensation (DCC) at pressure lower than atmospheric value is not sufficiently studied and documented in open literature, in which only condensation at atmospheric pressure is investigated [2]-[11]. In this scenario, the DCC effectiveness was investigated in an extended experimental activity carried out at Department of Civil and Industrial Engineering (DICI) of the University of Pisa, aimed to contribute to fill the literature gap and supporting a knowledge advancement and dissemination. An experimental Small Scale Test Facility (SSTF) was arranged to experimentally simulate, in a Reduced Condensation Tank (RCT) of 4.5 m^3, steam condensation in a water pool at sub-atmospheric pressure (up to 5 kPa abs) [12]-[14]. In particular, the postulated transient scenario occurring in one (Small LOCA Tank, SLT) of the four tanks of the VVPSS, as consequence of the ICE cat. IV event (multiple double ended break of >100 first wall cooling pipes), was experimentally simulated and described, for characterizing the occurring condensation regimes at vacuum conditions. The experimental boundary conditions were defined adopting a scaling law, to coherently reproduce the postulated transient event in the SLT.

Aiming to characterise the heat transfer coefficient (*HTC*) in condensation regimes along the whole transient, a wide set of images of the steam plume were elaborated at several transient instants. The acquired images were treated for pointing out the pure steam boundary profile and to compute its averaged length and steam-water heat transfer area. The *HTC* and steam jet length were determined and compared with correlations currently available in literature.

This activity is part of an extensive experimental and numerical University of Pisa effort for supporting the ITER VVPSS design and verification [15]-[19] and for literature improvement about DCC at sub-atmospheric conditions.

2. ITER VVPSS

The VVPSS of ITER machine aims to mitigate the postulated pressurization event of the VV, discharging pressurizing steam and air from VV into four Vapour Suppression Tanks (VSTs) of 100 m^3 each partially filled with water (left of Figure 1). Precisely, three Large LOCA tanks (LLTs) have 60 m^3 each of liquid water and one Small LOCA Tank (SLT) has 40 m^3 of pool, working with a cover gas at sub-atmospheric conditions, for allowing steam discharge from VV at pressure lower than the maximum allowed value of 0.15 MPa. A postulated ICE cat. IV transient events in ITER facility was numerically simulated [1] for the safety function evaluation of the suppression system. The piping lines connecting ITER VV to the VVPSS were designed to discharge steam only into the SLT (centre of Figure 1 and orange line in Figure 1 right) in case of Small LOCA event, whereas in case of larger LOCA, pressure is suppressed by condensation also in the three LLTs (blue line in Figure 1 right). The adopted sparger in all the VSTs has 1000 holes having a diameter of 10 mm, disposed on 20 rows of 50 holes each.

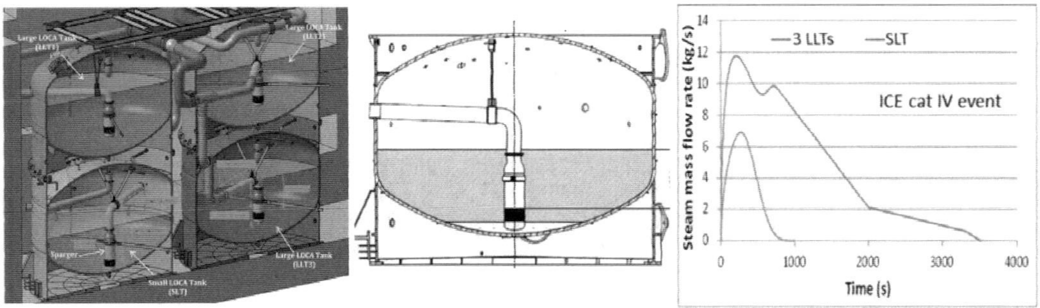

Figure 1: ITER VVPSS (left), SLT (centre) and postulated ICE cat. IV transient scenarios in the Small (orange) and Large (blue) LOCA Tanks (right)

3. Small Scale Test Facility and Scaling laws

The steam DCC at vacuum conditions, relevant for the ICE cat. IV scenario in SLT of ITER VVPSS, was experimentally investigated in the SSTF. Based on about 400 tests [13][14] carried out in the SSTF, proper scaling laws were defined, already published [15] and here summarized. It was found out that condensation regime is mainly function of water pool temperature (T_W), water pool pressure in front of the injection holes (P_W, i.e., downstream pressure) and steam mass flux (G_S, mass flow rate per unit of hole area). $G_S=q_S/A$ (kg s^{-1} m^{-2}) with q_S in kg/s is the steam mass flow per hole and A (m^2) is the flowing area of one hole (10 mm of diameter).

Defining the geometrical scale factor S as the ratio of VST to RCT volume, $S=V_{VST}/V_{RCT}$=100/4.5 is about 22. The volume ratio of cover gas to water is the same in VST and RCT. In this way, RCT has a reduced water pool mass (M_W) and consequently, the whole mass of injected steam (M_S) has to be scaled by S (M_S/S). The steam mass flow rate per hole q_S is conserved (to obtain the actual steam plume condensation regime) and for considering the different number of sparger holes, the amplified time factor $K=(N_{VST}/N_{RCT})/S$ was adopted (with N_{VST} the holes number of the sparger of VSTs equal to 1000 and N_{RCT} is the holes number of the sparger of RCT equal to 6). This ensures that the same water temperature increase (ΔT_W) and downstream pressure (conserving cover gas pressure and water head above the injection hole) are obtained in the RCT and VST, but at different time instants ($K\ t$ and t, respectively).

Assuming as reference transient scenario the postulate ICE cat. IV event [1] in the SLT (shown in orange in Figure 1 and on the left of Figure 2), with a sparger having 1000 holes the maximum mass flow rate of about 7 kg/s (i.e., 7 g/s per hole) was numerically predicted by ITER Organization adopting MELCORE code and provided as boundary conditions to UNIPI experiments. The scaled transient in the RCT, applying the abovementioned scaling laws, is shown on the right of Figure 2, in which the steam mass flow rate per hole is conserved and the time is amplified by $K=(N_{VST}/N_{RCT})/S=(1000/6)/22=7.58$. Indeed, the transient lasting about 1000 s is amplified to a time trends duration of 7580 s.

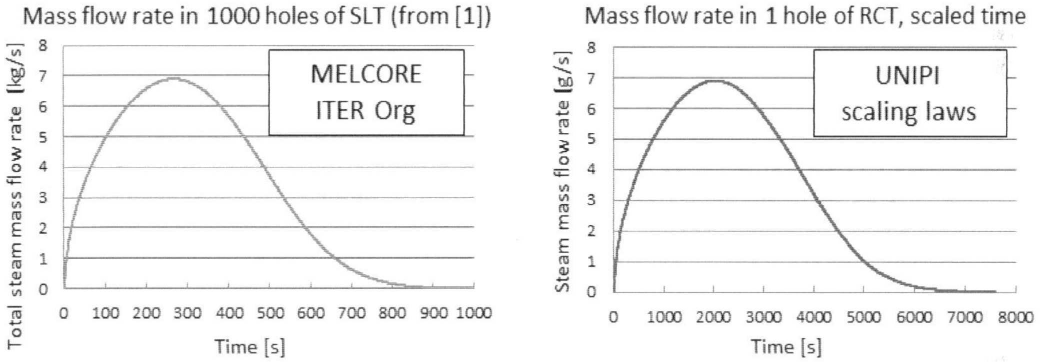

Figure 2: ICE cat. IV postulated transient scenario in the SLT (left, from [1]) and in the RCT with scaled duration (right).

4. Small LOCA transient

The postulated ICE cat. IV transient occurring in the SLT (left of Figure 2) was experimentally carried out in the RCT of the SSTF, adopting the mentioned scaling laws (right of Figure 2). The imposed steam mass flow rate time trend per hole (on the right of Figure 2) is shown in blue on the left of Figure 3 with the low oscillating experimental red mass flow rate time trend. In the same figure, the average water temperature and downstream pressure in front of the injecting holes are depicted in blue and red for the numerical and experimental data, respectively. A good agreement was achieved. On the right of Figure 3 the numerical (blue) and experimental (red) transient scenario is depicted in the condensation regime map, with average water temperature and steam mass flux divided by downstream pressure on ordinate and abscissa axis, respectively. In Figure 3 the six main condensation regimes are highlighted [14][15]: Chugging (C), Transitional Chugging (TC), Condensation Oscillation (CO), Bubbling Condensation Oscillation (BCO), Interfacial Oscillation Condensation (IOC), Stable Condensation (SC).

Figure 3 shows C, TC and CO occurring below 114 s (at lower mass flow rate), SC exists between 114 and 2888 s, IOC appears in the time interval 2888-4476 s, BCO between 4476 and 5000 s and C after 5000 s up to the end of the test.

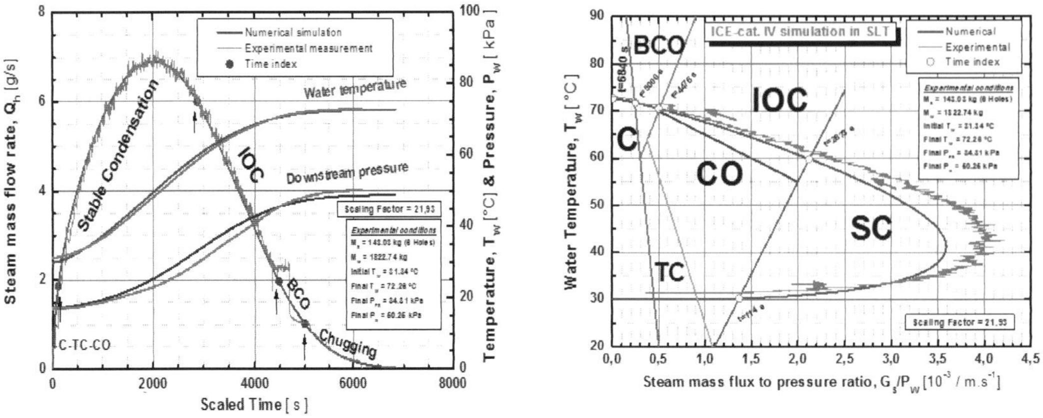

Figure 3: Experimental (red) and numerical (blue) mass flow rate, average water temperature and downstream pressure time trend of the postulate Small LOCA event (left) and transient scenario in the condensation regime map (right)

5. Results analysis

For studying and understanding the sub-atmospheric steam DCC, the steam-water heat transfer coefficient (*HTC*) was experimentally evaluated at several time instants of the transient test by Eq. 1.

$$HTC = \frac{q_S\,(H_S - H_F)}{A_{INT}\,(T_S - T_W)} \qquad (1)$$

Where q_S is the steam jet mass flow rate (kg/s); A_{INT} is the interfacial steam-water area (m^2); H_S and H_F are the steam and water enthalpy (J/kg) respectively; T_S and T_W are the steam and water temperature (K), respectively.

The A_{INT} parameter was evaluated at several steam mass flow rate (i.e., condensation regimes), analysing the recorded videos by a lateral camera GoPro HERO 4 with an open-source software for image elaboration (ImageJ-National Institute Health-USA). The camera has a sensor 1/2.3 inch and standard f stop (aperture) f/2.8, the focal length is 2.92 mm and acquired full HD 1080p videos at 120 fps. The correct and precise geometrical acquisition of the steam plumes is guaranteed by dimensional references. Indeed, in each video of the sparger, the acquisition calibration was performed evaluating the mm corresponding to a pixel on the basis of the known dimensions of holes diameters and interrexes. It proved that the video registered area of interest, recording six steam plumes, has linear characteristic and fish-eye effect or other deformation did not occur. The higher uncertainty lies in the axisymmetric assumption of the steam plume and the evaluation of its external surface as a surface of revolution.

The experimental run performed with the SSTF allowed to highlight the condensation regimes occurring during the transient evolution: initially, Chugging (C), Transition Chugging (TC) and Condensation Oscillation (CO) occured, then increasing the steam mass flow rate Stable Condensation (SC) appeared and successively Interfacial Oscillation Condensation (IOC), Bubbling Condensation Oscillation (BCO) took place and at the end of the transient (lower steam mass flow rate) Chugging occurred again. These condensation regimes are highlighted in Figure 4.

The entire transient test was discontinuously video recorded by 21 files (labelled from 7606 to 7626). The interfacial area was computed on the basis of the image analysis carried out at five different instants in SC (orange points in Figure 4), one time at IOC (green point in Figure 4) and an attempt was also done in C condition (red point in the same figure). Closely to these points, the label of the corresponding video file is written in the same figure. The chosen time instants are listed in Table 1. The video elaboration consisted to: 1) extract 30 frames (adopting FFMPEG software) from the 120 frames of the analysed instant (i.e. second); 2) apply the CLAHE filter (contrast limited adaptive histogram) increasing local contrast to each one of the 30 frames; 3) perform the median of these 30 images; 4) convert it from RGB colour image (with 8 bit for each one of the three channels) to 32 bit grayscale image, necessary to apply a 5) denoising R.O.F. filter (CLAHE produces a limited noise); and finally 6) use the EDGE filter to find the boundary of the steam plumes. In particular, the video analysis was carried out on 1 s of the transient at the 1st, 2nd, 3rd, 4th and 8th minute of the file 7606, at the 20th second of file 7616 and at the 18th second of video file 7622, as shown in Table 1. In the third column of this table the time instant of the elaborated images is referred to the whole transient time.

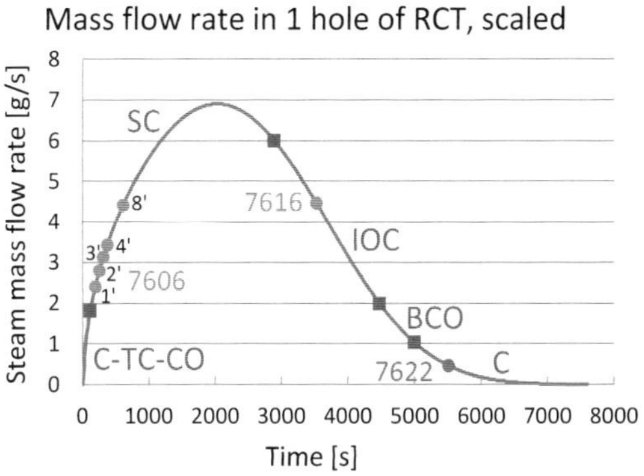

Figure 4: Condensation regimes in the ICE cat. IV event in the Small LOCA tank and time instants of image elaborations (orange and green) in the RCT

Table 1: Identification of the transient time instants of the analysed videos

Video file label	Time instant of the video file	Total transient time instant [s]
7606	1 min	192
	2 min	252
	3 min	312
	4 min	372
	8 min	612
7616	20 s	3502
7622	18 s	5700

In Figure 5 the plumes image is shown on the left, in the centre is provided the image editing result highlighting the plume edge and the blue points on the right provides the profile coordinates of the six plumes vertical cross section, for the 4th minute of the video file 7606 (Stable Condensation). For computing the average interfacial area (A_{INT}) of each plume, the asymmetric vertical dotted profile was divided by a horizontal axis in an upper- and a lower-profile, each one was rotated (360°) around this horizontal axis providing two surfaces of revolution, the average value of which provided the pursued interfacial area. The length of the plume (L) was also measured. Image analysis for IOC regime is shown in Figure 6. The abrupt oscillations in chugging conditions of steam-liquid water interface impeded to acquire an average profile and the heat transfer area was not possible to be evaluated, for this reason the 7622 data were excluded from the performed analysis. The most relevant measured parameters are listed in Table 2. The image analysis provided the averaged plume length (L) and interfacial area (A_{INT}) for the 6 steam jets at 6 transient time instants (5 of 7606 and 1 of 7616). The *HTC* of each steam plume was determined by Eq.1, assuming the steam mass flow rate per hole (q_S) as the total steam mass flow rate, measured by a vortex flow meter, distributed on the 6 holes proportionally to the square root of the pressure drop across the orifices. The latter was computed as $P_S - (P_W + \Delta P)$, with P_S the steam pressure inside the sparger, P_W the water pressure in front of the upper (first) orifice and ΔP the pool water head between two sparger holes.

Table 2: Main experimental parameters of the two condensation regimes considered

Test n.		T_W (°C)	P_W (kPa abs)	H_W (m)	q_{TOT} (g/s)	P_S (kPa abs)	T_S (°C)
7606	1 m	30.49	17.21	0.48	11.85	23.37	131.5
	2 m	30.73	17.23	0.48	14.22	25.8	132.4
	3 m	31.05	17.18	0.48	17.36	27.95	132.5
	4 m	31.43	17.27	0.48	18.61	29.45	132.3
	8 m	33.27	17.71	0.48	26.16	38.49	132.2
7616	18 s	66.32	36.22	0.48	27.49	52.56	111.8

Journal of Physics: Conference Series 2177 (2022) 012041 doi:10.1088/1742-6596/2177/1/012041

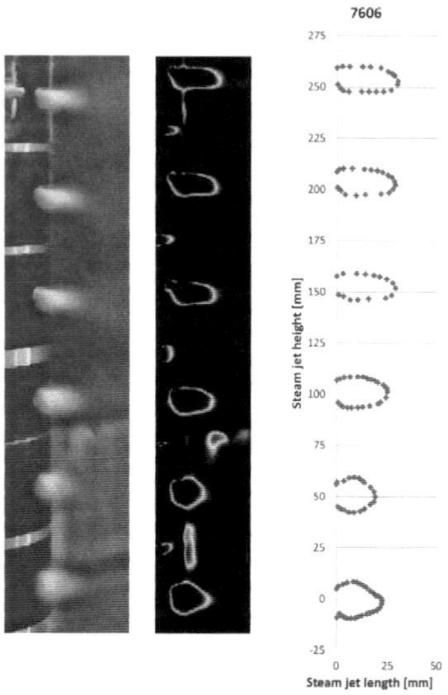

Figure 5: Median plume shape at 4th minute of video file 7606 (372 s of the whole transient) (left), average plume edge (centre) and dotted plume profile acquired (right)

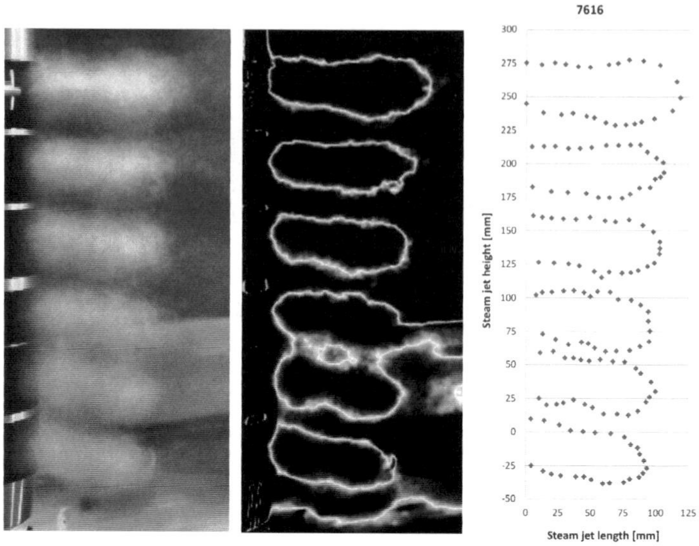

Figure 6: Median plume shape at 20 s of video file 7616 (3502 s of the whole transient) (left), average plume edge (centre) and dotted plume profile acquired (right)

Both *HTC* and *L/D* (with D the steam injection orifice diameter of 0.01 m) experimental data are plotted against the condensation driving potential *B*, as shown in Figure 7 and Figure 8, respectively. *B* is defined as the water specific heat capacity (c_P at water pool temperature and pressure) times the water subcooling ($\Delta T_{SUB}=T_{SAT}(P_W) - T_W$) divided by the latent heat $H_{GF}(P_W)$ at water pressure.

$$B = \frac{c_P \, \Delta T_{SUB}}{H_{GF}} \tag{2}$$

The measured values of videos 7606 and 7616 are plotted with already published *HTC* and *L/D* data [14], related to single steam plume in stable condensation regime (8 blue points in Figure 7 and Figure 8, respectively). Each point in the below figures represents a steam plume, an overall number of 44 plumes is therefore analysed in this work.

Several empirical correlations available in literature [2]-[4] provide *L/D* and *HTC* parameters for direct contact condensation at atmospheric pressure. The aim of this work is to elaborate and assess an analogous correlation for sub-atmospheric condensation (i.e., vacuum conditions), based on the abovementioned experimental data measured at Lab. Guerrini of the University of Pisa. The available data set [14] was expanded adding 30 datapoint (7606) at *B* about equal to 0.5 (orange in the above figures) and 6 datapoint (7616) at lower *B* (i.e., about 0.15, green in the same figures). In particular, the *HTC* correlations of Hwan [4] and Yeon [3] and the *L/D* correlations of Chul [2] and Yeon [3] were taken as reference, having the closer range of validity (even if not precisely suitable) to the performed experimental tests.

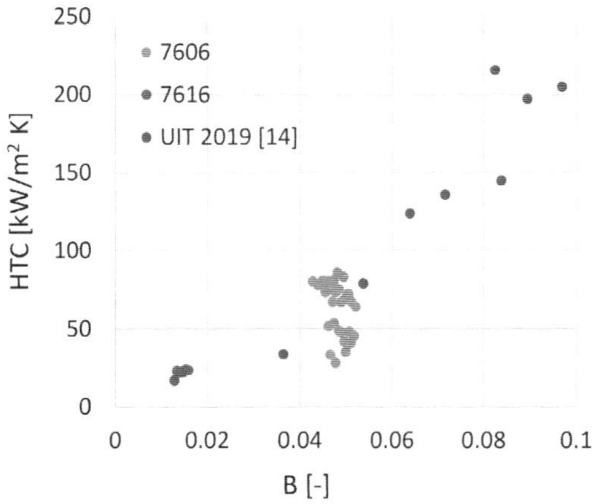

Figure 7: Average *HTC* of the steam jet plumes in transient 7606 (orange), 7616 (green) and data from [14] (in blue) as function of the condensation driving potential

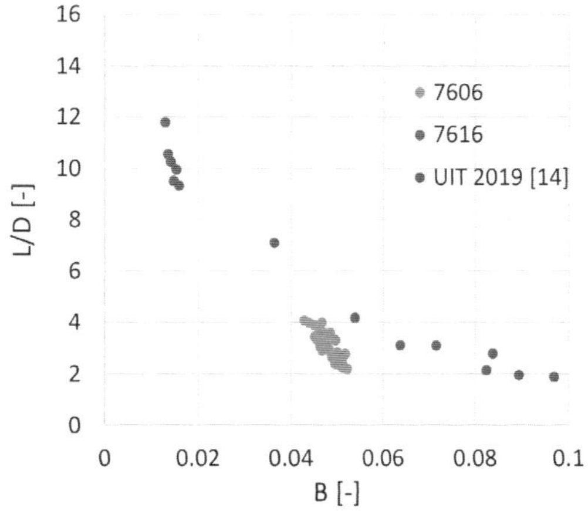

Figure 8: *L/D* of the steam jet plumes in transient 7606 (orange), 7616 (green) and data from [14] (in blue) as function of the condensation driving potential

$$HTC_{AVE} = 1.4453\ c_P G_m B^{0.03587} \left(\frac{G}{G_m}\right)^{0.13315} \quad \text{Hwan [4]} \qquad (3)$$

$$HTC_{AVE} = 1.3583\ c_P G_m B^{0.0405} \left(\frac{G}{G_m}\right)^{0.3714} \quad \text{Yeon [3]} \qquad (4)$$

$$\frac{L}{D} = 0.503\ B^{-0.70127} \left(\frac{G}{G_m}\right)^{0.47688} \quad \text{Chul [2]} \qquad (5)$$

$$\frac{L}{D} = 0.5923\ B^{-0.66} \left(\frac{G}{G_m}\right)^{0.3444} \quad \text{Yeon [3]} \qquad (6)$$

G_m is the critical steam mass flux (kg/m^2 s) computed by: $G_m = C_D \left[P_S \rho_S \gamma \left(\frac{2}{\gamma+1}\right)^{\frac{\gamma+1}{\gamma-1}}\right]^{0.5}$

Where C_D is the efflux coefficient ranging from 0.8 to 1 and assumed to be equal to 1, P_S and ρ_S are the pressure and density of the steam respectively and γ is the ratio between the specific isobaric and isochoric heat capacity of steam.

Aiming to determine correlations suitable for steam condensation at vacuum conditions, having parameters dependence coherent to the abovementioned correlations, the *HTC* and *L/D* experimental data were normalized for evaluating the dependence only on *B*, dividing by $c_P G_m \left(\frac{G}{G_m}\right)^x$ and $\left(\frac{G}{G_m}\right)^y$, respectively. In Figure 9 and Figure 10 the normalized *HTC* experimental data are shown with reference to the Hwan [4] (x = 0.13315) and Yeon [3] (x = 0.3714), respectively, moreover the corresponding trend line is provided.

The same analysis was carried out for *L/D* with reference to correlations (5) (y = 0.47688) and (6) (y = 0.3444), the normalized *L/D* experimental data (only dividing by $(G/G_m)^y$) are shown in Figure 11 and Figure 12, respectively.

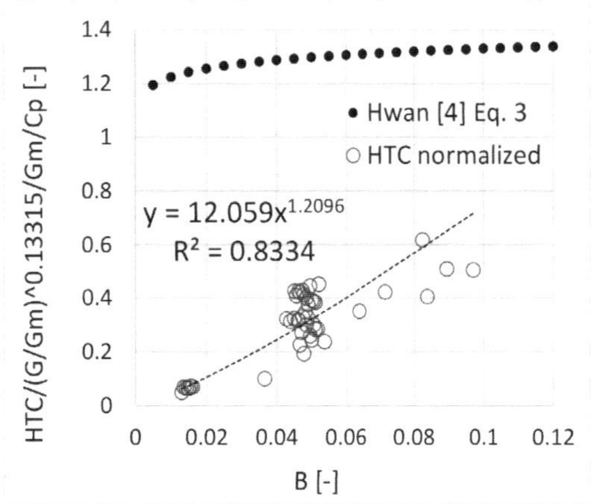

Figure 9: *HTC* normalized on the basis of Eq. 3 and trend line, underlining *HTC* dependence to only *B* parameter

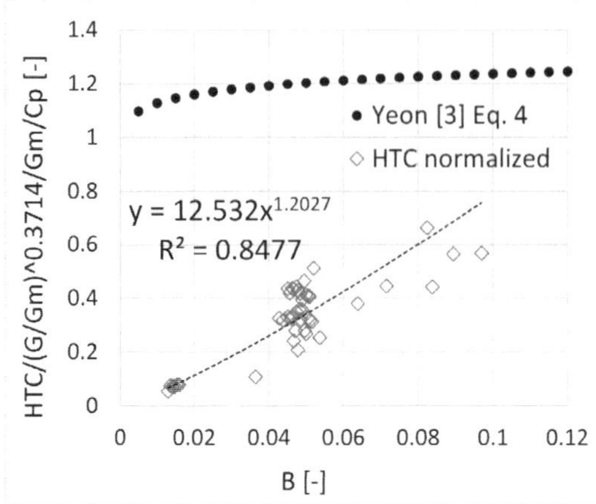

Figure 10: *HTC* normalized on the basis of Eq. 4 and trend line, underlining *HTC* dependence to only *B* parameter

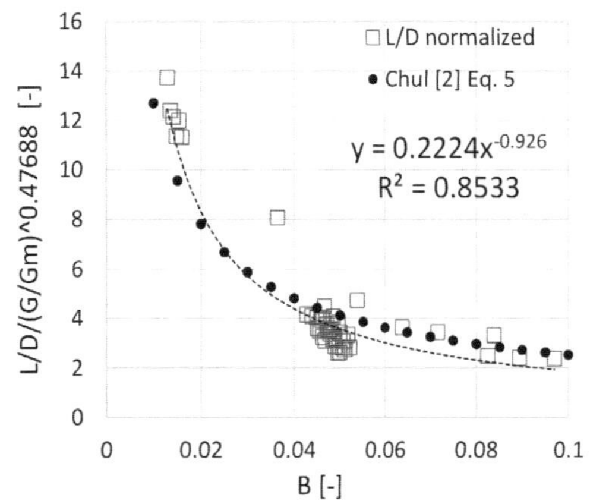

Figure 11: *L/D* normalized on the basis of Eq. 5 and trend line, underlining *L/D* dependence to only *B* parameter

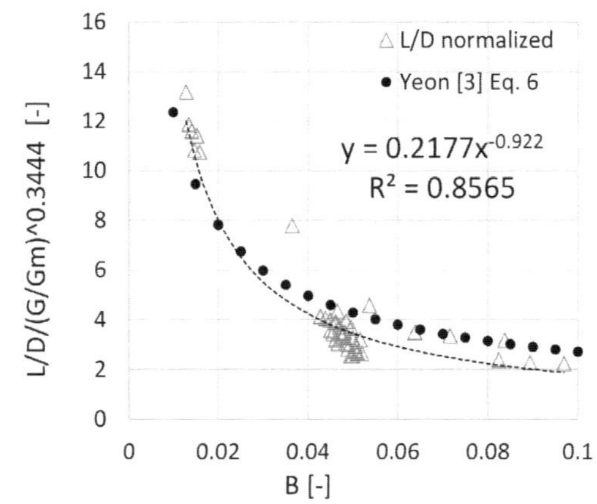

Figure 12: *L/D* normalized on the basis of Eq. 6 and trend line, underlining *L/D* dependence to only *B* parameter

The obtained trend line equations allowed to compose modified correlations for DCC at sub-atmospheric conditions. They are listed below, labelled by the original correlation number followed by "mod" (modified).

$$HTC_{AVE} = 12.059\ c_P G_m B^{1.2096} \left(\frac{G}{G_m}\right)^{0.13315} \tag{3mod}$$

$$HTC_{AVE} = 12.532\ c_P G_m B^{1.2027} \left(\frac{G}{G_m}\right)^{0.3714} \tag{4mod}$$

$$\frac{L}{D} = 0.2224\ B^{-0.926} \left(\frac{G}{G_m}\right)^{0.47688} \tag{5mod}$$

$$\frac{L}{D} = 0.2177\ B^{-0.922} \left(\frac{G}{G_m}\right)^{0.3444} \tag{6mod}$$

It is possible to highlight the significantly higher dependence of the *HTC* on *B* parameter at vacuum conditions, observing the exponent of *B* equal to about 1.2 instead of about 0.04. Instead, the *L/D* dependence on *B* increases slightly at sub-atmospheric conditions (0.9 instead of 0.7). This difference explains the significant prediction improvement of HTC correlations and the not relevant for L/D (see black dotted line from Figure 9 to Figure 12).

The *HTC* and *L/D* modified correlations are compared to the experimental data in Figure 13 and Figure 14, respectively. The high number of 7606 datapoint constitutes a weighting factor for the *HTC* trend line that entails the overestimation of about 30% at higher *B* values. A lower discrepancy was achieved for the modified *L/D* correlations as shown by the ± 15% lines.

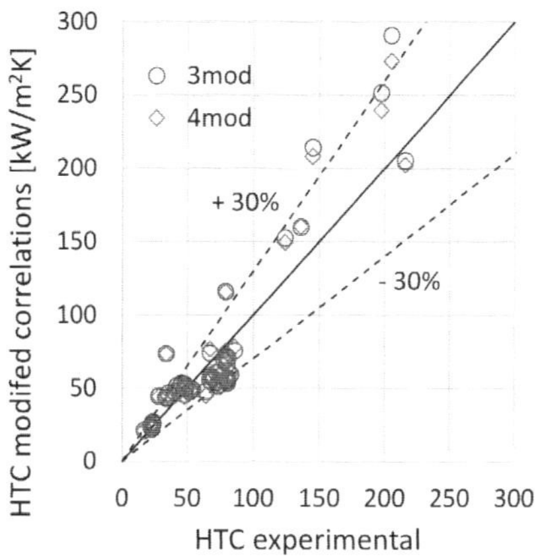

Figure 13: *HTC* modified correlations 2mod and 3mod compared to the experimental data

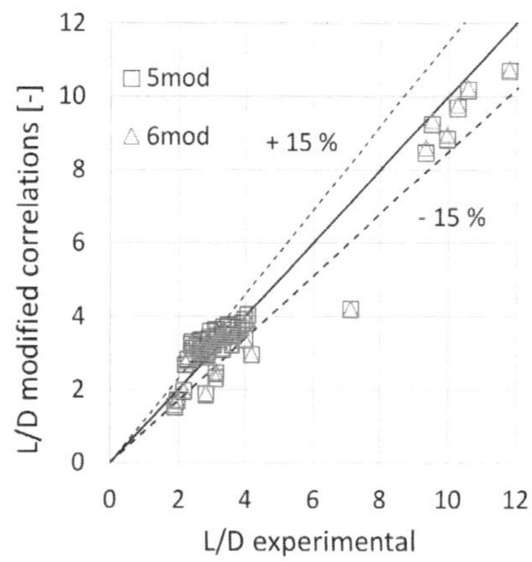

Figure 14: *L/D* modified correlations 4mod and 5mod compared to the experimental data

6. Conclusions

In ITER machine, as consequence of the postulated Small LOCA transient scenario, steam discharge occurs at sub-atmospheric conditions from the VV to the VVPSS Small LOCA Tank. The DCC of steam injected into the SLT was experimentally simulated at Lab. Guerrini of the University of Pisa, in a reduced scale facility, adopting a condensation tank of 4.5 m^3 implementing a sparger with 6 holes. The steam mass flow rate per hole was conserved and transient time was amplified, based on a developed and described similitude analysis, for taking into account the reduced tank volume (4.5 instead of 100 m^3) and lower sparger holes number (6 instead of 1000).

In this experimental activity, the characterisation of the steam DCC regimes at vacuum conditions was carried out, supported by video recording and steam plume image analysis, aiming to fill the gap of lacking literature and provide engineering feedbacks to ITER designer. The length (20-120 mm) and interfacial area (about 1000-20000 mm^2) of steam jets were measured in stable and interfacial oscillation condensation. On the basis of the elaborated area the average *HTC* was computed (20-200 kW/m^2K).

Correlations of steam plume *HTC* and length divided by orifice diameter were derived from the measured dataset (44 steam plumes were analysed), maintaining the parameter dependence of the atmospheric empirical correlations available in literature. In this work, the significantly higher dependence of the *HTC* on the condensation driving potential was clearly highlighted by its exponent, equal to about 1.2 at vacuum conditions and 0.04 at atmospheric pressure. Instead, for *L/D*, the driving potential exponent changes from about -0.7 to -0.9 passing from atmospheric to vacuum conditions and the prediction capability is not modified.

In the future work, the data set will be increased performing further images analysis at several steam mass flow rates (transient instants).

References

[1] ITER Internal Report, "VVPSS Interim Design Report", SDJNS7, 29 Jan 2016 / 2.2.
[2] Chul, Hwa Song, Seok Cho, Hwan Yeol Kim, YooN Young Bae, Moon Ki Chung – Korea Atomic

Energy Research Institute -"Characterization of Direct Contact Condensation of Steam Jets discharging into subcooled water"- Taejon, Republic of Korea

[3] Yeon Sik Kim, Moon Ki Chung, Jee Won Park, Moon Hyun Chun - "An experimental investigation of Direct Condensation of Steam Jet in Subcooled Water"- Journal of the Korean Nuclear Society Vol. 29, Number 1, pp. 45-57, (1997)

[4] Hwan Y. Kim, Yoon Y. Bae, Chul H. Song, Jong K. Park and Sang M. Choi - "Experimental study on stable steam condensation in a quenching tank" - International Journal of Energy Research Vol. 25, pp. 239-252, (2001)

[5] Qiang Xu,Liejin Guo, Suifeng Zou, Juanwen Chen, Ximin Zhang - State Key Laboratory of multiphase flow in Power Engineering, Xi'an Jiaotong University, Xi'an 710049, China - "Experimental study on direct contact condensation of stable steam jet in water flow in a vertical pipe" - International Journal of Heat and Mass Transfer pp.808-817, (2013) F

[6] Simpson, M.E. and Chan, C.K. - "Hydraulics of a subsonic Vapor Jet in Subcooled Liquid" - Journal Heat Transfer Vol. 104, pp. 271-278, (1982)

[7] J.C. Weimer, G.M. Faeth, D.R.Olson - Department of Mechanical Engineering Pennsylvania State University, University Park, Pennsylvania 16802 - "Penetration of Vapor Jets Submerged in Subcooled Liquids" - AlChE Journal Vol.19, pp.552-558, (1973)

[8] P.J. Kerney, G.M. Faeth, D.R.Olson - The Pennsylvania State University, Pennsylvania 16802 - "Penetration characteristics of a Submerged Steam Jet" - AlChE Journal Vol.18, pp. 548-553

[9] Xin-Zhuang Wu, Jun-Jie Yan, Wen-Jub Li, Dong-Dong Pan, Dao-Tong Chong - State Key Laboratory of multiphase flow in Power Engineering, Xi'an Jiaotong University, Xi'an 710049, China - "Experimental study on sonic steam jet condensation in quiescent subcooled water" - Chemical Engineering Science Vol.64, pp. 5002-5012, (2009)

[10] Hwan Y. Kim, Yoon Y. Bae, Chul H. Song, Jong K. Park and Sang M. Choi - "Experimental study on stable steam condensation in a quenching tank" - International Journal of Energy Research Vol. 25, pp. 239-252, (2001)

[11] Yeon-Sik Kim, Jee-Won Park and Chul-Hwa Song - Korea Atomic Energy Research Institute - "Investigation of the steam-water Direct Contact Condensation heat transfer coefficients using interfacial transport models" - International Comm. Heat Mass Transfer Vol. 31, pp. 397-408, (2004)

[12] MAZED, D., et al., Experimental investigation of functional performance of a vacuum vessel pressure suppression system of ITER, Fusion Engineering and Design 122 (2017) 42-46.

[13] MAZED, D., et al., Experimental Investigation of Steam Condensation in Water Tank at Sub-atmospheric Pressure, Nuclear Engineering and Design 335 (2018) 241-254.

[14] GIAMBARTOLOMEI, G., et al., "Direct Condensation of Steam in a Water Tank at Sub-Atmospheric Pressures", Journal of Physics: Conference Series, volume 1599, issue 1 (2020) article number 012024.

[15] PESETTI A. et al., "Direct steam condensation at sub-atmospheric pressure: experimental tests and similitude analysis", Journal of Physics: Conference Series, 1868 (2021) 012018. Doi:10.1088/1742-6596/1868/1/012018.

[16] PESETTI, A., et al., "Large Scale Experimental Facility for Performance Assessment of the Vacuum Vessel Pressure Suppression System of ITER", Fusion Engineering and Design 171 (2021) 112523.

[17] PESETTI A. et al., "Full Scale Test Facility for Operability Verification of Tokamak Pressure Suppression System at Sub-Atmospheric Conditions", Fusion Engineering and Design, Article number 112639, accepted for publication. Doi: under release.

[18] PESETTI, A., et al., "Numerical Analysis of Steam Condensation at Sub-Atmospheric Pressure in Water Suppression Tank", Proceedings of 27th International Conference on Nuclear Energy, ICONE, Tsukuba, Japan, 19-24 May 2019.

[19] PESETTI, A., et al., Numerical Analysis of Sub-Atmospheric Steam Condensation in Suppression Tank with Simmer IV Code, Fusion Engineering and Design 158 (2020) 111746.

Assessment of predictive methods for the heat transfer coefficient during flow boiling of R50 and R170 mixtures

William Ferretto, Andrea Lucchini, Luigi P.M. Colombo

Politecnico di Milano, Dipartimento di Energia, Via Lambruschini 4, 20156 Milano

luigi.colombo@polimi.it

Abstract. Liquefied natural gas (LNG) is a binary mixture of methane (R50) and ethane (R170). Proper evaluation of the heat transfer coefficient during flow boiling of such a mixture is very important to design LNG facilities. Pure fluid correlations have been tested with and without the adoption of corrections to account for the lowered performance due to mass transfer resistances arising both in the nucleate boiling and in the convective boiling regimes. In total, the prediction of 13 correlations has been compared with a new data set from the literature. Results show that the correction related to nucleate boiling has the greatest impact though, in general, the adoption of the corrections leads to under predict the data.

1. Introduction

Liquefied natural gas (LNG) is a binary mixture of methane (R50) and ethane (R170). The boiling characteristic of LNG have high importance not only for storage and transport, such as in vaporization section, but also for liquefaction, since R50 and R170 are the fundamental components in mixed-refrigerants for Joule-Thomson refrigeration cycle, widely used in LNG industry. The study of boiling heat transfer of R50/R170 binary mixtures is then very important to properly design LNG facilities, which requires reliable methods to compute the heat transfer coefficient.

Boiling heat transfer of multi-component mixtures has been extensively studied in the past and most of the approaches model the degradation of the heat transfer coefficient due to the arising mass transfer resistance.

Part of the models are developed for specific fluid mixtures under specific flow patterns. As significant examples, the Mishra [1] correlation is developed for R12/R22 mixtures in annular flow. Jung et al. [2], Granryd [3], Takamatsu et al. [4] work with R22/R114 and/or R12/R152a sharing at least part of the database. Accordingly, generalization of the results is hard to be achieved.

A more general approach consists of the correction of pure fluid correlations to account for performance degradation both in the nucleate boiling dominant and in the convective boiling dominant regime. Among the most recent contributions, starting from the Liu and Winterton [5] correlation, Zou et al. [6] modified the part describing the boiling effect, whereas Guo et al. [7] refitted both the convective enhancement factor and the boiling suppression factor using data for an R134a/R245fa mixture. On the other hand, Shah [8] proposed a general approach starting from any pure fluid correlation and adopting the Thome and Shakhir [9] correction for pool boiling of zeotropic mixtures to update the prediction in the boiling dominant regime, and the Bell and Ghaly [10] method to correct the prediction in the convective dominant regime. This approach was successful in predicting an 878 points database from 21 independent studies, except for LNG data taken from only one source [11].

One of the most recent approaches is provided by Zhang et al. [12] giving a physics-based correlation, in which a suitable formulation is derived according to the value of the ratio between the temperature glide and the saturation temperature. The correlation has been tested against 2091 data from 22 independent research groups, showing a mean absolute percentage deviation of 29.0%. However, validation does not seem to include explicitly LNG mixtures.

Data for LNG evaporation are relatively scarce [8]. Recently, Gong et al. [13] have published new measurements for R50/R170 mixtures at various composition, pressure, heat and mass flux. They reported 100 experimental points that can be used to check the so-called general prediction methods mentioned above. Accordingly, this paper presents a comparison between 13 correlation approaches applied to the prediction of the Gong et al. [13] LNG database as reported in the published work.

2. Correlations for the heat transfer coefficient during flow boiling in binary mixtures

As for a single-component system, a superheat is required to form a bubble of radius r. At equilibrium, from the Clausius-Clapeyron equation, it arises:

$$T_g - T_{SAT} = \frac{2\sigma}{(\partial p_{SAT}/\partial T)r} \tag{1}$$

For binary mixtures, the surface tension (σ), the bubble radius (r), and the slope of the vapour pressure curve ($\partial p_{SAT}/\partial T$) depend on the composition. An extension of Eq. 1 is provided by Stein [14]:

$$\frac{\partial p_{SAT}}{\partial T} = \left(\frac{\partial p}{\partial T}\right)_{\tilde{x}} + \frac{p}{\tilde{R}T}\left(\frac{\partial \tilde{x}}{\partial T}\right)_p (K-1)\tilde{x}\left(\frac{\partial^2 g}{\partial \tilde{x}^2}\right)_{T,p} \tag{2}$$

where K is the equilibrium constant and g the Gibbs free energy. It can be shown [15] that the second term in the right hand side is always negative, hence ($\partial p_{SAT}/\partial T$) will be always lower than the value for a pure fluid system. However, it is remarkable that this change is lower than the changes in both the surface tension and the active nucleation site radius. The maximum size of nucleation site is determined by the penetration of liquid into pre-existing surface cracks, a phenomenon influenced both by contact angle and surface tension changes. Shock [16] showed that for ethanol/water mixtures these changes can be very high and with a small addition of ethanol the superheating required drastically increases. There is a large difference between activation and deactivation superheating, which are affected by composition. Shakir and Thome [8] showed in a study about heterogeneous nucleation that Eq. 1 cannot predict the experimental variation between activation and deactivation of nucleation sites, which is considerably influenced by mass diffusion between the liquid and bubble.

2.1. *The correction model*

A model derived from an existing pure fluid correlation by applying corrections in order to take into account mixture effects is referred to as a correction model. Corrections apply to both nucleate boing and convective boiling regimes.

2.1.1. *Nucleate boiling*

Nucleate boiling heat transfer is strongly related to bubble dynamics and is the dominant mechanism at low qualities and high heat fluxes. Considering a binary mixture, at the liquid-vapor interface, the concentration of the less volatile component in the liquid is higher than in the bulk liquid due to the evaporation of the more volatile one. Hence, a higher bubble point is found, causing a decrease of the effective superheating. Moreover, the more volatile component from bulk liquid has to diffuse through the depleted region surrounding the bubble to ensure bubble growth, resulting in an additional mass transfer resistance. To account for these deviations compared to the pure fluid system, different corrections have been developed for pool boiling. Schlünder [17] developed a boiling model including mass diffusion shell resistance in an *n*-component mixture, derived from film layer analysis, which was able to predict the heat transfer coefficient over a wide range of reduced pressures:

$$F_c = \frac{h}{h_{id}} = \left\{1 + \frac{h_{id}}{q} \sum_{j=1}^{n-1} \left(T_{SAT,n} - T_{SAT,j}\right) \cdot \left(\tilde{y}_i - \tilde{x}_i\right) \cdot \left[1 - exp\left(-\frac{Bq}{\rho_l H_{lg} \beta_l}\right)\right]\right\}^{-1} \tag{3}$$

According to Shah [8] the ideal heat transfer coefficient is conveniently calculated from the Cooper correlation [18]:

$$h_{Cooper} = 55 p_r^{0.12} \left(-log\, p_r\right)^{-0.55} M^{-0.5} q^{2/3} \tag{4}$$

Moreover, taking the scaling factor B as unity implicitly assumes that all the heat transferred to the bubble interface is converted into latent heat. The liquid phase mass transfer coefficient β_l is set to the recommended value of 3×10^{-4} m/s. The Schlünder's model considers each component at equilibrium, but $\tilde{x}_i$ and $\tilde{y}_i$ are evaluated at their bulk values, which are not representative of the vapor-bubble interface. Moreover, at high and low reduced pressure some component may be under the freezing point or above the critical point, causing an impossible determination of saturation temperatures. Finally, Schlünder assumed for the sake of simplicity a fixed slope of the bubble-point curve for all composition, which does not describe well phase equilibria.

Thome and Shakir (TS) [9] refined the model to include the glide temperature and overcome the above mentioned issues. Their formulation reads:

$$F_c = \frac{h}{h_{id}} = \left\{1 + \frac{h_{id}}{q} \cdot \left(T_{dew} - T_{bub}\right) \cdot \left[1 - exp\left(-\frac{Bq}{\rho_l H_{lg} \beta_l}\right)\right]\right\}^{-1} \tag{5}$$

It is stated that these assumptions hold for mixtures with temperature glide up to 30 K.

2.1.2. Convective boiling

Convective boiling is governed by evaporation mechanism that takes place at liquid-vapour interface and is dominant at low heat fluxes and high qualities. Once nucleation is suppressed, the mass transfer resistance effect is present in both vapour and liquid films but with a smaller influence [19]. The main heat transfer degradation is here caused by the additional sensible heat transfer to the liquid and vapour phases due to increase of bubble point temperature that occurs along the flow direction. Furthermore, physical mixture properties are smaller compared to the ideal values. Finally, the temperature glide affects heat transfer in the sense that the higher the glide, the higher the mass resistance such that a full suppression of nucleation may happen.

The correction for convective boiling follows the method proposed by Bell and Ghaly (BG) [10], based on a study on condensation. In their approach, a vapour-side thermal resistance was included in the overall thermal resistance:

$$\frac{1}{h} = \frac{1}{h_c} + \frac{Y}{h_g} \tag{6}$$

$$Y = x c_{p,g} \frac{\partial T_{dew}}{\partial H} \tag{7}$$

The single-phase vapour convective heat transfer coefficient is estimated through the Dittus-Boelter correlation [20] assuming the vapour flowing alone (i.e. with the superficial velocity):

$$h_g = 0.023 Re_g^{0.8} Pr_g^{0.4} k_g / D \tag{8}$$

Since in both multicomponent convective boiling and multicomponent condensation sensible heat transfer and mass transfer resistance are present, an analogy is found. The method can be applied to boiling by replacing the condensation heat transfer coefficient h_c with the convective boiling heat transfer coefficient, and T_{dew} with T_{bub}. Finally, it was shown that:

$$\frac{\partial T_{dew}}{\partial H} \approx \frac{\partial T_{bub}}{\partial H} \approx \frac{T_{dew} - T_{bub}}{H_{lg}} \tag{9}$$

2.1.3. Pure fluid correlations

The two corrections reported above are applied to pure fluid correlations. Four of them, which are well-known in the literature, have been selected for this work: the Gungor and Winterton's issued in 1986 (GW86) [21], the Gungor and Winterton's issued in 1987 (GW87) [22], the Liu and Winterton's (LW) [5], the Shah's (SH) [8] correlations have been evaluated with both corrections (TSBG), with only one correction (TS) and without corrections at all. The equations are summarized in Table 1. When needed, the Dittus-Boelter correlation [20] has been adopted to compute the convective heat transfer coefficient for the liquid-alone flow (see Eq. 8 replacing the properties of the liquid).

Denomination	Correlation	Additional information
GW86 [21]	$h = Sh_{Cooper} + Fh_l$	$F = 1 + 24000 Bo^{1.16} X_{tt}^{-0.86}$ $S = \left(1 + 1.15 \cdot 10^{-6} F^2 Re_l^{1.17}\right)^{-1}$
GW86-TS	$h = F_c Sh_{Cooper} + Fh_l$	F and S as above
GW86-TSBG	$h = F_c Sh_{Cooper} + \left[(Fh_l)^{-1} + Y/h_g\right]^{-1}$	F and S as above
GW87 [22]	$h = Fh_l$	$F = 1 + 3000 Bo^{0.86} + 1.12(1/x - 1)^{-0.75}(\rho_l/\rho_g)^{0.41}$
GW87-TS	$h = F_{mix} h_l$	$F_{mix} = 1 + 3000(F_c Bo)^{0.86} + 1.12(1/x - 1)^{-0.75}(\rho_l/\rho_g)^{0.41}$
GW87-TSBG	$h = \left[(F_{mix} h_l)^{-1} + Y/h_g\right]^{-1}$	F_{mix} as above
LW [5]	$h = \left[(Fh_l)^2 + (Sh_{Cooper})^2\right]^{1/2}$	$F = \left[1 + x Pr_l(\rho_l/\rho_g - 1)\right]^{0.35}$ $S = \left(1 + 0.055 F^{0.1} Re_l^{0.16}\right)^{-1}$
LW-TS	$h = \left[(Fh_l)^2 + (F_c Sh_{Cooper})^2\right]^{1/2}$	F and S as above
LW-TSBG	$h = \left[(1/Fh_l + Y/h_g)^{-2} + (F_c Sh_{Cooper})^2\right]^{1/2}$	F and S as above
SH [8]	$h = max\left[h_{cb}; h_{nb}; h_{bs}\right]$ $h_{cb} = \left(J^{0.8}/1.8h_l\right)^{-1}$ $h_{nb} = h_l\left(230 Bo^{0.5}\right)$ $h_{bs} = 0.064 h_{nb} \exp\left(aJ^n\right)$	$J = Co$ if $Fr_l > 0.04$; $J = 0.38 Fr_l^{-0.3}$ if $Fr_l < 0.04$ $a = 2.47; n = -0.15$ if $J < 0.1$; $a = 2.47; n = -0.1$ if $J > 0.1$
SH-TS	$h = max\left[h_{cb,mix}; h_{nb,mix}; h_{bs,mix}\right]$ $h_{cb,mix} = \left(J^{0.8}/1.8h_l\right)^{-1}$ $h_{nb,mix} = h_l F_c\left(230 Bo^{0.5}\right)$ $h_{bs} = 0.064 h_{nb,mix} \exp\left(aJ^n\right)$	J, a and n as above
SH-TSBG	$h = max\left[h_{cb,mix}; h_{nb,mix}; h_{bs,mix}\right]$ $h_{cb,mix} = \left(J^{0.8}/1.8h_l + Y/h_g\right)^{-1}$ $h_{nb,mix} = h_l F_c\left(230 Bo^{0.5}\right)$ $h_{bs} = 0.064 h_{nb,mix} \exp\left(aJ^n\right)$	J, a and n as above

Table 1. Summary of the correction models.

2.2. *A recent correlation for binary mixtures*

Recently, Zhang et al. [12] introduced a new correlation specifically designed for binary mixtures. The model considers that the temperature glide plays a major role since a large value leads to a great suppression of nucleate boiling, hence to a dominant contribution of convective boiling. Besides, the saturation temperature affects heat transfer since its increase determines a lowered surface tension, a higher slope of vapour pressure curve thus increasing the size of active nucleation sites. Hence, at higher saturation temperatures the suppression of nucleate boiling is weaker.

It is then introduced the dimensionless temperature glide $T^*=(T_{dew}-T_{bub})/T_{bub}$. The approach considers a region ($T^*<0.06$) in which both convective boiling and nucleate boiling coexist, and a region where only convective boiling is present ($T^*>0.06$). Accordingly, the heat transfer coefficient is formulated as:

$$h_{TP}=\left[\left(0.7\cdot F_c h_{Cooper}\right)^2+h_{Mishra}^2\right]^{0.5} \quad \text{for} \quad T^*<0.06 \tag{10}$$

Where

$$h_{Mishra}=21.75\left(\frac{1}{X_{tt}}\right)^{0.29} Bo^{0.23} h_l \tag{11}$$

$$h_{TP}=\left[1+3000\left(F_c Bo\right)^{0.86}+1.12\left(\frac{x}{1-x}\right)^{0.75}\left(\frac{\rho_l}{\rho_g}\right)^{0.41}\right] h_l \quad \text{for} \quad T^*>0.06 \tag{12}$$

3. Data analysis

LNG flow boiling data are relatively scarce. Shah [8] based on the currently available database from Chen and Shi [11] concluded that unexpectedly mixture effects were not observed, thus correlations with no correction were able to achieve the best prediction.

Recently, Gong et al. [13] provided a new experimental dataset (in total 100 data point reported in the paper) for flow boiling of R50-R170 mixtures in 50 mm long, 6 mm inner diameter copper tubes, varying composition, pressure, heat and mass flux. The operating conditions range as follows: pressure, 0.3-0.7 MPa; heat flux, 5-62 kW m^{-2}; mass flux, 150-250 kg m^{-2} s^{-1}; component R50/R170 (mole fraction), 0.76/0.24, 0.43/0.57, 0.31/0.69, 0.15/0.85. Fluid properties have been computed through REFPROP [23]. As a representative sample, the phase diagram at 0.3 MPa is reported in Figure 1.

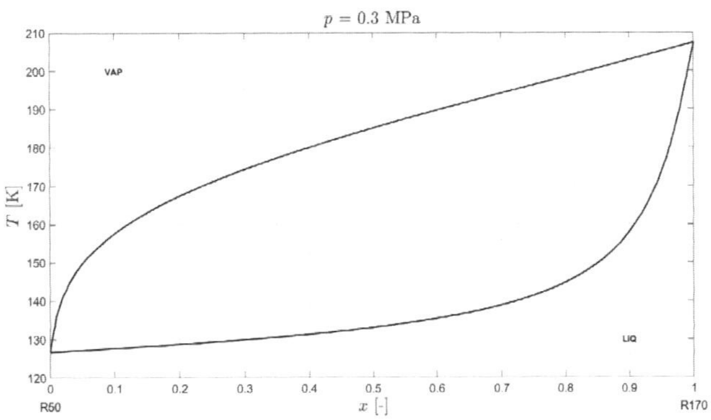

Figure 1. Phase diagram for R50-R170 mixtures.

All the 13 correlations listed in Table 1 have been tested against this database by evaluating the mean absolute relative deviation (MARD), the mean relative deviation (MRD), the percentage of data points within ±15% deviation (PCT15), the percentage of data points within ±30% deviation (PCT30) and the percentage of data points within ±50% deviation (PCT50):

$$MARD = \frac{1}{n}\sum_{i=1}^{n}\frac{\left|h_{i,\text{model}} - h_{i,\text{exp}}\right|}{h_{i,\text{model}}} \tag{13}$$

$$MRD = \frac{1}{n}\sum_{i=1}^{n}\frac{h_{i,\text{model}} - h_{i,\text{exp}}}{h_{i,\text{model}}} \tag{14}$$

It is worth noting that all the experimental conditions are characterized by $T^{*}>0.22$, which would imply only convective boiling according to Zhang et al. [12] (see Eq. 12). However, bubble point temperatures for R50-R170 mixtures are very low, hence assuming $T^{*}=0.06$ to discriminate between boiling dominant and convective dominant evaporation might be misleading. Accordingly, the threshold value of T^{*} has been set to ensure the best correlation performance. Reference is then made to the "modified" Zhang et al. model.

4. Results

The statistical parameters are summarized in Table 2. Figure 2 reports the predicted heat transfer coefficients against the experimental values for the three best predictors. The following considerations arise from the comparisons.

Model	MARD [%]	MRD [%]	PCT15 [%]	PCT30 [%]	PCT50 [%]
GW86 [21]	94.0	93.3	2.0	10.9	22.8
GW86-TS	42.6	38.3	25.7	38.6	59.4
GW86-TSBG	41.8	37.3	24.8	39.6	60.4
LW [5]	57.2	55.4	13.9	26.7	44.6
LW-TS	30.1	-24.9	12.9	52.5	88.1
LW-TSBG	42.0	-42.0	8.9	19.8	69.3
GW87 [22]	**28.0**	**23.8**	**34.7**	**52.5**	**85.2**
GW87-TS	29.1	-27.0	25.7	38.6	59.4
GW87-TSBG	47.7	-47.7	1.0	13.9	48.5
SH [8]	**18.8**	**4.7**	**37.6**	**80.2**	**100.0**
SH-TS	39.4	-38.3	11.9	22.8	75.3
SH-TSBG	53.8	-53.8	1.0	11.9	27.7
Zhang et al. [12]	**21.4**	**-19.5**	**36.6**	**73.3**	**99.0**

Table 2. Statistical indicators of prediction performance. The three best predictors are highlighted.

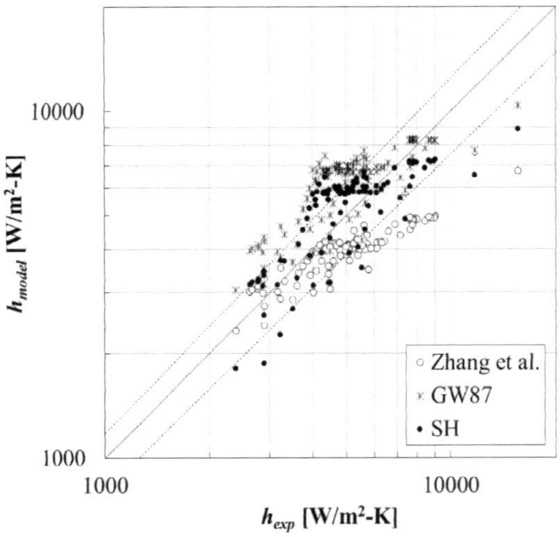

Figure 2. Parity plot for the three best predictors.

The GW87 [22], the SH [8] and the modified Zhang et al. [12] correlations are the best ones from the MARD point of view. The GW87 [22] underperforms slightly considering the PCT30 and PCT50 values compared to the others. For the Zhang et al. correlation [12], a threshold value T^*=0.39 has been determined as the one leading to the best performance for all the statistical parameters, which suggests that nucleation is a very important factor in determining the heat transfer performance. In any case, the SH [8], showing the best statistical indicators, and the GW87 [22] are pure fluid correlations, which confirms that mixture effects for R50 and R170 are possibly overestimated using either the Thome and Shakir [9] or the Bell and Ghali [10] corrections.

The Thome-Shakir correction [9] can either improve the heat transfer coefficient prediction as in GW86 [21] and LW [5] correlations or worsen it as in GW87 [22] and SH [8] correlations. Such a correction is mostly related to bubble nucleation, hence it can be inferred that nucleation plays a significant role in the data set, which confirms the above reported consideration about the effectiveness of the Zhang et al. [12] model. It can be noted from the MRD values that, as expected, the correction always lowers the prediction, leading in most cases to underestimate the data. However, it has to be stated that the validity range of the correction is up to 30 K glide, while for the dataset the glide ranges between 40 K and 55 K.

The TSBG correction generally worsens the prediction since the Bell and Ghali [10] model introduces further reduction in the heat transfer coefficient. Accordingly, strong data underestimation is observed. The only exception is seen with the GW86 [21] model since it over predicts so strongly the data that it benefits from the double correction.

5. Concluding remarks

A new data set including R50-R170 mixture flow boiling data has been analysed. 13 correlations have been tested to derive general considerations. The Shah [8] and the modified Zhang et al. [12] models showed the best MARD, respectively 18.8% and 21.4%.

There is evidence that on this data set the correction method by Thome and Shakhir [9] provides a general improvement compared with the basic pure fluid correlation. However, it may lead to significant overestimation of the heat transfer coefficient, which might be due to a glide temperature range higher than the applicable range.

The Bell and Ghali correction [10] always lowers considerably the prediction. If it is applied together with the Thome and Shakhir's [9], the results strongly under predict the data.

Nucleate boiling appears as a key mechanism in the heat transfer performance in LNG mixtures, hence extensive experimental study is still needed to extend the database useful to validate the models.

6. Acknowledgements

The authors are grateful to prof. Luca Molinaroli for his valuable advice.

References

[1] Mishra M, Varma H and Sharma C, 1981, "Heat transfer coefficients in forced convection evaporation of refrigerants mixtures", Letters in Heat and Mass Transfer, 8, 127 – 136

[2] Jung D, McLinden M, Radermacher R and Didion D, 1989, "A study of flow boiling heat transfer with refrigerant mixtures", Int. J. Heat Mass Transf. 32, 1751 – 1764

[3] Granryd E, 1991, "Heat transfer in flow evaporation of non-azeotropic refrigerant mixtures – A theoretical approach", Proc. 18th Int. Congress of Refrigeration, 1330 – 1334

[4] Takamatsu H, Momoki S and Fujii T, 1993, "A correlation for forced convective boiling heat transfer of nonazeotropic refrigerant mixture of HCFC22/CFC114 in a horizontal smooth tube", Int. J. Heat Mass Transf. 36, 3555 – 3563

[5] Liu Z and Winterton R H S, 1991, "A general correlation for saturated and subcooled flow boiling in tubes and annuli, based on a nucleate pool boiling equation", Int. J. Heat Mass Transf. 34, 2759 – 2766

[6] Zou X, Gong M Q, Chen G F, Sun Z H, Zhang Y and Wu J F, 2010, "Experimental study on

saturated flow boiling heat transfer of R170/R290 mixtures in a horizontal tube", Int. J. Refrig. 33, 371 – 380
[7] Guo C, Wang J, Du X and Yang L, 2016, "Experimental flow boiling characteristics of R134a/R245fa mixture inside smooth horizontal tube", Appl. Therm. Eng. 103, 901 – 908
[8] Shah M M, 2015, "A method for predicting heat transfer during boiling of mixtures in plain tubes", Appl. Therm. Eng. 89, 812 – 821
[9] Thome J R and Shakir S, 1984, "A new correlation for nucleate pool boiling of aqueous mixtures", AIChE Symp. Ser. 83, 46 – 51
[10] Bell K J and Ghaly A M, 1973, "An approximate generalized design method for multicomponent/partial condenser", AIChE Symp. Ser. 69, 72 – 79
[11] Chen D and Shi Y, 2013, "Two-phase heat transfer and pressure drop of LNG during saturated flow boiling in a horizontal tube", Cryogenics 58, 45 – 54
[12] Zhang J, Mondejar M E and Haglind F, "General heat transfer correlations for flow boiling of zeotropic mixtures in horizontal plain tubes", Appl. Therm. Eng. 150 (2019) 824–839
[13] Gong M, Song Q, Chen G, Zhuang X, Yang Z and Yao Y, 2020, "Boiling heat transfer characteristics for methane, ethane and their binary mixtures", Heat Transfer Engineering 41, 1 – 16
[14] Stein H, 1979, In Stralen S, Cole R, Eds., "Boiling phenomena: physicochemical and engineering fundamentals and applications", Hemisphere, Vol. 2, 535 – 553
[15] Collier J G, Thome J R, 1994, "Convective Boiling and Condensation", Chapter 12. Multi – component boiling and condensation, Clarendon Press
[16] Shock R, 1977, "Nucleate boiling in binary mixtures", Int. J. Heat Mass Transf., 20, 701 – 709
[17] Schlünder E U, 1983, Heat transfer in nucleate boiling of mixtures, Int. Chem. Eng., 23, 589 – 599
[18] Cooper M, 1984, In Hartnett J P and Irvine T F, Eds., "Advances in Heat Transfer", Elsevier, 16, 157 – 239
[19] Thome J, 1996, "Boiling of new refrigerants: A state-of-the-art review", Int. J. of Refrig., 19, 435 – 457
[20] McAdams W H, 1954, "Heat Transmission", McGraw-Hill, Chapter 9
[21] Gungor K and Winterton R, 1986, "A general correlation for flow boiling in tubes and annuli", Int. J. Heat Mass Transf. 29, 351 – 358
[22] Gungor K and Winterton R, 1987, "Simplified general correlation for saturated flow boiling and comparisons of correlations with data", Chem. Eng. Res. Des. 65, 148 – 156
[23] Reference Fluid Thermodynamic and Transport Properties-REFPROP, Version 9.1 National Institute of Standards and Technology, Standard Reference Data Program (Gaithersburg, MD, USA)

Numerical modelization of contact angle hysteresis of falling droplet under enhanced lubrication approximation

Nicola Suzzi[1] **and Giulio Croce**[1]

[1]DPIA - Dipartimento Politecnico di Ingegneria e Architettura - Universitá di Udine - Via delle Scienze - 33100 - Udine (UD) - Italy

E-mail: `nicola.suzzi@uniud.it`, `giulio.croce@uniud.it`

Abstract. Moving contact lines are involved in several engineering applications: in in-flight icing phenomenon, the eventual transition from droplet to rivulet or continuous film regime is crucial for the prediction of ice accretion over the aircraft surface; absorption process through structured packing is also characterized by a thin film flowing over the corrugated sheets. Disjoining pressure together with the assumption of a thin precursor film is largely used in numerical simulations of thin films and moving droplets in order to model the dynamics of moving contact lines and the surface wettability properties, in terms of imposed static contact angle. The disjoining pressure model was largely validated in case of falling films with the well known Voinov-Hoffman-Tanner law. On the other side, the capability of the disjoining pressure to model the contact angle hysteresis, which is a crucial parameter for predicting moving droplets behavior, has not been discussed yet. Here, numerical simulations of both falling films and moving droplets under lubrication approximation are conducted and the disjoining pressure model is used to predict the contact line dynamics. After verification of the full curvature implementation for a 1D falling film, the effective contact angle hysteresis is estimated for a moving droplet under different flow conditions and the transition from droplet to rivulet regime detected.

1. Introduction

The hydrodynamics of moving wetting lines still represents a complex phenomenon, characterized by physical discontinuities and different flow regimes. Indeed a thin liquid layer may evolve as a continuous film, an ensemble of rivulets, a droplet population or a combination of the mentioned regimes, with the occurrence of instability phenomena leading to a sudden change in the liquid layer configuration. Indeed, finger instability and moving dry patches have been largely analyzed in the Literature experimentally [1, 2] and, under the assumption of lubrication theory, both numerically and analytically [3–7]. Lubrication approximation, which allows to investigate hydrophilic surfaces, characterized by static contact angle $\theta_s < 90^\circ$, has the advantage to reduce the physical 3D problem to a 2D mathematical problem, ensuring a consistent reduction in computational cost when numerical computations are required. However, the numerical studies are mainly limited to even lower values of the static contact angle, with $\theta_s < 30^\circ$ usually investigated, due to the small slope approximation. Furthermore, the disjoining pressure model, usually adopted to model the contact line singularity, has been validated in the framework of moving film front, while, excluding the precursor work of Schwartz and Eley [8], its capability to properly model the droplet dynamics and the receding contact lines has not been tested yet. On the other side, the evolution of a liquid layer over a solid surface is involved in a number of engineering applications, such as in-flight icing on the aircraft surface, where the ice accretion is driven by the evolution of a droplet population [9], or absorption/distillation processes

through structured packing [10, 11].
Here the governing lubrication equations are numerically solved using an in-house code, previously developed and validated by the Authors [10, 12]. The small slope approximation is abandoned in favor of the full implementation of the film capillary pressure, allowing to investigate static contact angle up to $\theta_s = 60°$. Thus, the dynamics of a falling film and the occurrence of film rupture are studied, verifying consistency of the computed dynamic contact angle with the well known Hoffman-Tanner-Voinov law and discussing the benefit deriving from full curvature implementation instead of the small slope approximation. Then, a stand-alone droplet down a vertical plate is investigated and the apparent contact angle traced over the wetted perimeter. After the validation of the actual contact angle along the whole droplet with a literature, empirical correlation, the contact angle hysteresis is traced as a function of the flow characteristics, defined by the droplet Bond number, and the surface wettability, when the droplet regime is supported.

2. Mathematical model

Consider a thin liquid film flowing down a vertical plate. Applying Nusselt film theory, the undisturbed film thickness h_0 and the film velocity u_0 can be expressed as

$$u_0 = \frac{\rho g h_0^2}{3\mu}. \tag{1}$$

Lubrication theory allows to integrate continuity equation across the film thickness. Further assuming negligible film inertia, the momentum equation gives a parabolic velocity profile, which can be averaged across the film thickness and substituted in the continuity equation, leading to

$$\frac{\partial h}{\partial t} + \nabla \cdot \left(\frac{\rho \mathbf{g} h^3}{3\mu} - \frac{\nabla p}{3\mu} h^3 \right) = 0, \tag{2}$$

where h is the unknown film thickness and p is the local film pressure,

$$p = -2\sigma\kappa - \Pi, \tag{3}$$

with 2κ being the free surface curvature, required to calculate capillary pressure, and Π the disjoining pressure. The small slope approximation,

$$2\kappa = \nabla^2 h, \tag{4}$$

is usually adopted to estimate the free surface curvature. However, Eq. (4) is not accurate when the contact angle exceeds $\theta_s = 30°$ [13]. Following [10, 12, 14], the free surface curvature is estimated as

$$2\kappa \simeq \frac{\frac{\partial^2 h}{\partial x^2} + \frac{\partial^2 h}{\partial y^2}}{\left[1 + \left(\frac{\partial h}{\partial x}\right)^2 + \left(\frac{\partial h}{\partial y}\right)^2\right]^{3/2}}, \tag{5}$$

which allows to investigate contact angles up to at least $\theta_s = 60°$.
The disjoining pressure, which is related to intermolecular forces arising between solid and liquid surfaces, was integrated with lubrication theory by Schwartz and Eley [8],

$$\Pi = b\left[\left(\frac{d}{h}\right)^n - \left(\frac{d}{h}\right)^m\right] \tag{6}$$

$$b = f\,\frac{(n-1)(m-1)}{n-m}\,\frac{\sigma}{d}\,(1-\cos\theta_s),\; n > m > 1 \tag{7}$$

where d is the precursor film thickness and θ_s is the static contact angle. The correction coefficient f,

$$f = 1 + 6.069\,\delta + 161.7\,\delta^2 - 1547\,\delta^3 + 5890\,\delta^4 \tag{8}$$

with $\delta = d/h_0$ being the non-dimensional precursor film, was introduced by Zhao and Marshall [3], in order to extend disjoining pressure to non-infinitesimal values of the precursor film thickness, as it usually happens in numerical simulation.
Introducing the following non-dimensional quantities,

$$H = \frac{h}{h_0},\, \mathbf{X} = \frac{\mathbf{x}}{l_0},\, T = \frac{t}{(u_0/l_0)},\, \Pi^* = \frac{1}{1-\cos\theta_s}\frac{\Pi}{(\sigma/h_0)},\, \Gamma = \frac{1-\cos\theta_s}{Bo^{2/3}} \tag{9}$$

with Bo and l_0 being the film Bond number and the characteristic length scale [3] respectively,

$$Bo = \frac{\rho\, g\, h_0{}^2}{\sigma},\, l_0 = \left(\frac{\sigma\, h_0}{\rho\, g}\right)^{1/3}, \tag{10}$$

the governing lubrication equations, Eqs. (2), (3) and (5) can be recast as

$$\frac{\partial H}{\partial T} + \nabla \cdot \left(H^3 - \nabla P H^3\right) = 0 \tag{11}$$

$$P = -2K - \Gamma\Pi^* \tag{12}$$

$$2K \simeq \frac{\frac{\partial^2 H}{\partial X^2} + \frac{\partial^2 H}{\partial Y^2}}{\left[1 + Bo^{2/3}\left(\frac{\partial H}{\partial X}\right)^2 + Bo^{2/3}\left(\frac{\partial H}{\partial Y}\right)^2\right]^{3/2}}. \tag{13}$$

Eqs. (11), (12) and (13) are numerically solved on a structured computational grid. To do so, an in-house Finite Volume Method (FVM) solver, developed in FORTRAN, parallelized through OpenMP library for a shared memory machine and previously validated by the Authors [10,12,14] for contact angles up to 60°, is used. The Alternating Direction Implicit (ADI) method proposed by [15] is implemented for marching in time.

3. Problem statement

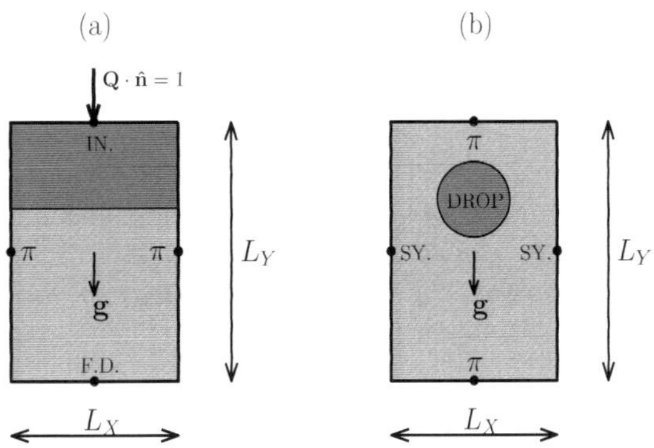

Figure 1. Investigated test cases and imposed boundary conditions (π – periodic; SY. – symmetry; IN. – inlet; F.D. – fully developed): 1D falling film (a); 2D stand-alone, moving droplet (b).

Two different test cases are investigated:

1. First, a 1D film falling down a vertical plate is investigated, Figure 1(a). Thus, periodic conditions are applied through the lateral boundaries, $X = 0$ and $X = L_X$ (giving an infinitely wide plate), while the liquid is uniformly injected through the inlet section, located at $Y = 0$, where $\mathbf{Q} \cdot \hat{\mathbf{n}} = 1$ and $H = 1$ are implemented, with $\mathbf{Q}$ being the non-dimensional film flux,

$$\mathbf{Q} = H^3 - \nabla P H^3. \tag{14}$$

Fully developed flow condition, $\nabla P \cdot \hat{\mathbf{n}} = 0$ and $\nabla H \cdot \hat{\mathbf{n}} = 0$, is imposed through $Y = L_Y$. The vertical plate is initialized with the precursor film thickness, $H|_{T_0} = \delta$. Simulations are conducted computing the free surface curvature using both the small slope approximation, Eq. (4), and the full expression provided by Eq. (13), which gives the exact curvature in case of a 1D film. The computational domain dimensions are set to $L_X = 1, L_Y = 50$ in order to ensure that the moving liquid front becomes self similar before reaching the bottomest plate section.

2. Then, a 2D droplet over a vertical plate of dimension $L_X \times L_Y$ is investigated, Figure 1(b). Periodic conditions are imposed through both the topmost section, $Y = 0$, and the bottomest section, $Y = L_Y$, giving an infinitely long plate, while symmetry condition, $\mathrm{Q} \cdot \hat{\mathbf{n}} = 0$ and $\nabla h \cdot \hat{\mathbf{n}} = 0$, is imposed at the lateral boundaries, $X = 0$ and $X = L_X$. The droplet is initialized as a spherical cap with prescribed contact angle θ_s at the center of the computational domain; the remaining initially dry region is initialized with the precursor film δ. The droplet Bond number is defined as the Bond number of the equivalent film, which is in turn defined by the average liquid thickness covering the spherical cap base area. Thus, the equivalent film thickness h_0 can be reduced to the radius r_0 of the initialized spherical cap,

$$\frac{h_0}{r_0} = \frac{(1 - \cos\theta_s)(2 + \cos\theta_s)}{3(1 + \cos\theta_s)}. \tag{15}$$

The free surface curvature is estimated through Eq. (13), allowing to simulate contact angles up to 60°.

4. Result

First of all, a film falling over a vertical plate (test case 1.) is investigated. Parametric computations are run and the dynamic contact angle, which is estimated from the maximum value of the film slope over the apparent contact line, traced as a function of the Bond number and the surface wettability, defined by the imposed static contact angle. Furthermore, the critical Bond number, leading to film instability, is also traced at different surface wettabilities. The spatial discretization step is set to $\Delta X = \delta/2$, in order to ensure grid independency, while the precursor film thickness and the disjoining pressure coefficients are set to $\delta = 5\times10^{-2}$ and $n = 3, m = 2$, according to Literature suggestions [3, 8, 16].

It is important to point out that, when small slope approximation is assumed, such a restrictive case (gravity driven film down a vertical plate) allows to reduce the two non-dimensional input parameters, Bo and θ_s, in a single parameter Γ, defined in Eq. (9), meaning that a single computation can be extended to a family of solutions in terms of Bo and θ_s. In fact, once the critical condition leading to film rupture is detected, it can be extended to any surface wettability. However, this is no more valid if the exact free surface curvature is computed via Eq. (13). The 1D computations are run both assuming small slope approximation and using the full curvature model, Eq. (13).

As a prove, the free surface profiles of the stable film solution and of the self similar droplet, the latter generated due to film rupture, are shown in Figures 2(a) and 2(c) at 3 different values of the static contact angle, $\theta_s = 30^\circ, 45^\circ, 60^\circ$, but the same Γ (with $\Gamma = 3.36$ leading to a stable film and $\Gamma = 3.92$ leading to film rupture), showing that different free surface profiles are obtained. Indeed, higher slopes are predicted at increasing θ_s, for both the stable film and the self similar droplet. The small slope approximation would hide this effect, leading to a single solution for both the stable film and the self similar droplet: Figures 2(b) and 2(d) demonstrates that the interface solution, obtained via full curvature modelization, progressively diverges from the one obtained assuming small slope approximation as $\theta_s > 30^\circ$.

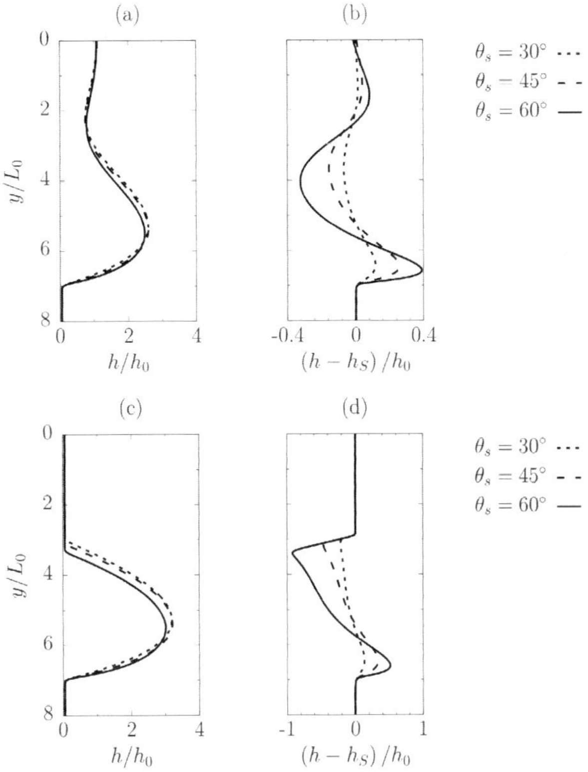

Figure 2. Self similar solutions of a 1D film falling over a vertical plate at different surface wettabilities ($\theta_s = 30°, 45°, 60°$): stable film at $\Gamma = 3.36$ (a); difference between small slope solution, h_S, and full curvature implementation at $\Gamma = 3.36$ (b); droplet generated due to film rupture at $\Gamma = 3.92$ (c); difference between small slope solution, h_S, and full curvature implementation at $\Gamma = 3.92$. Test case 1.

The dynamic contact angle in the stable film range is shown in Figure 3 as a function of the Bond number. Numerical results are consistent with the modified Hoffman-Voinov-Tanner law [17],

$$\left(\frac{\theta}{\theta_s}\right)^3 = C + Bo\left[D + E\log\left(Bo^{1/3}\right)\right] \tag{16}$$

with C being a constant coefficient introduced by Zhao and Marshall [3], D and E being constant coefficients as well. Furthermore, the dynamic contact angle always collapse to the corresponding static contact angle at Bond number approaching zero (i.e. static contact line).

The dashed line in Figure 3, which refers to small slope approximation, is obtained by reducing the detected dynamic contact angle of the continuous film at incipient film instability. It is worth to point out that the small slope approximation is not able to accurately predict the actual contact angle even at $\theta < 30°$. However, the film instability occurrence (i.e. the critical Bond number) is still correctly predicted.

The 2D computations of the stand-alone droplet (test case 2.) are run imposing spatial discretization step equal to $\Delta X, \Delta Y \leq 2/3\,\delta$, while the disjoining exponents are set again to $n = 3, m = 2$. Two different values of the precursor film thickness, $\delta = 5\times10^{-2}$ and $\delta = 2.5\times10^{-2}$, are investigated. The equivalent film thickness h_0, which defines the droplet Bond number and is required to initialize the spherical cap, is estimated via Eq. (15). Again, parametric computations are run and the contact angle hysteresis is traced as a function of the Bond number and the surface wettability. Since the disjoining pressure model

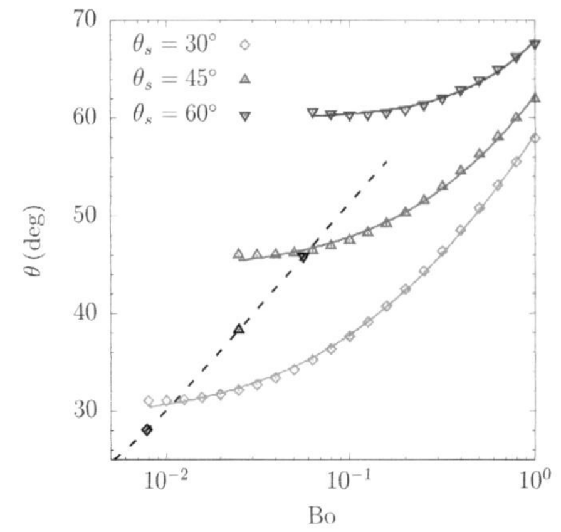

Figure 3. Dynamic (advancing) contact angle for a 1D stable film falling over a vertical plate as a function of Bond number and surface wettability: numerical results (markers) vs Eq. (16) (continuous lines). Dashed line predicts dynamic contact angle at incipient film rupture through small slope approximation. Test case 1.

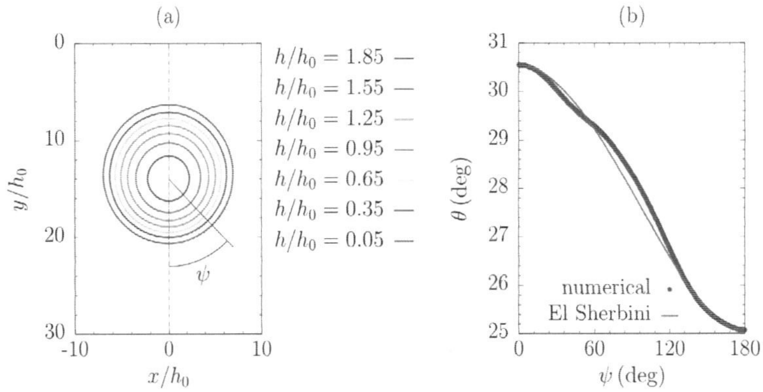

Figure 4. Self similar droplet falling down a vertical plate (a); numerical (markers) vs empirical [18] (continuous line) apparent contact angle as a function of azimuthal coordinate (b). Test case 2, $Bo = 10^{-2}$, $\theta_s = 30°$, $\delta = 5\times10^{-2}$.

have been largely validated and mostly used for the moving contact line of a liquid film [3, 4, 16], i.e. focusing on the advancing liquid front, a further validation is attempted in case of a falling droplet, i.e. looking at the phenomenology of a receding liquid boundary. To do so, once the slowly moving droplet becomes self similar, its apparent contact line, located where the local maximum slope is observed, is tracked over the computational domain. Thus, the droplet center of gravity is estimated and the apparent contact angle expressed as a function of the azimuthal coordinate ψ (with $\psi = 0°$ at the advancing front of the droplet and $\psi = 180°$ at the receding front). The numerical results are thus validated with the following empirical correlation [18],

$$\theta(\psi) = 2\,\frac{\cos\theta_{max} - \cos\theta_{min}}{\pi^3}\,\psi^3 - 3\,\frac{\cos\theta_{max} - \cos\theta_{min}}{\pi^2}\,\psi^2 + \cos\theta_{max}. \tag{17}$$

Figure 4(b) shows a good agreement between numerical results and empirical correlation, with a maximum uncertainty of about $\Delta/(\theta_{max} - \theta_{min}) \simeq 6\%$, meaning that we can be confident in modeling

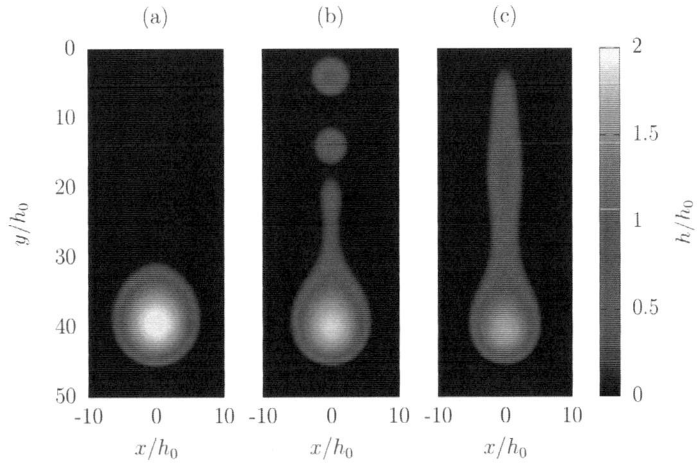

Figure 5. Liquid layer distribution at $T = 15$: $Bo = 2.00\times10^{-2}$ (a); $Bo = 3.16\times10^{-2}$ (b); $Bo = 5.01\times10^{-2}$ (c). Test case 2, $\theta_s = 30°$, $\delta = 5\times10^{-2}$.

droplet contact line dynamics along the whole droplet perimeter through disjoining pressure model. The corresponding droplet shape, shown in Figure 4(a), was obtained from a computation at relatively low Bond number, $Bo = 10^{-2}$, and $\theta_s = 30°$.

Increasing the droplet Bond number at fixed θ_s, the initial spherical cap deforms due to gravity. As

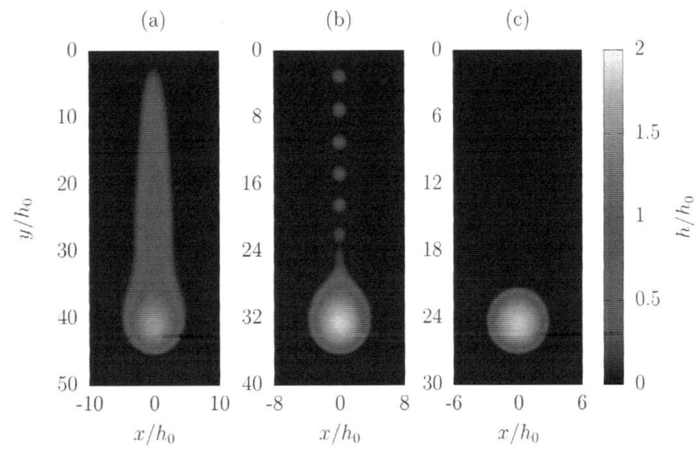

Figure 6. Liquid layer distribution at $T = 15$: $\theta_s = 30°$ (a), $\theta_s = 45°$ (b), $\theta_s = 60°$ (c). Test case 2, $Bo = 7.94\times10^{-2}$, $\delta = 5\times10^{-2}$.

shown in Figure 5, the moving, self similar droplet gets progressively stretched, until a critical condition is reached. In fact, when $Bo = 3.16\times10^{-2}$, the initial droplet breaks due to capillary instability, inducing formation of multiple droplets slowly draining down the vertical plate. Thus, a bifurcation can be observed for the droplet solution, as previously detected for the 1D falling film. Even something more similar to a long rivulet, not yet broken into smaller droplets, can be observed at higher Bond number, $Bo = 5.01\times10^{-2}$. Snapshots of Figure 5 are referred to the same non-dimensional time, $T = 15$, and the same surface wettability, $\theta_s = 30°$, but increasing Bond number.

The effect of the surface wettability is also presented in Figures 6(a), 6(b) and 6(c), which refer to the same Bond number, $Bo = 7.94\times10^{-2}$, but different surface wettabilities. As expected, droplet regime is incentivated at lower surface wettability, with bifurcation occurring at progressively higher Bond number. It is important to point out that such a behavior of the receding contact lines was analytically reported by Eggers [19], who demonstrated through asymptotic analysis that, under lubrication approximation, the receding contact line is no more sustainable and disappears at high contact line speed, which is exactly what is here observed. This may identify the transition between droplet and rivulet regime.

The contact angle hysteresis, defined as the difference between the maximum and the minimum contact

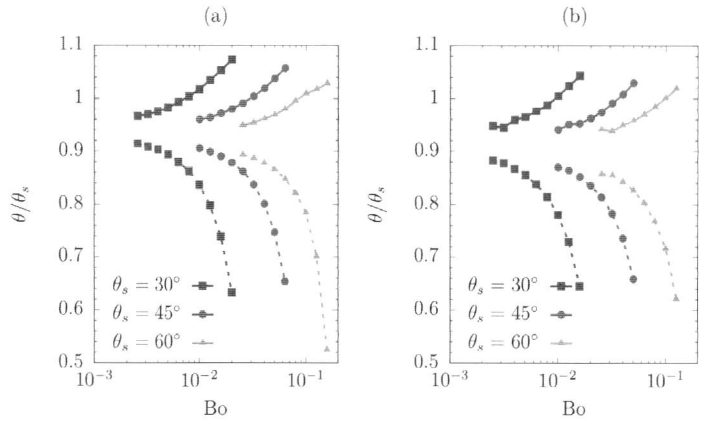

Figure 7. Normalized advancing contact angle (continuous lines) and receding contact angle (dashed lines) of a droplet down a vertical plate as a function of Bond number and surface wettability: $\delta = 5\times10^{-2}$ (a); $\delta = 2.5\times10^{-2}$ (b). Test case 2.

angle detected over the apparent contact line, is plotted in Figure 7 as a function of both the Bond number and the surface wettability. Results giving drop break up or an elongated droplet (which is more similar to a rivulet regime) are not considered since they do not provide information about the receding contact angle. Two different precursor film thickness are investigated, with Figure 7(a) referring to $\delta = 5\times10^{-2}$ and Figure 7(b) referring to $\delta = 2.5\times10^{-2}$, in order to verify its influence on the contact line dynamics. It can be observed that the contact angle hysteresis approaches zero for a sufficient low Bond number, i.e. negligible gravity forces, independently on the imposed static contact angle, meaning that the droplet progressively tends to the initialized spherical shape as its velocity tends to zero. On the other side, the contact angle hysteresis progressively increases with the Bond number and, thus, with the contact line velocity, until the critical condition leading to break up is detected. The effect of the solid surface wettability just seems to shift the hysteresis curves, with higher Bond numbers required to ensure an evident contact angle hysteresis. However, both the advancing and the receding contact angles seems to collapse to a value, which is a bit lower than the imposed static contact angle (with a discrepancy of about $\sim$5% observed for both $\delta = 5\times10^{-2}$ and $\delta = 2.5\times10^{-2}$). Such a small discrepancy is probably due to an inaccurate scaling in the definition of the Bond number, which is estimated through the equivalent film thickness, Eq. (15).

It can be also noticed that the effect of the precursor film thickness on the contact angle hysteresis is not relevant and even the critical condition (i.e. disappearance of the receding contact line) occurs at similar Bond numbers.

5. Conclusion

The governing lubrication equations were numerically solved in case of both a film down a vertical plate and a moving droplet driven by gravity. It was proved that the small slope approximation, usually adopted in literature and valid for low film slope, does not provide accurate results when $\theta_s \geq 30°$.

Indeed, computing the interface curvature via Eq. (5) ensures a proper modelization of the contact line dynamics, both for a flowing film and a moving droplet, at contact angle up to $\theta_s = 60°$. Furthermore, the disjoining pressure was validated for the first time in the framework of moving droplets, which involves a receding contact line. Indeed, the actual contact angle was traced along the whole droplet contact line and validated with a literature, empirical correlation. Thus, important information were provided about the contact angle hysteresis, which was traced as a function of the flow characteristics and the surface wettability. New instability phenomena, including droplet breakage and switch from moving droplet to rivulet regime, which are consistent with the physics of moving contact lines, were also detected. As a future work, the effect of the disjoining pressure exponents on the contact angle hysteresis should be investigated, in order to fully characterize the surface wettability properties. Furthermore, tracing the occurrence of elongated droplets or rivulets, may be crucial in the understanding the droplet behavior in practical applications as in-flight icing, which requires the definition of the threshold between droplet to rivulet regime to characterize the ice accretion model.

Nomenclature

b	disjoining coefficient
d	precursor film thickness
g	gravity acceleration
h	film thickness
H	non-dimensional film thickness
h_0	undisturbed film thickness
K	non-dimensional mean curvature
l_0	characteristic length scale
$L_{X,Y}$	non-dimensional plate size
m, n	disjoining exponents
$\hat{\mathbf{n}}$	normal outward versor
p	film pressure
P	non-dimensional film pressure
$\mathbf{Q}$	non-dimensional film flux
t	time
T	non-dimensional time
u	film velocity
u_0	undisturbed film velocity
x, y	Cartesian coordinates
X, Y	non-dimensional Cartesian coordinates
Greek symbols:	
Γ	non-dimensional disjoining coefficient
δ	non-dimensional precursor film thickness
θ	dynamic contact angle
θ_s	static contact angle
κ	mean curvature
μ	dynamic viscosity
Π	disjoining pressure
Π^*	non-dimensional disjoining pressure
ρ	density
σ	surface tension
ψ	azimuthal coordinate
Dimensionless numbers:	
$Bo = \rho g h_0^2/\sigma$	Bond number

References

[1] T Podgorski, J M Flesselles and L Limat 1999 *Physics of Fluids* **11** 845
[2] E Rio and L Limat 2006 *Physics of Fluids* **18** 032102
[3] Y Zhao and J S Marshall 2006 *Journal of Fluid Mechanics* **559** 355
[4] U Thiele, L Brusch, M Bestehorn and M Bar 2003 *The European Physical Journal E* **11** 255
[5] V Y Shkadov and G M Sisoev 2005 *Computers & Fluids* **34** 151
[6] J C-T Kao, A A Golovin and S H Davis 2006 *Journal of Colloid and Interface Science* **303** 532
[7] S K Wilson, B R Duffy and H Davis 2001 *European Journal of Applied Mathematics* **12** 233
[8] L W Schwartz and R R Eley 1998 *Journal of Colloid and Interface Science* **202** 173
[9] G Croce, E De Candido, W G Habashi, J Munzar, M Aube, G S Baruzzi and C Aliaga 2010 *Journal of Aircraft* **47** 1283
[10] N Suzzi and G Croce 2019 *Physics of Fluids* **31** 122106
[11] R K Singh, J E Galvin and X Sun 2016 *Chemical Engineering Science* **142** 244
[12] N Suzzi and G Croce 2017 *Journal of Physics: Conference Series* **923** 012020
[13] C A Perazzo and J Gratton 2004 *Journal of Fluid Mechanics* **507** 367
[14] N Suzzi and G Croce 2021 *Journal of Physics: Conference Series* **1868** 012010
[15] T P Witelski and M Bowen 2003 *Applied Numerical Mathematics* **45** 331
[16] M Sellier 2015 *International Journal of Multiphase Flow* **71** 66
[17] V S Ajaev 2012 *Interfacial Fluid Mechanics* (Springer)
[18] A I El Sherbini and A M Jacobi 2004 *Journal of Colloid and Interface Science* **273** 566
[19] J Eggers 2005 *Physiscs of Fluids* **17** 082106

Journal of Physics: Conference Series 2177 (2022) 012044 doi:10.1088/1742-6596/2177/1/012044

Experimental investigation of R744 transcritical flow and mixing in the two-phase ejector

Jakub Bodys, Michal Palacz, Michal Haida, Jacek Smolka, Damian Dziurowicz, Michal Majchrzyk, Andrzej J. Nowak

Silesian University of Technology, Department of Thermal Engineering, 44-121 Gliwice, Konarskiego 22, Poland

andrzej.j.nowak@polsl.pl

Abstract. In this work, the laboratory test rig for visualisation of the CO2 ejector is shortly presented. The internal flow structures in the premixing chamber and expansion cone are visualised using two different measurement arrangements, i.e. Particle Image Velocimetry (PIV) system and high-speed camera system. Both systems allow to determine some characteristic features of the R744 flows and propose some important quantities (i.e. expansion and mixing angles) describing expansion and mixing processes. Potentials of those preliminary results for field validation data are also pointed out.

1. Introduction

Recent studies on refrigeration cycles with carbon dioxide as refrigerant (denoted also as R744) has brought substantial enhancement in their technical solutions and particularly in their energy efficiency. As a result the modern R744-based refrigeration systems are commonly installed in the large-scale installations, for instance in supermarkets. Last commercial reports [1] showed that the number of R744 refrigeration supermarket installations is above 14,000 in Europe, above 3,100 in Japan and around 500 in North America. Continuous development of the R744 technology that concerns global features such as the improved layout of the systems as well as detailed investigation of the regulation methods and design tools for the particular components, allowed to extend the application range of this type of refrigeration systems into the hot and tropical climate zones [2].

The high efficiency two-phase carbon dioxide ejectors dedicated for the R744 systems need to be individually designed using numerical methods as discussed for instance in Elbel and Lawrence [3] and Besagni et al. [4]. It is also important to underline that all developed numerical flow models should be validated. The most common procedure is limited to a comparison of model results with relatively accurate measurements of the mass flow rates and all thermodynamic parameters at the ejector ports at different operating conditions. These kind measurements can be found relatively easy in the literature. However, if developing numerical model also adapted turbulent model should be validated as well, some measurement results within internal ejector ducts have to available. The turbulent model is absolutely crucial to describe properly flow and mixing processes inside premixing and mixing sections as well as diffuser. Unfortunately, these measurements inside ejector ducts are generally very difficult and their results are extremally rare in the literature. One possibility is a building transparent ejector and using flow visualisation techniques.

Depending on the flow characteristics, different flow visualisation methods could be used. In this context the shadow-graph visualisation [5], particle image velocimetry [6] or the high-speed video camera [7] should be mentioned. In the case of the R744 particularly interesting is the work by Zhu et al. [8] where flow in the two-phase ejector with rectangular cross-section is visualised. Those authors applied the light film and a single-lens reflex camera to observe the liquid and vapour flows. Special attention was given to the expansion angle within the mixing chamber. Based on the captured snapshots authors approximated the vapour quality in the fluid flow.

It should however be noticed that the majority of the currently used R744 ejectors have ducts with circular cross-section while a suction port is designed as radial (i.e. tangent to the cross-section). Such solution intensifies mixing of the secondary and primary streams due to the swirled flows [9]. To the best authors' knowledge, the only preliminary visualisation results for such R744 ejector (i.e. the circular cross-section with radial suction port) working in a transcritical refrigeration unit was presented quite recently by Haida et al. [10]. Current paper is a continuation of those experimental investigations with the circular transparent R744 two-phase ejector focusing on the analysis of local flows in the premixing chamber and expansion cone. The internal flow structures are visualised using two different measurement arrangements, i.e. Particle Image Velocimetry (PIV) system and high-speed camera system. Both systems allowed to record some different characteristic features of the R744 flows and propose some important quantities describing expansion and mixing processes. Potentials of those preliminary results for field validation data are also pointed out.

2. Short description of the test rig

The scheme of a laboratory R744 vapour compression test rig dedicated to two-phase ejector tests is presented in Fig. 1a. The ejector itself with the transparent mixing and diffuser sections was designed and manufactured in Gliwice, Poland. They can be seen in Fig. 1b where the ejector line with ejector ports are also shown.

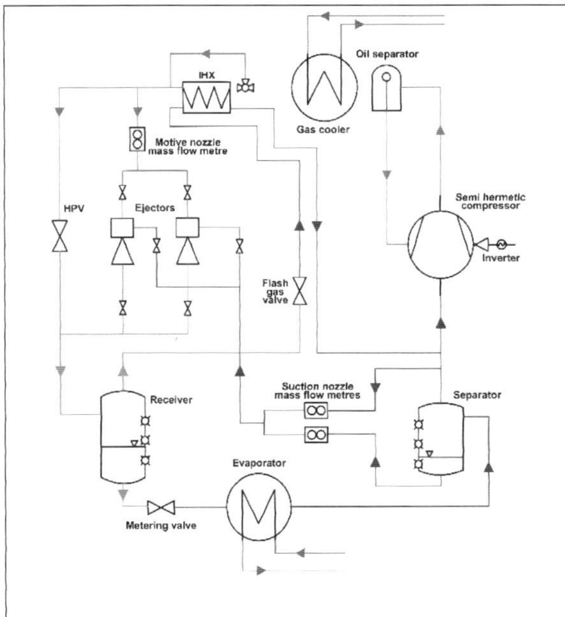

Fig. 1a. Scheme of the laboratory R744 refrigeration unit dedicated to ejector tests.

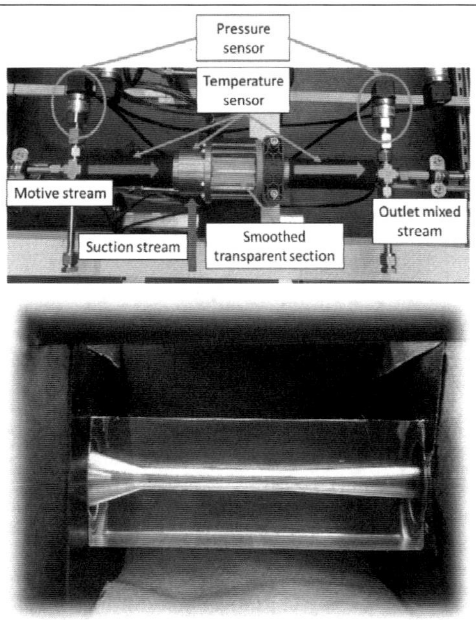

Fig. 1b. Ejector line with ports as well as transparent ejector.

The ejector section is located at separate rack than the other components to provide an easy access to complex measurements of each investigated ejector. The test rig was designed for the cooling capacity of 50 kW operated in the refrigeration and air-conditioning application at subcritical and transcritical conditions. The system layout is a standard transcritcal system with a compressor manufactured by DORIN. The brazed-plate type heat exchangers produced by SWEP were used to control heat absorption and rejection. The suction accumulator tank located at the evaporator outlet allows for the operation of the two-phase ejector either as the vapour ejector or the liquid ejector. The expansion process is also supported by the Danfoss CCMT type electronic expansion valve and the flash gas from the separator is expanded by the flash valve.

The test rig is equipped with the glycol loops (auxiliary loops) to control the cooling capacity and heat rejection in the gas cooler. Both loops are additionally integrated by the internal heat exchanger to minimise heat rejection to the ambient and control the operating conditions in an efficient way.

In order to carry out the test campaign at different operating conditions and cooling capacity the AK-PC-782A Danfoss control system is adopted. All measurements are recorded and stored by the Danfoss StoreView with a recording frequency of 5 seconds. The quasi-steady state conditions of the investigated operating point was set by constant ejector parameters or with small deviations within the period longer than 10 minutes (approx. 120 recorded points). The NIST guidelines were followed to evaluate and calculate the uncertainties from experimental tests [11].

The test rig is fully equipped with measurement instruments to control the operating conditions and cooling capacity of the system. The pressure transmitters Danfoss AKS 32R and AKS 2050 are used to measure and control the pressure levels within each section of the R744 loop up to the pressure level of 160 bar. The temperature is measured by the temperature sensors AKS-21 PT1000 that allow for the temperature range measurements from -70°C up to 180°C. Additional pressure and temperature sensors have been installed directly at the ejector inlets and the outlet. The refrigerant mass flow rate at the motive and suction ports of the ejector was measured by three precise Coriolis type mass flow meters manufactured by Endress+Hauser.

3. PIV vs high-speed camera approach to flow visualisation

As already mentioned, flow and mixing processes within premixing chamber are visualised using two systems: the high-resolution PIV system and system of the high-speed camera. It should be stressed at this moment that the R744 flow is a two-phase flow and application of any seeds would spoil considerably visualisation of flow structure. In other words, the PIV system with its laser positioning structure was used only to visualise flow but not to measure velocity field.

The PIV system was equipped with Dantec Dynamics Flow Sense 4M MK4 II camera which has a 2048x2048 CCD sensor with a pixel size of 7.4 μm. The maximum frame rate for the mentioned camera was 14 frames per second. A Nikon Nikkor AF 200 mm F/4D ED-IF Micro lens was paired with the specified camera. It is also worth mentioning that the macro reproduction ratio for that lens is equal to 1:1. In order to illuminate the analysed ejector section, a double-pulse Quantel Twin BSL 200 laser was used. The maximum energy output for that device is 200 mJ @ wavelength of 532 nm. All the devices of the utilised PIV system were synchronised using a Dantec Dynamics Synchronizer controlled with commercial software Dantec Dynamic Studio v. 3.41. Moreover, the mentioned software was used to postprocess the captured frames.

The general arrangement of the PIV system is presented in Fig. 2a, while the zoomed view at the ejector is presented in Fig. 2b. The high accuracy of the laser positioning is ensured on the basis of a dedicated frames presented in Fig. 2c. For the camera similar frame is shown in Fig. 2d. The last figure, i.e. Fig. 2e, presents the safety box. To highlight the core of the ejector, laser stream (marked by the green dashed line in Fig. 2a) is pointed perpendicularly on the transparent section. It should also be noted that the camera is pointed vertically and perpendicularly to the laser stream. The ejector case is covered with the black paper in order to minimise distractions and refraction of the laser stream. Moreover, the safety box is covered in order to reduce undesirable external light sources.

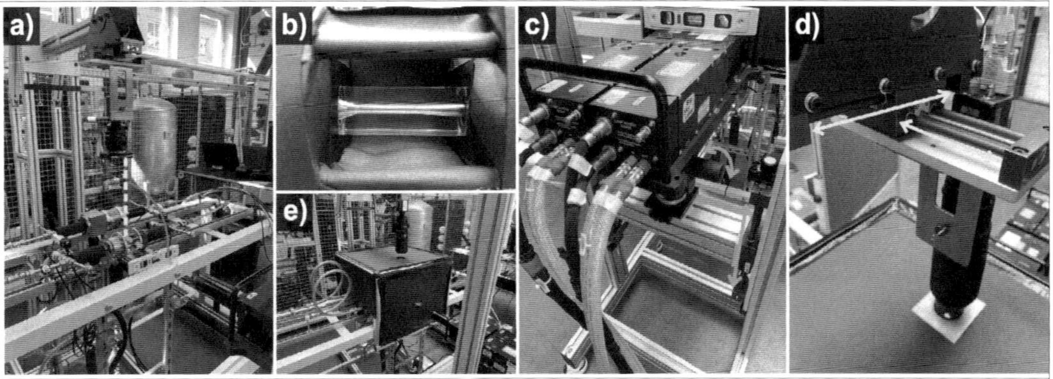

Fig.2. Arrangement of measurement stand for PIV approach.

Arrangement of the equipment in the second approach (i.e. high-speed camera approach) is less demanding. Positioning of the camera sensor (shown in Fig. 3) and compliance with angles between the glass not necessarily to be as precise as previously. On the other hand, the strong light source pointed on front and back sides of the transparent section was crucial in the visualisation process.

The high-speed camera type VEO340L produced by Photron was used with the same lens as described in PIV arrangement. The camera was able to record up to 200,000 frames per second with the resolution of 128x128px. The PCC Studio software was utilised for the recording set up and frames selection. The visualisation of the motive nozzle opening and steady state operation in off- and on-design conditions were considered as well as the ejector steady-state operation.

Fig. 3. High-speed camera and led lights arrangement

4. Selected results

Carried out experiments consists of visualisation of the steady-state operation of the ejector and the motive nozzle opening. The efficiency of the ejector [10] is equal to 19.99% ±1.0% with the mass entrainment ratio of 0.36 ±0.01. Absolute pressure of 80 bar and temperature equal to 32°C was set for the motive port. Suction and outlet ports were supplied by fluid having 28 bar and 8°C and 32 bar and -1°C respectively. The resulting motive and suction nozzle mass flow rates were at the level of 206 kg/h and 75 kg/h respectively. Due to the radial suction port resulting steady-state two-phase flow is a swirl and highly turbulent flow. Fluid having low value of the vapour quality is sucked along ejector wall and mixed with motive flow. This is clearly visible in Fig. 4 obtained with the help of the high-speed camera.

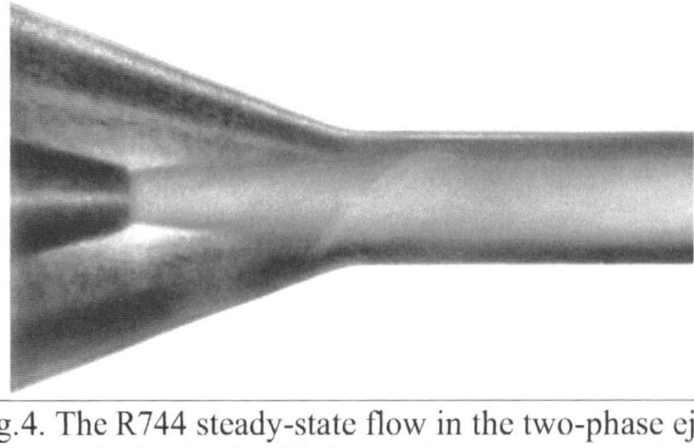

Fig.4. The R744 steady-state flow in the two-phase ejector obtained using high-speed camera approach

The same structure of the R744 flow has been also obtained using numerical modelling and is shown in Fig.5a. It is worth noting expansion of the motive flow (highlighted by orangish colour) leaving the suction nozzle and mixing with the sucked fluid (its flow is represented by path lines). In these perspective two angles describing flow and mixing processes are defined: expansion angle α (maximal and minimal ones) as well as mixing angle β. These two angles are clearly defined in Fig.5b.

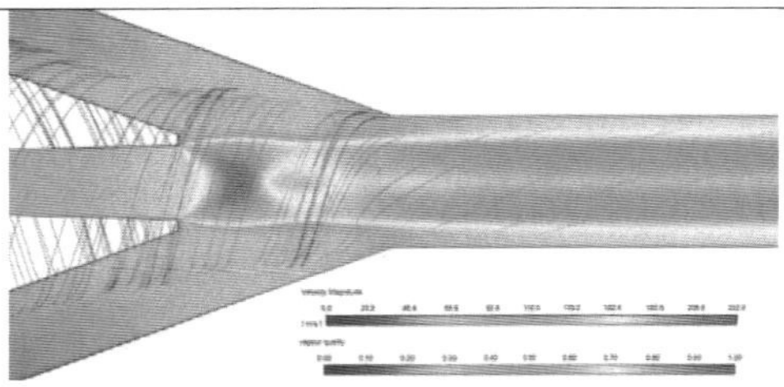

Fig. 5a. Typical investigated structure of the R744 flow in the two-phase ejector

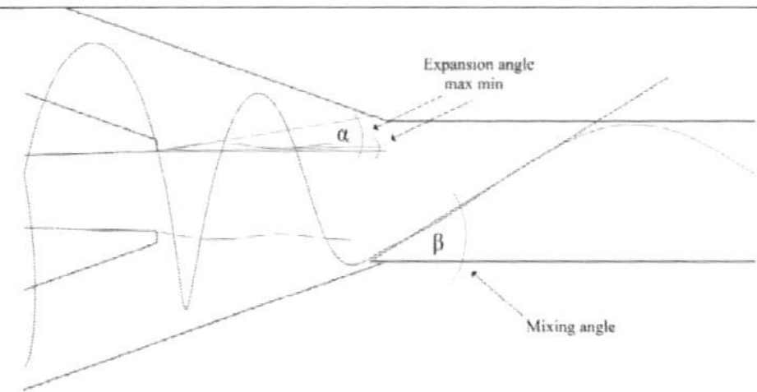

Fig. 5b. Definition of expansion and mixing angles

Processing recorded images the following results have been obtained: $\alpha \approx 5^{o}$ and $\beta \approx 36^{o}$. Although the recordings of the same steady-state flow obtained with the PIV system (see Fig.6) were more sharpen than those from the high-speed camera, the whole arrangement is more demanding mainly due to the small area where the laser and camera sensor should be precisely positioned.

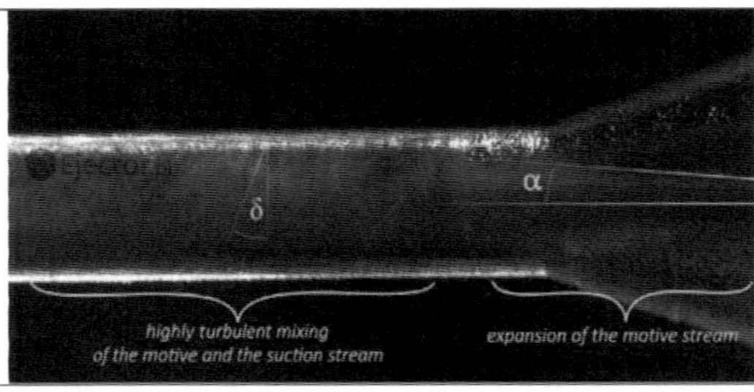

Fig.6. The R744 steady-state flow in the two-phase ejector obtained using PIV system

Colour map in Fig. 4a shows also position and shape of the first compression shock after opening of the motive nozzle valve. The high-speed camera allowed also to visualise this phenomenon at certain time moments. Fig. 7 presents image obtained at 25 milliseconds after opening of the motive nozzle valve.

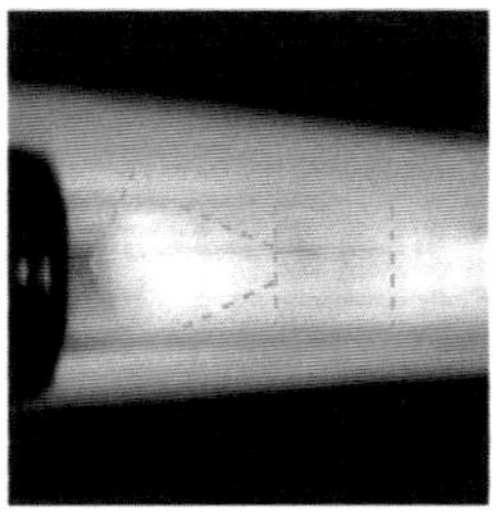

Fig.7. Frame from the motive nozzle opening process obtained high-speed camera

5. Final remarks

In this paper two possible approaches to visualise flow and mixing processes in the R744 ejector are presented and discussed. Comparing results from both techniques it seems that at this stage of the investigation the high-speed camera offers higher potential for quantitative analysis of the results. First of all, the equipment set up in the PIV system is very demanding due to the small area where the laser and camera sensor should be arranged keeping perpendicular orientation to certain surfaces. Secondly, the nature of the analysed flows gives a superiority to the high-speed camera with its large frames per second capability what is not easily achieved with the utilised PIV equipment. However, the recordings obtained with the PIV approach were more sharpen than those recorded with the high-speed camera.

The analysis of the high-speed camera recording requires support from the algorithms for grey-scale analysis to select frames with visible internal structures. Nevertheless, in both cases the analysis of the selected frame could deliver, based on the shape of the internal flow structures, a data for a choice of the proper turbulence model and validate numerical model.

Further work should be focused on the software development for reliable selection and analysis of the recordings. Analysis of the steady-state conditions with an opened suction nozzle should be investigated using high-speed camera, especially in the low speed region of the suction nozzle.

Acknowledgment

The research project realised under grant No. 2017/27/B/ST8/00945 funded by the National Research Center.

References

[1] Schecco. (2018). World Guide to CO2 Transcritical Refrigeration (Tech. Rep.).

[2] Gullo, P., Hafner, A., & Banasiak, K. (2018). Transcritical R744 refrigeration systems for supermarket applications: Current status and future perspectives. International Journal of Refrigeration, 93, 269310.

[3] Elbel, S., & Lawrence, N. (2016). Review of recent developments in advanced ejector technology. International Journal of Refrigeration, 62, 1-18.

[4] Besagni, G. (2019). Ejectors on the cutting edge: The past, the present and the perspective. Energy, 170, 9981003.

[5] Little, A. B., & Garimella, S. (2016). Shadowgraph visualization of condensing R134a flow through ejectors. International Journal of Refrigeration, 68, 118–129. Retrieved from http://dx.doi.org/10.1016/j.ijrefrig.2016.04.018 doi: 10.1016/j.ijrefrig.2016.04.018

[6] Gagan, J., Smierciew, K., Butrymowicz, D., & Karwacki, J. (2014). Comparative study of turbulence models in application to gas ejectors. International Journal of Thermal Sciences,

78, 9–15. doi: 10.1016/j.ijthermalsci.2013.11.009

[7] Hernández Cely, M. M., Baptistella, V. E., & Rodriguez, O. M. (2018). Study and characterization of gas-liquid slug flow in an annular duct, using high speed video camera, wiremesh sensor and PIV. Experimental Thermal and Fluid Science, 98, 563–575.

[8] Zhu, Y., Wang, Z., Yang, Y., & Jiang, P.X. (2017). Flow visualization of supersonic twophase transcritical flow of CO2 in an ejector of a refrigeration system. International Journal of Refrigeration, 74, 354–361. doi: 10.1016/J.IJREFRIG.2016.11.012

[9] Bodys, J., Smolka, J., Palacz, M., Haida, M., Banasiak, K., Nowak, A. J., & Hafner, A. (2016). Performance of fixed geometry ejectors with a swirl motion installed in a multiejector module of a CO2 refrigeration system. Energy, 117, 620–631. doi: 10.1016/J.ENERGY.2016.07.037

[10] Haida M., Fingas R., Szwajnoch W., Smolka J., Palacz M., Bodys J., Nowak A.J. (2020). A R744 Transcritical Refrigeration Test Rig Equipped with a Transparent Two-phase Ejector for Flow Visualisation: Experimental Study. IIR Rankine International Conference on Advances in Cooling, Heating and Power Generation, Rankine 2020, pp 539-546, doi: 10.18462/iir.rankine.2020.1215

[11] Taylor, B., & Kuyatt, C. (2001). Guidelines for Evaluating and Expressing the Uncertainty of NIST Measurement Results: Appendix D1. Terminology. Gaithersburg: National Institute of Standards and Technology, Standard Reference Data Program.

Adiabatic frictional pressure gradient during flow boiling of pure refrigerant R1233zd and non-azeotropic mixtures R448A, R452A and R455A

A Arcasi[1], R Mastrullo[1], A W Mauro[1*] and L Viscito[1]

[1] Department of Industrial Engineering, Federico II University of Naples, P.le Tecchio 80, 80125 Naples (Italy)
*Corresponding author e-mail: wmauro@unina.it

Abstract. The research on two-phase flow characteristics of refrigerants is of primary importance in several fields, such as air conditioning and refrigeration systems. Therefore, the determination of the pressure drop during flow boiling is important for the correct design of evaporators and heat spreaders systems. This paper presents a collection of experiments on flow boiling pressure drop using pure refrigerant R1233zd and new low-GWP refrigerant mixtures R448A, R452A and R455A. All tests were performed in adiabatic conditions, in a smooth horizontal stainless-steel tube having an internal diameter of 6.0 mm and a thickness of 1.0 mm. The effect of operating parameters, such as (bubble) saturation temperature (from 25 to 65 °C) and mass flux (from 150 to 600 kg/m^{2}s) is investigated and discussed, and the performance of the chosen fluids is also compared. Finally, an assessment of existing prediction methods is carried-out to find the most suitable correlations for two-phase pressure drop evaluation.

Keywords: *Flow boiling, low-GWP refrigerants, pressure drop, assessment*

1. Introduction

Nowadays, due to anthropic activities, the global warming issue is widely discussed. Particularly, since the early 2000s, the use of chlorofluorocarbon and hydro-chlorofluorocarbon refrigerants (CFCs and HCFCs) has been banned in several fields, with hydrofluorocarbon ones (HFCs) now being used thanks to their zero-ozone depletion potential (ODP). Nevertheless, HFC refrigerants are planned to be replaced due to their strong greenhouse effect. On this regard, the European F-Gas Regulation [1] restricts the global warming potential (GWP) value of the employed refrigerants in several sectors. For the abovementioned reasons, in the refrigerated transport sector as well as in the commercial refrigeration systems at low/medium temperature level the R404A refrigerant (GWP 3922) will be replaced with new low-GWP mixtures such as R448A (GWP 1390), R452A (GWP 2141) and R455A (GWP 148) having similar reduced pressure values. On the other hand, for low-pressure centrifugal chillers and for high temperature level systems such as organic Rankine cycles (ORC) – up to 120 °C - pure refrigerant R1233zd can be used thanks to its low reduced pressure, non-toxicity, non-flammability and GWP value equal to 1. In this context, a correct evaluation of the frictional pressure drop is useful both for a proper matching of the temperature level of the refrigerant and the secondary fluid in the evaporator and for a correct design of the system piping.

After a bibliographic research, for the abovementioned refrigerants there is a lack of frictional pressure gradient data in open literature. Some experimental studies provide two-phase pressure drop data of pure refrigerant R1233zd [2, 3, 4] during evaporation and condensation processes, whereas Jacob et al. [5] propose a comparison between the condensation pressure drop data of R404A and the substitutes R448A and R452A mixtures. Finally, Mauro et al. [6] provide frictional pressure drop data for the R455A mixture for different saturation temperatures and mass velocities. However, we found no assessment of the existing prediction methods for all the mentioned fluids and a thorough experimental campaign for pressure drop data of these new refrigerants has not been carried-out yet. For these reasons, this paper presents frictional pressure drop data of R1233zd, R448A, R452A and R455A refrigerants. The effect of the operating parameters in terms of bubble saturation temperature (from 25 to 65 °C, close to the working conditions of heat pumps for sanitary hot water production and to ORC systems, and corresponding to reduced pressures from 0.03 to 0.81) and mass flux (from 150 to 600 kg/m^2s) is discussed. The performance of the employed fluids in the same operating conditions is then compared. Finally, an assessment of the prediction methods is carried-out to find, among those implemented, the most appropriate correlation for the frictional pressure gradient.

2. Experimental apparatus

2.1. Refrigerant loop, test section and measurement instrumentation

The experimental apparatus is represented in Figure 1. In the primary loop (black line) the working fluid is pumped by a gear pump into a Coriolis mass flow meter. Then, the refrigerant passes through an electric pre-heater section where it partially evaporates and the desired vapor quality value is achieved. Successively, it passes through the test section where the pressure drop is measured; finally, the saturated fluid condenses into a plate heat exchanger and goes into a liquid receiver and a sub-cooler before closing the loop with the pump suction head. More information about the whole experimental test rig and its components can be found in previous publications of the same authors [7,8].

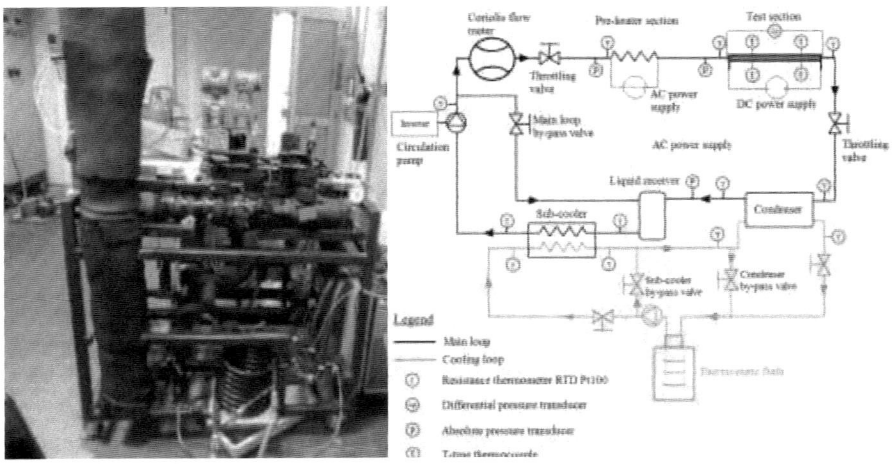

Figure 1. Photograph of the experimental apparatus and its schematic representation.

Table 1. Summary of operative range and accuracy of the measurement instrumentation.

Measurement	Range	Instrumental accuracy
Temperature	-80/250 °C	± 0.180 °C
Inlet absolute pressure (test section)	0/35 bar	± 0.1% reading
Absolute pressure (test rig)	0/50 bar	± 0.3% reading
Differential pressure (test section)	0/14.51 kPa	± 0.06 kPa
Flow meter	2.3/115.7 g/s	± 1.0% reading
Electric power	0/3.6 kW	± 1.0% reading

The test section is made of a horizontal stainless-steel tube (AISI SS316) having an inner diameter of 6.00 ± 0.05 mm and an external diameter of 8.0 ± 0.05 mm. Two pressure taps at a distance of 237 ± 0.91 mm are used both for the evaluation of the pressure drop across the test tube and the inlet absolute pressure value. Table 1 provides a summary of the instrument specifications of transducers and sensors employed for this experimental campaign.

3. Method

3.1. Data reduction and uncertainty analysis

The frictional pressure gradient across the test section is directly obtained by the pressure drop measurement. In fact, the gravitational contribution has not been considered due to the horizontal test section layout and the momentum contribution can be neglected since all tests are performed in adiabatic conditions, in order to obtain local values of the frictional pressure gradient. All the thermodynamics and transport properties of the chosen fluids are evaluated with the software Refprop 9.1 [9], developed by NIST, while the data reduction has been implemented with a MATLAB [10] code. The post processing of data has been integrated with the uncertainty analysis for both measured and derived parameters. Table 2 provides a summary of the operating parameters uncertainties (expanded uncertainty with a coverage factor equal to 2) for the whole database.

Table 2. Maximum uncertainty of operating parameters.

Parameter	Maximum uncertainty
(Bubble) saturation temperature $T_{sat,(B)}$	± 0.17 °C
Mass flux G	± 1.3%
Vapor quality x	± 0.3
Pressure gradient	± 29.0%

4. Experimental results

4.1. Experimental procedure and operating conditions

Each test of the experimental campaign for the employed fluids was performed in adiabatic and steady state conditions by setting mass flux, (bubble) saturation temperature and vapor quality values. Particularly, since the new low-GWP mixtures R448A, R452A and R455A have a considerable temperature glide during the evaporation process, the bubble saturation temperature has been fixed by adjusting the thermostatic bath set point. Pure refrigerant R1233zd and new low-GWP mixtures R448A, R452A and R455A have been studied in the same operating conditions such as (bubble) saturation temperature from 25 to 55 °C, mass flux from 150 to 600 kg/m^{2}s in order to compare their performance. Thermodynamic and transfer properties are shown in detail in Table 3 for the chosen fluids at the same bubble saturation temperature of 45 °C (common value for all refrigerants in this experimental campaign). The three mixtures vapor density values are very similar to each other as well as the liquid viscosity ones, whereas R1233zd vapor density value is significantly lower than the others and at the same time its liquid viscosity is substantially higher than the mixtures ones.

Table 3. Thermodynamic and transfer properties for the chosen fluids at the same bubble saturation temperature.

$T_{sat} = 45\ °C$	p_{red}	$\rho_V\ [kg/m^3]$	$\rho_L [kg/m^3]$	ρ_V/ρ_L	$\mu_V\ [\mu Pa-s]$	$\mu_L\ [\mu Pa-s]$	μ_V/μ_L
R1233zd	0.07	13.6	1212	0.01	11.8	364.2	0.03
R448A	0.45	82.6	1003	0.08	13.8	108.6	0.12
R452A	0.53	109.4	1020	0.1	14.4	103.7	0.14
R455A	0.45	68.9	965	0.07	13.2	106.2	0.12

4.2. Effect of the operating parameters on the frictional pressure gradient

Figures 2-5 show the frictional pressure gradient as a function of increasing vapor quality for pure refrigerant R1233zd and R448A, R452A, R455A mixtures, respectively. The effect of operating conditions in terms of mass flux and (bubble) saturation temperature is illustrated. In all cases the pressure gradient firstly increases with the increasing vapor quality until a peak value is reached and then it decreases. As regards the pure refrigerant R1233zd, Figure 2a shows the effect of the mass flux value from 150 to 500 kg/m^{2}s at a saturation temperature of 64.9 °C: due to the increasing flow velocity, the pressure gradient significantly increases with the mass flux. Figure 2b shows the effect of the saturation temperature from 25 to 65 °C at a mass flux value of 304 kg/m^{2}s: the pressure gradient decreases with the increase of the saturation temperature, due to an increased vapor-to-liquid density ratio. This phenomenon is more evident for high vapor qualities, in which the differences between the liquid and vapor velocities are more substantial. By fixing the vapor quality of 0.7, it passes from 20.0 kPa/m at a T_{sat} of 25 °C to 13.4 kPa/m at T_{sat} of 45 °C (-33%) and to 8.4 kPa/m at T_{sat} of 65°C (-58%). For low vapor qualities, instead, all values blend one another and small differenced fall within the uncertainty values.

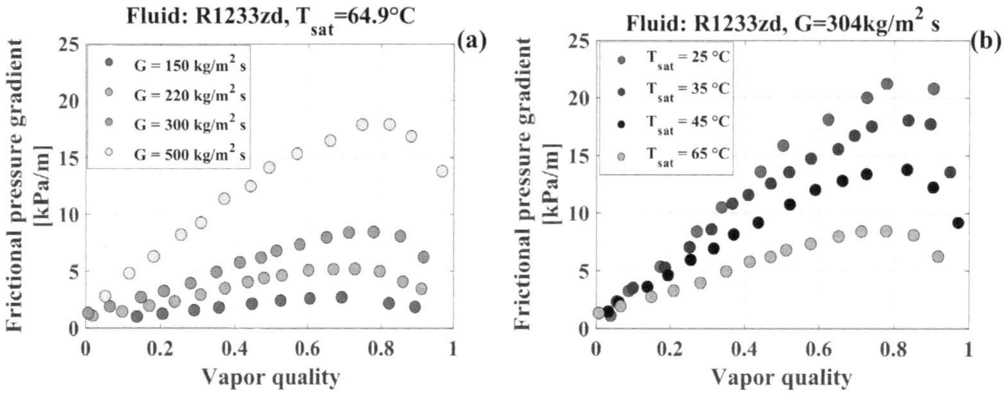

Figure 2. Frictional pressure gradient as a function of vapor quality for R1233zd. Effect of operating parameters. (a) Effect of mass flux for a saturation temperature of 64.9 °C; (b) Effect of the saturation temperature for a fixed mass flux value of 304 kg/m^{2}s.

The same trend in terms of pressure gradient can be observed for the three employed mixtures: as regards the refrigerant R448A, Figure 3a shows the effect of the mass flux value from 150 to 600 kg/m^{2}s at a bubble saturation temperature of 25 °C: the pressure gradient increases with increasing mass flux and it raises from 1.0 kPa/m at a mass flux of 150 kg/m^{2}s to 3.95 kPa/m at 400 kg/m^{2}s (+295%) and to 6.97 kPa/m at 600 kg/m^{2}s (+597%), when the vapor quality is fixed to 0.3. Figure 3b shows the effect of the bubble saturation temperature from 25 to 55 °C at a fixed mass flux of 404 kg/m^{2}s. At a vapor quality of 0.5, the pressure gradient decreases from 6.1 kPa/m at $T_{sat,B}$ of 25°C to 5.0 kPa/m at $T_{sat,B}$ of 35 °C (-18%) and to 1.9 kPa/m at $T_{sat,B}$ of 55 °C (-69%).

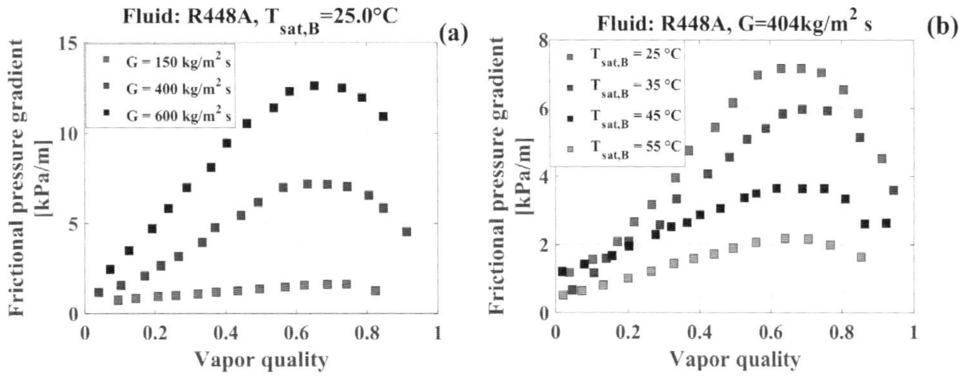

Figure 3. Frictional pressure gradient as a function of vapor quality for R448A. Effect of operating parameters. (a) Effect of mass flux for a bubble saturation temperature of 25.0 °C; (b) Effect of the bubble saturation temperature for a fixed mass flux value of 404 kg/m^{2}s.

For the R452A mixture, three experimental tests were carried-out at the same bubble saturation temperature at a mass flux of 150, 400 and 600 kg/m^{2}s respectively. Figure 4a shows the effect of the mass flux value on the pressure gradient: the pressure gradient increases with the mass flux and it passes from 0.9 kPa/m at a mass flux of 150 kg/m^{2}s 5.7 kPa/m at 600 kg/m^{2}s (+533%) when the vapor quality is set to 0.3. Figure 4b shows the effect of the bubble saturation temperature from 25 to 45 °C at a mass flux value of 402 kg/m^{2}s: the pressure gradient decreases with the increasing bubble saturation temperature. By fixing the vapor quality to 0.7, it reduces from 5.5 kPa/m at $T_{sat,B}$ of 25 °C to 2.6 kPa/m at $T_{sat,B}$ of 45 °C (-111%).

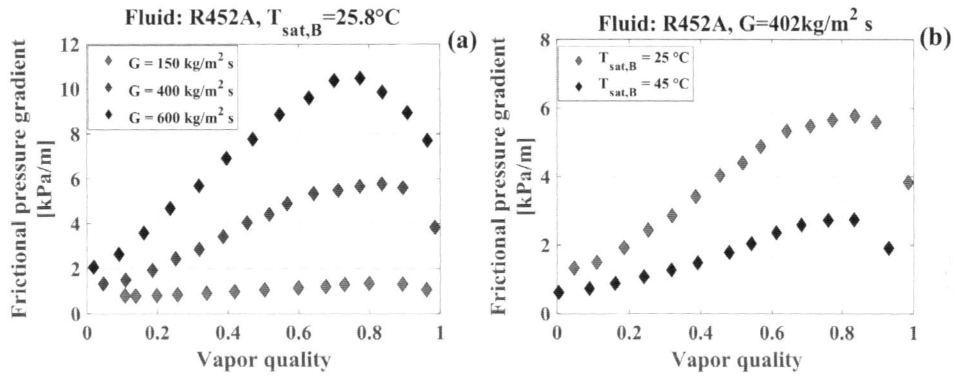

Figure 4. Frictional pressure gradient as a function of vapor quality for R452A. Effect of operating parameters. (a) Effect of mass flux for a bubble saturation temperature of 25.8 °C; (b) Effect of the bubble saturation temperature for a fixed mass flux value of 402 kg/m^{2}s.

Finally, for R455A mixture Figure 5a shows the mass flux effect on the frictional pressure gradient from 300 to 500 kg/m^{2}s at a bubble saturation temperature of 35 °C. Particularly, the frictional pressure gradient grows from 3.3 kPa/m at a mass flux of 150 kg/m^{2}s to 7.6 kPa/m at 500 kg/m^{2}s (+130%) when vapor quality is 0.6. Figure 5b shows the effect of the bubble saturation temperature from 35 to 55 °C at a mass flux value of 495 kg/m^{2}s: the frictional pressure gradient decreases with the bubble saturation temperature. At a vapor quality of 0.5, indeed, it decreases from 6.8 kPa/m at $T_{sat,B}$ of 35 °C to 5.3 kPa/m at $T_{sat,B}$ of 45 °C (-22%) and to 3.9 kPa/m at $T_{sat,B}$ of 55 °C (-42%).

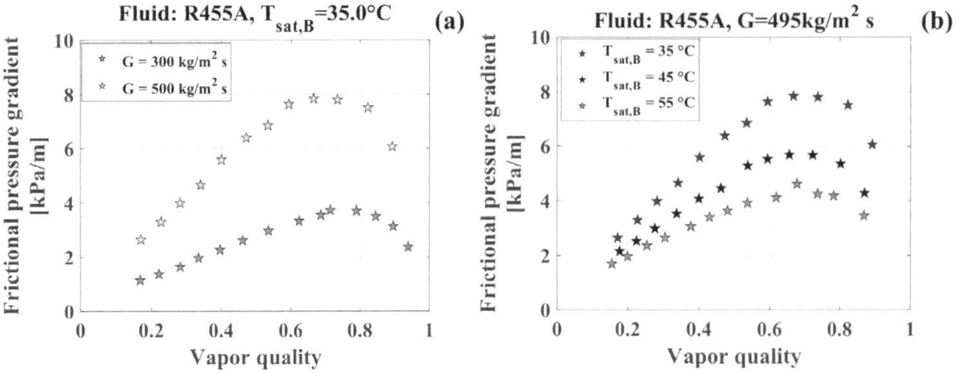

Figure 5. Frictional pressure gradient as a function of vapor quality for R455A. Effect of operating parameters. (a) Effect of mass flux for a bubble saturation temperature of 35.0 °C; (b) Effect of the bubble saturation temperature for a fixed mass flux value of 495 kg/m²s.

4.3. Effect of the working fluid

The frictional pressure gradient of the chosen fluids is shown in Figure 6 for the same bubble saturation temperature of 25 °C and mass flux value of 152 kg/m²s. It is worth noting that, in these conditions, the frictional pressure gradient for pure refrigerant R1233zd is significantly higher than mixtures ones. This behavior can be explained since the low reduced pressure value of R1233zd leads to a low vapor-to-liquid density ratio and consequently a considerable shear stress. As a result, the frictional pressure gradient of R1233zd is more than 100% higher than that of R448A, R452A and R455A. Specifically, when the vapor quality is fixed to 0.3 the frictional pressure gradient value of pure refrigerant R1233zd is 2.9 kPa/m whereas the ones for R448A, R452A and R455A are 1.0 kPa/m, 0.9 kPa/m and 0.85 kPa/m respectively. By fixing the vapor quality value to 0.7 the frictional pressure gradient of R1233zd is 6.6 kPa/m while the ones for R448A, R452A and R455A are 1.6 kPa/m, 1.3 kPa/m and 0.7 kPa/m respectively.

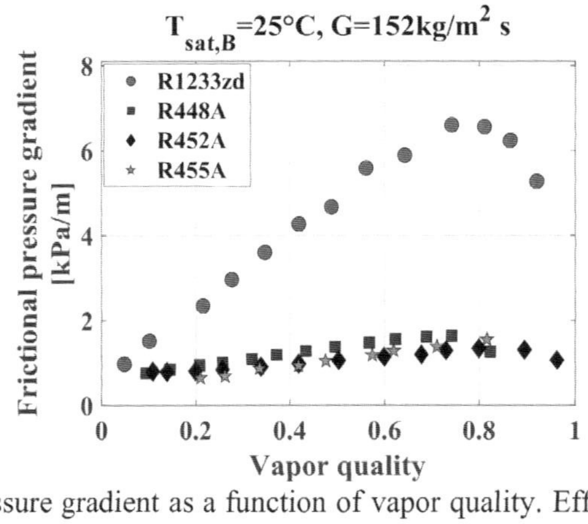

Figure 6. Frictional pressure gradient as a function of vapor quality. Effect of the chosen fluids for a bubble saturation temperature of 25 °C and a mass flux value of 152 kg/m²s.

5. Assessment of the existing prediction methods

More than 10 two-phase pressure drop prediction methods have been implemented, including both homogeneous and drift flux approaches. In this assessment, only the four correlations providing the best statistics (MAPE lower than 40%) are presented, regardless their rationale or the original operating conditions for which they have been developed. Müller-Steinhagen and Heck (1986) [11] method evaluates the pressure drop as a geometrical average on vapor quality of the all-liquid and all-vapor frictional pressure drop. This method produces a slight underestimation for the refrigerants R1233zd, R448A and R452A, whereas it slightly overestimates the R455A database, with more of 52% of the whole data points falling into a range of error of ±30%. Cioncolini-Thome (2017) [12] and Friedel (1979) [13] correlations consider a separated flow approach of the two-phases for the prediction of the frictional pressure drop: it is found that these methods provide a very good agreement with the whole dataset, with more of 70% and 73% of data points falling into a range of error of ±30% respectively. It is worth noting that the Cioncolini-Thome [12] correlation was originally developed only for annular flow: for this reason, the assessment is carried-out by excluding points that are likely (a visualization section is not available) to fall into the bubbly flow region and in the stratified region. Practically, Figure 7b includes only experiments with vapor qualities higher than 0.3 and at the same time mass flux values higher than 150 kg/m^{2}s. The statistics of this annular flow correlation slightly worsens when applied to the whole dataset, with a MAPE of approximately 30% for all the investigated substances. Finally, Cicchitti et al (1960) [14] method was developed for homogenous flow and displays a slightly underestimation for refrigerants R1233zd, R448A and R452A whereas it provides a very good agreement for the R455A dataset.

Figures 7a-d show the assessment results for the abovementioned correlations and a summary of the statistical parameters (MRPE, MAPE, STD and percentage of data falling into an error range of ±30%, defined as $\delta_{\pm 30\%}$) is provided in Table 4.

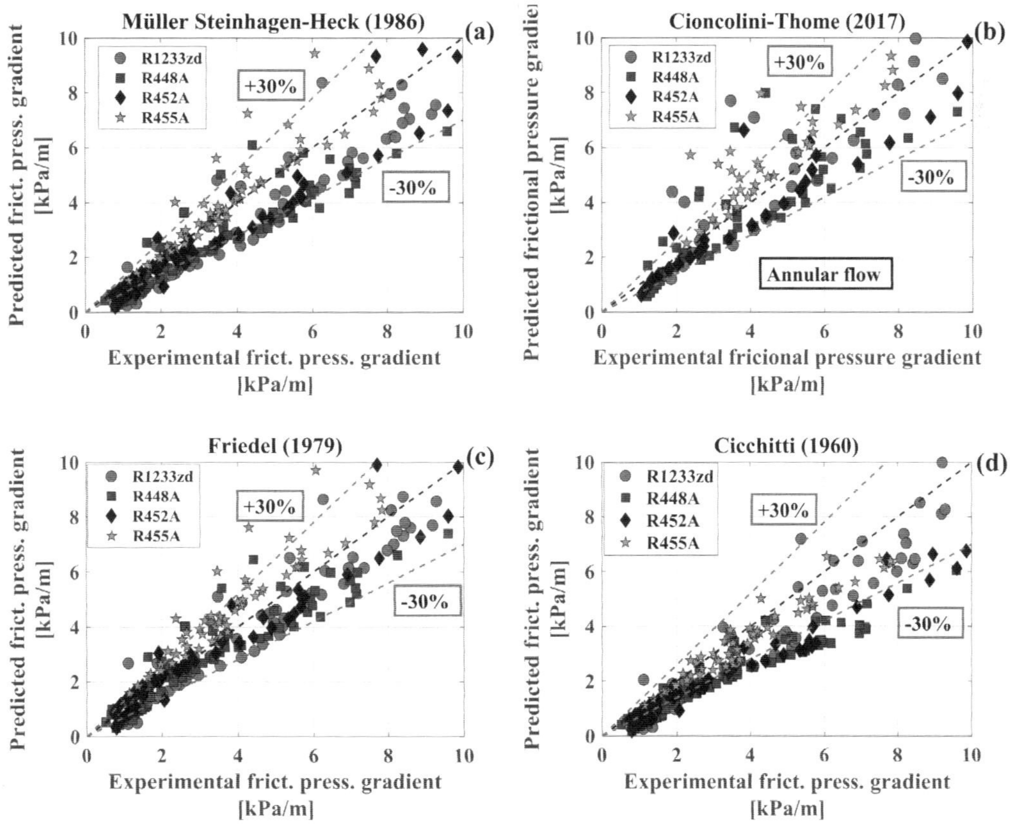

Figure 7. Experimental vs predicted frictional pressure drop. (a) Müller-Steinhagen and Heck (1986) method; (b) Cioncolini-Thome (2017) correlation; (c) Friedel (1979) method; (d) Cicchitti (1960) correlation.

Table 4. Assessment results for the chosen frictional pressure drop predictive methods.

	R1233zd [%]	**R448A [%]**	**R452A [%]**	**R455A [%]**
Müller-Steinhagen and Heck (1986) [11]	MAPE: 25.0 MRPE: -16.6 STD: 24.4 $\delta_{\pm30\%}$: 65.6	MAPE: 28.8 MRPE: -23.5 STD: 23.2 $\delta_{\pm30\%}$: 53.9	MAPE: 26.6 MRPE: -23.7 STD: 21.2 $\delta_{\pm30\%}$: 65.4	MAPE: 11.0 MRPE: 8.9 STD: 18.1 $\delta_{\pm30\%}$: 92.2
Cioncolini-Thome (2017) [12]	MAPE: 25.2 MRPE: 14.8 STD: 36.4 $\delta_{\pm30\%}$: 75.8	MAPE: 24.5 MRPE: -6.2 STD: 29.9 $\delta_{\pm30\%}$: 81.8	MAPE: 18.7 MRPE: -6.9 STD: 22.6 $\delta_{\pm30\%}$: 81.2	MAPE: 26.4 MRE: 26.4 STD: 29.3 $\delta_{\pm30\%}$: 71.4
Friedel (1979) [13]	MAPE: 17.8 MRPE: -6.0 STD: 24.3 $\delta_{\pm30\%}$: 90.6	MAPE: 21.4 MRPE: -7.0 STD: 25.2 $\delta_{\pm30\%}$: 78.6	MAPE: 18.7 MRPE: -7.6 STD: 22.2 $\delta_{\pm30\%}$: 80	MAPE: 23.1 MRPE: 13.4 STD: 29.0 $\delta_{\pm30\%}$: 73.4
Cicchitti et al (1960) [14]	MAPE: 21.3 MRPE: -15.1 STD: 21.9 $\delta_{\pm30\%}$: 72.9	MAPE: 35.3 MRPE: -35.1 STD: 15.1 $\delta_{\pm30\%}$: 33.7	MAPE: 35.1 MRPE: -35.1 STD: 15.3 $\delta_{\pm30\%}$: 29.1	MAPE: 12.0 MRPE: -9.0 STD: 10.4 $\delta_{\pm30\%}$: 100

6. Conclusions

Experimental adiabatic frictional pressure gradient data of pure refrigerant R1233zd and new low-GWP mixtures R448A, R452A and R455A have been obtained in a 6.00 mm horizontal stainless-steel tube. The main outcomes of this work are summarized as follow:

- The frictional pressure gradient increases with vapor quality up to a peak value before dropping. As regards the effect of the other parameters, it increases with the mass velocity whereas it decreases with the rise of the bubble saturation temperature. This trend has been observed for any operating condition investigated and for any fluid.
- When compared to the three mixtures data, pure refrigerant R1233zd frictional pressure gradients are significantly higher due to its lower reduced pressure and therefore low vapor-to-liquid density ratio: at a mass flux value of 150 kg/m^{2}s and a bubble saturation temperature of 25 °C the percentage difference reaches 560%.
- The assessment of the two-phase frictional pressure gradient predictive methods has shown that the correlation of Friedel [13] provides the best agreement with the experimental data, with more of 73% of data falling into an error band of ±30% MAPE index respectively equal to: 23.1% (R455A), 18.7% (R452A), 21.4% (R448A) and 17.8% (R1233zd).

References

[1] The European Parliament and the Council of the European Union 2014 *Off. J. Eur. Union* **150** 195-230
[2] Huang H, Borhani N, Thome J R 2016 *Int. J. Heat Mass Transfer* **98** 596-610
[3] Xiao J, Hrnjak P 2018 *Int. J. Heat Mass Transfer* **122** 442-450
[4] Zhang J, Kaern M R, Ommen T, Elmegaard B, Haglind F 2019 *Int. J. Heat Mass Transfer* **128** 136-149
[5] Jacob T A, Matty E P, Fronk B M 2020 *Int. J. Refrigeration* **116** 9-22
[6] Mauro A W, Napoli G, Pelella F, Viscito L 2020 *Int. J. Refrigeration* **119** 195-205
[7] Mastrullo R, Mauro A W, Revellin R, Viscito L 2018 *Appl. Therm. Eng.* **145** 251-263
[8] Lillo G, Mastrullo R, Mauro A W, Pelella F., Viscito L 2019 *Appl. Therm. Eng.* **161** 114146
[9] Lemmon E W Mc Linden M O, Huber M L, 2009 *REFPROP NIST* Database **23**
[10] MATLAB 2019a Release, *Mathworks*
[11] Müller-Steinhagen H, Heck K 1986 *Chem. Eng. Process* **20** 297-308
[12] Cioncolini A, Thome J R 2017 *Int. J Multiphase Flow* **89** 321-330
[13] Friedel L1979 *Proceedings*
[14] Cicchitti A, Lombardi C, Silvestri M, Soldaini G, Zavalluilli R 1960 *Energia Nucl.* **7** 407-425

Mechanisms of dropwise condensation on aluminum coated surfaces

Matteo Mirafiori[1], Marco Tancon[1], Stefano Bortolin[1], Alessandro Martucci[1], Davide Del Col[1*]

[1] University of Padova, Department of Industrial Engineering
Via Venezia 1, 35131 Padova, Italy

*Corresponding author e-mail: davide.delcol@unipd.it

Abstract. Dropwise condensation (DWC) is a complex phase-change phenomenon involving the formation of randomly distributed droplets on the condensing surface. The promotion of DWC instead of the traditional filmwise condensation (FWC) is a promising solution to enhance the efficiency of heat exchangers by increasing the condensation heat transfer coefficient. The interaction between the condensing fluid and the surface (wettability) is important in defining the condensation mode. On metallic surfaces widely employed in heat transfer applications, the condensing process occurs in filmwise mode. Ideally, an engineered surface designed to achieve high DWC heat transfer coefficients should present low contact angle hysteresis and low thermal resistance. Among the different available techniques to modify the surface wettability, hybrid organic-inorganic sol-gel silica coatings functionalized with hydrophobic moieties (phenyl or methyl groups) have been identified as a feasible solution to promote DWC on metallic surfaces. In the present paper, different aluminum sol-gel coated surfaces have been tested during DWC of steam in saturated conditions. The realized coatings have been characterized by means of dynamic contact angles and coating thickness measurements. Condensation tests have been performed using a two-phase thermosyphon loop operating in steady-state conditions that allows visualization of the condensation process and simultaneous heat transfer measurements. Heat transfer coefficients have been measured by varying the heat flux, at 106 °C saturation temperature and with vapor velocity equal to 2.7 m s^{-1}. A high-speed camera is used for the visualization of the DWC process.

1. Introduction

Water vapor condensation is a two-phase heat transfer process encountered in many engineering applications as thermal power plants, heat pipes, sea water desalinization and buildings heating/cooling systems. An improvement in the efficiency of the condensation process can have a significant effect on the reduction of energy consumptions. Researchers have identified dropwise condensation (DWC) as a potential solution for enhancing the efficiency of the filmwise condensation (FWC) heat transfer process. During classic FWC of pure steam, the presence of a continuous liquid film which forms in correspondence of the heat exchanger walls contributes to the main thermal resistance. On the contrary, DWC involves the formation of discrete droplets and their rapid removal is entrusted by an external force (e.g. gravity or vapor drag force). The heat transfer coefficient (HTC) during DWC is usually 5–7 times higher as compared to the one measured during filmwise condensation [1].

The interaction between the surface and the condensing fluid plays a crucial role in determining the condensation mode (filmwise or dropwise). Coatings that are able to promote DWC over metallic substrates usually present the following characteristics: low surface energy compared to the surface tension of the condensing fluid (for this reason is more difficult to achieve DWC with low surface tension fluids) and high droplet mobility. For the evaluation of droplet mobility, the sessile drop method can be used to measure dynamic contact angles (advancing θ_a and receding θ_r contact angle): a small difference between the advancing and the receding contact angle, named contact angle hysteresis $\Delta\theta$, is associated to high droplet mobility.

DWC is a quasi-cyclic process with characteristic timescales, fraction of the surface covered by droplets, and drop-size distribution [2]. DWC begins at the molecular scale with droplets nucleation in preferred sites. Drops grow by direct condensation at first and later by coalescence until they reach the critical dimension at which the external forces (e.g., gravity, vapor drag) overcome the retention forces and they start to move. Sliding droplets sweep the surface and make new nucleation sites available. Since the renewal mechanism is directly affected by the droplet mobility, promoting droplets sweeping was found to increase the heat transfer coefficient of the DWC process [3].

Thanks to the promising heat transfer enhancement given by DWC and to the availability of new surface coatings, research is nowadays focused on both experimental investigation (heat transfer measurements, visual observations of the droplet population) and mathematical modelling of the DWC. The models are aimed at describing all the processes involved during DWC: the nucleation of a drop, its growth until its departure, the heat exchanged by the drop during its lifetime, the droplet removal and the resulting population of the droplets on the surface. In particular, droplets population can be divided in two parts: the small droplet population characterized by droplets growing by direct steam condensation and the large droplet population in which the coalescence between neighbors drops is the dominant growing mechanism [4]. For the large droplets population, the droplet sizes distribution was proposed by Rose[1].

As demonstrated by several authors [4], the heat transferred through a single droplet is the result of several thermal resistances. In particular, the presence of a coating used to modify the wettability of the metallic surface introduces an additional thermal resistance that in turn depends on the thickness of the coating and on the thermal conductivity (and thus on the chemistry of the coating). When the coating thermal resistance is not negligible, the resulting overall heat transfer coefficient will be lower and the advantage given by DWC can be reduced as reported in Parin et al. [5].

In the present work, DWC of steam has been experimentally investigated on three different aluminum sol-gel coated surfaces. Heat transfer coefficients measurements and visualizations of droplets population have been performed at a constant vapor velocity of 2.7 m s^{-1} and varying the heat flux between 70 and 550 kW m^{-2}. The recorded high-speed videos have been analyzed by using a home-made MATLAB® program with the aim to link the droplet dynamic to the heat transfer coefficient for the different coated substrates.

2. Experimental setup

The experimental apparatus used for condensation tests is a two-phase thermosiphon loop which consists of a boiling chamber, a test section, a cooling water circuit, and a post-condenser. The layout of the experimental test rig is shown in Fig. 1a. Steam, which is generated in the boiling chamber by means of electrical heaters (with a maximum power of 4 kW), enters the test section in saturated conditions and it is partially condensed over a vertical aluminum sample (condensing area 50 mm high by 20 mm wide). The surface of the sample exposed to the condensing steam (front side) is coated with a sol-gel silica layer. To obtain continuous condensation in the test section, heat is removed from the back side of the sample by using chilled water provided by an external thermal bath at a controlled flow rate and inlet temperature. Vapor condensation inside the pipeline between the boiling chamber and the test section is avoided by using electrical resistance wires to balance the heat losses through the ambient. Downstream the test section, the two-phase mixture enters a post-condenser where the vapor is fully condensed and subcooled. The subcooled liquid returns to the boiler chamber driven by the buoyancy force due to the difference between the liquid and the vapor density and it closes the circuit.

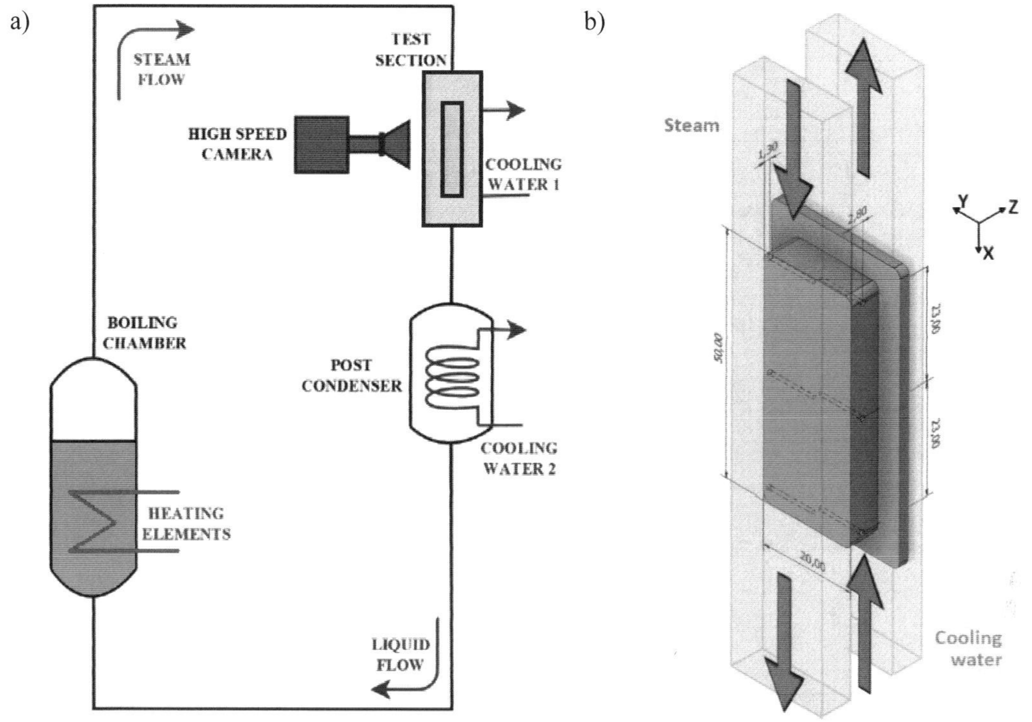

Figure 1. Experimental apparatus: a) schematic view of the thermosiphon loop; b) 3D model of the test section and the aluminum sample (dimensions in mm). Orientation of *x-y-z* axis is also shown.

The average condensation heat flux over the surface is obtained from the coolant mass flow rate $\dot{m}_{cool}$ and the temperature difference measured at the inlet and outlet of the test section ΔT_{cool} as follows:

$$q_{mean} = \frac{\dot{m}_{cool}\, c_{cool}\, \Delta T_{cool}}{A} \tag{1}$$

In Eq. (1), c_{cool} is the specific heat capacity of the coolant and A is the area of the condensing surface. Six thermocouples are placed inside the metallic specimen at two different depths from the condensing surface (z_1 = 1.3 mm and z_2 = 2.8 mm) and in correspondence of three longitudinal positions x along the sample (Fig. 1b). Starting from the thermocouples temperature measurements, the Fourier's law can be applied to obtain the surface temperature ($T_{wall,loc}$) for each of the three longitudinal positions:

$$T_{wall,loc} = T_{z_1} + \left(T_{z_1} - T_{z_2}\right)\frac{z_1}{z_2 - z_1} \tag{2}$$

In Eq. (2), T_{z_1} and T_{z_2} are the temperatures measured at depths z_1 and z_2 respectively. The three local values of surface temperature $T_{wall,loc}$ are then averaged over the whole condensing area to obtain the mean value of the HTC:

$$\text{HTC} = \frac{q_{mean}}{T_{sat} - T_{wall}} \tag{3}$$

where T_{sat} is the saturation temperature of the steam. The thermodynamic and transport properties of the steam are evaluated by means of NIST Refprop version 10.0 [6]. Each experimental data reported in this work is the mean of 480 readings acquired at 1 Hz and the uncertainty bars are calculated with a coverage factor k = 2. Additional information about the test rig, the measuring technique and the procedure for the evaluation of the experimental uncertainty are reported in Parin et al. [7].

3. Experimental results and discussion

Condensation tests have been performed in steady-state conditions and varying the heat flux from 70 to 550 kW m^{-2}, while maintaining a constant steam velocity (≈ 2.7 m s^{-1}), saturation temperature (≈ 106 °C) and coolant mass flow rate (0.11 kg s^{-1}). High-speed videos have been recorded at a constant value of heat flux (≈ 440 kW m^{-2} as show in Fig. 2a) in order to investigate the effect of the coating on droplets mobility. Experimental data have been obtained on three aluminum samples treated with different sol-gel coatings: coating #1 (Parin et al. [5]), coatings #2 and #3 (Parin et al. [7]). The three surfaces have been characterized by means of dynamic contact angle measurements using the sessile drop method and by coating thickness measurements. The results of surface characterization are summarized in Table 1. After functionalization, the three surfaces showed a nearly hydrophobic behavior with advancing contact angle θ_a ≈ 90° and receding contact angle θ_r ≈ 60°. According to the work proposed by Cha et al. [8], the reduced contact angle hysteresis (Table 1) improves droplet mobility and it allows to promote DWC even on surfaces with relatively low advancing contact angle (e.g. below 90°). With the aim to avoid possible changes of surface wettability due to the exposure in harsh environment, the samples have been tested for approximately 2 hours, which is a time interval shorter if compared to the coating lifetimes.

Table 1. Advancing and receding contact angles, contact angle hysteresis and coating thickness for the three investigated sol-gel coatings.

	θ_a [°]	θ_r [°]	$\Delta\theta$ [°]	δ_{coat} [nm]
Coating #1 [5]	89 ± 1	64 ± 3	25 ± 2	250 ± 5
Coating #2 [7]	83 ± 3	57 ± 6	26 ± 5	300 ± 25
Coating #3 [7]	87 ± 3	64 ± 2	23 ± 3	420 ± 35

In Fig. 2, the effect of the heat flux on the HTC during DWC is presented for the three coated aluminium samples. As shown in Fig. 2a, the saturation-to-wall temperature differences (ΔT) increases almost linearly with the heat flux and therefore the HTC remains nearly constant over the whole range of wall subcooling here investigated (Fig. 2b). The performed experiments showed that dropwise condensation can be sustained on the aluminum samples realizing thermal fluxes up to 500 kW m^{-2} with a temperature difference between saturation and wall below 6 K. On the contrary, in order to obtain a similar heat flux (≈ 400 kW m^{-2}) during filmwise condensation (FWC), the temperature difference would be around 30 K [9]. With coatings #2 and #3, the HTC is almost 5 to 6 times higher as compared to the HTC measured during FWC in similar conditions.

The sample treated with coating #1 exhibits the highest values of the HTC (between 180 and 190 kW m^{-2} K^{-1}) while, for coatings #2 and #3, the measured HTCs are significantly lower, with values that are in the range 85-100 kW m^{-2} K^{-1}. It must be pointed out that, considering the present experimental technique, the measured HTC takes into account the thermal resistances due to condensation process [4] plus the thermal resistance of the coating. The latter is strongly influenced by the thickness of the coating as deeply discussed in [5]. In particular, the thermal resistance associated with the conduction through the coating increases with the thickness of the coating. As it will be shown in the following analysis, the difference in the HTCs measured with the three coatings can only be partially explained by the different thickness of the DWC promoter layer (considering the same value of thermal conductivity for all the three sol-gel coatings). In order to predict the effect of the coating layer thermal resistance on the HTC, the dropwise condensation model by Kim and Kim [4] was considered. The thermal resistances per unitary area were calculated as the ratio of the film thickness (see Table 1) to the coating thermal conductivity λ supposed equal to 0.2 W m^{-1} K^{-1}. Considering that the wettability is almost the same for the three coatings (Table 1) as well as the other input parameters needed for the application of the model, the calculated HTC is affected only be the coating thermal resistance, which is higher for coating #2 and #3 (30% and 70% respectively as compared with coating #1 thermal resistance ≈ 1.32 K MW^{-1}). The HTC predicted for coating #1 is respectively increased by 10% and 25% as compared to the HTC calculated for coatings #2 and #3. Therefore, as shown in Fig. 2, the HTC increase measured with coating #1 is not only due to the lower thickness of the coating.

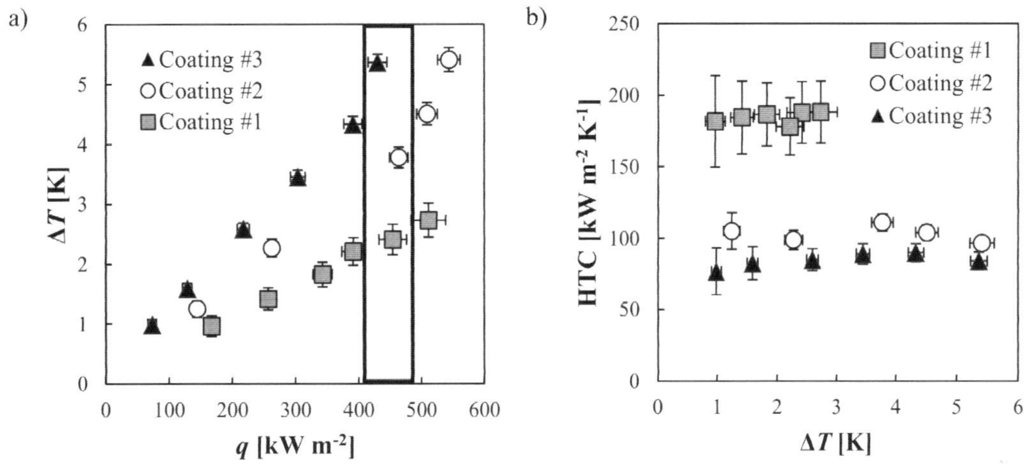

Figure 2. a) Saturation-to-wall temperature difference ($\Delta T = T_{sat} - T_{wall}$) versus heat flux ($q$) and b) heat transfer coefficient (HTC) versus temperature difference ΔT measured during DWC tests on the three aluminum samples (named coating #1, coating #2 and coating #3).

As a second step, heat transfer measurements are analysed considering also droplets dynamics. In order to study the droplets mobility, in particular to investigate the surface renewal time, the recorded videos were analysed to evaluate the fraction f of the surface area covered by droplets, which was firstly introduced by Rose [1]. Considering the droplets detection threshold of the present optical technique, the fraction f is determined from Eq. 4 as reported in Parin et al. [7]

$$f = \frac{A_c}{A_{tot}} \tag{4}$$

where A_c is the area of the condensing surface occupied by drops with radius greater than 200 μm (which represents the recognition threshold) and A_{tot} is the total investigated area which is the same for the three samples analysed (≈ 2.85 cm^{-2}). It must be noted that, in the present work, the parameter f is evaluated considering only a fraction of the large droplet population since the video acquisition system (which consists of a LED plus a high-speed camera Photron FASTCAM UX100 coupled with a macro-lens) allows to detect droplets down to 200 μm radius [7]. When f approaches to unity it means that the inspected area is almost entirely covered by drops whereas, when f is equal to zero, it means that there are no visible drops in the inspected area with radii larger than 200 μm. With the aim of evaluating the f parameter, each recorded frame extracted from the high-speed videos has been analysed using a home-made MATLAB® program for droplet recognition. The program allows to detect both the position and the radius of the droplets within the inspected area. A detailed description of the MATLAB® program can be found in [7].

The three graphs in Fig. 3a show the evolution against time of the fraction area f measured considering the three different samples. The analysis has been performed at a constant heat flux of ≈ 440 kW m^{-2} and with a fixed time interval equal to 1 s. For coating #1, the evolution of the fraction area with time (graph on the left in Fig. 3a) has been linked to the images of the droplet population reconstructed using the MATLAB® program (Fig. 3b). From the experimental data, it results that f can be described as a periodic function with its characteristic amplitude and frequency. Higher the amplitude and the frequency of the f function, shorter the surface renewal time and more efficient the sweeping mechanism that makes new nucleation sites available. The surface renewal frequency for coating #1 is around 1.8 Hz and this value results respectively 20% and 35% higher than the measure one for coating #2 and #3 (Fig. 3a). As expected, when the fraction area f reported in Fig. 3a is compared against HTC measurements (Fig. 2b), a higher surface renewal frequency corresponds also to higher values of the HTC. Looking at the amplitude of the f function, it can be observed that, despite of the similar surface renewal time, the amplitude for coating #2 is higher compared to the amplitude measured for the coating #3: this means that the sweeping mechanism cleans larger portions of the investigated area in the case of the coating #2, making the droplet renewal process more efficient.

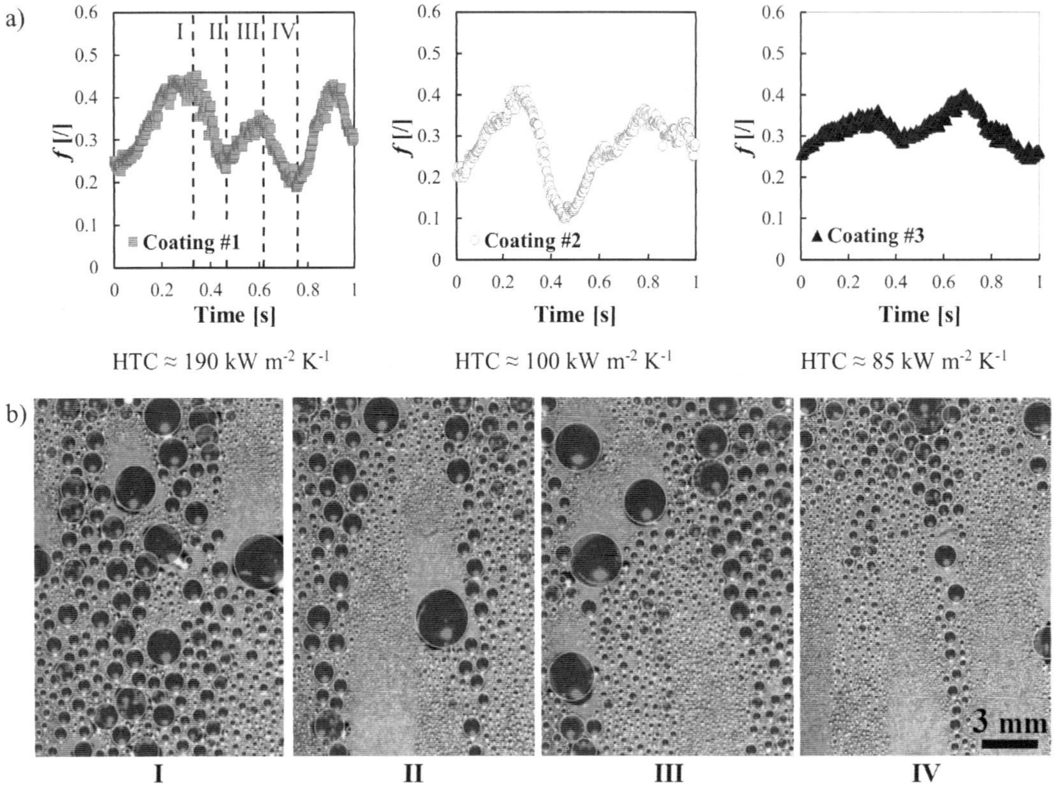

Figure 3. a) Fraction area f occupied by droplets with radii higher than 200 μm plotted versus time for the coatings #1, #2 and #3 (heat flux ≈ 440 kW m^{-2}, inspected area ≈ 2.85 cm^2). For each coating, the corresponding value of the measured HTC is also reported. b) Time evolution of the droplets population on the sample with coating #1. The resulting image after video analysis and droplet recognition by MATLAB® program (red transparent circles) is superimposed to the original image taken by the high-speed camera.

From the video analysis, it is possible to obtain the sweeping period (τ) that, according to the work by Kim and Kim [4], can be expressed as the ratio of the selected area to the sweeping rate at which the surface is renewed by falling drops. Therefore, it indicates the time necessary for renewing the whole surface. The sweeping period was calculated evaluating the maximum and minimum peak of the fraction area f. In fact, the variation Δf between a maximum of the parameter f (e.g. at time step I in Fig. 3a) and the following minimum (at time step II in Fig. 3a) corresponds to a certain portion of the sample surface that was renewed up to that time instant. Thus, the sweeping period can be calculated as the ratio between the observation time interval (e.g. 1 s) and the sum $\sum \Delta f$ of the fraction area variations between a maximum and next minimum, measured in the same observation time interval. The results obtained with the three coatings show that the sweeping period for coating #1 is 1.7 s, whereas DWC on coatings #2 and #3 is respectively characterized by sweeping periods of 2.5 s and 4.2 s. As expected, the highest HTC was measured with the coating exhibiting the lowest sweeping period. It must be considered that the sweeping period calculated with the present method accounts for surface renewal due to coalescence with sliding droplets (droplets that have reached the departure diameter) whereas it does not account for surface renewal due to coalescence between static droplets (droplets that do not have yet reached the departing diameter) as reported in the work by Lethuilleir et al. [10].

4. Conclusions

In the present paper, HTC measurements and droplets dynamic analysis during DWC of steam have been presented for three different sol-gel coated aluminum samples. The three coatings displayed similar wettability ($\theta_a \approx 90°$ and $\theta_r \approx 60°$) but different coating thickness (in the range between 250 ÷ 420 nm). Condensation tests have been performed at constant saturation temperature (≈ 106 °C) and steam velocity (≈ 2.7 m s^{-1}), while varying the heat flux between 70 kW m^{-2} and 550 kW m^{-2}.

All the three coated samples sustained DWC with HTCs considerably higher than the values measured during FWC. In particular, the HTC measured with coating #1 was the highest (≈ 185 kW m^{-2} K^{-1}), while the HTCs measured with coatings #2 and #3 were respectively around 100 and 85 kW m^{-2} K^{-1}. The effect of the additional thermal resistance due to the thickness of the coating was estimated using the DWC model by Kim and Kim [4]. The predicted overall HTC for coating #1 results to be increased by 25% compared to coating #3 due to the lower thickness of the sol-gel coating. However this does not fully explain the higher HTC values measured with coating #1.

Images of the droplets population evolution taken with a high-speed camera have been analysed by a home-made software in MATLAB® code and the following parameters have been determined: the time evolution of the fraction area f occupied by droplets and the sweeping period τ. From the measurements, the sweeping period measured over coating #1, which exhibited the highest HTC, was found to be lower than the value measured with the other two coatings.

Acknowledgements

The authors acknowledge the financial support of the Department of Industrial Engineering, University of Padova through the BIRD187743 project. The support of the European Space Agency through the MAP Condensation program ENCOM (AO-2004-096) is also acknowledged.

References

[1] J. W. Rose, "Dropwise condensation theory and experiment: A review," *Proceedings of the Institution of Mechanical Engineers, Part A: Journal of Power and Energy*, vol. 216, no. 2, pp. 115–128, Jan. 2002, doi: 10.1243/09576500260049034.

[2] B. S. Sikarwar, K. Muralidhar, and S. Khandekar, "Effect of Drop Shape on Heat Transfer During Dropwise Condensation Underneath Inclined Surfaces," *Interfacial Phenomena and Heat Transfer*, vol. 1, no. 4, pp. 339–356, 2014, doi: 10.1615/interfacphenomheattransfer.v1.i4.30.

[3] M. Tancon, R. Parin, S. Bortolin, A. Martucci, and D. Del Col, "Effect of steam velocity during dropwise condensation," *International Journal of Heat and Mass Transfer*, vol. 165, p. 120624, 2021, doi: 10.1016/j.ijheatmasstransfer.2020.120624.

[4] S. Kim and K. J. Kim, "Dropwise Condensation Modeling Suitable for Superhydrophobic Surfaces," *Journal of Heat Transfer*, vol. 133, no. 8, p. 081502, 2011, doi: 10.1115/1.4003742.

[5] R. Parin, M. Sturaro, S. Bortolin, A. Martucci, and D. Del Col, "Heat transfer during dropwise condensation of steam over a mirror polished sol-gel coated aluminum substrate," *International*

Journal of Thermal Sciences, vol. 144, pp. 93–106, Oct. 2019, doi: 10.1016/j.ijthermalsci.2019.05.017.

[6] E. W. Lemmon, I. H. Bell, M. L. Huber, and M. O. McLinden, "NIST Standard Reference Database 23: Reference Fluid Thermodynamic and Transport Properties-REFPROP, Version 10.0, National Institute of Standards and Technology." 2018, doi: https://dx.doi.org/10.18434/T4JS3C.

[7] R. Parin, M. Tancon, M. Mirafiori, S. Bortolin, L. Moro, L. Zago, F. Carraro, A. Martucci, and D. Del Col, "Heat transfer and droplet population during dropwise condensation on durable coatings," *Applied Thermal Engineering*, vol. 179, no. January, p. 115718, 2020, doi: 10.1016/j.applthermaleng.2020.115718.

[8] H. Cha, H. Vahabi, A. Wu, S. Chavan, M. K. Kim, S. Sett, S. A. Bosch, W. Wang, A. K. Kota, and N. Miljkovic, "Dropwise condensation on solid hydrophilic surfaces," *Science Advances*, vol. 6, no. 2, 2020, doi: 10.1126/sciadv.aax0746.

[9] A. Bisetto, S. Bortolin, and D. Del Col, "Experimental analysis of steam condensation over conventional and superhydrophilic vertical surfaces," *Experimental Thermal and Fluid Science*, vol. 68, pp. 216–227, Nov. 2015, doi: 10.1016/J.EXPTHERMFLUSCI.2015.04.019.

[10] J. Lethuillier, P. Lavieille, and M. Miscevic, "About the Role of Falling Droplets' Sweeping in Surface Renewal during Dropwise Condensation," *Langmuir*, vol. 36, no. 43, pp. 12877–12886, Nov. 2020, doi: 10.1021/acs.langmuir.0c02092.

Journal of Physics: Conference Series 2177 (2022) 012047 doi:10.1088/1742-6596/2177/1/012047

Experimental study on evaporation of droplets in microgravity and in the presence of electric field

Alekos Ioannis Garivalis, Paolo Di Marco*

DESTEC, University of Pisa, Largo L. Lazzarino, 1, 56122, Pisa, Italy

Sam Dehaeck, Alexey Rednikov, Pierre Colinet

TIPs, Universitè libre de Bruxelles (ULB), C.P. 165/67 Avenue F.D. Roosevelt, 50 1050 Brussels, Belgium

*p.dimarco@ing.unipi.it

Abstract. One of the key limiting factors in the evaporation rate of sessile droplets is the evacuation of the generated vapour away from the liquid-air interface. In the absence of convection, the main driving mechanism is the vapour diffusivity: as mass diffusion is quite a slow process, this generally leads to small evaporation rates. For liquids with a heavy vapour, natural convection induced by density gradients in the vapour cloud can increase the evaporation rate significantly on ground. Nevertheless, this effect is absent in space and one therefore has to find alternative mechanisms to accelerate evaporation. The electric field may induce some convection and influence evaporation. This work describes experiments regarding evaporation of droplets in microgravity conditions performed during a sounding rocket campaign in June 2019. Microgravity allows to exclude the contribution of natural convection that otherwise prevails on the other effects. The electric field produces a clear change in the vapour concentration distribution surrounding the droplet and an increase of the evaporation rate with respect the case with no electric field. Moreover, in the absence of electric field, the Marangoni convection in the liquid has an effect on the vapour cloud too, preventing the pure diffusive behaviour.

1. Introduction

A sessile droplet is a small isolated volume of liquid deposited on a flat solid substrate, which can be heated or kept at environment temperature, generally in a gas (air) ambient. Drop evaporation was studied since the time of Maxwell in the late 19th century. During the following century, the diffusive isothermal model of the evaporating sessile drop was developed [1]. More recently, the theory was refined to consider other effects than diffusion, such thermocapillary flows and instabilities, convection, wettability, substrate properties and complex fluids. Comprehensive reviews are given by Cazabat and Guéna [2] and Kovalchuk et al. [3]. Among others, some authors investigated the evaporation process in the presence of an external electric field [4], with the double aim to control the droplet shape and influence the evaporation rate. The evaporation of a sessile drop is present in daily life phenomena and became important in the last decades because it is a common theme in many industrial contexts as spray cooling [5], inkjet printing [6], nano-materials production and deposition of solid particles [7], to cite but a few. One of the key limiting factors in the evaporation rate of liquids is the evacuation of the generated vapour away from the liquid-air interface. In the absence of natural or forced convection, diffusion of the vapour into the stagnant air is the controlling mechanism; a comprehensive analytical model in the limit of a dilute vapour is due to Popov [8]. As mass diffusion is quite a slow process, this generally leads to small evaporation rates. For liquids with a heavy vapour, natural convection induced by density gradients in the vapour cloud, can increase the evaporation rate significantly on ground [9]. This beneficial effect is absent in space, however, and one therefore has to find alternative mechanisms to accelerate evaporation.

The change in evaporation rate is reflected in variations of the vapour concentration surrounding the liquid-air interface of the droplet relative to the pure diffusive behaviour; Kelly-Zion et al. [10] measured by means of infrared tomography the vapour distribution in the presence of natural convection showing that the vapour cloud is flattened compared with the pure diffusion one. With the present contribution, we intend to highlight two phenomena that can modify the vapour cloud surrounding the droplet, that are also related to a change in evaporation rate:

1. Entrainment of the vapour cloud due to internal droplet recirculation (Marangoni effect);

2. Electrohydrodynamic forces acting on the vapour cloud and its dielectric permittivity gradients.

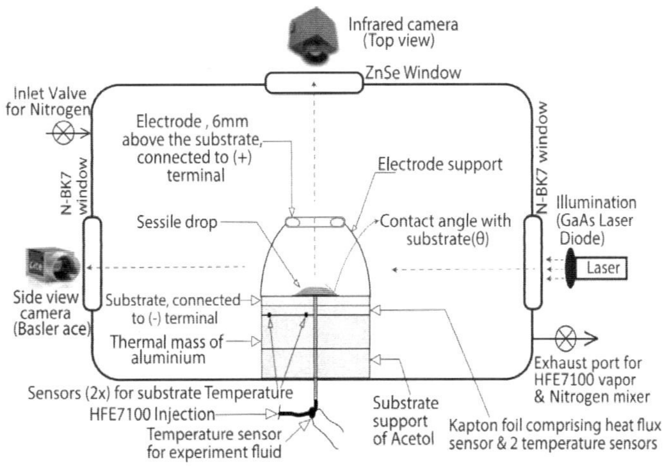

Figure 1 – ARLES module picture of cell interior (top) and schematics (bottom, picture by Kumar et al. [11]).

These effects were investigated in microgravity conditions with experiments performed during a sounding rocket campaign (named ARLES: *Advanced Research of Liquid Evaporation in Space*) in June 2019, funded by the European Space Agency (ESA), which allowed for 6.4 uninterrupted minutes of a good-quality microgravity (< 10^{-4} g). Microgravity allows to exclude the contribution of natural convection that otherwise prevails over the other effects. Both effects will be shown to modify strongly the vapour concentration field around the droplet, the electric field giving rise to a non-negligible enhancement of the evaporation rate. Understanding those phenomena may open the field for advanced vapour manipulation techniques.

2. Experimental apparatus ad microgravity facility

ARLES is a part of the payload of the sounding rocket MASER 14 that was successfully launched on Monday, 24 June 2019 from the Esrange Space Center in northern Sweden. It is a collaborative experiment involving several teams from universities and research institutions, under the umbrella of ESA (financial support) and Swedish Space Corporation (experiment manufacturing and launch management). It consists of several chambers in which some droplets of HFE-7100 (3M™ Novec™ 7100 Engineered Fluid, a hydrofluoroether also known as methoxy-nonafluorobutane $C_4F_9OCH_3$) were evaporated during the flight. We will focus on the main cell only, which has more extended diagnostics and the electrode. A more detailed description of the entire apparatus and its complete functionalities is given by Kumar et al. [11].

The apparatus is shown in Figure 1: a silicon substrate of 25.4 mm diameter is fixed in the middle of a 4-liter cell initially filled with pure Nitrogen at 1 bar; silicon is thermally connected to a heater embedded in a large metal block. A heat flux sensor measures the mean flux through the silicon, several temperature sensors and one pressure transducer measure the cell conditions. During the experiments, the nominal temperature of the cell environment is set to 26 °C while the temperature of the substrate is set and kept to 28 °C. The droplet is produced by injecting pure HFE-7100 through a 0.7 mm diameter opening in the middle of the substrate. A geometrical barrier in the form of a groove etched into the silicon (50 μm wide and deep, etched by Deep Reactive Ion Etching) ensures the pinning of the droplet at exactly 4 mm diameter for most of the evaporation process. An external electrostatic d.c. field can be applied via a washer-shaped electrode, placed axisymmetrically at 6 mm distance above the substrate (the silicon is grounded). Due to the electrode large central opening (13 mm), its influence on the evaporation rate of the droplet is assumed to be negligible.

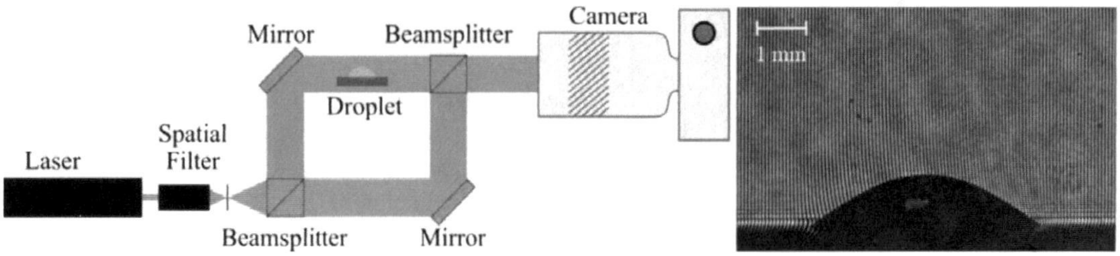

Figure 2 – Interferometer schematics: the laser beam is divided in the beamsplitter, then the two paths converge in a second beamsplitter and their union is recorded by a BW camera.

In the main cell the droplet evolution in time is detected by an infrared camera from the top and an interferometer from the side; we will treat the interferometer only, which allows to evaluate the droplet volume and observe the vapour concentration field surrounding the droplet. A Mach-Zehnder interferometer using a red diode laser (633 nm) generates a homogeneous system of vertical fringes. The scheme of the interferometer components is shown in Figure 2; fringes (visible all over the gas phase in the BW image, Figure 2 on the right) come from the interference of two laser beams, the first following a direct optical path to the camera (Basler Ace ac2440-gm20, resolution 1500x1500), the second crossing the region above the droplet and the silicon substrate. The two beams issue from a spatially filtered laser beam passing through the first beamsplitter. They are then redirected by the mirrows and finally recombined in the camera by means of the second beamsplitter. The fringes are distorted when the vapour cloud is present, which changes the refractive index of the gas. It is possible to correlate the fringes deviation from the vertical position to the optical phase difference accumulated when the light ray traverses the vapour cloud surrounding the droplet. In principle, under the assumption of axial symmetry, an inverse Abel transform algorithm permits to perform the tomographic reconstruction, extracting the refractive-index field around the droplet (n), which is then related to the vapour mole-fraction field χ. The detailed procedure and algorithms used are described by Dehaeck et al. [12]. However, due to an extended vapour cloud and a limited field of view in the present case, a full tomographic reconstruction cannot be realized. Yet, the isolines of the optical phase difference will be well representative of the form of the isolines of the vapour concentration (cf. below). By isolating the black profile of the droplet form the background (the images from the camera look like Figure 2 on the right) it is possible to calculate the volume of the droplet (assumed symmetrical, using Pappus-Guldinus theorem) as a function of time.

The entire flight took less than 15 minutes and the rocket reached an apogee height of 245 km. The microgravity period lasted 6 minutes and 22 seconds. During this time, five droplets in total were injected and left to evaporate for about 35 seconds one after another. The contact line was pinned on the groove and so the volumes decreased nearly linearly; the remaining droplet volume was flushed away before the injection of a new droplet. Four droplets were composed of pure HFE while the last had some graphene added (nanofluid). The latter will be not considered here. Two of the pure droplets evaporated without electric field and two in the presence of electric field. However, for two of these droplets some mishaps disturbed the evaporation process (a fluctuation in the injection of the first droplet without electric field and a Taylor cone formed on the first droplet with electric field), so we will focus on the two remaining droplets only in the following section.

3. Results and discussion

Before the rocket launch, some experiments were performed on ground in normal gravity (following the same procedure of the flight). In Figure 3 four tests are reported; for each test, droplets 0 and 2 are without electric field and droplets 1 and 3 with electric field (with 6.5 to 8 kV applied to the electrode). The initial volume of the injected droplets varies from 4 to 10 μl. The experimental error of the volume measurement is ±0.2 μl. Despite the different initial volumes, the evaporation rate is nearly constant for most of the time; only at the very end of the process it deviates from the linear trend, because to the depinning of the contact line and the departure from the Constant Contact Radius evaporation regime [13]. Noticeably, the evaporation rate does not change in the presence of the electric field, even if the droplets shapes are clearly modified. Considering the linear part, the mean measured evaporation rate is 0.221 μl/s.

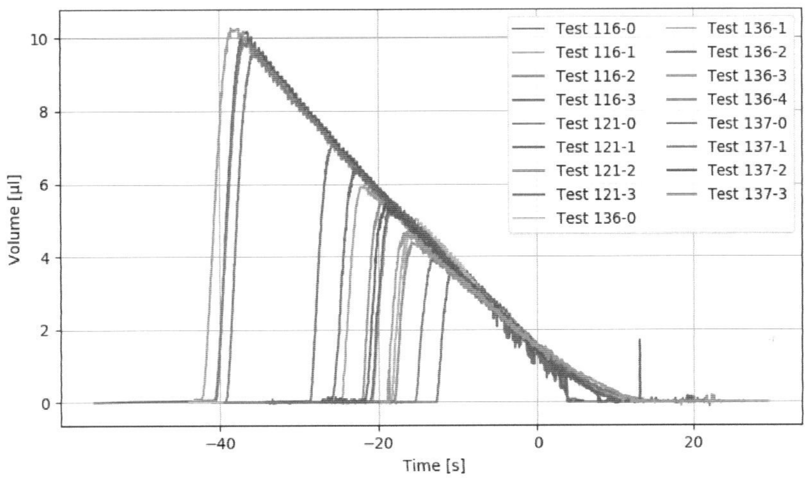

Figure 3 – Evolution of droplets volume vs time on ground.

We now consider the two earlier mentioned flight droplets (see Figure 4), one without electric field and the other with 5.7 kV applied to the electrode; they did not reach a volume small enough for the contact line to detach before flushing, so the volume decreases nearly linearly with time. The initial volume is slightly different, but on ground it has been shown that the evaporation rate is not affected so much. Evaporation rates in microgravity are quite smaller than ground tests (approximately by half), but there emerges a difference between the case without and the case with electric field: the rates are 0.088 μl/s and 0.110 μl/s respectively (Thus, in the presence of the electric field the droplet evaporates 25% faster.

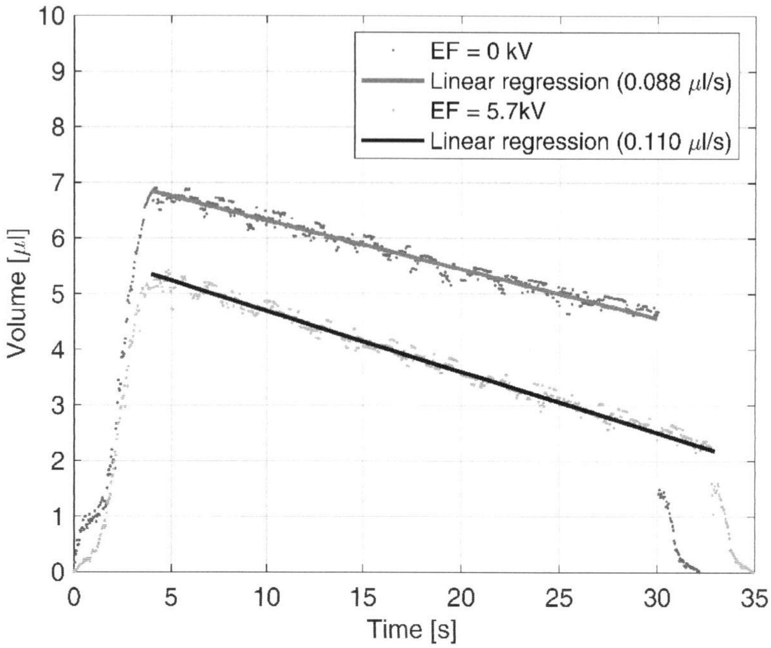

Figure 4 – Volume vs time of the two droplets with and without electric field in flight.

The difference between the two above droplets is confirmed by the interferometric measurements. As earlier explained, the optical phase difference field, shown (modulo 2π) in Figure 5, is representative of the vapour concentration isolines, and gives an idea of the vapour cloud distribution. The unavailability of a full tomographic reconstruction will be compensated with a comprehensive set of numerical simulations, that will confirm the vapour distribution trends and allow to analyse the evaporation process more in detail. Vapour cloud in pure diffusive evaporation exhibits a hemispherical shape. Thus, the effect of natural convection is evident on ground, because the (heavier) vapour cloud is flattened by gravity (see Figure 5a). In microgravity, gravity-driven convection is absent, but even in this case the vapour cloud shape does not look like a pure diffusive one. It appears elongated vertically (see Figure 5b), which is explained as follows: it is known that the temperature gradients in the liquid produce a gradient in surface tension, which induces an internal fluid recirculation directed from the bottom to the top of the droplet (Marangoni convection). Marangoni effect is masked by gravity on ground, but becomes evident in microgravity. The liquid internal recirculation that induces a movement of the vapour too explains the observed vapour distribution. Finally, adding an electric field in microgravity produces a great effect on vapour cloud, as the previously observed elongation disappears and the isolines go back to being flattened (see Figure 5c).

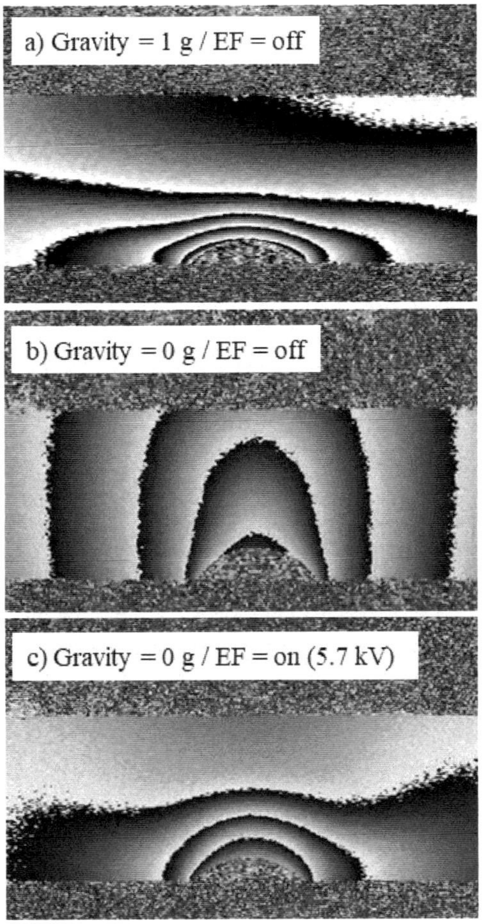

Figure 5 – Vapour clouds in different conditions: (a) test on ground without electric field, (b) test in flight without electric field, (c) test in flight with electric field (5.7 kV).

Indeed, it is known that the electric field acts on conductor and dielectric droplets, especially at the liquid-air interface. The expression of the volumetric vector force $\boldsymbol{f}_e'''$ (N/m^3) acting on a given volume of dielectric (as the gas-vapour mixture) in the absence of free charges and immersed in an electric field is given by [8]:

$$\boldsymbol{f}_e''' = -\frac{\varepsilon_0}{2}E^2\nabla\varepsilon_r + \frac{\varepsilon_0}{2}\nabla(bE^2) \tag{1}$$

with ε_0 being the electric permittivity of vacuum, ε_r the relative permittivity of the medium (function of temperature and density), E the electric field and b the electrostriction coefficient. The first term on the left-hand side is named dielectrophoresis, due to electric permittivity gradient, and the second term is the electrostriction. It is the first term that is of interest here, as the second one is irrotational and gives rise to a contribution into the pressure field, but not into a flow field. Usually, as a first approximation, electric permittivity of gases is considered to be close to the vacuum one, so the relative electric permittivity is $\varepsilon_r \approx 1$ and the electrostriction coefficient is $b = 1 - \varepsilon_r \approx 0$ (see Landau and Lifshitz [14]). However, surprisingly, it can be shown that the effect is here significant, in spite of $\varepsilon_r \approx 1$ (but not exactly one, the deviation turning out to be non-negligible for HFE-7100 vapour with its large molecules). This results in the observed electroconvection in the gas phase, flattening the vapour cloud similarly to gravity.

4. Conclusions

The experiments conducted during the sounding rocket campaign showed that in microgravity the vapour cloud does not follow a diffusion-controlled distribution, but a vertically elongated shape. This must be attributed (in the absence of other external actions) to the Marangoni-induced internal recirculation inside the droplet entraining a corresponding flow in the gas, which is consistent with the observations. Indeed, it is overshadowed by natural convection when gravity is present, but manifests itself in microgravity conditions. On the other end, when the droplet evaporates in the presence of an external electrostatic field, the vapour cloud is altered too, but this time it is flattened (resembling the shapes observed on ground). Thus, vapour concentration distribution is influenced by the electric field, and the electroconvection of the air-vapour phase is found to enhance the evaporation rate by about 25% for the observed droplets. Of course, further investigations are needed to confirm these limited observations, but hopefully they are currently on-going or planned in the short term.

5. Near future work

The ARLES experiment was conducted as a preparation for an experiment that is to be performed in a few years at the European Drawer Rack 2 of the International Space Station under the EVAPORATION project of ESA. During the ARLES campaign, some design and construction flaws prevented a perfect experimental outcome: concerning the above-mentioned work, the injection system was no able to ensure initial droplet volume repeatability, the electrode was not perfectly aligned and a quantitative measure of the vapour cloud concentration was not possible. However, important evidences of evaporations rates and the qualitative shape of concentration vapour isolines were successfully collected, and the vapour cloud trend was well reproduced with numerical simulations (still on-going). Moreover, a second flight of the upgraded module was granted: ARLES-II will flight in June 2022 and all the design issues found were successfully rectified. Another campaign will provide stronger evidences allowing to deepen the analysis, opening the way to the extensive test campaign on board ISS to come.

Acknowledgments

The present work was carried out in the framework of the European Space Agency research project AO-1999-110: EVAPORATION.

References

[1] Erbil H Y 2012 Evaporation of pure liquid sessile and spherical suspended drops: A review *Adv. Colloid Interface Sci.* **170** 67–86
[2] Cazabat A M and Guéna G 2010 Evaporation of macroscopic sessile droplets *Soft*

Matter **6** 2591–612

[3] Kovalchuk N M, Trybala A and Starov V M 2014 Evaporation of sessile droplets *Curr. Opin. Colloid Interface Sci.* **19** 336–42

[4] Vancauwenberghe V, Di Marco P and Brutin D 2013 Wetting and evaporation of a sessile drop under an external electrical field: A review *Colloids Surfaces A Physicochem. Eng. Asp.* **432** 50–6

[5] Liang G and Mudawar I 2017 International Journal of Heat and Mass Transfer Review of spray cooling – Part 1 : Single-phase and nucleate boiling regimes , and critical heat flux *Int. J. Heat Mass Transf.* **115** 1174–205

[6] Park J and Moon J 2006 Control of Colloidal Particle Deposit Patterns within Picoliter Droplets Ejected by Ink-Jet Printing *Langmuir* **22** 3506–13

[7] Bertola V 2008 Some Applications of Controlled Drop Deposition on Solid Surfaces *Recent Patents on Mec. Eng.* **1** 167–74

[8] Popov Y O 2005 Evaporative deposition patterns: Spatial dimensions of the deposit *Phys. Rev. E - Stat. Nonlinear, Soft Matter Phys.* **71** 1–17

[9] Carle F, Sobac B and Brutin D 2013 Experimental evidence of the atmospheric convective transport contribution to sessile droplet evaporation *Appl. Phys. Lett.* **102**

[10] Kelly-Zion P L, Pursell C J, Hasbamrer N, Cardozo B, Gaughan K and Nickels K 2013 Vapor distribution above an evaporating sessile drop *Int. J. Heat Mass Transf.* **65** 165–72

[11] Kumar S, Medale M, Marco P Di and Brutin D 2020 Sessile volatile drop evaporation under microgravity *npj Microgravity* **6** 1–8

[12] Dehaeck S, Rednikov A and Colinet P 2014 Vapor-based interferometric measurement of local evaporation rate and interfacial temperature of evaporating droplets *Langmuir* **30** 2002–8

[13] Picknett R G and Bexon R 1977 The evaporation of sessile or pendant drops in still air *J. Colloid Interface Sci.* **61** 336–50

[14] Landau L D and Lifshitz E M 1984 Electrostatics of Dielectrics *Electrodynamics of Continuous Media* (Elsevier) pp 34–85

AUTHOR INDEX